INTERNATIONAL SERIES OF MONOGRAPHS IN
ORGANIC CHEMISTRY

GENERAL EDITORS: D. H. R. BARTON and W. DOERING

VOLUME 5

APPLICATIONS OF NUCLEAR MAGNETIC RESONANCE SPECTROSCOPY IN ORGANIC CHEMISTRY

Applications of Nuclear Magnetic Resonance Spectroscopy in Organic Chemistry

2nd Edition

BY

L. M. JACKMAN

Professor of Organic Chemistry
Pennsylvania State University
U.S.A.

AND

S. STERNHELL

Reader in Organic Chemistry
University of Sydney
Australia

THE QUEEN'S AWARD
TO INDUSTRY 1966

PERGAMON PRESS

OXFORD · LONDON · EDINBURGH · NEW YORK
TORONTO · SYDNEY · PARIS · BRAUNSCHWEIG

Pergamon Press Ltd., Headington Hill Hall, Oxford
4 & 5 Fitzroy Square, London W.1
Pergamon Press (Scotland) Ltd., 2 & 3 Teviot Place, Edinburgh 1
Pergamon Press Inc., Maxwell House, Fairview Park, Elmsford, New York 10523
Pergamon of Canada Ltd., 207 Queen's Quay West, Toronto 1
Pergamon Press (Aust.) Pty. Ltd., 19a Boundary Street, Rushcutters Bay,
N.S.W. 2011, Australia
Pergamon Press S. A. R. L., 24 rue des Écoles, Paris 5e
Vieweg & Sohn GmbH, Burgplatz 1, Braunschweig

First edition 1959
Second edition 1969

Library of Congress Catalog Card No. 68 – 18524

PRINTED IN GERMANY
08 012542 5

To Marie and Alice

CONTENTS

Preface   xii

Preface to the 1st Edition   xiv

Editorial Preface to the 2nd Edition   xvi

PART 1

**An introduction to the theory and practice of nuclear magnetic resonance
spectroscopy**   1

Chapter 1–1. THEORY OF NUCLEAR MAGNETIC RESONANCE   1

    A. Dynamic and magnetic properties of atomic nuclei   1

    B. Nuclear resonance   2

    C. Relaxation processes   5

    D. Chemical effects in n.m.r.   11

        (i) Chemical shift   11

        (ii) Absorption intensities   13

        (iii) Spin-spin coupling   13

        (iv) Time-dependent phenomena   17

Chapter 1–2. THE EXPERIMENTAL METHOD   21

    A. The nuclear magnetic resonance spectrometer   21

        (i) The magnet   22

        (ii) The radiofrequency oscillator   25

        (iii) Scanning devices   26

        (iv) The detector   26

        (v) Field-frequency locks   29

    B. Experimental factors which influence resolution and the shapes
of absorption lines   30

        (i) Homogeneity of the magnetic field   30

        (ii) Spinning side bands   33

        (iii) Radiofrequency phase   33

        (iv) Sweep rate   33

        (v) R.f. power level   36

        (vi) Sample preparation   36

    C. Experimental factors which influence sensitivity (signal-to-noise ratio) 37

        (i) Frequency response   38

        (ii) R.f. power level   38

        (iii) Concentration   38

        (iv) Availability of compounds—small capacity cells   40

        (v) Enhancement of sensitivity by time-averaging devices   41

    D. Measurement of line positions and determination of the chemical shift   42

    E. Measurement of intensities   49

    F. Solvents   52

    G. Deuterium exchange   53

PART 2

Theory of chemical effects in nuclear magnetic resonance spectroscopy   55

Chapter 2–1. TIME-DEPENDENT EFFECTS IN NUCLEAR MAGNETIC RESONANCE
      SPECTROSCOPY   55

Chapter 2–2. THEORY OF THE CHEMICAL SHIFT   61

  A. Classification of shielding effects   61
  B. Local diamagnetic proton shielding   63
    (i) Direct electrostatic effects   67
    (ii) Inductive effects   69
    (iii) Mesomeric effects   70
    (iv) Hybridization   70
    (v) Van der Waal effects   71
  C. Long range shielding   72
    (i) General principles   72
    (ii) Shielding by carbon–carbon single bonds and carbon–hydrogen bonds   78
    (iii) Shielding by halogens   80
    (iv) Shielding by oxygen atoms in ethers and alcohols   80
    (v) Shielding by nitrogen atoms   81
    (vi) Shielding by carbon–carbon double bonds   83
    (vii) Shielding by the carbonyl group   88
    (viii) Shielding by the carbon–carbon triple bond   92
    (ix) Shielding by the nitrile group   93
    (x) Shielding by the nitro group   94
    (xi) Shielding by aromatic rings   94
    (xii) Shielding by three-membered rings   98
  D. Contact shifts   102
  E. The hydrogen bond   103
  F. Solvent effects   104

Chapter 2–3. THEORY OF SPIN–SPIN MULTIPLICITY   114

  A. Recognition of spin–spin multiplets   115
  B. The mechanism of spin–spin coupling   115
  C. Analysis of n.m.r. spectra   118
    (i) Some definitions   119
    (ii) The exact analysis of n.m.r. spectra   122
    (iii) The first-order theory   124
    (iv) Recognition of second-order spectra   128
    (v) Discussion of individual spin systems   128
      a. The two-spin system (AB)   129
      b. The three-spin systems: The AB_2 to AX_2 cases   130
      c. The three-spin systems: The AMX, ABX and ABC cases   132
      d. The multispin systems related to ABX: The ABX_n system   134
      e. The symmetrical four-spin systems: A_2X_2, A_2B_2, AA'BB', AA'XX'   134
      f. More complex systems   137
    (vi) Field variation as an aid in the analysis of spectra   139
    (vii) Analysis of proton spectra with the aid of ^{13}C satellite bands   140
    (viii) Deuterium substitution as an aid in the analysis of proton spectra   142
    (ix) Solvent effects as an aid in the analysis of spectra   144
    (x) Contact shifts as an aid in the analysis of spectra   144
    (xi) Analysis of spectra by spin decoupling   145
    (xii) Virtual coupling and deceptively simple spectra   147

Chapter 2–4. THEORY AND APPLICATIONS OF MULTIPLE IRRADIATION   151

  A. Theory   151
  B. Applications   153
     (i) Simplification of complex spin–spin multiplets   153
     (ii) Determination of relative signs of spin–spin coupling constants   154
     (iii) Highly accurate determination of chemical shifts   155
     (iv) Location of "hidden" absorptions   155
     (v) Interrelation of two or more coupled groups of nuclei   157
     (vi) Investigation of moderately rapid chemical exchange reactions   157

PART 3

Applications of the chemical shift   159

Chapter 3–1. GENERAL CONSIDERATIONS   159

Chapter 3–2. PROTONS BONDED TO NON-CYCLIC sp^3 CARBON ATOMS   163

  A. H—C—C   167
  B. H—C—X   174
  C. Effect of multiple substitution at the α-carbon atom   181

Chapter 3–3. PROTONS BONDED TO NON-AROMATIC sp^2 CARBON ATOMS   184

  A. Olefinic protons H—C=C   184
  B. Formyl protons H—C=X   192

Chapter 3–4. PROTONS BONDED TO sp CARBON ATOMS   193

Chapter 3–5. PROTONS BONDED TO sp^3 CARBON ATOMS IN NON-AROMATIC
    CYCLIC STRUCTURES   195

Chapter 3–6. PROTONS BONDED TO AROMATIC AND HETEROCYCLIC
    CARBON ATOMS   201

  A. Benzene derivatives   201
  B. Polynuclear and non-benzenoid carbocyclic aromatic compounds   204
  C. Heterocyclic compounds   207

Chapter 3–7. PROTONS BONDED TO ELEMENTS OTHER THAN CARBON   215
  A. —OH, —NH and —SH groups   215

Chapter 3–8. STEREOCHEMISTRY AND THE CHEMICAL SHIFT   219

  A. Symmetry arguments   220
     (i) Structural applications   220
     (ii) Stereochemical applications   222
  B. Geometrical isomerism in ethylene derivatives   222
  C. Geometrical isomerism in systems containing C=N and N=N bonds   226
  D. Cyclopropanes and heterocyclopropanes   227
  E. Multi-ring structures   229
  F. *Cis–trans* stereochemistry in four-, five-, and six-membered rings   234
  G. The chemical shifts of axial and equatorial protons and groups attached to
    six-membered rings in chair conformation   238
  H. Chemical shift—stereochemistry correlations in some complex natural
    products   241
  I. Conformation   245
  J. Stereochemistry and the solvent shift   246

Chapter 3–9. CARBONIUM IONS, CARBANIONS AND RELATED SYSTEMS   249

    A. Carbonium ions   249

        (i) Alkyl and alicyclic carbonium ions   250
        (ii) Aryl and arylalkyl carbonium ions   251
        (iii) Alkenyl and cycloalkenyl carbonium ions   252
        (iv) Alkynyl carbonium ions   252
        (v) Cyclopropyl carbonium ions   253
        (vi) Arenonium ions   254
        (vii) Non-classical carbonium ions   256
        (viii) Aromatic cations   261
        (ix) Acyl and related cations   262
    B. Carbanions   262
        (i) Grignard reagents and alkylmagnesium compounds   263
        (ii) Lithium alkyls and related compounds   264
        (iii) Aromatic anions   265
        (iv) Miscellaneous carbanions   266

PART 4

Spin–spin coupling   269

Chapter 4–1. GEMINAL INTERPROTON COUPLING   270

    A. Geminal coupling across an sp^3 carbon atom   270
        (i) Influence of the H—C—H angle   273
        (ii) Influence of neighbouring π-bonds   273
        (iii) Influence of ring size   275
        (iv) Influence of substituent electronegativity   276
        (v) Effect of orientation of α-substituents   277
        (vi) Miscellaneous effects   277
    B. Geminal coupling across an sp^2 carbon atom   277
    C. Geminal coupling across a heteroatom   279

Chapter 4–2. VICINAL INTERPROTON COUPLING   280

    A. Vicinal coupling across three single bonds   280
        (i) The relation between vicinal coupling and dihedral angle—the Karplus
            rule   281
        (ii) Influence of substituent electronegativity   283
        (iii) Influence of hybridization   284
        (iv) Negative vicinal coupling constants   285
        (v) Influence of other factors   286
        (vi) Vicinal coupling in cyclic systems   286
        (vii) Vicinal coupling in flexible systems   289
        (viii) Uses and abuses of the Karplus equation   292
        (ix) Examples of application of the Karplus rule   294
        (x) Coupling in the system CH_3—CH   298
        (xi) Coupling in the system H—C—X—H where X is a heteroatom   298
        (xii) Vicinal coupling as an aid in structure determination   300
    B. Vicinal interproton coupling across one double and two single bonds 301
        (i) *Cis* and *trans* coupling   301
        (ii) Influence of electronegativity of substituents   302
        (iii) Influence of H—C=C angle   303
        (iv) Influence of bond-order of the double bond   303

Chapter 4–3. INTERPROTON COUPLING IN AROMATIC AND HETEROCYCLIC SYSTEMS 305

   A. Influence of ring size 310
   B. Influence of bond order 310
   C. Effects due to heteroatoms 310
   D. Effects due to substituents 310

Chapter 4–4. LONG-RANGE INTERPROTON COUPLING 312

   A. Observation of long-range coupling 312
   B. Theory of long-range coupling 315
   C. Allylic and homoallylic coupling 316
   D. Long-range coupling in acetylenes, allenes and cumulenes 328
   E. Benzylic coupling 330
   F. Coupling between protons separated by five bonds along an extended zig-zag path 333
   G. Coupling across four single bonds 334
   H. Long-range coupling in 1,3-butadiene derivatives 341
   I. Miscellaneous examples of long-range coupling 342

Chapter 4–5. COUPLING BETWEEN PROTONS AND OTHER NUCLEI 345

   A. Coupling between ^{1}H and ^{13}C 345
   B. Coupling between ^{1}H and ^{19}F 348
   C. Coupling between ^{1}H and ^{31}P 353
   D. Coupling between protons and other elements 356

PART 5

Applications of time-dependent phenomena 357

Chapter 5–1. GENERAL APPLICATIONS 358

   A. Experimental methods involved in the study of time-dependent phenomena in n.m.r. spectroscopy 358
   B. Hydrogen exchange and hydrogen bonding 359
   C. Ligand exchange 360
   D. Partial double bond character 361
   E. Valence-bond tautomerism 362
   F. Conformational changes 364
   G. Miscellaneous processes 366

Chapter 5–2. INTERNAL ROTATION AND EQUIVALENCE OF NUCLEI 368
Chapter 5–3. TAUTOMERISM 380

References 385

Index 441

Other Titles in the Series 457

PREFACE

WHEN the original edition of this book was written the pertinent literature consisted of approximately 400 references—the corresponding figure at present is well over 6000 references, and although the majority of these publications contain only passing mention of n.m.r., the total amount of information has clearly increased enormously. It is for this reason that an almost complete rewriting of the original work has been undertaken.

Our own experiences in discussing n.m.r. spectroscopy with organic chemists have made us acutely aware of the various stumbling blocks† to the adequate comprehension of the subject. Thus, while this edition, like its predecessor, is addressed to the entirely non-mathematical chemist, an attempt has been made to point out explicitly various pitfalls associated with the inadequate comprehension of the theoretical side of the subject. For those interested in more rigorous treatments, copious references are made to the many excellent and comprehensive papers, reviews and books covering the mathematics and physics of the subject.

The wealth of the available data is such, that in a work of this size only a small proportion of the contributions could receive adequate mention. The selection was not made on the authors' opinions of the relative merits of the work, but upon its suitability for adequate discussion within the framework of this book. Further, in chapters dealing with applications, the authors were inevitably tempted to emphasize the methods they had themselves found useful in the solution of organic chemical problems. It is hoped that our experience has been sufficiently wide for this tendency to be beneficial rather than harmful.

Specific coverage of n.m.r. spectroscopy of important and well investigated groups of natural products, such as steroids, sugars, flavanoids, porphyrins,

---

† The most commonly encountered misconceptions appear to be:

(i) Incorrect interpretations of spin multiplets due to over extension of first-order rules and an inadequate understanding of the concepts of chemical and magnetic equivalence. For instance, errors often arise with systems involving mixed strong and weak coupling ("virtual coupling", "deceptive simplicity") and in spin systems such as $X_3A—A'X_3'$ which are frequently confused with the AX_3 type.

(ii) Abuse of the Karplus relation.

(iii) Reliance on small changes in chemical shifts due to a lack of understanding of the complexity and multiplicity of factors governing shielding.

(iv) Lack of understanding of the nature of time-dependent phenomena. This leads to confusion about equivalence and symmetry, and difficulties in interpretation of spectra of species in which fast chemical exchange or conformational inversion take place.

Not included are the almost universal misconceptions about the nature of the n.m.r. phenomenon itself and of relaxation processes, as these seldom lead to misinterpretation of spectra.

di- and tri-terpenes, bitter principles and others, as well as of important groups of synthetic substances (norbornanes and related bicyclic derivatives, strained cage compounds, polymers, organometallic compounds, σ- and π-complexes, phosphorus derivatives, etc.) was not attempted as we feel that each of these topics merits a separate review or monograph. We have, however, listed a number of references to work in most of these fields wherever appropriate, in the hope of providing the interested reader with reasonably direct access to the pertinent literature.

After careful consideration we decided to omit specific discussion of the n.m.r. spectroscopy of ^{13}C, ^{19}F, ^{31}P and other nuclei. We feel that we lack the necessary expertise to discuss these techniques and that, in any case, an adequate treatment would require an extension of the text not commensurate with their current utility in organic chemistry. It is quite likely, however, that in the future ^{13}C spectroscopy will play an increasing role in structural organic chemistry.

In spite of the substantial changes introduced in this edition, the aims remain unaltered. They are to provide the *minimum* theoretical background necessary for the intelligent application of n.m.r. spectroscopy to common problems encountered in organic chemistry, to give a sufficient body of data for most routine general applications, and to illustrate applications with suitable examples. The utility of n.m.r. spectroscopy in organic chemistry is now beyond question and this work is intended for the practising organic chemist.

We thank Dr. John Grutzner for reading the entire manuscript and for innumerable helpful suggestions. We are grateful to Miss Jan Richards for typing the manuscript and assisting at the final stages of its preparation, to Donald Jackman for help in processing the bibliography, and to Mrs. Virginia Haddon for proof-reading the text.

Finally, we acknowledge authors who made available manuscripts prior to publication, and authors, the editors and publishers of the *Journal of the American Chemical Society, Australian Journal of Chemistry, Recueil des travaux chimiques des Pays-Bas, Pure and Applied Chemistry, Canadian Journal of Chemistry, Journal of Chemical Physics, Chemische Berichte, Tetrahedron, Proceedings of the Royal Society (London)* and *Journal of Organic Chemistry,* and to Varian Associates, Academic Press Inc., Institute of Physics and the Physical Society and Pergamon Press for permission to reproduce diagrams.

PREFACE TO THE 1st EDITION

NOWADAYS, the practice of pure organic chemistry requires the use of a number of physical methods, the fundamentals of which belong to the realms of chemical physics. Consequently, organic chemists have come to rely on standard treatises which provide simplified introductions to the theory of such methods, together with compilations of relevant data which can be used for the characterization of organic compounds and for the elucidation of molecular structure and stereochemistry. The most recent method to be adopted by the organic chemist is nuclear magnetic resonance spectroscopy and although the widespread use of this technique has only just commenced many workers are already alive to its considerable potential. For this reason there is a need for a text which provides a non-mathematical introduction to the theory and practice of n.m.r. and which provides such classified data as is at present available. In this book I have attempted to satisfy these needs. I have tried to keep the physical background to a minimum since physical concepts often constitute a "potential barrier" to the organic chemist. At the same time I consider that an understanding of the pertinent physical principles is vital if this powerful technique is to be used with maximum effect. Thus, I have tried to present these principles descriptively at the same time providing references for those readers who may wish to pursue the rigorously mathematical approach.

An unfortunate consequence of writing about n.m.r. at such an early stage in its development is that I have had to commit myself and my readers to one specific method of expressing the chemical shift of hydrogen, whereas in fact a number of systems are in current use and several committees are at present deliberating on such matters. The system I have used is that of τ-values introduced by G.V.D. Tiers. I have chosen the method of Tiers because I believe that its simplicity will appeal to organic chemists and because the values for a vast majority of protons in organic molecules lie between zero and ten and are thus readily committed to memory. If an alternative system is adopted by international agreement it must certainly be based on an internal reference so that the data in this book will be readily convertible to the new units.

I have drawn on the chemical literature up till May 1959 but as it appeared desirable to limit this book to its present length I have been able to quote only a few of the many examples of the application of n.m.r. in structural studies.

I consider that one of the principle features of this book is the summarized experimental data which it contains and I am therefore most happy to acknowledge my gratitude to Dr. G.V.D. Tiers of the Minnesota Mining and Manufacturing Co., and Dr. N.F. Chamberlain of the Humble Oil and Refining Company who provided me with their extensive compilations of un-published data.

Although this book is comparatively short, the list of colleagues to whom I am indebted is long and I am glad of this opportunity to acknowledge their help. I am particularly grateful to Professor D.H.R.Barton, F.R.S., for grant-ing me the opportunity to work in the field of n.m.r., and indeed it was at his suggestion that I undertook to write this book. He has followed its course with interest and has read the entire manuscript. I owe much to Dr. L.H.Pratt and Dr. D.F.Evans who provided my early education in the subject and who also read part of the manuscript. Dr. R.E.Richards, F.R.S., carefully read the manuscript from the point of view of an n.m.r. spectroscopist and kindly suggested several alterations which I feel have greatly improved the text, while Dr. E.S.Waight and Dr. B.C.L.Weedon read the book as organic chemists and were likewise able to draw my attention to many obscurities. I also received some much appreciated encouragement from Professor W. E.Doering who read several of the early chapters. Dr. D.W.Turner made helpful criticisms of Chapter 3. I am most grateful to Miss B.A.Harsant for typing the manuscript.

Finally, I am grateful to many authors, to the editors and publishers of the *Journal of the American Chemical Society, Helvetica Physica Acta,* the *Annals of the New York Academy of Science, Journal of Chemical Physics, Molecular Physics, Transactions of the Faraday Society,* and to Varian Associates for permission to reproduce diagrams.

July 1959   L.M.JACKMAN

Note added in proof. It is now conventional to present spectra with the field increasing from left to right. With the exception of Figs. 2.15, 5.4, 6.1, 6.2 and 6.16, spectra reproduced herein acccord with this convention.

EDITORIAL PREFACE TO THE 2nd EDITION

THE above Editorial Preface, written for the 1st edition of Professor Jackman's monograph on the *Applications of Nuclear Magnetic Resonance Spectroscopy in Organic Chemistry,* adumbrated clearly the fundamental position that nuclear magnetic resonance was to occupy in organic chemistry. Now, less than ten years later, nuclear magnetic resonance spectroscopy is the single most important physical tool available to the organic chemist. Not only does this technique characterize functional groups, but it describes the relationships of appropriate nuclei to each other in constitutional and stereochemical detail. It is a very powerful method for studying details of conformation in solution. No other method can give such useful detail in the liquid state. Increasingly the method is applied to nuclei other than hydrogen.

During the decade since the 1st edition of this book considerable progress in instrumentation has been made. Spectrometers have become more powerful, more easy to use and, for 60 mc/s instruments, cheaper. Almost all organic chemistry laboratories now have some sort of spectrometer available.

The 1st edition of this book was well received by the scientific public and has remained in use up to the present time. Nevertheless, the substantial developments in the subject outlined above do fully justify a new and expanded edition. It is fortunate that Professor Jackman has agreed to undertake this heavy task and fortunate also that another internationally known authority in the field, Dr. Sever Sternhell, has agreed to collaborate in the work. It is certain that this book will occupy the same important position in the personal libraries of organic chemists that the earlier edition did. It is warmly recommended to all organic chemists and to workers in the other disciplines as was the 1st edition.

Imperial College   D. H. R. BARTON
London, S.W. 7

xvi

PART 1

AN INTRODUCTION TO THE THEORY AND PRACTICE OF NUCLEAR MAGNETIC RESONANCE SPECTROSCOPY

CHAPTER 1–1

THEORY OF NUCLEAR MAGNETIC RESONANCE

A. DYNAMIC AND MAGNETIC PROPERTIES OF ATOMIC NUCLEI

Apart from the use of atomic numbers and isotopic weights, the organic chemist has largely developed his subject without any special knowledge of the properties of atomic nuclei. The recent advent of n.m.r. spectroscopy and, to a much lesser extent, microwave and pure quadrupole spectroscopy, has altered this state of affairs and organic chemists of the present generation have now to become acquainted with certain subjects hitherto the domain of the nuclear physicist and spectroscopist. Thus today a table of atomic weights of those elements commonly encountered by the organic chemist might usefully include other nuclear properties such as spin numbers, nuclear magnetic moments, and nuclear electric quadrupole moments.

Of these additional nuclear properties the spin number, I, and the nuclear magnetic moment, μ, are of particular interest; the nuclear electric quadrupole moment, Q, will enter only occasionally into our discussions.

The nuclei of certain isotopes possess an intrinsic spin; that is they are associated with an angular momentum. The total angular momentum of a nucleus is given by $(h/2\pi) \cdot [I(I + 1)]$ in which h is Planck's constant and I is the *nuclear spin* or *spin number* which may have the values $0, \frac{1}{2}, 1, \frac{3}{2}, \ldots$ depending on the particular isotopic nucleus ($I = 0$ corresponds to a nucleus which does not possess a spin). Since atomic nuclei are also associated with an electric charge, the spin gives rise to a magnetic field such that we may consider a spinning nucleus as a minute bar magnet the axis of which is coincident with the axis of spin.† The magnitude of this magnetic dipole is

---

† The neutron has $I = \frac{1}{2}$ and possesses a magnetic moment even though it has no net charge. This apparent paradox is resolved in terms of Yukawa's dissociation theory in which it is assumed that the neutron, for a fraction of its life-time, is partially dissociated to a proton and negative meson, the magnetic moment of the latter being larger than, and of opposite sign to, that of the former.

expressed as the nuclear magnetic moment, μ, which has a characteristic value for all isotopes for which I is greater than zero.

In a uniform magnetic field the angular momentum of a nucleus ($I > 0$) is quantized, the nucleus taking up one of $(2I + 1)$ orientations with respect to the direction of the applied field. Each orientation corresponds to a characteristic potential energy of the nucleus equal to $\mu \cdot H_0 \cdot \cos \theta$ where H_0 is the strength of the applied field and the angle θ is the angle which the spin axis of the nucleus makes with the direction of the applied field. The importance of I and μ in our discussion is that they define the number and energies of the possible spin states which the nuclei of a given isotope can take up in a magnetic field of known strength. A transition of a nucleus from one spin state to an adjacent state may occur by the absorption or emission of an appropriate quantum of energy.

Nuclei of isotopes for which $I > \frac{1}{2}$ are usually associated with an asymmetric charge distribution which constitutes an electric quadrupole. The magnitude of this quadrupole is expressed as the nuclear quadrupole moment Q.

B. NUCLEAR RESONANCE

Although the literature contains several detailed mathematical treatments of the theory of nuclear magnetic resonance which are based on microphysical[265,1929] or macrophysical[263,264,1929] concepts we will be content to develop the theory, as far as possible, in a purely descriptive manner by stating in words the results of the physicists' equations. In doing so we will no doubt lose the elements of exactness but as organic chemists we will gain tangible concepts of considerable utility, which would otherwise be lost to all but those possessing the necessary mathematical background.

The starting point of our discussion is a consideration of a bare nucleus, such as a proton, in a magnetic field of strength H_0. Later we will consider collections of nuclei. We will also add the extranuclear electrons and ultimately we will build the atoms into molecules. We have just seen that certain nuclei possess two very important properties associated with spin angular momentum. These properties are the spin number I and the magnetic moment μ. We are only concerned with the nuclei of those isotopes for which these two quantities are not equal to zero.

When such a nucleus is placed in a static uniform magnetic field H_0 it takes up one of $(2I + 1)$ orientations which are characterized by energies dependent on the magnitudes of μ and H_0. If the bare nucleus is a proton, which has a spin number I equal to one half, we can liken it to a very tiny bar magnet. A large bar magnet is free to take up any possible orientation in the static field so that there are an infinite number of permissible energy states. Quantum mechanics tells us that the tiny proton magnet is restricted to just two possible orientations $[(2I + 1) = 2]$, in the applied field and these can be considered to be a low energy or parallel orientation in which the magnet is aligned with the field and a high energy or anti-parallel orientation in which it is aligned against the field (i.e. with its N. pole nearest the N. pole of the static field). Since these two orientations correspond to two energy states it should be possible to induce transitions between them and the frequency, v, of the electromagnetic radiation which will effect such transitions is given by the equation

$$hv = \frac{\mu\beta_N H_0}{I} \tag{1-1-1}$$

where β_N is a constant called the nuclear magneton. Equation (1–1–1) may be rewritten as (1–1–2)

$$v = \gamma \cdot H_0/2\pi \tag{1-1-2}$$

where γ is known as the *gyromagnetic ratio*. The absorption or emission of the quantum of energy hv causes the nuclear magnet to turn over or "flip" from one orientation to the other. For nuclei with spin numbers greater than $\frac{1}{2}$ there will be more than two possible orientations (3 for $I = 1$, 4 for $I = \frac{3}{2}$, etc.) and in each case a set of equally spaced energy levels results. Again, electromagnetic radiation of appropriate frequency can cause transitions between the various levels with the proviso (i.e. selection rule) that only transitions between adjacent levels are allowed. Since the energy levels are equally spaced this selection rule requires that there is only one characteristic transition frequency for a given value of H_0. If we insert numerical values into our equation for v we find that, for magnetic fields of the order of 10,000 gauss, †the characteristic frequencies lie in the radiofrequency region (ca. 10^7–10^8 Hz). Thus our first primitive picture of n.m.r. spectroscopy may be summed up by stating that atomic nuclei of certain elements ($I > 0$) when placed in a strong magnetic field may absorb radiofrequency radiation of discrete energy.

We shall now develop a classical picture of the absorption process which takes us a little further in our understanding of nuclear magnetic resonance and which provides a useful model for discussing the experimental procedure. Let us consider our spinning nucleus to be oriented at an angle θ to the direction of the applied field H_0 (Fig. 1–1–1). The main field acts on the nuclear magnet so as to decrease the angle θ. However, because the nucleus is spinning, the net result is that the nuclear magnet is caused to precess about the main field axis. The angular velocity, ω_0, of this precessional motion is given by equation 1–1–3

$$\omega_0 = \gamma H_0 = 2\pi v \tag{1-1-3}$$

The precessional frequency ω_0 is directly proportional to H_0 and to the gyromagnetic ratio‡ and is exactly equal to the frequency of electromagnetic radiation which, on quantum mechanical grounds, we decided was necessary to induce a transition from one nuclear spin state to an adjacent level. We

---

† The gauss is strictly speaking the unit of magnetic induction and the field strength should be measured in oersteds. In air the two units are almost equivalent and the former is commonly used.

$\ddagger$ We may note that ω_0 is *independent* of the angle θ.

are now in the position to establish the exact character of the radiofrequency radiation necessary to do this. The act of turning over the nucleus from one orientation to another corresponds to an alteration of the angle θ. This can only be brought about by the application of a magnetic field, H_1, in a direction at right angles to the main field H_0. Furthermore, if this new

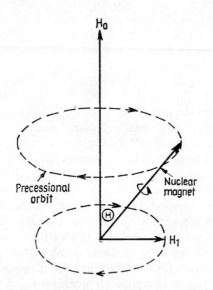

FIG. 1-1-1. The behaviour of a nuclear magnet in a magnetic field (in this figure, the applied field H_0, the rotating field H_1 and the nuclear magnetic dipole are represented as vectors).

field H_1 is to be continuously effective, it must rotate in a plane at right angles to the direction of H_0 in phase with the precessing nucleus. When these conditions are met the rotating magnetic field and the precessing nuclear magnet are said to be in resonance and absorption of energy by the latter can occur.

The important point to be derived from the classical model at this stage is that in order to obtain a physically observable effect it is necessary to place the nucleus in a static field and then to subject it to electromagnetic radiation in such a way that the magnetic vector component of the radiation rotates with the appropriate angular velocity in a plane perpendicular to the direction of the static field.

This is achieved by means of a coil, carrying an alternating current of the appropriate radiofrequency, the axis of which is at right angles to the direction of the applied magnetic field. Detection of absorption of energy is accomplished using the same coil or a separate coil which is orthogonal to both the first coil and the applied magnetic field. The spectrum is scanned by varying either the frequency or the strength of the applied magnetic field.

C. RELAXATION PROCESSES

We have now to enquire into the fate of the energy which has been absorbed by our nuclear magnet. Radiation theory tells us that the emission of energy in the form of electromagnetic radiation can take place either spontaneously or by a process stimulated by an electromagnetic field, and that the probability of occurrence of the latter process is exactly equal to the probability of absorption of energy from the field. Furthermore, the theory shows that the probability of spontaneous emission depends on the frequency of the emitted radiation in such a way that, at radiofrequencies, this probability is negligible. If we now consider a collection of nuclei of the same isotope which are equally divided between two adjacent spin states, we may conclude that the rate of absorption of energy by the lower state will exactly equal the rate of induced emission from the upper state and no observable effect is possible. The situation is saved by the fact that a collection of nuclei in a static magnetic field are not equally distributed between the various possible spin states but rather they take up a Boltzmann distribution with a very small but finite excess in favour of the lower levels. In other words the nuclei ($I = \frac{1}{2}$) all prefer to be aligned parallel to the main field (as they would be at equilibrium at $0°K$) but because of their thermal motions the best that can be managed is a slight excess of parallel spins at any instant. Small though this excess is, it is sufficient to result in a net observable absorption of radiofrequency radiation since the probability of an upward transition (absorption) is now slightly greater than that of a downward transition (emission).

If now we irradiate the collection of nuclei ($I = \frac{1}{2}$), the rate of absorption is initially greater than the rate of emission because of the slight excess of nuclei in the lower energy state. As a result, the original excess in the lower state steadily dwindles until the two states are equally populated. If we are observing an absorption signal we might find that this signal is strong when the radiofrequency radiation is first applied but that it gradually disappears. This type of behaviour is in fact sometimes observed in practice. More generally, however, the absorption peak or signal rapidly settles down to some finite value which is invariant with time. The reason for this behaviour is that induced emission is not the only mechanism by which a nucleus can return from the upper to the lower state. There exist various possibilities for radiationless transitions by means of which the nuclei can exchange energy with their environment and it can be shown[19][29] that such transitions are more likely to occur from an upper to a lower state than in the reverse direction. We therefore have the situation in which the applied radiofrequency field is trying to equalize the spin state equilibrium while radiationless transitions are counteracting this process. In the type of systems of interest to the organic chemist, a steady state is usually reached, such that the original Boltzmann excess of nuclei in the lower states is somewhat decreased but not to zero so that a net absorption can still be registered.

The various types of radiationless transitions, by means of which a nucleus in an upper spin state returns to a lower state, are called *relaxation processes*.

Relaxation processes are of paramount importance in the theory of nuclear magnetic resonance for not only are they responsible for the establishment and maintenance of the absorption condition but they also control the lifetime expectancy of a given state. The uncertainty principle tells us that the "natural" width of a spectral line is proportional to the reciprocal of the average time the system spends in the excited state. In u.v. and i.r. spectroscopy, the natural line width is seldom if ever, the limit of resolution. At radiofrequencies, however, it is quite possible to reach the natural line width and we shall therefore be very much concerned with the relaxation processes which determine this parameter.

We may divide relaxation processes into two categories namely *spin–lattice relaxation* and *spin–spin relaxation*. In the latter process a nucleus in its upper state transfers its energy to a neighbouring nucleus of the same isotope by a mutual exchange of spin. This relaxation process therefore does nothing to offset the equalizing of the spin state populations caused by radiofrequency absorption and is not directly responsible for maintaining the absorption condition. In spin–lattice relaxation, the energy of the nuclear spin system is converted into thermal energy of the molecular system containing the magnetic nuclei, and is therefore directly responsible for maintaining the unequal distribution of spin states. Either or both processes may control the natural line width.

Spin–lattice relaxation is sometimes called longitudinal relaxation.[263] The term *lattice* requires definition. The magnetic nuclei are usually part of an assembly of molecules which constitute the sample under investigation and the entire molecular system is referred to as the lattice irrespective of the physical state of the sample. For the moment we will confine our attention to liquids and gases in which the atoms and molecules constituting the lattice will be undergoing random translational and rotational motion. Since some or all of these atoms and molecules contain magnetic nuclei such motions will be associated with fluctuating magnetic fields. Now, any given magnetic nucleus will be precessing about the direction of the applied field H_0 and at the same time it will experience the fluctuating magnetic fields associated with nearby lattice components. The fluctuating lattice fields can be regarded as being built up of a number of oscillating components (in the same way as any complicated wave-form may be built up from combinations of simple harmonic wave-forms) so that there will be a component which will just match the precessional frequency of the magnetic nuclei. In other words, the lattice motions, by virtue of the magnetic nuclei contained in the lattice, can from time to time generate in the neighbourhood of a nucleus in an excited spin state a field which, like the applied radiofrequency field H_1, is correctly oriented and phased to induce spin state transitions. In these circumstances a nucleus in an upper spin state can relax to the lower state and the energy lost is given to the lattice as extra translational or rotational energy. The same process is involved in producing the Boltzmann excess of nuclei in lower states when the sample is first placed in the magnetic field. Since the exchange of energy between nuclei and lattice leaves the total energy of the sample unchanged, it follows that the process must always operate so as to establish the most probable distribution of energy or, in other words, so as to establish the Boltzmann excess of nuclei in lower states.

The efficiency of spin–lattice relaxation can, like other exponential processes, be expressed in terms of a characteristic relaxation time, T_1, which,

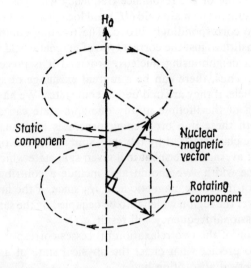

FIG. 1–1–2. The resolution of the magnetic vector of a nucleus into a static and a rotating component.

in effect, is the half-life required for a perturbed system of nuclei to reach an equilibrium condition. A large value of T_1 indicates an inefficient relaxation process. The value of T_1 will depend on the gyromagnetic ratio (or ratios) of the nuclei in the lattice and on the nature and rapidity of the molecular motions which produce the fluctuating fields. Because of the great restriction of molecular motions in the crystal lattice, most solids exhibit very long spin–lattice relaxation times, often of the order of hours. For liquids and gases the value of T_1 is much less, being of the order of one second for many organic liquids. We shall presently discuss certain conditions under which T_1 falls to even lower values.

The term spin–spin relaxation, sometimes called transverse relaxation,[263] usually embraces two processes which result in the broadening of resonance lines. One of these processes is a true relaxation in that it shortens the life of a nucleus in any one spin state, whereas the other process broadens a resonance line by causing the effective static field to vary from nucleus to nucleus. Both effects are best understood from a consideration of the interaction of two precessing nuclear magnets in close proximity to one another. The field associated with a nuclear magnet which is precessing about the direction of the main field may be resolved into two components (Fig. 1–1–2). One component is static and parallel to the direction of the main field H_0. The other component is rotating at the precessional frequency in a plane at right angles to the main field. The first component will be felt by a neighbouring nucleus as a small variation of the main field. As the individual nuclei in a system are not necessarily in the same environment, since the surrounding magnetic nuclei are in various spin states, each may experience a slightly different local field due to neighbouring nuclei. Consequently, there will be a spread in the value of the resonance frequency which is, of course, proportional to the sum of the main field (H_0) and local fields. The resonance line will therefore be correspondingly broad. The rotating component at right angles to H_0 constitutes just the correct type of magnetic field for stimulating a transition in a neighbouring nucleus, provided it is precessing with the same frequency. Thus, there can be a mutual exchange of spin energy between the two nuclei if they are in different spin states. We have already seen that the limiting of the lifetime in any one spin state can also cause line broadening. Both these line broadening processes are usually considered together and are characterized by the spin–spin relaxation time, T_2, corresponding to that average time spent in a given spin state which will result in the observed line width. We should also include a contribution from the inhomogeneity of the static magnetic field H_0 since, if the field varies from point to point over the region which is to be occupied by the sample, a spreading of the precessional frequency will result.

A consideration of the two relaxation processes corresponding to T_1 and T_2 enables us to predict what effect the physical state of a substance will have on the observed absorption line.

Many solids may be considered as more or less rigid assemblages of nuclei in which random movement of the lattice components is negligible. For this reason spin–lattice relaxation times may be very long. On the other hand, local fields associated with spin–spin interaction are large, with the result that absorption lines of solids are usually very broad (this is known as dipolar broadening). In fact, the broadening in solids is usually several powers of ten greater than the effects which are of interest to the organic chemist and for this reason we will consider solids no further. In liquids, molecules may undergo random motion (Brownian motion) and it can be shown[264] that, provided this motion is sufficiently rapid, the local fields average out to a very small value, so that sharp resonance lines can be observed. Indeed, with liquids other factors including the spin–lattice relaxation are of comparable importance in determining line width. A further consequence of random motion in liquids is a marked lowering of the spin–lattice relaxation time T_1. The observed value of T_1 depends, amongst other things, on the viscosity of the liquid. The dependence is not a simple one but assumes the form indicated in Fig. 1–1–3. The origin of this behaviour can be understood in terms of the fluctuating fields which effect relaxation. At high viscosities the molecular motions are relatively slow and the fluctuating fields are largely built up of lower frequency components so that the intensity of the component ν_0, which matches the precessional frequency of the nuclei and causes relaxation, will be relatively low. On the other hand the fluctuating fields in a very mobile liquid are comprised of a large range of frequencies so that any one frequency (in particular ν_0) makes only a small contribution. Evidently some intermediate viscosity will provide the maximum intensity of the correct frequency ν_0 (see Fig. 1–1–4) and hence the minimum value of T_1. In addition to the effect of viscosity on T_1, we should note that at very high viscosities it is possible that the molecules are moving about too slowly to effect

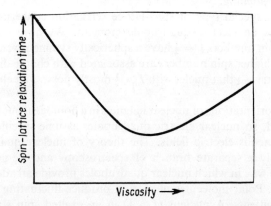

FIG. 1–1–3. The dependence of the spin–lattice relaxation time on viscosity.

complete time-averaging of local fields, so that even though T_1 is long T_2 becomes short and broadened lines are observed.

We need to know about two special types of spin–lattice relaxation which sometimes influence our observations. The first of these may be termed paramagnetic broadening and results from the presence of paramagnetic mole-

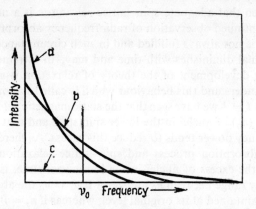

FIG. 1–1–4. The variation of the frequency distribution of fluctuating molecular magnetic fields with viscosity: (a) high viscosity; (b) intermediate viscosity; (c) low viscosity.

cules or ions in the sample under investigation. The electron magnetic moment is some 10^3 times larger than nuclear magnetic moments. Consequently the motions of paramagnetic lattice components will produce very intense fluctuating magnetic fields and greatly reduced spin–lattice relaxation times, T_1, result. Under these conditions T_1 becomes very short and

1a AN

makes a large contribution to line width, and the nuclear magnetic resonance lines of paramagnetic substances are usually very broad. Furthermore, the presence of even small quantities of paramagnetic impurities in a sample can cause line-broadening.

The second special type of spin–lattice relaxation concerns those nuclei which possess an electric quadrupole moment. We have seen (p. 2) that nuclei with spin numbers $I = \frac{1}{2}$ have a spherically symmetric charge distribution whereas higher spin numbers are associated with charge distributions of lower symmetry, so that nuclei with $I > \frac{1}{2}$ mostly possess an electric quadrupole moment.

Just as the orientations of nuclear magnets in a homogeneous magnetic field are quantized, so nuclear electric quadrupoles assume specific orientations in inhomogeneous electric fields. The theory of nuclear quadrupole resonance is really a separate branch of spectroscopy and we need only consider here the way in which nuclear quadrupoles provide an additional mode of relaxation. Polar molecules in motion produce fluctuating local electrostatic field gadients. A nucleus $(I > \frac{1}{2})$ in an excited spin state, by virtue of the interaction of its quadrupole with fluctuating field gradients, is thus offered an additional method of giving up its spin energy to the lattice. The essential feature of the electric interaction is that it is usually stronger and falls off less rapidly with distance than its magnetic counterpart which persists only over a very short distance (<4 Å). Consequently, nuclei with quadrupole moments frequently exhibit very short spin–lattice relaxation times and the observed absorption lines associated with these nuclei are correspondingly broad.

We have seen that adequate spin–lattice relaxation is a necessary condition for the continued observation of radiofrequency absorption. In practice this condition is not always fulfilled and in such circumstances the observed absorption signal diminishes with time and may, in extreme cases, vanish. The preceding development of the theory of relaxation is sufficient for us to be able to understand this behaviour which is called *saturation*. Considering nuclei with $I = \frac{1}{2}$ we have seen that the static magnetic field H_0 establishes a small excess (n_0) of nuclei in the lower spin state and that the absorption of radiofrequency power tends to reduce this excess. As there is competition between the absorption process and spin–lattice relaxation a new steady value (n_s) for the excess of nuclei in the lower spin state, is obtained. The value of n_s may range between n_0 and zero. If $n_s = n_0$ the absorption condition will be maintained at its original level, whereas if $n_s = 0$ the absorption of radiofrequency power will cease. Between these two extremes we have the situation where the absorption starts at some value and rapidly falls to a lower value. The ratio $n_s/n_0 = Z_0$ is known as the saturation factor and is of course a direct measure of the degree to which the absorption condition is maintained. It can be shown that for $I = \frac{1}{2}$ the saturation factor is given by equation (1–1–4).[263,265,1929] We note that low values of Z_0 correspond to a high degree of saturation.

$$Z_0 = [1 + \gamma^2 H_1^2 T_1 T_2]^{-1} \qquad (1\text{–}1\text{–}4)$$

Equation (1–1–4) tells us the conditions under which we may expect appreciable saturation. Firstly, the inclusion of the term H_1^2 expresses the self-evident conclusion that the greater the radiofrequency power applied to the sample the greater will be the degree of saturation. The term T_2 takes into account the width of the absorption line. A narrow absorption line corresponds to a high proportion of nuclei with precessional frequencies exactly equal to the applied radiofrequency field. Thus, taken together, $\gamma^2 \cdot H_1^2 \cdot T_2$ expresses the probability of radiofrequency induced transitions. The term T_1 is an inverse measure of the probability of spin–lattice transitions. Therefore, a combination of large radiofrequency fields, narrow absorption lines and long spin–lattice relaxation times may prevent the observation of a spectrum. Furthermore, when saturation is appreciable it is those nuclei with exactly the correct precessional frequency which are lost from the excess n_0 so that the new excess n_s will correspond to a broader resonance line. The strength of an observed resonance signal from a given type of nucleus is a function both of the radiofrequency power H_1 and of the concentration of those nuclei in the sample, so that the saturation factor Z_0 ultimately determines the minimum concentration which will give an observable spectrum (see Chapter 1–2 C).

D. CHEMICAL EFFECTS IN N. M. R.

The purpose of this section is to introduce, in the simplest terms, the effects which make n.m.r. spectroscopy important in chemistry. It is necessary to be acquainted with the phenomena described in this section in order to follow the detailed and separate accounts of each, given in subsequent Parts.

(i) CHEMICAL SHIFT

So far in our discussion of nuclear magnetic resonance we have more or less assumed that the resonance frequency of a nucleus is simply a function of the applied field and the gyromagnetic ratio of the nucleus. If this were indeed the case nuclear magnetic resonance would be of little value to the organic chemist. It turns out, however, that the observed resonance frequency is to a small degree dependent on its molecular environment. This is because the extranuclear electrons magnetically screen the nucleus so that the magnetic field felt by the nucleus is not quite the same as the applied field. Naturally, we might expect the efficiency of this shielding by the extranuclear electrons to bear some sort of relationship to the type of chemical bonding involved.

Thus, naïvely, we might predict that electron withdrawal from a given nucleus would decrease the shielding of that nucleus. To the extent to which this is true we can regard the magnetic nucleus as a tiny probe with which we may examine the surrounding electron distribution. Although the nuclear resonance picture of electron distribution is not nearly as simple as the one drawn above, it has been found that nuclear resonance frequencies, when properly determined, are remarkably characteristic of molecular structure.

We shall be concerned almost exclusively with the nuclear magnetic resonance of the hydrogen nucleus, that is with proton magnetic resonance, and a preliminary idea of the effect of structure on proton resonance frequencies can be gained by reference to Fig. 1–1–5. We see that protons in

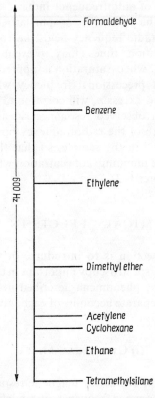

FIG. 1–1–5. Chemical shifts of protons in different environments.

the usual types of organic environments have relative frequencies spread over about 600 Herz (Hz) at a field strength of 14,092 gauss for which resonance frequency of the hydrogen nucleus $v_H = 60 \times 10^6$ Hz. Although this spread is only equivalent to about ten parts per million, *relative* values for proton signals can be readily determined with an accuracy of better than ± 1 Hz.

The separation of resonance frequencies of nuclei in different structural environments from some arbitrarily chosen line position is generally termed the *chemical shift*. If the arbitrary line position is that of a bare proton the chemical shift is equal to the screening constant which measures the difference between the applied field and the actual field felt by the nucleus. The chemical shifts of protons are amongst the smallest for all nuclei but, as both carbon (^{12}C) and oxygen (^{16}O) do not possess magnetic moments, the major application of n.m.r. spectroscopy to organic chemistry involves the study of proton shifts.

The theory of the chemical shift is dealt with in Part 2, Chapter 2, and Part 3 deals with applications.

(ii) Absorption intensities

The low resolution spectrum of ethanol (Fig. 1–1–6) consists of three bands as expected from the three types of protons in the molecule. The intensities of these bands, as measured by the areas they enclose, are in the ratios 1:2:3. This is to be expected, since the nature of the transition, namely the inversion of the proton spin, is the same irrespective of the chemical environment. Each proton therefore has the same transition probability and

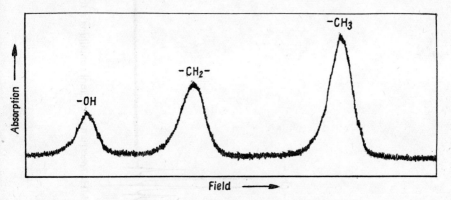

FIG. 1–1–6. The low resolution spectrum of ethanol.

the intensities of absorption bands are proportional only to the numbers of protons responsible for the absorption.† Thus, if a spectrum of a pure compound consists of a number of discrete bands arising from protons in different chemical environments, the intensities of the bands provide a measure of the number of protons in each environment.

(iii) Spin–spin coupling

If we examine a sample of ordinary ethanol (not highly purified) under higher resolution than that used to obtain Fig. 1–1–6, we find a somewhat more complicated picture (Fig. 1–1–7). The bands associated with the methyl and methylene groups now appear as multiplets, the total areas of which are still in the ratio of 3:2. The spacing of the three components of the methyl group triplet are found to be equal to that in the quartet from the methylene

† This is correct to at least a very high order of approximation, provided the experimental conditions are chosen to render differences in saturation unimportant.

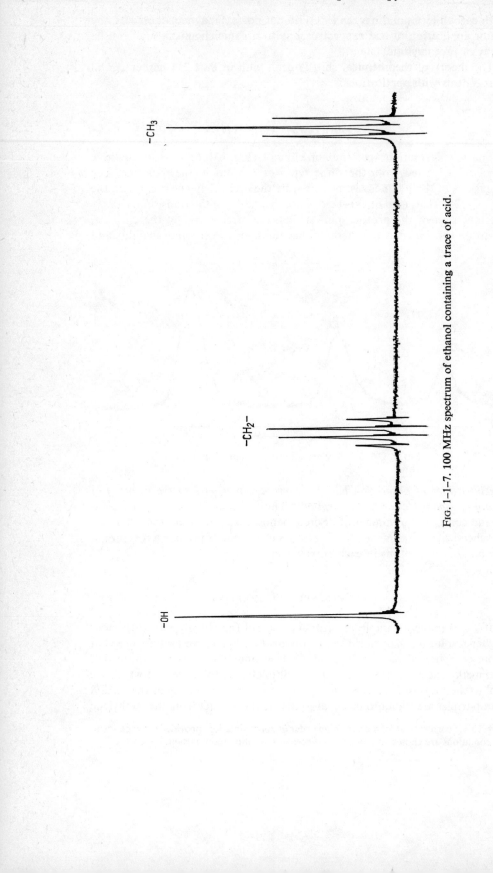

Fig. 1-1-7. 100 MHz spectrum of ethanol containing a trace of acid.

group. Furthermore the areas of the components of each multiplet approximate to simple integral ratios (1:2:1 for the triplet and 1:3:3:1 for the quartet). We can understand these observations if we imagine that the field experienced by the protons of one group is influenced by the spins of protons in the neighbouring group. Let us consider the methyl protons in relation to the possible spin arrangements of the two methylene protons. There

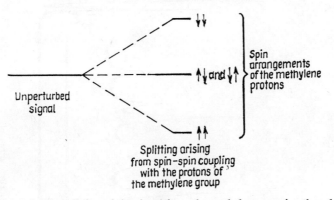

Fig. 1–1–8. The splitting of the signal from the methyl protons in ethanol by spin–spin interaction with the protons of the methylene group.

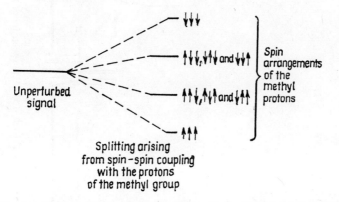

Fig. 1–1–9. The splitting of the signal from the methylene group protons in ethanol by spin–spin interactions with the protons of the methyl group.

are four possible spin arrangements for the methylene group. If we label the two protons A and B then we have (1) A and B both in parallel spin states, (2) A and B both in antiparallel spin states, (3) A parallel and B antiparallel, and (4) A antiparallel and B parallel. The arrangements (3) and (4) are equivalent. The magnetic effect of these arrangements is in some way (Chapter 2–3) transmitted to the methyl group protons so that these protons will experience one of three effective fields according to the instantaneous spin arrangement of the methylene group. Thus, for a collection of ethanol molecules there will be three equally spaced transition energies (frequencies) for the methyl protons. Since the probabilities of existence of each

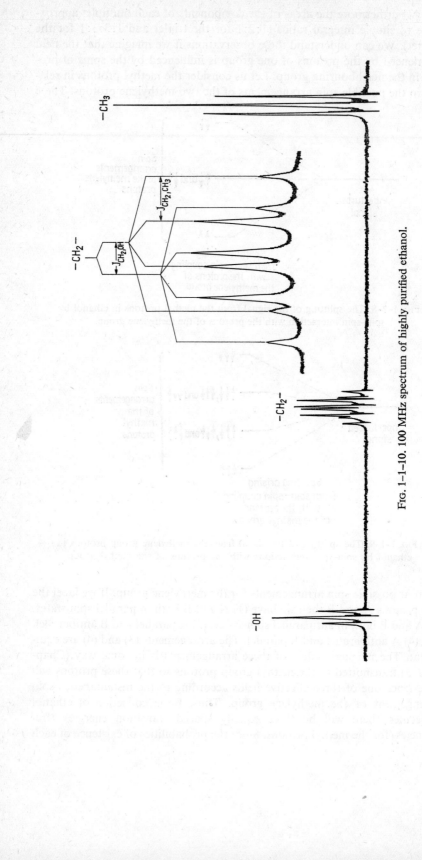

Fig. 1-1-10. 100 MHz spectrum of highly purified ethanol.

of the four spin arrangements are, to a high order of approximation, equal and taking into account the fact that (3) and (4) are equivalent it follows that the intensities of the three transitions will be 1:2:1. These results are illustrated schematically in Fig. 1–1–8. Similarly we find that the spins of the methyl group can be arranged in eight ways of which there are two sets of three equivalent arrangements (Fig. 1–1–9) thus accounting for the observed structure of the methylene multiplet.

In the example just considered, the spacing of adjacent lines in the multiplets is a direct measure of the *spin–spin coupling* of the protons of the methylene group with those of the methyl group, and is known as the *spin–spin coupling constant, J*.

The above argument is a greatly simplified one which only applies in rather special cases. A more general treatment of the interpretation of spin–spin multiplicity and of the determination of spin–spin coupling constants is given in Part 2, Chapter 3.

(iv) Time–dependent phenomena

The lack of multiplicity of the hydroxyl proton signal and further splitting of the methylene group signal in Fig. 1–1–7 is due to a time dependence which frequently influences n.m.r. spectra. If the spectrum of a highly purified specimen of ethanol is examined, the expected multiplicity of the hydroxylic proton signal is observed together with an increase in the multiplicity† of the band from the methylene group (Fig. 1–1–10). It has been found that acidic or basic impurities are responsible for the removal of the coupling between the hydroxylic and methylenic protons. The explanation of this phenomenon lies in the existence of a rapid chemical exchange of hydroxylic protons which is catalysed by acids or bases. As a result of this exchange any one hydroxylic proton, during a certain interval of time, will be attached to a number of different ethanol molecules and will thus experience all possible spin arrangements of the methylene group. If the chemical exchange occurs with a frequency which is substantially greater than the frequency separation of the components of the multiplet from the hydroxylic proton, the magnetic effects corresponding to the three possible spin arrangements of the methylenic protons are averaged, and a single sharp absorption line is observed. In other words, rapid chemical exchange causes spin decoupling of the hydroxylic and methylenic protons.

Chemical exchange can also affect the chemical shifts of nuclei. To illustrate this we shall consider the behaviour of a mixture of ethanol and water. In the absence of acidic or basic catalysts the spectrum of this mixture possesses bands characteristic of the protons of water and of the hydroxylic protons of ethanol⁺ (Fig. 1–1–11a). If a trace of an acid or base is added to the mixture these two bands coalesce to a single sharp line (Fig. 1–1–11b). It is

† Which now appears as a doublet of quartets (see insert in Fig. 1–1–10).
⁺ This is only true at room temperature if the concentration of water is comparatively low (<30 per cent v/v).

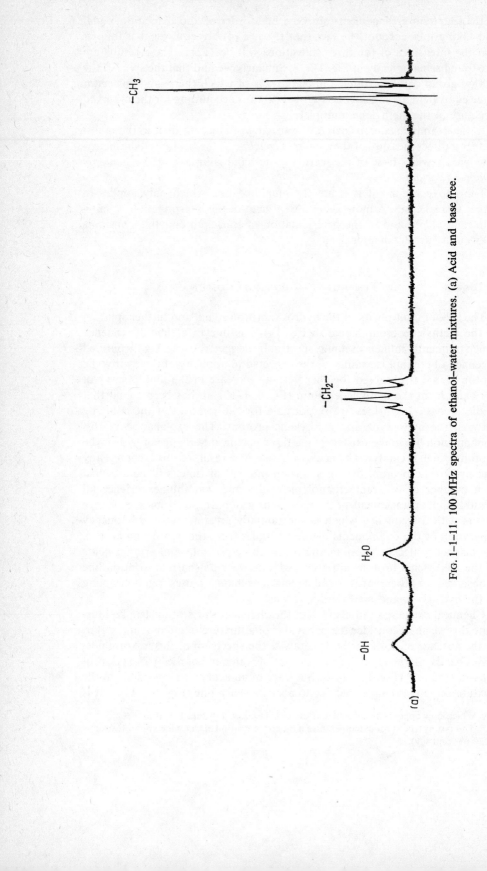

FIG. 1-1-11. 100 MHz spectra of ethanol–water mixtures. (a) Acid and base free.

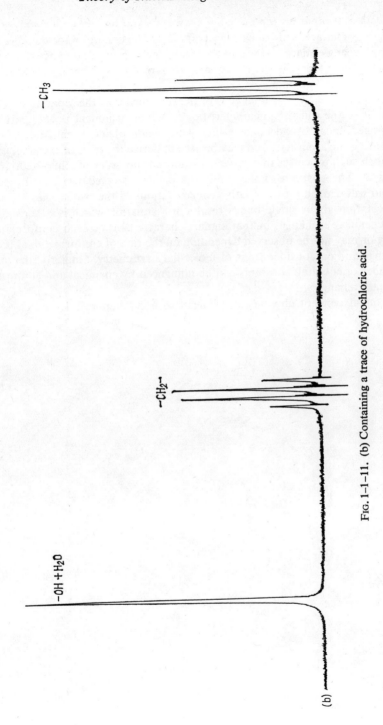

FIG. 1-1-11. (b) Containing a trace of hydrochloric acid.

(b)

evident that rapid proton exchange between water and ethanol averages the shielding characteristic of each environment. The condition for the observation of a single sharp line is that the frequency of proton exchange should be much greater than the frequency separation of the two lines observed in the absence of exchange.

If the frequency of exchange is of the same order as the separation of the two lines, the shielding characteristic of each environment is only partially averaged and a broad band results. This region of indeterminancy can be defined quite accurately and may be used to derive the rates of exchange even though such processes may have half-times of the order of a hundredth of a second. The position of the single sharp line observed when exchange is rapid will depend on the relative concentrations of the two species.

Rotation about single bonds results in a situation which is rather similar to chemical exchange, in that signals characteristic of each conformation may or may not be observed depending on the rate of conformational interchange and on considerations of molecular asymmetry. Similarly, the magnitudes of coupling constants can be influenced by chemical and conformational exchange.

Time-dependent phenomena are dealt with in Chapter 2–1.

THE EXPERIMENTAL METHOD

A HIGH resolution n.m.r. spectrometer in its present form is an instrument of considerable complexity and a detailed knowledge of the electronic equipment will not be of any great advantage to the organic chemist. We will therefore describe the apparatus in the broadest possible terms and discuss in detail only those points of the experimental technique which lie within the operator's control, and which are of importance in that they directly determine the character of the spectra obtained. Our particular concern in this chapter will be the calibration of spectra and the methods of expressing results, for from the organic chemist's point of view, n.m.r. is principally a spectroscopic method which, like infrared and ultraviolet spectroscopy, provides us with numbers characteristic of atoms and their arrangements in complex molecules. We will also consider the measurement of absorption intensities as this forms the basis of the use of n.m.r. spectroscopy for quantitative analysis in organic chemistry.

A. THE N.M.R. SPECTROMETER

The apparatus consists essentially of four parts:

(i) a magnet capable of producing a very strong homogeneous field;
(ii) a means of continuously varying either the magnetic field or frequency over a very small range;
(iii) a radiofrequency oscillator;
(iv) a radiofrequency receiver.

The magnet is necessary to produce the condition for the absorption of radiofrequency radiation. The remaining components then have analogues in other methods of absorption spectroscopy. Thus, the radiofrequency oscillator is the source of radiant energy. The device for varying the magnetic field or frequency over a small range corresponds to a prism or grating in as much as it permits us to scan the spectrum and determine the positions of absorption lines in terms of frequency or field strength. The radiofrequency receiver is the "detector" or device which tells us when energy from the source is being absorbed by the sample.

(i) THE MAGNET

Both permanent and electromagnets are employed in nuclear magnetic resonance spectroscopy. The essential feature of the magnet is that it should possess a region between the pole faces in which the magnetic field is homogeneous to a high order (1 in 10^9). By homogeneous we mean that the strength and direction of the field should not vary from point to point. Furthermore, it is desirable that the strength of this should be as high as practically possible, for we have seen that chemical shifts are proportional to field strength. Thus higher field strengths result in better separation of absorption bands from protons in different environments. High field strengths also have the advantage of giving rise to stronger absorption signals since the Boltzmann excess in the lower spin state depends on H_0. The strength of an absorption signal relative to the irreducible background of radiofrequency noise (signal-to-noise ratio) varies with approximately the one and one-half power of the field strength[86] and in the final analysis an increase in field strength can mean a reduction in the total amount of a compound needed for spectral determination, a consideration which is often of considerable importance to the organic chemist.

The factors which are important in the design of magnets for spectrometers are the *constancy* of the field strength, the *homogeneity* or uniformity of the field, and the maximum obtainable *strength* of the field.

A highly constant magnetic field strength is necessary for the accurate determination of relative line positions in a spectrum. Permanent magnets are capable of maintaining fields which are sufficiently constant provided they are very carefully thermostated and are located in situations which are well away from variable magnetic influences such as large moving iron objects. Conventional electromagnets are intrinsically less stable but the desired stability can be obtained by ancillary stabilizing devices. In these devices, coils wound around the pole pieces sense a change in flux which is then cancelled by passing a current of the correct sign and magnitude through a similar set of coils. Recently, electromagnets consisting of superconducting solenoids operating in liquid helium cryostats have been developed and these are highly stable. Electromagnets interact much less with external magnetic influences than do permanent magnets. In practice, minor variations in field strength can be tolerated if they produce corresponding changes in the irradiating frequency.

Devices which achieve this are called *field-frequency locks* and are incorporated in several commercial spectrometers. Field-frequency locks, which result in reproducibility of a very high order (better than 1 Hz/hr), will be discussed in more detail in section F.

In the preceding chapter we observed that, for liquids, the natural line width of a nuclear resonance signal was governed by the average lifetime of a nucleus in any one spin state. In practice, it is frequently though not always found that the homogeneity of the static magnetic field is the determining factor. Inhomogeneity of the magnetic field causes nuclei in different parts.

of the sample to experience different field strengths and, consequently, to precess at different frequencies. If the broadening of the signal due to this effect is greater than that governed by spin–spin and spin–lattice relaxation the inhomogeneity of the applied field constitutes the limit of resolution.

With both permanent magnets and conventional electromagnets highly homogeneous fields can be obtained by careful machining and alignment of the pole faces. Further improvement is effected by *current shims*. These consist of coils, located at the pole faces, which can be used to produce field

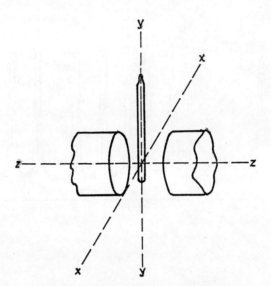

FIG. 1–2–1. Definition of co-ordinate axes.

gradients in various directions so as to cancel out gradients inherent in the main field. Their operation, which can be partially automatic in some instruments, will be discussed in Section B (i). The effect of the finite volume of the sample on attainable resolution can be very significantly reduced by using a cylindrical sample tube and spinning it about the Y-axis (Fig. 1–2–1). If the tube is spun at a rate which is substantially greater than the desired resolution (in Hz), nuclei lying on a given circle in the XZ plane will experience a field which is the average of the fields at each point on that circle so that the apparent homogeneity is increased. All high resolution spectrometers employ this technique.

As mentioned above, several advantages accrue from increasing field strength. Figure 1–2–2 exemplifies the increase in chemical shift which accompanies an increase in field strength. This is particularly desirable when a spectrum is complicated by second order spin–spin splitting (Chapter 2–3) as in the example shown.

The increased signal-to-noise ratio, which in going from 60 MHz to 100 MHz is *ca.* 1·6, is also evident. Permanent and conventional electromagnets are used in spectrometers operating at up to 100 MHz. Super-conducting

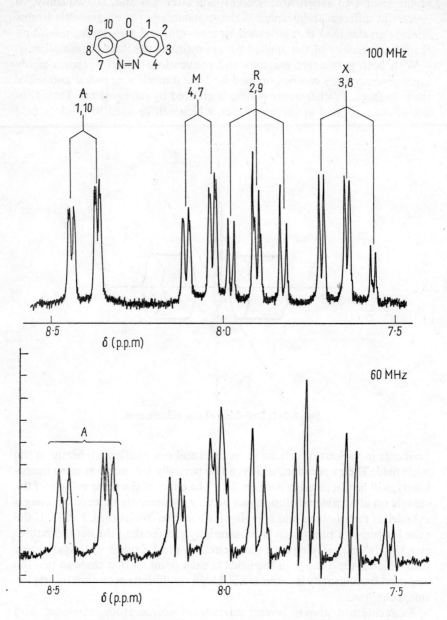

FIG. 1–2–2. 60 and 100 MHz spectra of dibenzo-[*c*,*f*] [1,2]-diazepin-11-one 8 per cent in CDCl₃ (by courtesy of R. G. Amiet and R. B. Johns, University of Melbourne). The 60 MHz and 100 MHz spectra were obtained on different horizontal scales, so that the dimensionless (ppm) scales coincide. The higher dispersion of the 100 MHz spectrum is evident from its first-order character (cf. Chapter 2–3 C).

magnets[1860] have realised fields corresponding to 230 MHz. An excellent example of the advantage of very high fields is provided by the determination at 200 MHz of the very similar chemical shifts of the aromatic protons in alkyl aromatic hydrocarbons, a task which would have been exceedingly difficult or even impossible at 100 MHz or less.[325]

(ii) THE RADIOFREQUENCY OSCILLATOR

We have seen (p. 4) that in order to induce a nuclear transition it is necessary to provide a rotating electromagnetic field, the magnetic component of which moves in a plane perpendicular to the direction of the applied magnetic field. Although the production of such rotating fields is feasible, it is more convenient to use a linearly oscillating field.

This achieves the same result because such a field may be regarded as being the resultant of two components rotating in phase but in opposite directions as is shown in Fig. 1–2–3. One of these components will be rotating in the same sense as the precessing nuclear magnet with which it will interact when the frequencies are the same. The other component will have no effect on the nucleus and need not be further considered.

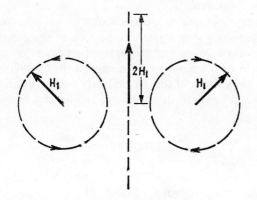

FIG. 1–2–3. The resolution of a linearly oscillating field into two fields rotating in opposite senses.

Linearly polarized radiofrequency radiation is precisely that generally used in radio communication. Consequently our source of radiant energy is merely a radio transmitter (r.f. oscillator) capable of generating a signal of constant frequency, the power of which can be varied if necessary. This signal is fed to a coil situated in the pole gap of the magnet and wound with its axis perpendicular to the direction of the magnetic field. Such an arrangement produces a magnetic component of the electromagnetic field rotating in a plane at right angles to the main field direction. Thus, if a sample is placed inside the coil it will be effectively submitted to a rotating magnetic field correctly oriented for the induction of nuclear magnetic transitions.

(iii) SCANNING DEVICES

The precessing magnetic nuclei can be brought into resonance with the rotating magnetic field by suitable variation of the frequency of either the former or the latter. A spectrum can therefore be scanned either by varying the applied field H_0 (*field sweep* method), which alters the precessional frequencies of the nuclei, or by varying the frequency of the r.f. oscillator (*frequency sweep* method) which varies the frequency of the rotating magnetic field, H_1.

The sweeping or scanning of the static field may be accomplished in two ways. Firstly, it is possible to apply a direct current to coils wound on the two pole pieces of the magnet. In the second method a direct current is fed to a pair of Helmholtz coils† which flank the sample with their axis parallel to the direction of the static field. Either method allows the effective value of H_0 to be varied over a small range without detriment to the homogeneity of the field. The first method is only suitable for relatively low sweep rates as high inductance limits the rate of response of H_0 to the superimposed sweep potential. In practice, both methods can be so operated that the rate of change of the field is constant with time; that is a linear sweep is employed. The slow sweep must operate through the stabilizing system (p. 22), if one is employed.

With certain types of field-frequency lock systems it is feasible to sweep the observing frequency over a small range. One of the great advantages of field-frequency locked systems or very stable permanent magnets is that the sweep (either field or frequency) can be ganged to the X-arm of an XY recorder so that the position of the X-arm bears a direct relation to the field/ frequency ratio and hence to the resonance frequencies of nuclei.

(iv) THE DETECTOR

The passage of radiofrequency radiation through the magnetized sample is associated with two phenomena, namely absorption and dispersion. The terms absorption and dispersion have exactly the same meaning here as they have in classical optics and the way in which each depends on the frequency of the radiation is the same in n.m.r. as in optics.

The line shapes associated with absorption and dispersion are shown in Figs. 1–2–4 and 1–2–5. It is clear that the observation of either dispersion or absorption will enable the resonance frequency to be determined. In practice it is usually easier to interpret an absorption spectrum than the corresponding dispersion spectrum.

Basically, the function of the detector is twofold. It must separate the absorption signal from the dispersion signal and from that of the r.f. oscillator. This second function is necessary because, although the amplitude of

---

† Each coil has the same number of turns and the separation of the coils is equal to their radius.

the applied r.f. signal remains constant, it is very much larger than the amplitude of the absorption signal. There are two principal methods of detection. The first involves the measurement of the effect of the absorption and/or dispersion signals at the transmitter coil of the r.f. oscillator.[265] This

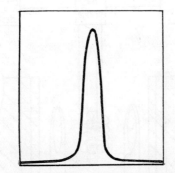

Fɪɢ. 1–2–4. The absorption signal.

requires the use of a radiofrequency bridge which functions in very much the same way as the more familiar Wheatstone bridge. The bridge network balances out the transmitter signal and allows the absorption and dispersion

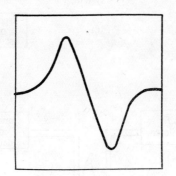

Fɪɢ. 1–2–5. The dispersion signal.

signals to appear as an out-of-balance e.m.f. across the bridge. In this method, the one coil which surrounds the sample serves as both a transmitter and a receiver coil (Fig. 1–2–6). The second method of detection employs a separate receiver coil (Fig. 1–2–7) and is sometimes called the crossed coil or nuclear induction method.[264] If the two coils have their axes at right angles to each other (and also the direction of the static magnetic field) they will not be effectively coupled. In this way the transmitter signal is separated from the absorption and dispersion signals. Provision is made for a variable degree of inductive and capacitive coupling between the two coils so that the transmitter signal can be accurately suppressed.

The separation of the absorption and dispersion signals is achieved by taking advantage of the fact that they differ in phase by 90°. The usual

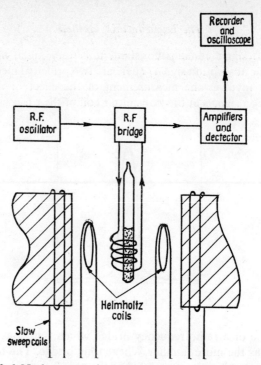

FIG. 1–2–6. Nuclear magnetic resonance spectrometer. The single coil system.

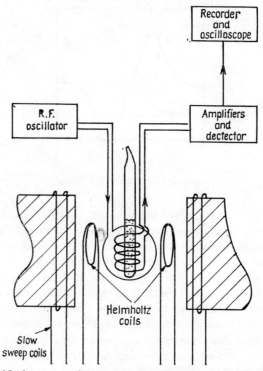

FIG. 1–2–7. Nuclear magnetic resonance spectrometer. The nuclear induction or crossed-coil system.

method is to employ a phase sensitive detector which permits the operator to select the phase of the signal to be detected.

The absorption (or dispersion) signal is extremely weak and requires considerable amplification before it is fed to a chart recorder or oscilloscope.

Figures 1–2–6 and 1–2–7 are block diagrams illustrating the two methods. A detailed understanding of the origins of the absorption and dispersion signals can be obtained from Bloch's original paper.[264]

(v) Field-frequency locks

Some field-frequency locks, as well as other ancillary techniques, depend on the phenomenon known as side-band modulation which is now described. Let us suppose we are examining the spectrum of water. Under normal conditions we observe a single line. If we now apply, say a 100 Hz sinusoidal e.m.f. from an audiofrequency oscillator to either the sweep coils or to the radio-frequency oscillator, we observe that the original absorption line is now flanked by pairs of lines as illustrated in Fig. 1–2–8. It can be shown[75] that the separation of the first pair of side bands from the main signal corresponds to the frequency of the sinusoidal e.m.f., in this instance 100 Hz. The intensities of the first side bands relative to the centre band depend on the ratio of the imposed sinusoidal e.m.f. to its frequency. Thus, modulation can be used to produce radiofrequency fields which differ from the original rotating field H_1 in both magnitude and frequency.

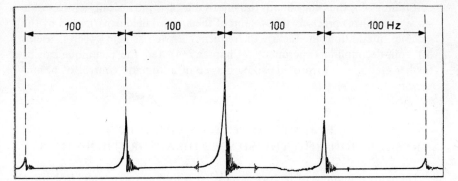

Fig. 1–2–8. The pattern of side bands produced by modulation of the applied field with a sinusoidal e.m.f. (100 Hz).

In practice, it is possible to modulate simultaneously at more than one frequency.

As we are only interested in the *relative* positions of resonance absorptions in a spectrum (p. 12) we wish to measure the position of a given absorption relative to a reference or control absorption which is observed at a *constant* field-to-frequency ratio. The control absorption can be derived from a separate sample (*external lock* system) or from a reference compound added to the sample under investigation (*internal lock* system).

In one method of external locking (e.g. Varian A 60), the sample and reference are simultaneously observed with a side band, the modulation frequency of which always assumes a value which maintains the control sample at resonance. The reference sample, *but not the control*, is field swept with Helmholtz coils. An alternative method (Varian HA 60 with external lock; JEOL JNM-60-C) uses the dispersion signal (which changes sign at its centre) from the control sample to sense a change in the field-to-frequency ratio and to supply a compensating direct current to the stabilizing system (p. 22).

Internal locking is achieved by using two side bands of different frequencies to observe the control absorptions of the internal reference and the signals of the sample (Varian HA 100; JEOL JNM-4H-100). The control signal is again used to compensate, via the stabilizing system, any variation in the applied field. A frequency sweep can be accomplished by linearly varying the frequency of the side band used for observing the sample. If the frequency of the control side band is varied, a compensating change in the applied field is produced and thus effects a field sweep of the spectrum. In either case, the difference in the absorption frequencies of the sample and the control is given directly by the difference in the frequencies of the two side bands. This frequency difference is then made equivalent to the travel of the *X*-arm of an *XY*-recorder.

The external lock has the advantage that it operates continuously, whereas the internal lock is lost when the sample is removed and has to be re-established for each subsequent sample, a process which can be somewhat time consuming. On the other hand, the internal lock is intrinsically more stable since it stabilizes precisely that region of the applied field experienced by the sample. In addition, the facility of frequency as well as field sweep is useful for spin-decoupling experiments (Chapter 2–4) and the technique can be further elaborated to provide some degree of automatic control of homogeneity.

B. EXPERIMENTAL FACTORS WHICH INFLUENCE RESOLUTION AND THE SHAPES OF ABSORPTION LINES

There are a number of factors associated with the experimental technique which can alter the shape of a n.m.r. absorption signal. It is important to be aware of these factors and to take them into account before attempting to assign lines in a spectrum.

(i) HOMOGENEITY OF THE MAGNETIC FIELD

We have mentioned that the line width of an absorption signal is frequently determined by the homogeneity of the magnetic field. In assessing any spectrum, it is necessary to know the resolution at the time of measurement and we therefore require an index of resolution. A convenient estimate of resolving power is provided by the measurement of line width. For this purpose a substance capable of giving a very narrow absorption line should be used.

Acetaldehyde or tetramethylsilane are recommended†. The line width is expressed as the width (in Hz) of the line, measured at half-height (see Fig. 1–2–9). Another criterion of resolution is introduced below (p. 36). The statement of an index of resolution with published spectra is a desirable

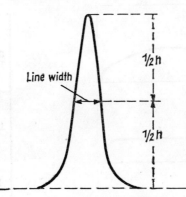

Fɪɢ. 1–2–9. The definition of line width at half-height.

practice. There are many cases where a singlet absorption is observed but where line width measurements reveal the existence of broadening by unresolved spin–spin splitting (Chapter 4–4A).

The homogeneity of the magnetic field between the parallel pole faces of a magnet is greatest in the central region and falls off towards the periphery. For this reason it is desirable that the diameter of the pole pieces should be large (of the order of 6–12 in.) in order to provide an adequate central region which is free from "edge" effects.

Clearly, the smaller the volume of homogeneous field required the higher will be the effective homogeneity. Thus considerations of resolution demand small samples. This requirement has to be balanced against the demands of sensitivity which is related to the number of nuclei of a given type in the sample under investigation. In other words, we have generally to compromise between the conflicting requirements of resolving power and signal strength.

We have seen that homogeneity in the X and Z directions (Fig. 1–2–1) are improved by spinning the sample about the Y-axis. Further control of gradients along these axes is effected by current shims (p. 23). However, the principal sources of inhomogeneity are gradients along the Y-axis and gradients of the type referred to as *curvature*, neither of which is effectively reduced by spinning.

Y-axis gradients lead to increased line width (decreased resolution) and are removed by current shimming. Some spectrometers are also provided

---

† It is important to remove atmospheric oxygen from the sample otherwise the line width may be controlled by paramagnetic broadening (p. 9).

with a mechanical shimming device which permits a minute control over the alignment of the pole faces in the Y-direction, and which constitutes a coarse adjustment of Y-gradients.

Since resolution is so sensitive to Y-gradients an automatic control of the current in the Y-shims is a feature of some instruments (e.g. the "Autoshim" provided with the Varian HA 100).

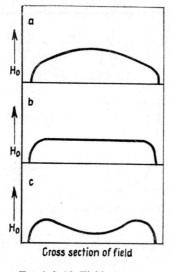

FIG. 1–2–10. Field contours.

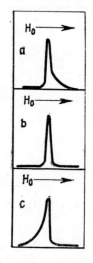

FIG. 1–2–11. Line shapes associated with different field contours.

Curvature consists of concentric gradients about the Z-axis. When an electromagnet is switched on and left for a period to attain an equilibrium state it is frequently found that the central regions of the poles are more heavily magnetized than their peripheries, so that the field has a dome-shaped contour (Fig. 1–2–10a) in directions normal to the Z-axis. In order to remove these undesirable gradients and obtain a flat contour (Fig. 1–2–10b), the magnetizing current is raised for a short period (1–5 min) and then returned to its operating value. This has the effect of increasing magnetization in the peripheral regions and reducing the contour. The flat contour can be approached by successive approximation, the process being termed "cycling". If the process is carried too far, i.e. if the magnet is over-cycled a dished contour results (Fig. 1–2–10c).

Cycling amounts to a coarse adjustment of curvature and a fine control is provided by current shims. Magnets operating at 100 MHz for protons (i.e. 23,487 gauss) cannot be easily cycled as they are saturated. In this case, all curvature correction must be made by current shims and a more elaborate system of shims is used.

Incorrect adjustment of curvature is readily recognized by its effect on line shapes. Coupling between the r.f. field and the cylindrical sample is greatest at the axis of the sample. Therefore, if the sample is situated at the top of a "dome" and the field swept from low to high fields through resonance, the centre of the sample, being at a slightly higher field, comes into resonance first, and as this part of sample is strongly coupled it gives rise to a strong signal. As the sweep continues the outer parts of the sample now resonate but as they are less strongly coupled the strength of the signal tails off. The resulting line shape is shown in Fig. 1–2–11 a. By a similar argument, it follows that a sample situated at the bottom of a "dish" will have the line shape depicted in Fig. 1–2–11 c. Ideally, the field contour should be flat so that symmetrical signals (Fig. 1–2–11 b) can be observed.

(ii) Spinning side bands

The practice of spinning samples to increase resolution was considered on p. 23. If the spinning frequency is too low, the averaging of the field is incomplete and the main absorption signal is accompanied by side bands as illustrated in Fig. 1–2–12. In practice, side bands can also result from uneven spinning of the sample tube so that care must be taken to use tubes which spin smoothly. Spinning side bands are often associated with abnormal field gradients. The spacing of side bands is symmetric about the main band and is equal to the spinning frequency or some integral multiple thereof. Spinning side bands can therefore be identified by comparison of spectra obtained by using different spinning frequencies. Very high spinning frequencies can cause the formation of a vortex which may extend into the operative region of the sample. This is to be avoided since it can cause a serious reduction in resolution.

(iii) Radiofrequency phase

A common source of distortion of absorption signals is incorrect adjustment of the phase sensitive detector (p. 29). This results in signals which are mixtures of the adsorption and dispersion modes and have the shapes indicated in Fig. 1–2–13.

Since the phase of a signal can vary with the volume magnetic susceptibility of the sample it should be adjusted for each sample. The adjustment can be made accurately by examining a single strong line (usually the internal frequency reference line) in the spectrum observed at high gain and power, and varying the phase until there is no negative component in the base-line at either side of the signal. The observation of the integral of the spectrum also provides a sensitive criterion of correct adjustment of phase (p. 50).

(iv) Sweep rate

The rate at which the magnetic field is varied for the purpose of scanning a spectrum can have a profound effect on the shape of absorption signals. Figure 1–2–14a shows the proton signal of chloroform measured with an

2 AN

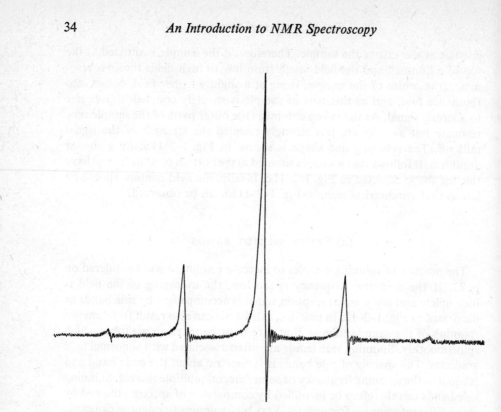

FIG. 1–2–12. The proton signal of chloroform with large spinning side bands caused by the use of an abnormally low spinning frequency.

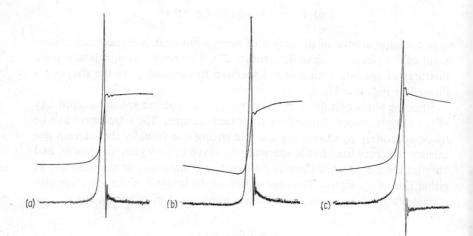

FIG. 1–2–13. The effect of phase adjustment on the shape of signals and their integrals; (a) correct phase adjustment; (b) +15° error; (c) —15° error.

unduly high sweep rate and it is seen that rapid sweeping is associated with considerable distortion of the signal. The distortion is described as a "wiggle" or "*ringing*" and it occurs after the magnetic field has passed through the resonance value. The ringing depicted in Fig. 1–2–14a decays exponentially with time. The origin of the effect can be described in the following way.

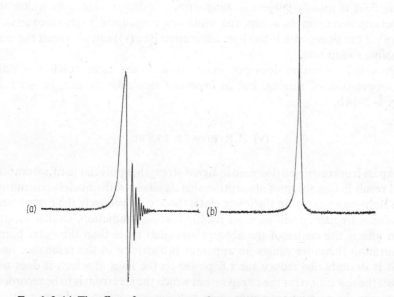

FIG. 1–2–14. The effect of sweep rate on the appearance of signals. Sweep rates: (a) 1·0 Hz per sec. (b) 0·1 Hz per sec.

The Boltzmann excess of nuclei in the ground state corresponds to a magnetization of the sample in the direction parallel to the applied field. At resonance, part of this magnetization appears as a rotating component of magnetization in the plane perpendicular to the applied field. It is in fact this rotating component of magnetization which induces the absorption and dispersion signals in the receiver coil. If at resonance the r.f. field, H_1, is suddenly removed, the rotating component of magnetization will not immediately vanish but rather it will decay exponentially at a rate determined by the relaxation times of the nuclei giving rise to the signal. A similar situation results from rapid sweeping through the resonance condition. In this circumstance, however, the frequency of the rotating component of magnetization varies with the changing sweep field, and as a result the induced signals are alternately in and out of phase with the applied r.f. field, the frequency of which is constant. The result is the characteristic ringing pattern illustrated in Fig. 1–2–14a. Evidently the shorter the relaxation times T_1 and T_2, the less will be the distortion due to ringing. Since T_2 includes the field inhomogeneity (p. 8) the observation of ringing is a good indication of a homogeneous field.

The obvious way of removing the unwanted ringing from recorded signals is to use a slow sweep. There are, however, several factors which place a lower limit on permissible sweep rates. If, as is invariably the case, we ultimately wish to determine the relative separation of bands in a spectrum, it is vital that the sweep rate should not vary throughout the recording of the spectrum. However, unless a field-frequency lock is employed the static magnetic field is usually subject to long-term, non-linear variations which are superimposed upon the sweep, and which can consequently introduce serious errors if the sweep rate is too low. Saturation [see (v)] can also limit the permissible sweep rate.

Provided saturation does not occur slow sweep rates result not only in suppression of ringing, but in *improved resolution* as can be seen in Fig. 1–2–14b.

(v) R.F. POWER LEVEL

Apart from causing a decrease in signal strength, appreciable r.f. saturation will result in distortion of absorption signals since, of the nuclei constituting the Boltzmann excess in the lower state, those with precisely the correct precessional frequency will be most readily lost on irradiation, so that saturation affects the centre of the absorption signal more than the outer parts. Saturation therefore causes an apparent broadening of the resonance line.

It is desirable to reduce the r.f. power to the level at which it does not affect the resolution for the sweep rate at which the spectrum is to be recorded. However, in practice, a compromise between resolution and sensitivity is usually necessary.

(vi) SAMPLE PREPARATION

Occasionally, samples are encountered which give poorly resolved spectra, even when the factors referred to above have been optimized. In these cases it is desirable to pay particular attention to the preparation of the samples.

Quite often poor resolution is associated with the presence of paramagnetic or ferromagnetic impurities in the sample, and, in certain cases, these can be readily removed.

The most common paramagnetic impurity is dissolved molecular oxygen, removal of which is mandatory if maximum resolution is to be obtained. Removal of dissolved oxygen can be effected by flushing with "oxygen-free" nitrogen or more effectively by out-gassing at the boiling point of the solvent. Degassing is especially indicated in dealing with compounds, such as olefins and aromatic hydrocarbons, which tend to form charge transfer complexes with oxygen. Soluble paramagnetic impurities can sometimes be removed from water immiscible solutions by shaking with dilute hydrochloric acid.

Ferromagnetic impurities are often present in samples in the form of dust particles from the atmosphere. A simple filtration suffices to remove this source of line-broadening. It is good experimental technique to filter each sample immediately prior to introduction to the sample tube. This can be conveniently effected by the simple device illustrated in Fig. 1–2–15. The need for this type of pre-treatment cannot be overstressed. Other suspended impurities are also undesirable.

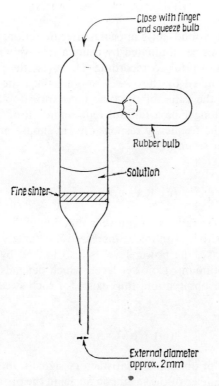

FIG. 1–2–15. Filtering device for sample preparation.

Sometimes poor resolution is associated with dipolar broadening (p. 8) which arises because of the formation of large molecular aggregates in solution. If all other attempts to obtain good resolution have failed it is worthwhile examining the compound at different concentration and in different solvents.

C. EXPERIMENTAL FACTORS WHICH INFLUENCE SENSITIVITY (SIGNAL-TO-NOISE RATIO)

The signal appearing at the receiver coil is extremely weak and requires considerable amplification prior to rectification and recording. The limit to which the signal can be amplified is determined by the ratio of its amplitude to the mean amplitude of the background radiofrequency noise. We may assume that commercial instruments have been designed to reduce the intrinsic noise factor to a minimum so that we may confine our attention to some general considerations over which we have some measure of control.

(i) FREQUENCY RESPONSE

Part of the background noise appears, after detection, as very short term fluctuations and can be eliminated by using a relatively long time constant (frequency response) prior to recording. However, the frequency response will also be determined by the sweep rate at which the spectrum is to be recorded since, if the value of the time constant is too long, the recorded spectrum will no longer be a faithful representation of the absorption signals. In general, the frequency response in Hz should not be less than the sweep rate (Hz per second).

(ii) R.F. POWER LEVEL

The signal-to-noise ratio is, of course, a function of the r.f. power. Because of saturation, sensitivity cannot be increased indefinitely by increasing H_1. As mentioned above, the power level is also limited by considerations of resolution. An optimum r.f. power level, which depends on the line width of the signal under observation, thus exists for each sweep rate.

(iii) CONCENTRATION

The easiest way of obtaining a satisfactory signal-to-noise ratio is to provide a sufficient number of the magnetic nuclei in the operative region of the sample; that is, the region of the sample which is effectively coupled with the r.f. field. We have already seen that the volume of this region is limited by considerations of resolution so that the only way of increasing the number of nuclei it contains is to increase the concentration of the sample in solution.

It is useful to introduce a convenient criterion of the signal-to-noise ratio (S/N). The most common criterion used relates to the largest peak of the quartet of a 1 per cent v/v solution of ethyl benzene recorded under specified conditions (Fig. 1–2–16). Spectrometers operating at 60 MHz detect this signal with a ratio to the RMS noise level of 5:1–15:1† and values of the order of 30:1 are obtained with 100 MHz instruments. If it is assumed that a peak can be identified at a signal-to-noise ratio of 2:1, then a 100 MHz instrument is capable of detecting one proton in a molecule if the concentration of the compound is ca. 0·005 M provided the proton signal is a singlet of line

---

† S/N is referred to the tallest peak of the quartet in the spectrum of ethylbenzene (Fig. 1–2–16). The RMS noise level is obtained by dividing the maximum peak-to-peak noise amplitude by 2.4. A meaningful value is the average of several determinations.

width less than 0·6 Hz. If the proton is involved in spin–spin coupling the resulting absorption is a multiplet, the peaks of which have intensities equivalent to fractions of a proton and very much higher concentrations may be needed in order to detect the components of the multiplets.

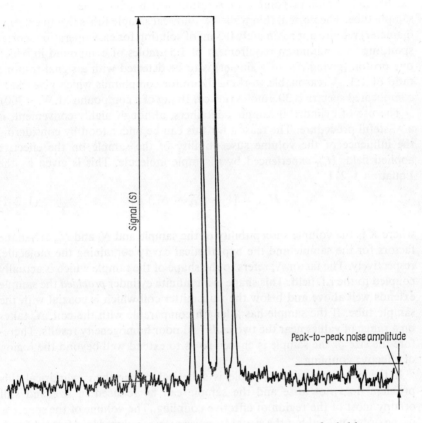

FIG. 1–2–16. 100 MHz spectrum of the methylene protons of ethyl benzene (1 per cent in CCl₄). The spectrum was recorded on a Varian HA 100 spectrometer under the following conditions: R.f. power level; optimized. Sweep rate; 1 Hz per second. Frequency response; 0·5 Hz.

As suitable solvents for nuclear magnetic resonance spectroscopy are rather limited (Section F) the solubility of a compound often determines whether or not its spectrum can be observed. Operation at elevated temperatures can sometimes overcome these difficulties. Some applications demand observations on dilute solutions in which solute–solute interactions are minimized and the desired concentrations may be below the value given above. In these circumstances recourse may be made to the methods of signal enhancement discussed below.

(iv) AVAILABILITY OF COMPOUNDS—SMALL CAPACITY CELLS

In structural organic chemistry, the quantity of material available may be exceedingly small. We are therefore interested in the minimum quantity of material necessary for the determination of a spectrum or, what is equivalent, the minimum volume of solution which has to be placed in the sample tube. The normal thin walled cylindrical sample tubes (5 mm external diameter) require approximately 0·3 ml of solution for easy operation, corresponding to a minimum requirement of 1·5 μmoles of compound in which one proton (giving rise to a singlet) is to be detected with a signal-to-noise ratio of 2:1. A reasonable working figure for compounds which give rise to complicated spectra is 30 μmoles (i.e. ca. 10 mg of a compound M.W. $\sim$300).

The use of cylindrical sample containers, although highly convenient, is a wasteful procedure. The reason for this can be understood by considering the influence of the volume susceptibility of the sample on the effective applied field, H_{eff} experienced by a sample molecule. This is given by the Equation 1–2–1

$$H_{eff} = [1 - (N_s - N_m) K] \qquad (1\text{--}2\text{--}1)$$

where K is the volume susceptibility of the sample and N_s and N_m are shape factors for the sample and the hypothetical cavity containing the molecule, respectively. The factor N_s refers to the shape of the sample which is actually coupled to the r.f. field. This shape is an infinite cylinder *provided* the sample extends well above and below the transmitter coil which is coaxial with the sample tube. If the sample has a length comparable with the coil, N_s takes on a range of values near the two ends and poor homogeneity results. Therefore the size of the sample is chosen so as to extend well beyond the region of effective coupling.

If the geometry of the sample is spherical rather than cylindrical this problem does not arise and the sample can be reduced in size and still occupy most of the region of effective coupling. The volume of the sphere is of the order of 25 μl but the signal-to-noise ratio is reduced by a factor of one-third. Nevertheless, this technique leads to a reduction by a factor of five for the absolute requirement of the sample so that 1–2 mg of a compound (M.W. $\sim$ 300) are sufficient even if the spectrum is complicated.

Several designs have been used for spherical sample containers. The Varian "Microcell" for instance comprises two nylon plugs with hemispherical cavities at each end. These fit together in a precision tube, to give a spherical cavity. The upper plug has a fine capillary outlet to facilitate filling. N.M.R. Specialities supply "all glass" tubes with a spherical cavity at the bottom of a wider capillary through which the sample can be introduced by means of a hypodermic syringe. Samples from gas–liquid chromatograms can be directly trapped into this device. The wider capillary results in slight deterioration in resolution. Successful operation with spherical sample containers requires considerable care with regard to the positioning of the cell in the probe of the spectrometer.

(v) Enhancement of sensitivity by time-averaging devices

The signal-to-noise ratio is a function of the time taken for the observation. Perhaps the easiest way to understand this is to consider the sum of a number of recordings of a peak in a spectrum. The absorption itself will give rise to a positive response which is proportional to its intensity. The summed intensity over a number of scans will simply be that number multiplied by the intensity. The signal arising from random noise at a particular frequency will vary in magnitude and *sign* from one observation to another and its sum over a number of observations will increase less rapidly than an absorption signal. In fact, the sum of the noise will be proportional to the square root of the number of observations. The two basic methods by which this principle can be applied to the enhancement of sensitivity will now be outlined.

The first method employs a computer of averaged transients and is known as the CAT technique.[53,1407,1270] The output voltage (which is normally applied to the recorder) at a series of equally spaced frequency intervals is converted to a digital equivalent and stored in a computer. In other words, if, say, a thousand equally spaced points (frequencies) along the abscissa are selected, the computer assigns a memory address to each point and stores in that address a number which is proportional to the signal strength at the corresponding frequency. The process is repeated a number of times, the results being *added* to those of previous scans. The contents of the stores are then converted back to analogue voltages and recorded as for a normal spectrum. The result is a spectrum with a signal-to-noise ratio which is improved by the square root of the number of scans. The frequency intervals into which the spectrum is divided must be less than the desired resolution. It is also essential that the data stored in a given memory address corresponds in each scan to the same point in the spectrum. For this reason, a field-frequency locked spectrometer, preferably with automatic homogeneity control, or a permanent magnet instrument must be used. Actually, it is only necessary to maintain high frequency stability for the period of one scan since the acquisition of data can be made to commence at the same point in the spectrum for each scan. Small computers designed specifically for n.m.r. spectrometers are commercially available but conventional computers, particularly small data processing machines, can be used, provided "on-line" access to them is available. The spectral data can be read out in digital form on to paper or magnetic tape, a procedure which can be useful in certain physico-chemical work, if it is intended to subject the data to some form of mathematical analysis.

The second method of time averaging has been facetiously called the DOG technique.[2018] In this method, the time averaging is effected by using extremely slow sweep rates and long receiver time constants (i.e. low frequency response). The signal-to-noise ratio will be proportional to the square root of the sweep rate. The method suffers from two disadvantages. It

2a AN

demands higher *long term* stability than the CAT technique. Because slow sweep rates increase the tendency towards saturation, the DOG method must usually be operated at r.f. power levels lower than with the CAT method. A modified DOG procedure which involves only a minor modification of the spectrometer has been described by Dehlsen and Robertson,[618] and can lead to useful improvements in sensitivity as shown in Fig. 1–2–17.

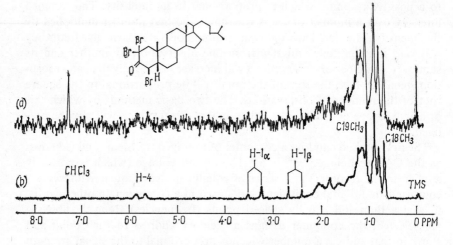

FIG. 1–2–17. Improvement in S/N obtainable with the DOG method[618]. (a) Normal scan—Varian A 60 spectrometer; (b) DOG scan.

The CAT is the better of the two methods although the ancillary equipment is costly. Overnight operation can result in a sensitivity enhancement of the order of 25:1 so that useful spectra can be obtained from ca. 100 μg of a compound (M.W. ~ 300) corresponding to concentrations of the order of 0·0002 M. It is necessary to check the purity of solvents used for measurements of this type since trace impurities from this and other sources can now give rise to observable peaks.

D. MEASUREMENT OF LINE POSITIONS AND DETERMINATION OF THE CHEMICAL SHIFT

The organic chemist is familiar with i.r. and u.v. spectroscopic methods in which the absolute line positions are obtained directly from the instrument. For a number of reasons this practice is not possible with n.m.r. spectroscopy. Indeed, the determination of absolute frequencies to ±1·0 Hz would require measurements to be made with an accuracy of 1 part in 10^8. Fortunately, *relative* line positions can be readily determined with an accuracy of ±1·0 Hz and often with even higher precision.

N.m.r. spectrometers which employ either permanent magnets or field-frequency locks are sufficiently stable to permit the use of pre-calibrated chart paper. The "zero" of the chart can be made to correspond to the position of any line in the spectrum and the positions of all other peaks can be read off directly. In general, the "zero" corresponds to the line position of a reference substance, usually tetramethylsilane (see below). Most spectrometers currently available use pre-calibrated charts.

If the long term stability of the spectrometer is of a lower order, alternative methods for the calibration of spectra are used. The most common procedure is the "side band" technique. As explained above (p. 29) modulation of the applied field or r.f. field produces side bands of known frequency. These side bands can be used to establish the relation between frequency and distance along the X-axis of the spectral chart. If we apply this technique to a many-line spectrum then, provided we can identify the side bands, we can again calibrate the chart and so determine the separations, in Hz, of the various lines in the spectrum. Commercial audiofrequency oscillators are suitable for this purpose. The frequency of the modulation will usually be within one Herz of the value set on the dial and this accuracy is adequate for most of our purposes. If higher accuracy is required the audiofrequency can be measured with a frequency counter. An error arises if the modulation frequency is less than about 20 Hz. The most serious source of error in the calibration made by this method arises from the non-linearity of the magnetic sweep caused by the superimposition of long-term fluctuations in the main magnetic field. Errors of this type can be eliminated in the following way. We measure the separation of two lines in a spectrum by varying the modulation frequency until the first side band of one signal is coincident with the second signal. The frequency separation can then be read off directly from the dial of the oscillator and the value thus obtained is now independent of the linearity of the sweep. The superimposition method is however difficult and tedious to apply to a complicated spectrum and an interpolation method is usually adopted. When using the interpolation method it is important to check for variations in the sweep rate. This can be done by comparing the results obtained for several determinations of the spectrum. Alternatively, a number of side bands can be introduced at regular intervals in the spectrum. The linearity of the sweep may then be established by showing that the distances between successive side bands are in the same ratio as their corresponding frequency separations. Even when this method is employed it is wise to calibrate a second spectrum, in which the field is swept in the opposite direction, and check the consistency of the results.

Another method of determining line separations, known as the "wiggle beat" technique, is ideal for measuring small separations (< 15 Hz) such as observed for many spin multiplets. We have seen that rapid sweeping through sharp signals produces a distortion known as ringing (p. 35). The ringing pattern from a single absorption peak decays exponentially. If two closely situated signals are rapidly traversed, the decay pattern of the first interferes with that of the second. The result is a decay envelope exhibiting a series of maxima and minima (Fig. 1–2–18). The number of maxima observed per second is equal to the separation (in Hz) of the two signals and can be determined by displaying the pattern on an oscilloscope with a known time base.

Alternatively, the pattern can be recorded on a rapidly moving chart, together with time intervals of one second marked off with a fiducial marker. The recorder used for this purpose must have a very rapid response. In suitable cases an accuracy of $\pm 0\cdot1$ Hz can be obtained.

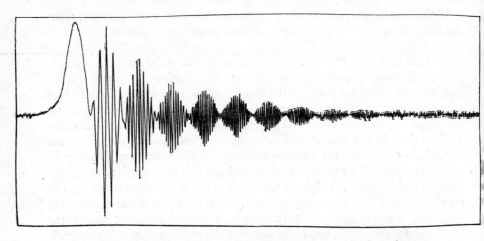

FIG. 1–2–18. The ringing pattern produced by sweeping rapidly through two or more equally spaced signals.

If we are to compare the positions of absorption lines of one compound with those of other compounds, it is necessary to refer all line positions to some standard position. The choice of the standard line position is arbitrary but we must be able to identify it in every spectrum. At first sight we might suppose that this could always be accomplished by adding to a sample a sealed capillary containing a standard substance giving a single sharp absorption signal, which we could then use as a standard line position. When used in this way the standard is called an *external reference*. However, this technique is only legitimate if the effective fields experienced by the sample and reference are the same or if the difference between them is constant for all samples. Now, from equation (1–2–1), the field experienced by a molecule in a sample will depend on the shape and magnetic susceptibility of the sample. If the molecular cavity is assumed to be spherical, the term $(N_s - N_m)$ reduces to $2\pi/3 = 2\cdot09$ for a cylindrical sample oriented with its axis perpendicular to the field (i.e. the normal experimental position of the sample). A similar expression holds for the capillary reference. Bothner-By and Glick[303] have shown that the line position of an external reference is independent of the susceptibility of the sample. Therefore, the field experienced by the sample is not equal to that experienced by the external reference. Furthermore, since K will vary from sample to sample the difference between H_{eff} for sample and reference will not be constant. For these reasons *the values of line positions obtained by the external reference technique will depend on the volume susceptibility of the sample and will not be a true measure of intramolecular shielding.*

In principle, it should be possible to correct for the susceptibility contribution either by determining K or by using spherical samples ($N_s = N_r$). However, Bothner-By and Glick[303] have shown empirically that the correction, $(N_s - N_r) K$, does not depend solely on the shape factors and that $(N_s - N_r)$ may have values between 2·3–3·0 instead of the theoretical 2·09. Insertion of numerical values into equation (1–2–1) reveals that the uncertainty of the correction could result in errors of an order of magnitude greater than we can tolerate.

The use of an external reference can be adapted so as to give reliable shielding values. To do this it is necessary to measure the line positions relative to the external reference from spectra of the compound in solutions of varying concentration, and to extrapolate to infinite dilution. Extrapolation to infinite dilution is, in effect, an extrapolation to the volume susceptibility of the solvent.

Therefore, if the spectra of all compounds are determined in the same solvent and the line positions extrapolated to infinite dilution in each case, the results obtained will permit a direct comparison of the relative shielding of nuclei. The obvious disadvantages of this procedure are that a number of measurements at different concentrations is required for each compound and that the same solvent must be used throughout.

The difficulties of obtaining a constant susceptibility difference between reference and sample are best avoided by using an *internal reference*. An internal reference is actually dissolved in the sample. The reference substance therefore experiences exactly the same field as the compound under investigation and the question of volume susceptibilities does not arise. Thus, although the absolute line position of the reference varies with the susceptibility of the sample–reference mixture, the relative positions of the sample and reference signals remain constant. There is, however, an inherent danger in this method of standardization which derives from the assumption that admixture of the reference compound with the sample does not alter the absolute line position of the former other than by virtue of the change in volume susceptibility. If there is any chemical interaction, however weak, between the reference and solvent or sample molecules, the above assumption is no longer necessarily valid. The data in Table 1–2–1 give some indication of the inconsistencies which may result from an injudicious choice of the internal reference. The data in Table 1–2–1 have been compiled in the following way: the positions of the absorption lines of the reference compounds relative to the single line of tetramethylsilane were determined with an external reference by extrapolation to infinite dilution in carbon tetrachloride; small quantities of each reference were added to samples of pure ethylbenzene and the separation, between the reference lines and the aromatic proton line of ethylbenzene were measured; these separations were then referred to the tetramethylsilane line which was given the arbitrary value of zero. The results should be compared with the last entry in the table which is the relative line position of the aromatic proton of ethylbenzene, obtained with an external reference by extrapolation to infinite dilution in carbon tetrachloride. It is evident that compounds such as chloroform, methylene dichloride and dioxane are unsuitable for use as internal references, presumably because they are specifically solvated by the sample (see Chapter 2–2F).

We can now outline the properties which are demanded of a satisfactory internal reference.

(a) It must be chemically inert to a high degree.
(b) It must be magnetically isotropic or nearly so.
(c) It should give a single, sharp, and readily recognizable absorption signal.
(d) It should be readily miscible with a wide variety of solvents and other organic liquids.
(e) It should be relatively volatile in order to facilitate the recovery of valuable sample material.

TABLE 1–2–1. A COMPARISON OF INTERNAL REFERENCE COMPOUNDS

Reference	Line position of the aromatic protons of ethylbenzene (Hz from tetramethylsilane at 40 MHz)
Chloroform	325·5
Methylenedichloride	317·5
Dioxane	290·5
Cyclohexane	281·5
Tetramethylsilane	282·0
Tetramethylsilane †	284·5
External reference ‡	284·4

† Added to a 10 per cent solution of ethylbenzene in CCl_4.
‡ By extrapolation to infinite dilution CCl_4.

The most suitable reference compound at present available for proton spectroscopy is tetramethylsilane, $[(CH_3)_4Si]$. Its use as an internal reference was first proposed by Tiers,[2403] and is now standard practice. Tetramethylsilane is chemically very inert and, as the twelve protons are spherically distributed, it is magnetically isotropic. In addition, it is volatile (b.p. 27 °C) and miscible with organic solvents. It gives a single absorption line. Tiers has shown that the use of tetramethylsilane as an internal reference corresponds closely to the method employing an external reference at infinite dilution. The data in Table 1–2–2 demonstrate the equivalence of the two methods. Tetramethylsilane has the important advantage that its line position is at higher fields than absorptions of all the common types of organic protons. Its absorption can therefore be readily identified † in all spectra and, further-

† Beware of silicones!

more, it provides a convenient signal for the operation of an internal field-frequency lock.

The solubility of tetramethylsilane in water is far too low for it to be used as an internal reference for samples in aqueous solution. Several compounds have been suggested as suitable internal references for aqueous systems. Tiers and Coon[2410] have described the use of sodium 4,4-dimethyl-4-silapentane-1-sulphonate $[(CH_3)_3SiCH_2CH_2CH_2SO_3^-Na^+]$, which, however, has several disadvantages. It gives rise to absorptions in a region of the

TABLE 1–2–2. THE EQUIVALENCE OF LINE POSITIONS OBTAINED USING TETRAMETHYLSILANE AS AN INTERNAL REFERENCE WITH THOSE OBTAINED EMPLOYING AN EXTERNAL REFERENCE AND EXTRAPOLATION TO INFINITE DILUTION IN CARBON TETRACHLORIDE

The line positions are given as values of δ defined below

Compound†	Internal‡	External††
C_6H_6	7·27	7·26
$C_6H_5C_2H_5$	7·11	7·11
$(C_6H_5CH_2)_2$	7·11	7·11
$p\text{-}C_6H_4(CH_3)_2$	6·95	6·95
Cyclo-octatetraene	5·69	5·74
CH_3NO_2	4·28	4·31
$C_6H_5OCH_3$	3·73	3·69
CH_3OH	3·38	3·40
$(C_6H_5CH_2)_2$	2·87	2·87
$C_6H_5CH_2CH_3$	2·62	2·58
$C_6H_5CH_3$	2·34	2·33
$(CH_3CO)_2O$	2·19	2·19
CH_3I	2·16	2·19
CH_3COCH_3	2·085	2·09
CH_3CO_2H	2·07	2·10
CH_3CN	1·97	1·90
Cyclohexane	1·44	1·49

† The protons to which the values refer are printed in boldface type.

‡ All measurements were made on dilute (1–6% v/v) solutions in CCl_4.

†† These values were obtained from various sources (see Ref. 2403 for references).

spectrum which is frequently of interest and it appears to be sensitive to the presence of aromatic solutes.[1103] Acetonitrile, dioxan and t-butanol have also been recommended,[1304] the first two being preferred, since t-butanol appears to be more affected by aromatic solutes. Tetramethylammonium chloride has been used as an internal reference for concentrated sulphuric acid solutions[631,794] and is probably satisfactory for aqueous solutions. The absorption signal of water itself is an unsatisfactory reference as its position is sensitive to both pH and temperature.

In order to obtain the most reliable shielding values with an internal reference it is important *to make the measurements on a dilute solution* (< 5 per cent w/v) *of a compound in a suitable solvent* (see Section F). The actual concentration employed is usually determined by the signal-to-noise ratio, but extrapolation to infinite dilution may be necessary for some investigations.

Finally, it is necessary to decide on the most convenient way of expressing shielding values. We have seen (p. 23) that the chemical shift is proportional to the field strength or, what is equivalent, to the r.f. oscillator frequency. It is thus desirable to express line positions in a form which is independent of the field strength and frequency. This can be done by using the chemical shift parameter δ' defined by equation (1–2–2) where H_s and H_r are the field strengths

$$\delta' = (H_s - H_r)/H_r \qquad (1–2–2)$$

corresponding to resonance at constant frequency for a particular nucleus in the sample, and the reference, respectively. As we usually calibrate our spectra in c/s we may rewrite equation (1–2–2) as (1–2–3).

$$\delta' = (\nu_s - \nu_r)/\nu_r = (\nu_s - \nu_r)/\text{Spectrometer Frequency} \qquad (1–2–3)$$

where $(\nu_s - \nu_r)$ is negative if the sample nucleus is less shielded than the reference nucleus.† The line position of tetramethylsilane has now been accepted as the standard for proton spectra but unfortunately there are two methods of expressing chemical shift values. The first is known as the *delta scale* in which the line position of tetramethylsilane is zero. The delta value is given by equation (1–2–4).

$$\delta = (\nu_{\text{TMS}} - \nu_s) \cdot 10^6 / \text{Spectrometer Frequency} \qquad (1–2–4)$$

The factor 10^6, which reduces the values to parts per million (ppm), is introduced for convenience since most proton chemical shifts then lie in the range 0–10 ppm. Note that protons less shielded than those of tetramethylsilane have positive δ-values and that *increasing δ-values correspond to decreasing shielding.*

The second expression for proton chemical shifts is the *tau-scale*[2403] defined by equation (1–2–5).

$$\tau = 10 - \delta \qquad (1–2–5)$$

Here the line position of tetramethylsilane is taken as 10 ppm and *increasing τ-values correspond to increasing shielding.* The two scales are widely used in the literature. It is an incredible circumstance that no international agreement to adopt one or other of the scales has been forthcoming so that the

---

† This expression follows from the direct proportionality between ν and H_0 (equation 1–1–2) and applies to a field sweep experiment. For frequency sweep at constant field strength, $(\nu_s - \nu_r)$ will be positive if the sample nucleus is less shielded than the reference nucleus.

literature is now about equally populated with the two symbols with δ possibly gaining ascendancy. *The δ-scale will be used throughout this book.* Table 1–2–3 gives the δ-values of some reference compounds.

TABLE 1–2–3. δ-VALUES OF REFERENCE COMPOUNDS

Tetramethylsilane (T.M.S.)	(0·00)
Sodium 4,4-dimethyl-4-silapentane	
sulphonate	0·00
Acetonitrile	2·00
Dioxan	3·64
t-Butanol	1·27
Tetramethylammonium chloride	3·10

It should be noted that δ-values refer only to the chemical shifts of nuclei. *The line positions of components of multiplets should be quoted in* Hz from tetramethylsilane, together with the spectrometer frequency.[2295] Spin–spin coupling constants are simply given in Hz as they are independent of field strength.

E. MEASUREMENT OF INTENSITIES

The intensity of a n.m.r. absorption is proportional to the area under the absorption curve and, under certain limiting conditions, to the concentration of nuclei which produce the signal. We will therefore be concerned with the factors which control the accuracy of intensity measurements and with the relation of intensity to concentration.

The integral of an absorption spectrum is readily obtained by electronic integration, a facility which is now available with all commercial spectrometers. Figure 1–2–19 is a typical absorption spectrum with its integral recorded on the same chart. The integral is a step function, the heights of the steps being proportional to the intensities of the corresponding absorption bands. The heights of the steps can be determined accurately with a digital voltmeter or by measurement from the chart. The actual signal which is fed to the integrator may be a mixture of the absorption and dispersion signals together with a leakage signal due to imperfect suppression of the carrier signal (p. 27). If the integral is to be a true measure of intensity it is necessary that the dispersion signal is completely rejected and that the leakage signal remains constant.

The integral of the dispersion signal is not a step function and furthermore it extends some distance either side of the resonance position, so that in many spectra there will usually be significant overlapping with the integrals of nearby signals. For these reasons it is necessary to remove completely the dispersion signal by careful adjustment of the phase control of the phase sensitive detector (p. 28). This may be done while observing the absorption spectrum, or alternatively by adjusting the phase to give the maximum value of the absorption integral. The effect of incorrectly adjusted detector phase is illustrated in Fig. 1–2–13.

The carrier leakage can be cancelled electronically. The appropriate adjustment simply consists of balancing the detector to give zero output in a region of the spectrum where there is no absorption. However, the leakage

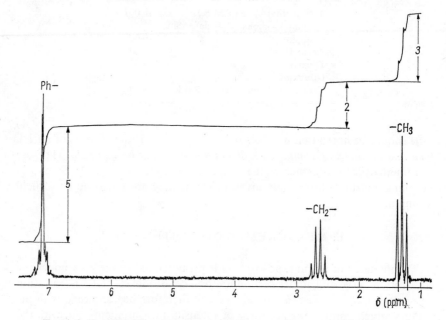

FIG. 1–2–19. 100 MHz spectrum of ethylbenzene (10 per cent in CCl₄) with integral. (Varian HA 100 spectrometer.)

level is subject to fluctuations due to small changes, presumably mechanical in origin, in inductive and capacitive coupling within the probe and such changes can seriously affect the integrals. Errors arising in this way can be minimized by careful adjustment of the probe to minimum leakage. Some instruments incorporate a modulation technique which greatly reduces the effect of fluctuations of this type. The applied magnetic field is modulated at audio-frequency and the absorption signals are amplified at audio-frequency and detected with an audio-phase detector. Since the absorption signals, but not the carrier leakage, is modulated only the former is detected and fluctuations in leakage are considerably reduced.

The value of the integral of an absorption band (i.e. the height of a step in Fig. 1–2–19) is inversely proportional to the sweep rate. The accuracy of integrals as a measure of intensities will therefore depend on the linearity of the sweep. If a field-frequency lock is used or the spectrometer employs a permanent magnet this source of error is unimportant. With less stable equipment it is necessary to use rather high sweep rates in order to minimize errors arising in this way. However, a compromise is necessary as the absolute value of the integral, and hence the accuracy with which it can be measured, decreases with increasing sweep rate.

The distortion of signals by ringing (p. 35) is generally assumed not to contribute to the value of the integral.

We now turn to a consideration of the relation between intensity and concentration of protons. In this connection it should be noted that we are invariably concerned with the measurement of relative intensities within the same sample, in order to determine the relative numbers of protons of various types in a molecule or the relative concentrations of components in a mixture. The determination of absolute concentrations involves calibration of the instrument with a reference compound present at known concentration. The reference compound may, in fact, be incorporated in the sample under investigation.[1588,1761,171,1294,2314,1316] Some suitable reference compounds are 1,3,5-trinitrobenzene and disodium terephthalate for non-aqueous and aqueous solutions, respectively.

The intensity of absorption is directly proportional to the Boltzmann excess of nuclei in the ground state. In the absence of the radiofrequency field, H_1, this will have the same value for all protons irrespective of their chemical environment. However, the application of the radiofrequency field reduces the excess by an amount which is a function of H_1, T_1 and T_2, the relaxation times for the protons. The function is of the form:

$$\text{Intensity} \propto [1 + \gamma^2 H_1^2 T_1 T_2]^{-1/2}$$

Now T_1 and T_2 vary for protons in different environments even within the same molecule. Thus, unless the second term is much smaller than unity, observed intensities may not be proportional to concentrations. It is therefore necessary to establish that the observed relative intensities in a spectrum are independent of the r.f. power (H_1) before equating these values to concentrations or relative numbers of protons. The tendency towards saturation is also dependent on the sweep rate, in such a way that lower sweep rates accentuate errors arising from differential saturation of protons in different environments.

Several further precautions are necessary in quantitative work. Solvents should be examined in case impurities are present at levels which could interfere in analyses. For highly accurate work, it is sometimes necessary to take into account the presence of ^{13}C nuclei. This isotope, which has $I = \frac{1}{2}$, is present at a natural abundance of 1 per cent. Protons directly attached to ^{13}C nuclei give rise to doublets with a spacing of 120–130 Hz. These absorptions appear as satellites of the main absorption band, each component having 0·5 per cent of the intensity of the main band. If one component overlaps a weak absorption whose intensity is to be determined it is clear that an appreciable error can result.

The following is a summary of the factors which are important in the use of n.m.r. spectroscopy as a quantitative analytical technique:

(i) Careful adjustment of r.f. phase to suppress the dispersion signal.

(ii) Use of a properly designed integrator which employs the audio-frequency modulation technique.

(iii) Selection of a sweep rate which is commensurate with the frequency stability of the instrument and with the desired sensitivity of measurement of the integral.

(iv) Examination at several r.f. power levels to establish the independence of the relative values of the integrals.

(v) Performance of a sufficient number of observations under optimum conditions to establish a precision index for the final answer.

(vi) Examination of solvent "blanks".

(vii) Correction for ^{13}C satellites.

Careful attention to the above factors can lead to a precision of better than 1 per cent. The method thus compares favourably with combustion methods for the analysis of hydrogen in organic molecules, and has the distinct advantage that it is non-destructive. It is also a most convenient method for studying equilibria and the kinetics of reactions.

F. SOLVENTS

The choice of suitable solvents for proton spectroscopy of organic compounds is very limited. As well as being capable of giving fairly concentrated solutions of organic compounds, a satisfactory solvent should be chemically inert, magnetically isotropic and preferably devoid of hydrogen atoms. Carbon tetrachloride is the ideal solvent and should be used whenever possible. However, many compounds are insufficiently soluble in carbon tetrachloride and other solvents have to be used. Carbon disulphide, cyclohexane and deuterochloroform† are often suitable and the δ-values measured in these solvents usually correspond closely to measurements in carbon tetrachloride. Sometimes other solvents must be used. These include pyridine, acetone, acetonitrile, dimethylformamide and their perdeutero derivatives. Dimethyl sulphoxide-d_6 is a particularly useful solvent. Frequently, addition of a few per cent of this solvent to deuterochloroform is sufficient to solubilize many poorly soluble compounds and this avoids the line broadening associated with the high viscosity of neat dimethyl sulphoxide. Trifluoroacetic acid has also found wide use as a solvent. With many of these solvents, the observed δ-values differ appreciably from those obtained in carbon tetrachloride, deuterochloroform and carbon disulphide. This is particularly true of aromatic solvents, such as pyridine, where "solvent shifts" of the order of 0·5 ppm are often observed.

Advantage is sometimes taken of the "solvent shifts" to remove the overlap of absorption bands and thus aid the interpretation of the spec-

---

† However, Laszlo has quoted examples in which shifts as great as 0·3 ppm are associated with the use of deuterochloroform as a solvent.[1503]

trum.[2298,2430] This is well illustrated by the spectra in Fig. 1–2–20 taken from the work of Slomp and MacKellar.[2298] Solvent effects are discussed in some detail in Chapters 2–2F, 2–3C and 3–8J.

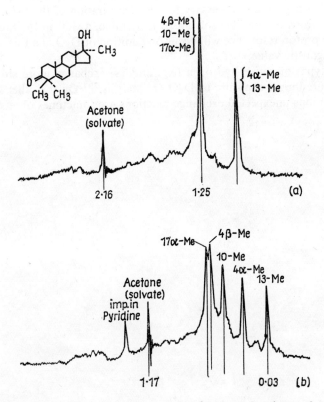

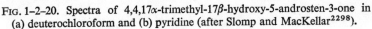

FIG. 1–2–20. Spectra of 4,4,17α-trimethyl-17β-hydroxy-5-androsten-3-one in (a) deuterochloroform and (b) pyridine (after Slomp and MacKellar[2298]).

G. DEUTERIUM EXCHANGE

Deuterium exchange is an extremely useful technique in conjunction with n.m.r. spectroscopy, as it results in the effective removal of protons and hence in simplification of spectra (see, e.g. Fig. 2–3–13 on p. 143). We shall discuss the spectra of deuterated compounds in several places but in the present context a mention must be made of the technique of *in situ* exchange of labile protons, e.g. those of the hydroxyl and amino groups.[791]

If the spectra of such compounds are determined in heavy, water-immiscible solvents (CCl_4 or $CDCl_3$), a brief vigorous shaking† with a few drops of D_2O generally results in complete exchange and the protium thus liberated

† The temptation to use the index finger as a stopper of sample tubes in general is almost irresistible, but should be suppressed when solvents such as trifluoroacetic acid are used, as surprisingly severe burns will result.

is, of course, present in the aqueous layer which is outside the coupled region of the sample. The identical technique can be used with solvents such as benzene, but the amount of D_2O must be very small, otherwise the aqueous layer may extend into the effective region of the sample. In either case centrifuging is generally necessary for clean separation of the two layers.

Even with water-miscible solvents e.g. pyridine, a shift in the position of the labile proton resonance will occur on addition of D_2O and this may also be of diagnostic value.

Some types of active hydrogen (e.g. amides) exchange more slowly and may require longer exposure to D_2O or catalysis.[791] On the other hand, in D_2O solutions unexpected exchange reactions are sometimes observed.

PART 2

THEORY OF CHEMICAL EFFECTS IN NUCLEAR MAGNETIC RESONANCE SPECTROSCOPY

TIME-DEPENDENT EFFECTS IN NUCLEAR MAGNETIC RESONANCE SPECTROSCOPY

THE nature of all n.m.r. spectra depends on the rates of various processes such as inter- and intra-molecular motions and chemical exchange. As these processes can influence the values of spectral parameters, such as line width, chemical shifts and spin–spin coupling constants, it is important that the underlying principles should be fully understood before we proceed with a detailed treatment of the relation between n.m.r. spectra and chemical constitution. In particular we will see that the spectra of some molecules can change radically if the rates of certain processes are changed by altering the conditions of observation, e.g. temperature or pH of solution.

We have already considered one process, namely intermolecular interactions with neighbouring dipoles (p. 8), the effect of which is time-dependent. Here, we noted that in solids a particular nucleus experiences a field from neighbouring magnetic nuclei and the magnitude of this field depends on the spin orientations of these nuclei. Furthermore, we observed that this effect was absent in the liquid state because of rapid random motion of molecules. What we really meant was that in solids the magnetic environment of a particular nucleus exists for a period which is long compared with the time of survival of its environment in the liquid state. In the latter situation the nucleus experiences an *average* environment (in this example the average contribution is, for all practical purposes, zero). What we need to know now is how rapidly the environment must change in order for its effect to be equal to the average over all instantaneous values.

It will be convenient at this stage to consider an example in which the environment can have only two values rather than the general case where many values may be possible. Such an example is provided by a system in which a proton is undergoing chemical exchange between two sites. For instance, we could consider the exchange of a proton between two bases (equation 2–1–1). In a system of this type,

$$AH^+ + B: \rightleftharpoons A: + BH^+ \qquad (2\text{–}1\text{–}1)$$

the two environments are characterized by the chemical shifts (ν_A and ν_B) in Hz of the proton relative to some arbitrary reference. If the life-time of AH^+ and BH^+ are τ_A and τ_B seconds, we define the life-time of the system as $\tau = \tau_A\tau_B/(\tau_A + \tau_B)$. To begin with we will consider the special case in which the concentrations of AH^+ and BH^+ are equal. In this case $\tau = \tau_A/2 = \tau_B/2$. We also assume that the transverse relaxation times (T_2) for the proton in the two species are the same and are large compared with $1/(\nu_B - \nu_A)$.

We can now state, in qualitative terms, the conditions for which the proton gives rise to a single line at $(\nu_A + \nu_B)/2$ and a pair of lines at ν_A and ν_B. The former situation arises if $\tau \ll 1/(\nu_B - \nu_A)$ and similarly two lines at ν_A and ν_B are observed if $\tau \gg 1/(\nu_B - \nu_A)$, it being assumed that $\nu_B > \nu_A$. Of particular interest is the situation in which τ and $1/(\nu_B - \nu_A)$ are of the same order. The behaviour of the system under this condition was first deduced theoretically by Gutowsky, McCall and Slichter[1071] and later, and more simply, by McConnell.[1616] The predicted behaviour is presented in Fig. 2–1–1. We note that as the life-time decreases through the critical range the two lines broaden and move together until they coalesce to a broad singlet. As the life-time decreases still further the broad singlet sharpens and eventually reaches a line width which is determined only by T_2. The life-time corresponding to coalescence is given by equation (2–1–2)

$$\tau = \sqrt{2}/2\pi\,(\nu_B - \nu_A) \qquad (2\text{–}1\text{–}2)$$

Since the line shapes can be computed for any value of τ and $(\nu_B - \nu_A)$, and since $(\nu_B - \nu_A)$ can be determined directly from the spectrum observed under conditions of slow exchange, it follows that this phenomenon provides a method of determining the rates of quite fast reactions. Since $(\nu_B - \nu_A)$ is generally in the range 10–1000 Hz, life-times of the order of 10^{-2}–10^{-4} sec can be investigated. For the particular case with which we are dealing and with $T_2 \gg 1/(\nu_B - \nu_A)$, four approximate methods are available for extracting τ from spectra corresponding to partial coalescence. The first, due to Gutowsky and Holm,[1065] takes advantage of a simple relation between the separation, $\Delta\nu_{obs}$, of the two maxima in partially coalesced spectra (equation 2–1–3)

$$\frac{\Delta\nu_{obs}}{(\nu_B - \nu_A)} = \left[1 - \frac{1}{2\pi^2\tau^2\,(\nu_B - \nu_A)^2}\right]^{1/2} \qquad (2\text{–}1\text{–}3)$$

The second method[2131] relates life-time to the ratio of the peak maxima to the minimum between them (equation 2–1–4)

$$\tau = \pm\frac{[2r \pm 2(r^2 - r)^{1/2}]^{1/2}}{2\pi(\nu_B - \nu_A)} \qquad (2\text{–}1\text{–}4)$$

where $r = $ Max. Intensity/Min. Intensity. In the region where the two peaks are broadened but do not overlap, equation (2–1–5),[1053,1975] which is independent of peak separation, can be applied

$$\frac{1}{\tau} = 2\left[\frac{1}{T_2^{exch}} - \frac{1}{T_2^0}\right] \qquad (2\text{–}1\text{–}5)$$

where T_2^{exch} and T_2^0 are found by multiplying π by the line widths found for exchange and in the absence of exchange, respectively.

Line width measurements of coalesced spectra can also yield life-times through equation (2–1–6)[1975,66]

$$\frac{1}{\tau} = \pi^2 (\nu_B - \nu_A)^2 \left[\frac{1}{T_2^{\text{exch}}} - \frac{1}{T_2^0} \right]^{-1} \qquad (2\text{--}1\text{--}6)$$

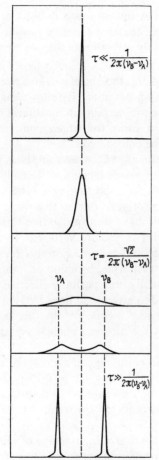

FIG. 2–1–1. Theoretical line shapes for nuclei exchanging between two equally populated sites.

It is stressed that the last four equations do not hold in the region of coalescence, i.e. when $\tau \sim \sqrt{2}/2\pi\,(\nu_B - \nu_A)$. In this region, and when $1/T_2$ and $(\nu_B - \nu_A)$ are of similar magnitude, the life-time can be determined by comparison with theoretically computed spectra.[1049]

In addition to chemical exchange processes, such as equation (2–1–1), there are other interactions which can provide two or more magnetic environments for a proton in a molecule, and which can therefore be associated with time-dependent phenomena. One of the more important of these is spin–spin coupling (p. 13). In a molecule such as (I) the proton A will have two magnetic environments depending on the spin state of proton B. Whether

$$CH_ABr_2 \cdot CH_B(CN)_2$$
(I)

or not the resonance of A appears as a doublet or a singlet will therefore depend on the life-time of the spin states of proton B.

This situation is precisely the same as the one we have just discussed. We can therefore conclude that if $\tau \gg \sqrt{2}/2\pi J$, where J is the spin–spin coupling constant between H_A and H_B, the nuclei will give rise to doublets. As in most cases we observe spin–spin coupling, it follows that this condition usually prevails. However, there are certain circumstances in which the spin state life-time is reduced to a value for which complete or partial averaging occurs. These will now be considered.

Mechanisms for the exchange between spin states which are always operative are spin–lattice and spin–spin relaxation. Usually these are too inefficient to average resolvable spin–spin splitting. One system in which this is not the case is that in which the proton is coupled to a nucleus which has a non-zero quadrupole moment. We have seen (p. 10) that this provides efficient spin–lattice relaxation and this is usually sufficient to average, completely or partially, the coupling to the proton. For this reason we do not observe coupling of protons with chlorine, bromine or iodine nuclei. In the case of ^{14}N, the averaging is usually only partially effective so that protons directly attached to nitrogen tend to give broad lines.[2408] An exception is the ammonium ion in strongly acidic media.[1053] In this case there are no strong electric field gradients at the nitrogen nucleus and quadrupole relaxation is therefore inefficient.

Averaging of spin–spin coupling can also result from chemical exchange. The example of ethanol has been given in the opening chapter (Figs. 1–1–7 and 1–1–10). In effect, the exchange process results in a rapid interchange in the two magnetic environments (provided by the hydroxylic protons) of the methylene protons.

A third way in which the rate of interchange of spin states of a nucleus can be effected is by strong irradiation at the resonance frequency of the nucleus. This causes averaging of the coupling and is the principle of a very valuable technique known as spin-decoupling or multiple irradiation, which will be discussed in Chapter 2–4.

The majority of organic molecules can exist in more than one conformation. Since conformational interchange at room temperature is generally rapid on the n.m.r. time scale, the spectra we observe are averages of all conformations. Rapid conformational equilibria average spin–spin coupling as well as chemical shifts.

In some cases, the process causes protons, which would have different chemical shifts and coupling constants in any one conformation, to become magnetically eqivalent (i.e. same chemical shifts and equally coupled to other protons in the molecule; cf. Chapter 2–3 C). This is always the case with the three protons of the methyl group.

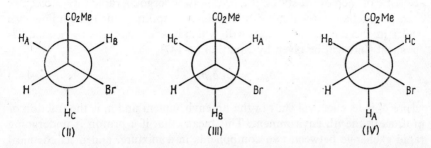

For instance, if we arbitrarily label as H_A one of the protons in methyl α-bromopropionate, we see that it has a different chemical environment in each of the three conformations (II), (III) and (IV). Furthermore, in each conformation the coupling between H_A and the α-proton will be different. However, interchange between these three conformations is so rapid that H_A has an average chemical shift and an average coupling constant to the α-proton which are of course exactly the same for the other two protons, H_B and H_C.

It is clear that, in order to predict the appearance of the spectrum of a compound, we need to have a feeling for the rates of processes which the molecules can undergo under the conditions of measurement. The following approximate rules will serve as a rough guide for the usual types of organic molecules under normal conditions.

(i) Rapid chemical exchange will usually only occur with "acidic" protons; e.g. —CO_2H, —OH, —NHR. Hydroxylic protons and protons of amines frequently do not undergo rapid exchange at low concentrations in neutral solutions. However, these exchange processes are catalysed by minute concentrations of acids and bases.

(ii) Conformational changes involving rotation or partial rotation about single bonds are almost always sufficiently rapid to result in averaging at room temperature.

(iii) Rotation about double bonds is of course too slow to cause averaging.

(iv) Rotation about partial double bonds (e.g. the C—N bond in amides) may be in the range which corresponds to partial averaging.

These and other phenomena are discussed in Chapter 5–1.

Our initial treatment of time-dependent phenomena was restricted to equal population of states having equal and long relaxation times. It is clear that many processes, such as conformational interchange and spin–spin coupling resulting in multiplicity greater than a doublet, will require a more elaborate theory. Techniques for dealing with more complicated systems are available and have been the subject of some excellent reviews.[1286,2079] We will however deal explicitly with the particular case of very rapid exchange between a number of *unequally* populated states.

Very rapid exchange between two or more unequally populated states will give rise to completely averaged spectra and our only concern will be the relation of the average chemical shifts and coupling constants to those characteristic of each of the states. If a proton is undergoing rapid intermolecular chemical exchange it will not exhibit spin–spin splitting due to coupling with nuclei in the molecules or ions with which it is momentarily associated. Its chemical shift will be given by equation (2–1–7)

$$\delta_{obs} = \sum_i \delta_i p_i \qquad\qquad (2–1–7)$$

where δ_i is its chemical shift in the ith environment and p_i is the fraction of protons in the ith environment. This means that if a proton is undergoing rapid exchange between two components in a mixture, and if its chemical shift in each component is known, the relative concentrations of the two components can be determined directly from its chemical shift in the mixture.

Equation (2–1–7) also holds for rapid intramolecular exchange processes, such as conformational interconversion. In this case a similar expression, equation (2–1–8), holds for coupling constants, since such changes do not result in an effective change of the spin state of a nucleus.

$$J_{obs} = \sum_i J_i p_i \qquad\qquad (2–1–8)$$

THEORY OF THE CHEMICAL SHIFT

A. CLASSIFICATION OF SHIELDING EFFECTS

It has already been pointed out (Chapter 1–1 D) that the field experienced by a nucleus in an atom or molecule is not precisely equal to the applied field, H_0, because the nucleus is to some extent shielded by the extra-nuclear electrons associated with it and with neighbouring nuclei. In principle, electrons can influence the field at a nucleus in one or both of two ways. Firstly, the presence of unpaired electron spins can, in certain circumstances, give rise to a magnetic component at the nucleus. This effect is confined to paramagnetic molecules and is rarely encountered (see Section D). The second way in which electrons shield nuclei is common to all atoms and molecules and is associated with the motions of electrons in a magnetic field. This latter effect is, indeed, the origin of atomic and molecular diamagnetism as well as the chemical shift. An initial insight into the mechanism of this second effect is best gained by a consideration of an atom.

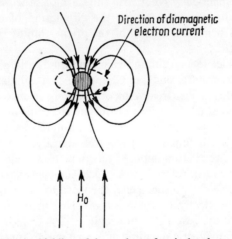

FIG. 2–2–1. Diamagnetic shielding of the nucleus of an isolated atom.

For an atom in a magnetic field, Larmor's theorem[2469] states that the motion of electrons differs from that in the field-free system by a common precession of frequency $H_0 e/4mc$. The physical significance of this theorem is now explained by reference to the Bohr model of the hydrogen atom shown in Fig. 2–2–1. In the presence of the magnetic field, H_0, the electron circulates (with a frequency $H_0 e/4mc$) in the direction shown (clockwise). The motion of the electron is equivalent to an electric current flowing in a closed loop and as such it is associated with a secondary magnetic field which is also depicted in Fig. 2–2–1. The secondary magnetic field opposes the applied field, H_0, at the nucleus which consequently experiences a total field which is slightly less than H_0. Thus, the observed resonance frequency of a proton appears to be slightly less than that predicted from the value of H_0 and the magnetic constants of the proton. Equation (1–1–2) can therefore be modified to

$$\nu = \gamma H_0 (1 - \sigma)/2\pi \qquad (2\text{–}2\text{–}1)$$

where σ is the shielding constant for the hydrogen atom. Figure 2–2–1 is based on the Bohr model but the qualitative extension of the argument to the orbital model is obvious. An important result which follows from the above treatment is that the frequency of the Larmor precession is proportional to H_0; hence the strength of the secondary magnetic field and consequently the chemical shift (measured as a frequency) are likewise proportional to H_0.

Extension of the treatment of shielding in free atoms to that of atoms in molecules is complicated by several factors. An understanding of these factors is particularly important as they are related to the way in which molecular structure affects the chemical shift. When hydrogen is chemically bound, as in H—X, the circulations of electrons are modified in two ways. Firstly, the electron density at the proton will not be the same as in the hydrogen atom. Secondly, the distribution of electrons around the proton is no longer spherically symmetric and Larmor's theorem does not hold. Chemical bonding can also restrict the circulation of electrons, the degree of restriction being a function of the orientation of H—X with respect to the applied field.

The associated shielding, which in liquids and gases is a time-average of all possible orientations (see Chapter 2–1), is similarly affected.

In the theoretical treatment of shielding in molecules it is customary to separate the shielding arising from the electrons directly associated with the nucleus, into two terms. The first term relates to normal circulation in atoms and is a function of electron density. The second term takes into account the restriction imposed on circulation of electrons by chemical bonding. The two are referred to as *local diamagnetic shielding* and *local paramagnetic shielding*, respectively, the word local being included to specify the shielding by electrons which are involved in the bonding of the proton in question to the rest of the molecule.

Local diamagnetic shielding makes an important contribution to the shielding of protons in organic molecules. Its magnitude will differ from that of the shielding in the hydrogen atom by virtue of changes in electron density in the immediate vicinity of the nucleus, which result when the hydrogen atom is involved in chemical bonding.

For the proton, local paramagnetic shielding is probably much less important than the diamagnetic term. In the case of the hydrogen molecule it is approximately an order of magnitude smaller,[2046] and its variation with molecular structure probably parallels that of the local diamagnetic shielding. For other nuclei, e.g. ^{19}F, ^{13}C, ^{14}N etc., local paramagnetic shielding dominates the chemical shift and, as predicted by theory,[2155] the shielding of such nuclei decreases with increasing electronegativity of the elements or groups to which they are bound.

In addition to the local shielding effects just considered, significant contributions to the shielding of protons may arise from the circulations of electrons associated with other atoms or groups in the molecule. Shielding of this type is called *long-range shielding*. It will be shown below (Section C) that long-range shielding arises only if the magnetic polarizability of the electrons is anisotropic, i.e. if the circulations induced by the applied field are greater for some orientations of the molecule in the applied field than for others. Long-range shielding may arise from electron circulations about a neighbouring atom in the molecule in which case it is referred to as *neighbouring paramagnetic shielding* or, alternatively, circulations may occur about two or more atomic centres and this effect is known as *interatomic diamagnetic shielding*. An important feature of long-range shielding is that it can be either positive or negative, depending on the geometric relation between the group of electrons responsible for the effect and the nucleus under consideration. The magnitude of the effect is such that it is important in discussions of the chemical shifts of protons (and deuterons) but of less significance for heavier nuclei for which the overall scale of chemical shifts is dominated by local paramagnetic shielding.

Chemical shifts of protons in organic compounds are almost invariably measured in the liquid phase, usually as solutions in suitable solvents. Any discussion of the theory of the chemical shift must therefore include a consideration of intermolecular effects. A solvent can contribute to the shielding of a proton in a molecule by its influence on the electronic structure of the molecule (an effect which can be substantial in the case of strong solvation) and by direct magnetic field contributions in the case of magnetically anisotropic solvents. The final section in this chapter deals with solvent effects.

In summary, the shielding (σ) of a nucleus can be expressed as the sum of five terms

$$\sigma = \sigma_d^L + \sigma_p^L + \sigma_p^N + \sigma_d^N + \sigma_s \qquad (2\text{--}2\text{--}2)$$

which are the local diamagnetic and paramagnetic, neighbouring paramagnetic and interatomic diamagnetic, and solvent shielding contributions, respectively, it being realized that σ_s is only the direct shielding contribution by the solvent and that the solvent may also influence the magnitudes of the other four terms.

B. LOCAL DIAMAGNETIC PROTON SHIELDING

The relation between the local diamagnetic shielding of a proton and chemical structure involves the effect of electric fields on the circulations of electrons around the proton, and of the electron density at the proton.

The influence of an electric field (E) on the hydrogen atom (electron spin being neglected) has been shown[1679] quantum mechanically to result in a reduction in shielding proportional to $E.^2$ The dependence is such that a partial charge of 0·2 of an electron at a distance of 3 Å causes a down-field shift of 0·08 ppm. The effect is thus small and in most cases it is negligible compared with a term directly proportional to the component of E along the X—H bond, which applies to protons in molecules and which is discussed below. It does however assume importance in the discussion of solvent effects (Section F). The term in E contributes to both σ_d^L and σ_p^L, although the contribution to the latter is small.[1679]

The much more important factor determining the magnitude of local diamagnetic shielding is the local electron density. Indeed, the relation between σ_d^L and the polarity of the C—H bonds, as inferred from several general criteria, forms the basis of perhaps the most useful means of predicting approximate proton chemical shifts in organic molecules.

Various attempts have been made to correlate proton chemical shifts in substituted alkanes with the electronegativities of the substituents. The earliest attempts utilized data for monosubstituted methanes,[591,64,451,1099,62,1619] and ethanes.[591,64,1853,374] It was noted that there was an approximately linear relation between the shifts for the methanes and the differences (δ_{internal}) in chemical shifts between the methyl and methylene protons of the ethanes and that both of these quantities were linearly related to the electronegativities of the substituents. Accordingly, it was suggested by Dailey and Shoolery[591] that (δ_{internal}) could be used to measure electronegativities, the relation between the two being given[451] by equation (2–2–3)

$$\text{electronegativity} = 0\cdot684\delta_{\text{internal}} + 1\cdot78 \qquad (2\text{–}2\text{–}3)$$

where δ_{internal} is in ppm.

The apparent success of the above correlations suggests that the shielding in alkanes is due solely to the local diamagnetic term (σ_d^L) which in turn is controlled by the inductive ($\pm$I) effect of the substituent. However, it is now clear that this is not the case. Inspection of the data for substituted methanes shows that chlorine, bromine, iodine and sulphur are anomalous[2340] (Fig. 2–2–2). It has also been shown that the shifts of the α-protons in isopropyl chloride, bromide and iodide are in the *opposite* order to that found for the corresponding methyl halides and the α-protons of the ethyl halides.[310] Furthermore, the internal shifts (δ_{internal}) in substituted hexachlorobicyclo[2.2.1]-heptenes(I)[2588] and bicyclo[2.2.1]heptene(II)[1508] depend rather critically on stereochemistry. These findings cannot be explained simply in terms of the inductive effect of substituents. Spiesecke and Schneider[2340] have considered the proton chemical shift data for substituted methanes and ethanes together with data for ^{13}C chemical shifts in the same molecules. They concluded that long-range shielding (Section C) is largely responsible for the anomalous results for the halogens and sulphur, and that the success of the Dailey–Shoolery relation is due to a fortuitous cancellation of the long-range shielding contributions at the α- and β-positions. The dependence of δ_{internal} in (I) and (II) on stereochemistry is also consistent with the existence of a long-range shielding effect, although it may also relate to the influence of the C—X dipole on the local shielding (see below) of the two

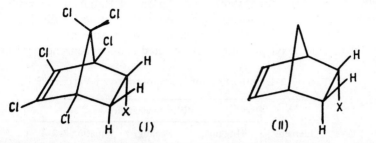

β-protons. The observation by Bothner-By and Naar-Colin[310] that the α-protons in isopropyl halides show a slight *increase* in shielding with the electronegativity of the halogen substituent is puzzling, although it has been suggested that it could arise from an interplay of the two long range shielding terms σ_p^N and σ_d^N which are predicted to have opposite signs for the α-protons in alkyl halides.[310]

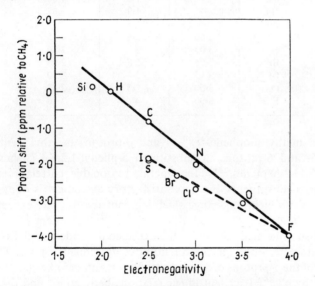

FIG. 2–2–2. Proton chemical shifts of CH_3X compounds vs. the electronegativity of X. (After Spiesecke and Schneider[2340].)

It appears that, with substituents for which long-range shielding effects are small (see Section C), there is a fair correlation between the shielding of α-protons and electronegativity. However, equation (2–2–3) would seem to offer a better basis for the estimation of electronegativities of groups. Using more extensive data, Muller[1802] has modified the parameters in equation (2–2–3). In Table 2–2–1 the electronegativities calculated with the new equation (2–2–4) are compared with values obtained by other methods.

$$\text{Electronegativity} = 0.0564 \, \delta_{\text{internal}} + 2.00 \qquad (2–2–4)$$

The proton chemical shift has also been correlated with parameters which reflect the effective electronegativity of the group to which the proton is attached. For instance, correlations of Hammett σ-constants for substituents have been attempted for aromatic protons in substituted benzenes,[2306,1494,1201,2341,2367] aldehyde protons in substituted benzaldehydes[1408] the methoxyl protons in substituted methoxybenzenes,[1141] methyl

TABLE 2–2–1. COMPARISON OF ELECTRONEGATIVITY VALUES[1802]

Element	Electronegativity		
	Huggins	Pauling	Equation 2–2–4
Li	—	1	0·77
Al	—	1·5	1·4–1·60
Zn	—	1·6	1·52
Ga	—	1·6	1·73
Si	1·90	1·8	1·77
Sn	1·90	1·8	1·80
Ge	—	1·8	1·84
Hg	—	1·9	1·87
Pb	—	1·8	2
S	2·60	2·5	2·75
N	3·05	3	2·86
Br	2·95	2·8	2·98
Cl	3·15	3	3·20
O	3·50	3·5	3·42
F	3·90	4	3·76

protons in methylthiophenes[1030] α- and β-protons in ethyl benzenes[2590] and the 5- and 6-protons in substituted 5-phenyl-1,2,3,4,7,7-hexachlorobicyclo[2.2.1]-2-heptenes.[2590] In general, reasonable correlations are reported for p-substituents but less satisfactory agreement is obtained with m-protons, for which long-range shielding contributions are expected to be more important. The chemical shifts of the methyl esters of substituted acetic and succinic acids show[1327] the expected dependence on Taft's $\sigma*$ inductive parameter although the quantitative agreement is poor. The chemical shifts of the β-protons of 1-substituted adamantanes likewise are not well correlated by $\sigma*$.[862] Excellent linear relations between σ^- and the shifts of the hydroxylic protons of substituted phenols in dimethylsulphoxide[1921] and amine protons of anilines in acetonitrile[717] have been found. However, in these examples, the principal contribution of the substituent is to control the strength of the hydrogen bond with the solvent, which in turn has a profound effect on the chemical shift of the weakly acidic protons (Section E). The chemical shifts of the hydroxylic protons of phenols at infinite dilution in carbon tetrachloride show a complex dependence on the nature of p-substituents.[1949] An excellent linear correlation of the shifts of the ethynyl protons with a combination σ^0 of inductive and resonance parameters has been reported for a series of substituted phenylacetylenes (Fig. 2–2–3)[532] in which long-range shielding contributions by substituents are likely to be negligible. Similarly, the chemical shifts of the α- and β-protons of p-substituted *trans* β-phenyl-, sulphonyl-, sulphinyl-, and thio-acrylic acids correlate well with σ.[1201]

The dependence of local diamagnetic shielding on electronegativity and Hammett-type substituent parameters clearly involves polarization of the C—H bond, the shielding being reduced as the C^-H^+ character of the bond

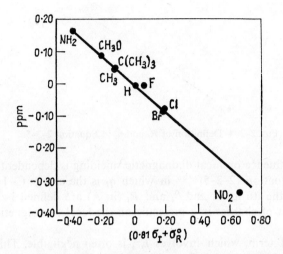

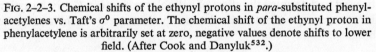

FIG. 2–2–3. Chemical shifts of the ethynyl protons in *para*-substituted phenyl-acetylenes vs. Taft's σ^0 parameter. The chemical shift of the ethynyl proton in phenylacetylene is arbitrarily set at zero, negative values denote shifts to lower field. (After Cook and Danyluk[532].)

increases. A more detailed understanding of the role of local diamagnetic shielding therefore requires an analysis of the factors which affect the polarity of C—H bonds. These factors are:

   (i) Direct electrostatic effects of charges and dipoles
  (ii) Inductive effects
 (iii) Mesomeric effects
 (iv) Hybridization of the carbon to which the proton is attached.
  (v) Van der Waal effects.

(i) DIRECT ELECTROSTATIC EFFECTS

We have already seen that the local diamagnetic shielding of the hydrogen atom is influenced in an electric field by a term proportional to the square of the field (E^2). In a molecule, the fields can arise from the existence of dipoles or formal charges. Such fields may also have a component in the direction of the C—H bond and will therefore contribute to the polarity of the bond. This effect is linearly dependent on the component of field (E_z) in the direction of the C—H bond so that $\sigma_d^L = k \cdot E_z$. The constant, k, has been estimated theoretically by Buckingham[388] and Musher[1821] to have a value between 2 and 3.4×10^{-12}. An empirical value of 2.6×10^{-12} has been obtained by Schaefer and Schneider[2192] from a consideration of the proton chemical shifts in aromatic anions and cations. Since E_z will depend on the orientation of dipoles and charges with respect to the C—H

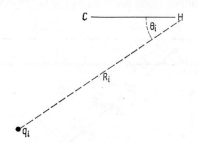

FIG. 2–2–4. Definition of R_i and θ_i in Equation 2–2–5.

bond, their influence on local diamagnetic shielding is dependent on stereochemistry. Equation (2–2–5)[2228] in which q_i is the charge (-1.00 for one electron) on the ith atom and θ_i and R_i (in Å) are defined in Fig. 2–2–4 gives the effect of charges in a molecule on the local diamagnetic shielding of a proton.

The second term, which involves E^2, is often negligible. This equation can be used to determine the effect of a dipolar group, e.g. C—F, by assigning partial charges equivalent to the bond dipole moment on the two atoms of the bond.

$$\Delta\sigma = 12.5 \times 10^{-6} \left(\sum_i \frac{q_i \cos \theta_i}{R_i^2} \right) - 17.0 \times 10^{-6} \left(\sum_i \frac{q_i}{R_i^2} \right)^2 \qquad (2\text{--}2\text{--}5)$$

In neutral molecules, shifts arising in this way will usually be in the range -0.5 to $+0.5$ ppm. Equation (2–2–5), or variants differing only in the magnitude of the parameters, have proved useful for correlating chemical shifts of protons in aromatic ions, non-alternant aromatic hydrocarbons and aromatic heterocycles, with π-electron charge densities calculated by various quantum mechanical techniques.†

One particularly interesting investigation of the effect of dipolar groups on the shielding of angular methyl groups in steroids has been reported by Zürcher.[2644] He has written the equation for the electric field effect of a group, with a dipole moment μ, in the form

$$-\Delta\sigma = k\mu e_z \times 10^{-12} = \varkappa e_z \times 10^{-12}$$

where e_z is the component of E along the C—H bond due to a hypothetical substituent with unit dipole moment. The variable, $\varkappa$, was then evaluated from data for eight ketosteroids, twelve chlorosteroids, twelve hydroxysteroids and five cyanosteroids, account being taken of the long-range shield-

† Pertinent references: 872, 960, 981, 1255, 1391, 1437, 1634, 2192, 2222, 2223, 2228, 2307, 2342, 2616, 2617. A crude, but useful approximation is that a unit positive charge on a carbon atom deshields a proton directly attached to it by 10 ppm.

ing of the substituents. It was found to vary linearly with the dipole moments of the groups investigated, (Fig. 2–2–5) thus establishing the validity of the method. The value of k thus obtained is 4·07 which lies just outside the

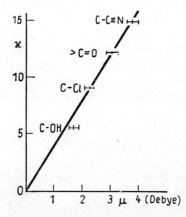

FIG. 2–2–5. The relationship between $\varkappa$ and the electric dipole moment of substituents. (After Zürcher[2644].)

range quoted by Musher.[1821] It is possible that Zürcher's value is appropriate for long-range effects whereas the lower value (2·60) is better for shorter range interactions, for which the point dipole and the point charge approximations, inherent in the derivation of both values, are less adequate.

(ii) INDUCTIVE EFFECTS

The inductive effect differs from the direct field effect just discussed in that it is relayed by the successive polarization of the σ-bonds separating the substituent from the proton under investigation. Its effect on local diamagnetic shielding is therefore expected to be independent of the spatial relation between the proton and the substituent. To what extent this mechanism contributes over and above the direct field effect is by no means clear.[1821] Perhaps the best evidence for inductive control of shielding is provided by the Hammett σ-correlations for the system (III),[2590] in which the

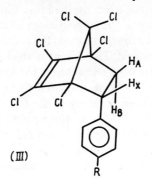

(III)

chemical shifts of protons A and B exhibit the *same* sensitivity to changes in the σ-constant of R. This would not be the case if the direct field effect made a significant contribution, since the bonds C—H_A and C—H_B are very differently oriented with respect to the partial charges developed in the aromatic ring by the substituents, R. However, the total spread of the chemical shifts of the protons A and B is only 0·17 ppm and further definitive investigations are clearly needed. The failure of Taft's inductive constants, σ^*, to correlate closely with chemical shifts (p. 66) has little significance, as much of the scatter can be attributed to long-range shielding contributions. In any case, the inclusion of direct field effects is implicit in the derivation of σ^* values.

(iii) MESOMERIC EFFECTS

Mesomeric effects can cause relatively large variations in local diamagnetic shielding as illustrated by the shifts (relative to ethylene) in the substituted ethylenes shown below.[159] Although long-range shielding is important in

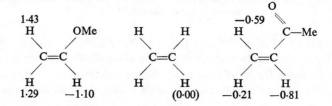

these examples,[2193,1259] it is clear that the mesomeric effect influences the electron density at the β-protons and that the observed shifts are in qualitative agreement with predictions made on the basis of the electronic theory of organic chemistry. Some progress has been made in calculating the mesomeric effect on local diamagnetic shielding in aromatic molecules and ions, using quantum mechanical methods in conjunction with equation (2–2–5).

(iv) HYBRIDIZATION

The hybridization of the carbon atom to which a proton is attached is expected to have a profound effect on the local diamagnetic shielding of the proton. As the *s*-character of the carbon atom of the C—H bond increases, the pair of electrons will be drawn towards the carbon atom and the shielding of the proton will be decreased. The change in shielding in going from ethane (*sp³*-hybridization; 25 per cent *s*-character) to acetylene (*sp*-hybridization; 50 per cent *s*-character) has been estimated to be 5–8 ppm,[1145,2073] after correcting for the long-range shielding of the triple bond. It might be anticipated that the chemical shift of a proton would be sensitive to the valence angles of the carbon atom to which it is bonded, and that in cyclic systems chemical shifts might be a function of ring size. However, ring size appears to have only a minor effect (except in cyclopropanes and aromatic systems in which long-range shielding is very important). The change in *s*-character in going from cyclohexane to cyclobutane, as judged from ^{13}C—H coupling constants (see however, Chapter 4–5A) is *ca.* 2 per cent so that the variation in chemical shift can only be of the order of 0·4 ppm (cf. cyclohexane $\delta = 1·44$ and cyclobutane $\delta = 1·96$)[2565] and hence similar in magnitude to changes in long-range shielding which might accrue from the alteration of geometry. However, Veillard has analysed the factors which influence the chemical shifts of protons in various aromatic heterocycles and, after allowing for the effects of long-range shielding and π-electron charge densities, he concludes that protons attached to planar five membered rings are deshielded by 0·35 ppm relative to those of planar six membered rings.[2477] It is probable that the decreased shielding of the protons of cubane (IV)[731] compared with the methine proton of isobutane (V) is associated with increased *s*-character of the C—H bonds in the former compound.

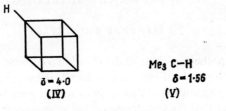

δ = 4·0
(IV)

Me$_3$ C—H
δ = 1·56
(V)

(v) VAN DER WAAL EFFECTS

There are a number of examples[2297,976,980,409,1836,2186,1207,2600] in which significant down field shifts have been observed for protons which are in very close steric proximity to other groups. Such shifts are often due in part to the direct field effect or to long-range shielding, both of which can assume large values at short distances. Excellent examples are provided by the half cage compound (VI a) and its corresponding anion (VI b).[2600]

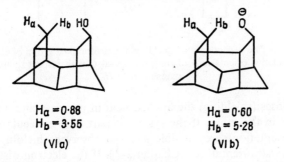

H$_a$ = 0·88
H$_b$ = 3·55

(VI a)

H$_a$ = 0·60
H$_b$ = 5·28

(VI b)

It is worth noting that, in addition to these causes, there is a contribution which is independent of polarity and magnetic anisotropy and which can be termed an intramolecular van der Waal's shift, since it is associated with the oscillating electric fields which give rise to London dispersion forces. Even for an isotropic electron distribution the mean squared value of its time dependent dipole moment is not zero and so it can provide an E^2 term in equation (2–2–5). The value of E^2 which arises in this way has a six-fold inverse dependence on the distance separating the proton under investigation from the neighbouring group.[1224] Calculations for two isolated hydrogen atoms reveal that the associated shielding contribution is negligible if the atoms are separated by more than the sum of their van der Waal radii but is significant for shorter distances. Thus, for two hydrogen atoms separated by 2·4 Å (sum of the van der Waal radii) the shielding contribution is $-0·01$ ppm, compared with $-0·2$ ppm at 2 Å and $-0·5$ ppm at 1·7 Å. The effect on protons in molecules is probably somewhat greater.[2356] The influence of other atoms will be in proportion to their polarizabilities and will therefore be greater at equivalent distances. In all cases, this mechanism leads to deshielding.

C. LONG RANGE SHIELDING

(i) GENERAL PRINCIPLES

Consider a hypothetical electron distribution (G) about one or more nuclei in the vicinity of a proton. Figure 2–2–6 depicts the secondary magnetic field for two possible orientations of the system with respect to the applied magnetic field. We see that the component of the secondary magnetic field at the

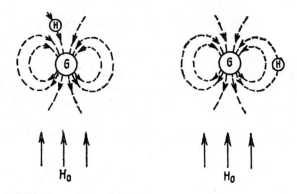

Fig. 2–2–6. The dependence of shielding of a proton by a group of electrons G on the relative orientations of the proton and G with respect to the applied magnetic field H_0.

proton opposes (shielding) the applied field in one case and augments (deshielding) it in the other. If the system is part of a molecule in the liquid state, all orientations are possible and the observed shielding will be the average over all orientations (Chapter 2–1). If the electron distribution (G) has spherical symmetry, i.e. if its magnetic polarizability is the same for all orientations, it can be shown that its shielding contribution to the proton averages to zero. Thus, long-range shielding can only arise from electron distributions which are *diamagnetically anisotropic*. Consideration of a specific example, *viz.* the proton shielding in acetylene will help to clarify this concept.

Figure 2–2–7 shows the effect of the applied magnetic field on the circulations of the π-electrons in acetylene for two orientations of the molecule. In acetylene, the cylindrical π-electron distribution about each carbon atom allows normal induced circulations provided the applied field and principal

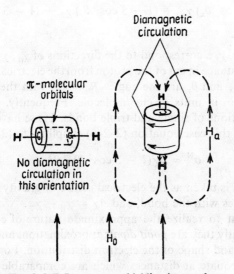

FIG. 2–2–7. Long range shielding in acetylene.

axis of symmetry are coincident. In this orientation the secondary magnetic field from the π-electrons shields the two protons. In the other orientation, in which the applied field is transverse to the molecular axis, no circulations are possible[1987] and there is no shielding contribution from the π-electrons. All other orientations are intermediate between these two and so the average shielding over all orientations will be positive. We may say that the long-range shielding by the π-electron system of acetylene is a consequence of its diamagnetic anisotropy.

If the triple bond is incorporated in a molecule, the situation may arise in which a particular proton is located immediately above the π-electron system (e.g. the proton H_a in Fig. 2–2–7). This proton is deshielded in the longitudinal orientation and its average shielding over all orientations is therefore negative. Long-range shielding may thus be either positive or negative depending on the geometric relation of the proton to the electron distribution. Our discussions of long-range shielding will therefore be concerned with the diamagnetic anisotropies of electron distributions associated with one or more atoms, and the methods by which we can determine, in a system of known geometry, the shielding of a proton by an electron distribution of known diamagnetic anisotropy.[1615]

In general, the diamagnetic anisotropy of an electron distribution (e.g. a chemical bond) is expressed in terms of its three principal magnetic susceptibilities, $\chi_{\parallel}$, $\chi_{\perp}$ and χ_T which are respectively parallel, perpendicular, and transverse to a convenient axis of the system chosen such that the origin is the electrical centre of gravity of the electron distribution. The shielding field can then be simulated by a point magnetic dipole located at the origin and the sign and magnitude of this field, averaged over all orientations, is then given by equation (2–2–6)[1851]

$$\sigma^N = [(1 - 3\cos^2\theta_x)\chi_{\|} + (1 - 3\cos^2\theta_y)\chi_{\perp} + (1 - 3\cos^2\theta_z)\chi_{T}]/3R^3$$
(2–2–6)

where the axes x, y, z correspond to the directions of $\chi_{\|}$, $\chi_{\perp}$ and χ_{T}, respectively. R is the distance in cm of the proton from the electrical centre of gravity of G and θ_x, θ_y, and θ_z are the angles R makes with the three axes. The susceptibilities are in units of $cm^3/molecule$. Frequently, we will consider electron distributions of single and triple bonds. These have axial symmetry and $\chi_{\perp} = \chi_{T}$. In this case equation (2–2–6) simplifies to (2–2–7)

$$\sigma^N = \Delta\chi(1 - 3\cos^2\theta)/3R^3$$
(2–2–7)

where the origin is taken as the electrical centre of gravity of the bond, θ is the angle R makes with the bond, and $\Delta\chi = \chi_{\|} - \chi_{T}$.

It is important to realize the approximate nature of equations (2–2–6) and (2–2–7). Firstly they are *point dipole* approximations and do not take into account the size and shape of the electron distribution. For this reason they are grossly inadequate at distances which are comparable with the dimensions of the electron distribution. Secondly, in the case of polar systems the electrical centre of gravity is not known. Nevertheless, these equations may provide a qualitative basis for the discussion of long-range shielding.

In certain systems, the point dipole treatment can be replaced by a somewhat more accurate approximation in which the electron distribution is regarded as a loop carrying the induced current (electron circulations). The current is calculated directly from Larmor's theorem and its associated magnetic field is thus derived. This method was developed by Waugh and Fessenden[2507] for the shielding associated with the π-electron system of benzene (p. 95). Its extension to triple bonds has not been investigated but would seem to offer advantages over the point dipole approximation. It has the advantage that it does not require *a priori* knowledge of diamagnetic anisotropies.

It is well known that, for organic compounds, the molar susceptibility is an additive and constitutive property. It is therefore reasonable to assume that it is possible to assign diamagnetic anisotropies to atoms, bonds or groups and that these anisotropies will then be approximately independent of molecular structure. There are various ways in which the determination of diamagnetic anisotropies can be attempted. These are: from quantum mechanical calculations, from molecular Zeeman effects, from measurements of spin–rotational interaction, from chemical shift data on oriented single crystals and molecules, from magnetic susceptibility measurements on oriented single crystals, from Cotton–Mouton constants, from the Gans–Mrowka equation and from the analysis of chemical shift data for suitably chosen model compounds.

The accurate calculations of diamagnetic anisotropies by quantum mechanical techniques are only feasible for very simple systems such as the hydrogen molecule.[1091] Nevertheless, treatments based on approximate methods have yielded useful results,[1996,1997,1074,2414] especially for aromatic systems (see below). In particular, they indicate the order of magnitude of $\chi_{\parallel}$, $\chi_{\perp}$ and χ_T, and provide a basis for the choice of suitable model compounds from which values of $\chi_{\parallel}$, $\chi_{\perp}$ and χ_T might be obtained from equations (2–2–6) and (2–2–7).

Rotational energy levels in diamagnetically anisotropic molecules can be split by a magnetic field. This is known as the molecular Zeeman effect and it can be observed as the doubling of lines in the microwave spectra of suitable simple molecules in the presence of a homogeneous magnetic field. The splitting factors (g-values) can be resolved into their three principal components along the inertial axes of the molecule and, by combining this information with the value for the mean magnetic susceptibility, it is possible to derive $\chi_{\parallel}$, $\chi_{\perp}$ and χ_T for the molecule.[834] Similar information is also available from microwave or molecular beam studies of spin–rotation interactions for simple molecules.[833]

In principle, molecular diamagnetic anisotropies are available from measurements of the dependence of the chemical shift on the orientation of single crystals in a magnetic field. In practice, the method has limited application as usually the changes in chemical shift are orders of magnitude smaller than the dipolar broadening of absorption lines in solids. The only example so far reported is due to Lauterbur who was able to measure the ^{13}C chemical shift of the carbonate ion in calcite.[1511a] In this case the only magnetic nucleus is ^{13}C which is present at its natural abundance of 1 per cent so that dipolar broadening is of a very low order. An interesting variant of this general method is to study molecules in the nematic phase (liquid crystals) in which there is some degree of orientation but in which dipolar broadening is no longer serious. Englert and Saupe[2173,2171,2174] have studied the anisotropies of aromatic molecules by this method.

Studies of single crystals using the oscillation technique,[2237] permit the determination of the differences between the molar susceptibilities along the three principal axes of polarizability. From a knowledge of the structure of the unit cell and of the molar susceptibilities of the crystal, it is possible to calculate the three principal susceptibilities of the *molecule*. The separation of the total anisotropy into atom, bond and group contributions is more difficult and requires the comparison of data for closely related series of compounds. Since an X-ray crystallographic investigation is required for each molecule studied, acquisition of data is difficult and comparatively little basic information on group anisotropies is, at present, available from this source.[2237,945] The method has been most successful in determining the anisotropies associated with inter-atomic diamagnetic circulations of π-electrons in aromatic compounds, although here again it is necessary to separate the total anisotropy into various contributions.[1176]

Molecular diamagnetic anisotropies are also available from studies of magnetic double refraction (Cotton–Mouton effect)[393,1527] of certain molecules. The Cotton–Mouton effect is the magnetic counterpart of the Kerr electro-optical effect. The magnitude of birefringence, which is proportional to the square of the applied magnetic field, is also a function of the three principal diamagnetic susceptibilities and of the three principal electric polarizabilities which can be independently determined.[1526] For axially symmetric molecules, $\chi_\perp = \chi_T$, and since the average molecular susceptibility, which equals $(\chi_\parallel + \chi_\perp + \chi_T)/3$, can be measured by conventional methods, the molecular diamagnetic anisotropy can be determined. The method has the advantage that it can be applied to simple molecules (in the gaseous or liquid state) for which the separation of molecular diamagnetic anisotropy into bond anisotropies is more readily achieved. Thus, Buckingham, Prichard and Whiffen[395] have studied the Cotton–Mouton effect for ethane and have found that $\Delta\chi^{C-C} - 2\Delta\chi^{C-H} = -8\cdot1 \times 10^{-30}$ cm³/molecule where $\Delta\chi^{C-C}$ and $\Delta\chi^{C-H}$ are the anisotropies of the carbon–carbon and carbon–hydrogen bonds, respectively. To date, little information has been available from this source but it is to be hoped that investigations will be extended to other molecules e.g. acetylene, hydrogen cyanide, ethylene, etc.

Gans and Mrowka[945] have derived an equation which relates the principal magnetic susceptibilities of an atom or molecule to its principal electric polarizabilities. In certain circumstances, this relation can be solved simultaneously with other equations involving these quantities, e.g. by using data from Kerr constant, light scattering, refractivity and mean molecular diamagnetic susceptibility measurements to yield values for the diamagnetic anisotropies of bonds.[2643] Bothner-By and Pople[314] have pointed out that the Gans–Mrowka equation requires the largest magnetic susceptibility to be at right angles to the largest electric polarizability and, for paraffinic hydrocarbons this is at variance with the experimentally observed negative sign of the Cotton–Mouton constant.[395,408] Consequently, there is doubt as to the validity of the Gans–Mrowka equation.

The use of chemical shift data for determining the diamagnetic anisotropies of atoms, bonds and groups is essentially the use of the proton as a probe for sensing the shielding field associated with the anisotropy. If an isolated proton could be placed anywhere in the vicinity of, say, a carbon–carbon double bond then an infinite set of equations of type (2–2–6) would be obtained from which $\chi_\parallel$, $\chi_\perp$ and χ_T could be evaluated within the limits of the point dipole approximation on which equation (2–2–6) is based. In practice, it is comparatively simple to find a molecule or molecules which have protons in a sufficient number of different positions, relative to a carbon–carbon double bond, to determine uniquely the three susceptibilities in equation (2–2–6). However, the major difficulty in applying this method is to separate from the total shielding of the proton the long-range shielding due to the double bond, remembering that the total shielding embodies *all* effects influencing local diamagnetic shielding, and the long-range contributions from *all* electron distributions in the molecule. The various approaches to the circumvention of this difficulty will now be summarized.

In many systems, in which the long-range shielding of a non-polar group is under consideration, effects of local diamagnetic shielding can be minimized by considering the shifts of protons which are attached to the same carbon atom but differently disposed to the shielding group. For instance, the difference in chemical shift of the olefinic protons in 2-methylbut-1-en-3-yne (VII) should reflect the difference between the long-range shielding of

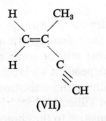

(VII)

the C—H bonds of the methyl group and that of the ethynyl residue. Similarly, the chemical shifts of axial and equatorial protons which can be observed in the low-temperature spectrum of cyclohexane can be expressed in terms of $\Delta\chi^{C-C}$ and $\Delta\chi^{C-H}$.

If polar groups are present, or if the group whose long-range shielding is being investigated is itself polar, allowance must be made for their effect. This can be done by application of equation (2–2–5). Zürcher[2644] has made use of this technique to investigate the anisotropies of several groups (see below).

Procedures of the type just discussed are, in principle, incapable of yielding unique values for anisotropies from a single example, since the final expressions, as in the case of cyclohexane, always involve two anisotropies. Ideally many examples should be examined in order to obtain values for both anisotropies, together with their precision indices.

If the effect of X on the local diamagnetic shielding of the protons in CH_3—X could be accurately predicted, information concerning the anisotropy of X would be available directly from the proton chemical shift. One interesting attempt to obtain information in this way has been made by Reddy and Goldstein.[2071] These workers have argued that, since there is a relation between J_{13C-H} and the electronegativity of X (see Chapter 4–5A), the former can be used to estimate the effect of X on the local diamagnetic shielding of the α-protons. In other words, in the absence of other shielding influences, there should be a linear relation between J_{13C-H} and chemical shifts in compounds of the type H—CXYZ. Reddy and Goldstein determined the parameters for the linear relation from data for compounds in which long-range shielding was assumed to be constant.[998] Deviations from this relation observed for other compounds then provided a measure of the long-range shielding contributions of the substituents. The interpretation of these results in terms of equation (2–2–7) is difficult because the values of R are in the range in which the point dipole approximation is invalid.[652]

In general, the evaluation of diamagnetic anisotropies from chemical shift data in terms of equations (2–2–6) and (2–2–7) will only be justified in cases where the anisotropy is large and the data correspond to large values of R. The calculations by Didry and Guy[652] for the hydrogen molecule indicate that the point dipole approximation may be invalid for values of R less than 6–7 Å.

(ii) SHIELDING BY CARBON–CARBON SINGLE BONDS AND CARBON–HYDROGEN BONDS

It is convenient to consider the long-range shielding of these two bonds together as, experimentally, their diamagnetic anisotropies are usually determined from the same systems and in some cases the values for the two quantities are expressed as dependent variables.

The first attempts to evaluate the anisotropies of the C—C and C—H bonds were made by Guy and Tillieu,[1074,2414,653] using quantum mechanical methods. Their results are sensitive to the choice of wave functions and parameters, and they give $\Delta\chi^{C-C} = 2\cdot0$† and $\Delta\chi^{C-H} = 1\cdot5$, as preferred values. There have been several investigations in which values of $\Delta\chi^{C-C}$ have been obtained from chemical shift data on the assumption that $\Delta\chi^{C-H}$ is negligible.[1828,1819,1085,1793,309,2236,1771,1200]

Values ranging from 5·5–25 are quoted. Reddy and Goldstein have reported a value of 16[2071] using ^{13}C—H coupling constant data to evaluate σ_d^L (cf. p. 77). Bothner-By and Pople[314] have pointed out that, since the mean susceptibility of the C—C bond is 5·3 any value of $\Delta\chi$ greater than 7·5 is unacceptable as it would require $\chi_{||}$ to be paramagnetic.

It is now evident that $\Delta\chi^{C-H}$ is by no means negligible. Davies[605] has pointed out that the crystal diamagnetic anisotropies of long chain fatty acids require $\Delta\chi^{C-C} - 2\Delta\chi^{C-H} \sim -6\cdot6$, which agrees well with $\Delta\chi^{C-C} - 2\Delta\chi^{C-H} = -8\cdot0$ derived from the Cotton-Mouton constant for ethane.[395] Davies has pointed out that $\Delta\chi$ for all single bonds probably has the same sign. This is certainly true for the quantum mechanically derived values for C—C, C—H[1074,2414] and H—H.[1091] It therefore follows that $\Delta\chi^{C-H} > 4\cdot0$.

Recognition of the significance of the long-range shielding due to the C—H bond means that in evaluating anisotropies from chemical shift data it is necessary to solve simultaneously for both $\Delta\chi^{C-C}$ and $\Delta\chi^{C-H}$. Several attempts to do this have been made.[605,2641,1847,2623,1252] However, analysis of available data indicates that the long-range shielding effects of C—C and C—H bonds probably cannot be accounted for solely in terms of their anisotropies. This is illustrated by a consideration of the data of Eliel, Gianni, Williams and Stothers[748] for a number of alkyl cyclohexanols (Table 2–2–2). The calculated values have been obtained[1252] by regression analysis using the geometric factors in equation (2–2–7) as independent variables. This treatment gives $\Delta\chi^{C-C} = 25\cdot8 \pm 7\cdot4$ and $\Delta\chi^{C-H} = 13\cdot7 \pm 6\cdot0$. These values are far too large. In other words, the observed shifts, such as the pronounced shielding of an equatorial proton by axial and equatorial 2-methyl groups are not explained. The conformation of the hydroxyl group introduces an uncertainty into the analysis, since the effect of its anisotropy and electric dipole moment on the α-proton will depend on the conformation about the C—O bond. However, this is probably not the origin of the observed dis-

† All anisotropies are quoted in the units "$\times 10^{-30}$ cm³/molecule" and can be directly inserted in equations (2–2–6) and (2–2–7) where R is in cm.

crepancies, as consideration of the spectra reported by Muller and Tosch[1812] clearly indicates that shifts of comparable magnitude occur in hydrocarbons.[292] In this connection it is significant that the values calculated from the difference in chemical shift of the axial and equatorial protons in cyclohexane[102] and the Cotton–Mouton constant for ethane are $\Delta\chi^{C-C} = 10\cdot5$ and $\Delta\chi^{C-H} = 7\cdot7$, if the electrical centre of gravity is taken as $0\cdot77$ Å from the carbon atom, and even greater for points closer to the carbon atom. It is likely that the basic assumption that the shielding field due to anisotropy can be simulated by a point magnetic dipole is the major source of error. A modified treatment in which the magnetic dipole is assigned a finite length does not materially improve the calculated anisotropies.[116a]

TABLE 2–2–2. REGRESSION ANALYSIS OF CHEMICAL SHIFT DATA FOR THE α-PROTONS IN ALKYLCYCLOHEXANOLS[1252]

Substituents	δ_{obs}^{748}	δ_{calc}	$(\delta_{obs} - \delta_{calc})$
cis-3-Methyl	3·45	3·49	−0·04
trans-2-Methyl	2·98	2·85	0·13
cis-3,3,5-Trimethyl	3·64	3·96	−0·32
trans-2-trans-6-Dimethyl	2·42	1·96	0·46
3,3,5,5-Tetramethyl	3·82	3·70	0·12
3,3-Dimethyl	3·65	4·03	−0·38
trans-2-Methyl-trans-4-t-butyl	2·93	2·73	0·20
cis-2-Methyl-trans-4-t-butyl	3·56	3·66	−0·10
2,2-Dimethyl-trans-4-t-butyl	3·12	3·46	−0·34
cis,cis-3,5-Dimethyl	3·48	3·49	−0·01
trans-4-t-butyl	3·36	3·49	−0·13
cis-4-t-butyl	3·94	3·93	0·01
cis-2-cis-6-Dimethyl	3·46	3·36	0·10
trans-2-Methyl-cis-4-t-butyl	3·56	3·41	0·15
trans-3-trans-5-Dimethyl	4·02	4·08	−0·06
2,2-Dimethyl-cis-4-t-butyl	3·30	3·14	0·16
cis-2-Methyl-cis-4-t-butyl	3·67	3·61	0·06

It has been suggested[1822] that the shifts observed in cyclohexyl and bicyclo[2.2.1]heptyl[1828] systems may arise, in part, from the electrostatic effects associated with polarity of C—H bonds. Inclusion of geometric factors associated with such electric effects equation (2–2–5) in the regression analysis of the data in Table 2–2–2 reveals no significant correlation. On the other hand, there is a correlation with the inductive effect of methyl groups if this is assumed to be attenuated by a factor of $0\cdot33$ for each bond separating a methyl group from the α-proton. The effect of a methyl group in the 2-position is an up-field shift of ca. $0\cdot2$ ppm. Narasimhan and Rogers[1847] have attempted to allow for changes in local diamagnetic shielding of the two types of protons in propane. They used two approaches; one based on equation (2–2–3) in which propane is considered as CH_3CH_2X where X is CH_3 with an electronegativity of $2\cdot32$; the other utilized quantum mechanically calculated values for the charges on the carbon atoms in propane. The two estimates agree quite closely, and, when used to correct the chemical shift difference in propane, they give the expressions $\Delta\chi^{C-C} - 1\cdot17\Delta\chi^{C-H} = 2$, and $0\cdot7$, respectively. Combining these expressions with that derived from the Cotton–Mouton constant of ethane[395] (see above) gives, respectively, the following values for the two anisotropies.

$$\Delta\chi^{C-C} = 16\cdot0; \quad \Delta\chi^{C-H} = 11\cdot6$$
$$\Delta\chi^{C-C} = 12\cdot4; \quad \Delta\chi^{C-H} = 10\cdot6$$

which again are far too large.

There seems little doubt that C—C and C—H do give rise to significant long-range shielding (the total range of shifts in Table 2–2–2 is *ca.* 1·6 ppm) but that an interpretation in terms of diamagnetic anisotropies is made difficult by the inadequacies of the point dipole approximation.

(iii) Shielding by halogens

McConnell[1615] has suggested that the anisotropies of the halogen atoms in molecules are approximately 20 per cent of the average susceptibilities of the corresponding anions, i.e. $\Delta\chi = -3, -7\cdot5, -10$ and -16 for F, Cl, Br and I, respectively. Pople[1997] has calculated a value of $-4\cdot8$ for fluorine and Itoh[1248] has derived a theoretical value for the long-range shielding by the C—Cl in methyl chloride which is in substantial agreement with the value predicted by McConnell. The value of $\Delta\chi = -33\cdot3$ has been obtained for the iodine–iodine bond, from measurements on single crystals of iodine.[638] This result suggests an upper limit of *ca.* -16 for the anisotropy of the carbon–iodine bond. Presumably, the lone pairs of electrons make the most important contribution, although as Bothner-By and Naar–Colin[310] have pointed out, the total anisotropy arises from two terms associated with the bonding electrons and with the lone pairs, respectively, which probably are opposite in sign.

Evaluation of $\Delta\chi$ for the halogens from chemical shift data is difficult because a large proportion of the long-range shielding effect arises from the contribution to local diamagnetic shielding by the electrostatic effect associated with the dipole moment of the C—X bond. Several attempts[2071,831, 2193,2340,2186] have been made to account for chemical shifts in terms of the electric effects and diamagnetic anisotropies of the carbon–halogen bonds. In particular, Zürcher[2644] has concluded from an analysis of chemical shift data for twelve different chloro-compounds (steroids and bicycloheptane derivatives) that the long-range shielding effect of the chlorine substituent arises entirely from electrical effects (p. 68).

(iv) Shielding by oxygen atoms in ethers and alcohols

Pople[1997] has considered the diamagnetic anisotropy of the oxygen atom in ethers and alcohols and has concluded that its largest axis of polarizability corresponds to the *y*-axis defined in Fig. 2–2–8, with the other two principal axes of polarizability being equal. The calculated anisotropy, $\Delta\chi$, is -5.3 and the associated long-range shielding is described by the cones in Fig. 2–2–8. However, no experimental test of this hypothesis has so far been made. Indeed, the derivation of $\Delta\chi$ from chemical shift data is made difficult, not only by the existence of large electrostatic contributions of oxygen groups to local diamagnetic shielding, but also by the fact that both sign and magnitude of the contributions of both the electric dipole moment and diamagnetic anisotropy will depend on the conformation about the bond between RO and the molecule.

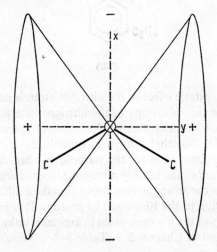

FIG. 2–2–8. Definition of axes for an ether oxygen atom and the sign of shielding predicted by the point dipole approximation.

Several investigators[434,2275,1383,1894,1548] have noted that an oxygen atom in close proximity to a proton causes a down field shift, but the rationalization of these results is not obvious. Zürcher[2644] has analysed the shifts produced by hydroxyl groups in steroids. By assuming the hydroxyl group to be freely rotating and axially symmetric about the C—OH bond, he concluded that the effect of diamagnetic anisotropy is unimportant since the shifts can be accounted for solely in terms of the electrostatic effect.

(v) SHIELDING BY NITROGEN ATOMS

Molecular orbital calculations[1997] of the diamagnetic anisotropy of the nitrogen atom bonded in the sp^3 hybridized configuration suggest that this system is isotropic. However, similar calculations give the same result for the tetrahedral carbon atom although it appears certain that the C—C bond is anisotropic. It is therefore possible that there is an anisotropy associated with an sp^3-hybridized nitrogen atom. Yamaguchi, Okuda and Nakagawa[2625] have attributed the down-field shift (0.35 ppm) of the 10-β-proton

relative to the 10α-proton in dihydrodeoxycodeine-D (VIII) to a long-range
shielding by the nitrogen atom. They claim that these two protons are symmetrically disposed with respect to the aromatic nucleus and that calculations

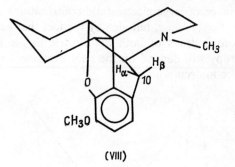

(VIII)

show that the electrostatic effect of the nitrogen atom is negligible. They calculate $\Delta\chi = 8\cdot9$ for the anisotropy of the nitrogen lone pair (the axis is that
of the lone pair) and note that this would account for the observed shifts of
axial and equatorial 2-methyl groups in quinolizidines.[1801] However, the
system (VIII) is far from ideal for the purpose of observing the long-range
shielding contribution of the tertiary nitrogen, since the argument depends on
the assumption that the aromatic ring does not make a differential long-range
shielding contribution to the 10α- and 10β-protons. It is probably fair to say
that the anisotropy of the nitrogen atom in amines is unknown at the present
time. Data presented in Chapter 3–8 (Table 3–8–10) could be of use in this
connection.

There is convincing evidence that the lone pair of an sp^2 hybridized nitrogen atom is associated with an appreciable anisotropy. This is predicted,
theoretically, as being associated with the low-lying excited state involved
in the n–π* transitions of such systems. Gil and Murrell[981] have calculated
the anisotropy of the nitrogen atom in pyridine to be $\Delta\chi = -7\cdot1$ (the axis is
perpendicular to the plane of the aromatic ring) and have found that this
value accounts quite well for the shift in the ^{14}N resonance, of pyridine on
protonation. This anisotropy gives rise to deshielding of protons in the plane
of the ring, an effect which is, of course, greatest at the α-position. The anisotropy is substantially reduced by protonation of the nitrogen atom so that
Gil and Murrell have satisfactorily explained why the chemical shift of the
α-proton is only slightly affected by protonation, whereas the β- and γ-pro-
tons are considerably shifted to lower fields. At the α-position the deshielding
effect of the positive charge is largely cancelled by the reduction in the anisotropy of the nitrogen atom. Similar effects are predicted for sp^2 nitrogen
atoms in other systems such as Schiff's bases, oximes, etc. (cf. data in Chapter 3–8 C).

(vi) SHIELDING BY CARBON–CARBON DOUBLE BONDS

Several attempts to predict the diamagnetic anisotropy of the carbon–carbon double bond have been made. Tillieu,[2414] using the variation method, has computed the values -3.6, -5.4 and -7.8 for the three principal susceptibilities in the x, y and z directions (Fig. 2–2–9); i.e. $\chi_x > \chi_y > \chi_z$. Conroy[531] has suggested that the magnitudes are in the order $\chi_x > \chi_z > \chi_y$. Pople's calculations[1997] give $\chi_y > \chi_x = \chi_z$ with $\Delta\chi = \chi_y - \chi_z = 7$ at each carbon atom. In the first edition of this book it was suggested on the basis

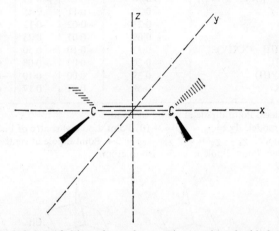

FIG. 2–2–9. Definition of axes for a carbon–carbon double bond.

of meagre experimental evidence that $\chi_x = \chi_y > \chi_z$ might approximate to the anisotropy of the double bond. This conclusion is similar to Tillieu's prediction. There are now many more data which can be used to test these predictions. The results are summarized in Table 2–2–3. It is evident that all models, except Conroy's, give a reasonable account of the sign of the long-range shielding effect of the double bond.

While no significance should be attached to the absolute magnitudes of the predicted values in Table 2–2–3, certain conclusions regarding the sign of the shielding effect seem justified. It is apparent that protons in the regions of space above and below the plane of the double bond and near the z-axis are shielded, while those in the plane and near the y-axis are deshielded. The theories of Tillieu and Pople differ in that the former predicts deshielding in the region in-plane and near the x-axis whereas the latter gives rise to shielding in this region. The shifts in the β-protons of the pair of isomeric triene diesters, (XXIX) and (XXX), are in accord with Tillieu's model. However, it is conceivable that this result may reflect differences in electron density at the β-protons, rather than a long-range shielding effect and, in any case, the distances of these protons from the source of long-range shielding are far too short to permit any confidence in the point dipole approximation. Data obtained from a study[2625] of codeine derivatives is claimed to support the model adopted in the first edition (and hence Tillieu's model). These con-

TABLE 2–2–3. SHIELDING EFFECTS OF THE CARBON–CARBON DOUBLE BOND

Compounds	Observed differences	Calculated difference			
		T^a	J^b	C^c	P^d
[(IXa)—(Xa)]—[(XI)—(XII)]	−0·03 to −0·13	−0·04	−0·18	0·05	−0·14
[(IXb)—(Xb)]—[(XI)—(XII)]	−0·22 to −0·25	−0·04	−0·18	0·05	−0·14
[(IXc)—(Xc)]—[(XI)—(XII)]	−0·13 to −0·16	−0·04	−0·18	0·05	−0·14
(XIII)—(XV)	−0·12	−0·03	−0·16	0·02	−0·09
(XIV)—(XVI)	0·18	0·00	0·03	−0·04	0·04
[(XIII)—(XIV)]—[(XV)—(XVI)]	−0·30	−0·03	−0·19	0·06	−0·13
(XVII)—(XVIII)^e	−0·22	−0·11	−0·45	0·02	−0·20
(XIX) *endo-exo*	−0·23	−0·02	−0·12	0·04	−0·11
(XX)—(XXI)	−0·09	−0·05	−0·23	0·07	−0·25
(XXII)[=(XXIII)]—(XXIV)	−0·33	−0·10	−0·50	0·04	−0·30
(XXV)—(XXVI)	−0·20	−0·03	−0·08	0·07	−0·15
(XXVII)—(XXVIII)	0·39	0·00	0·10	−0·11	−0·18
(XXIX)—(XXX)	0·58	0·07	0·17	0·18	−0·22

^{a} Tillieu's values. Point dipole at centre of bond.
^{b} Jackman's model. $\Delta\chi = \chi_z - \chi_y = -10$. Point dipole at centre of bond.
^{c} Conroy's model. $\chi_z - \chi_x = 4\cdot2$; $\chi_z - \chi_y = 2\cdot4$. Point dipole at centre of bond.
^{d} Pople's value. Point dipole at each carbon atom.
^{e} *cf.* ref. 2422.

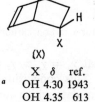

	X	δ	ref.		X	δ	ref.	δ	ref.	δ	ref.
^{a}	OH	3.71	1943	^{a}	OH	4.30	1943	3·58	1827	4.12	1827
	OH	3.78	613		OH	4.35	613	3.66	2614	4.17	2614
	OH	3.82	2614		OH	4.46	2614				
^{b}	CO_2H	2.14	613	^{b}	CO_2H	2.90	613				
^{c}	CN	2.18	613	^{c}	CN	2.85	613				

(IX) (X) (XI) (XII)

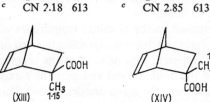

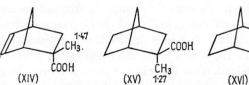

(XIII) (XIV) (XV) (XVI)

(ref. 1943)

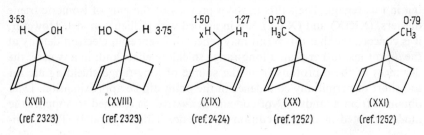

(XVII) (XVIII) (XIX) (XX) (XXI)

(ref. 2323) (ref. 2323) (ref. 2424) (ref. 1252) (ref. 1252)

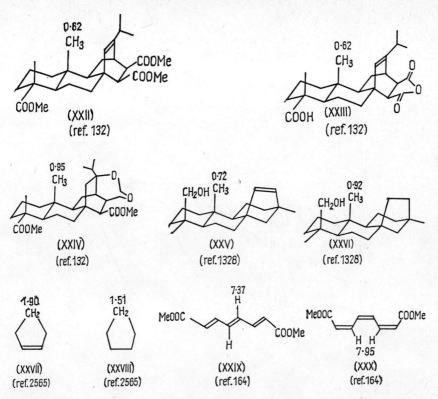

clusions are also subject to criticism as the molecules contain an aromatic ring which is associated with a large long-range shielding effect, so that even quite minor changes in molecular geometry could cause substantial changes in the shielding of neighbouring protons.

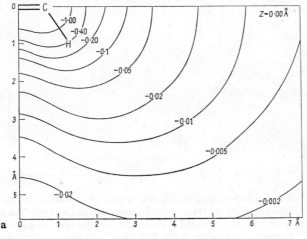

FIG. 2–2–10a. (Legend, see page 86.)

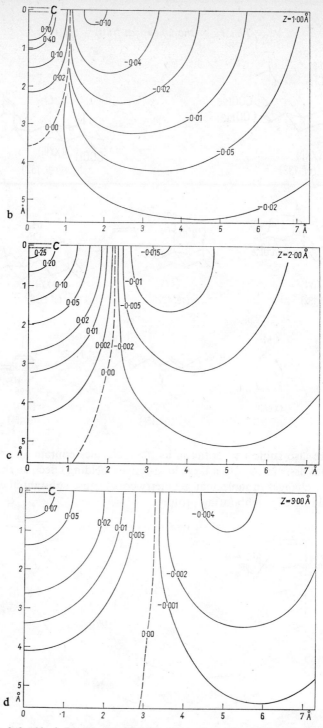

Fig. 2-2-10b–d. Long-range shielding due to the carbon–carbon double bond, based on Tillieu's values for the principal susceptibilities (see text). The figures on the contours represent shielding in ppm, negative values denoting deshielding. The four diagrams are sections in the X–Y plane with the Z coordinate equal to 0, 1, 2 and 3 Å. Only the lower right-hand quadrant is shown.

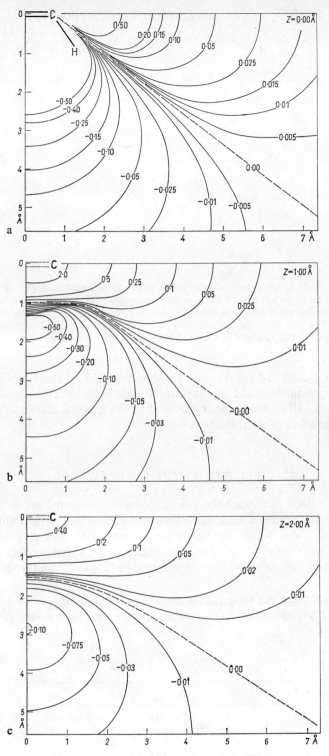

FIG. 2–2–11 a–c. (Legend, see page 88.)

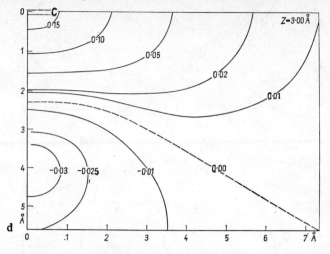

FIG. 2–2–11 d. Long-range shielding due to the carbon–carbon double bond, based on Pople's values for the principal susceptibilities (see text). The figures on the contours represent shielding in ppm, negative values denoting deshielding. The four diagrams are sections in the X–Y plane with the Z coordinate equal to 0, 1, 2 and 3 Å. Only the lower right-hand quadrant is shown.

Figures 2–2–10 and 2–2–11 show the shielding field predicted by the point dipole approximation and the anisotropies calculated by Tillieu and Pople, respectively. All models predict that olefinic protons are deshielded by the long-range effect of the π-electron systems.

(vii) SHIELDING BY THE CARBONYL GROUP

Pople[1997] has calculated the anisotropy of the carbonyl group and his values for the principal susceptibilities at the carbon and oxygen atoms are recorded in Table 2–2–4. The values for the total contribution of the paramagnetic term to the principal susceptibilities do not agree in either relative or absolute magnitude with those determined by Flygare[834] for formaldehyde

TABLE 2–2–4. THE PRINCIPAL SUSCEPTIBILITIES FOR THE CARBONYL GROUP
(These values do not include the isotropic diamagnetic term)

$$>\!C\!=\!O$$

		χ_x	χ_y	χ_z
C	\} (Pople[1997])	10·8	17·9	10·8
O		18·0	16·7	7·2
Total		28·8	34·6	18·0
Total (Flygare[834])		50·7	72·0	81·3

from a study of its molecular Zeeman effect. However, it will be shown that the anisotropy predicted by Pople gives a much better account of the long-range shielding effect of the carbonyl group.

Since the carbonyl group possesses a substantial electric dipole moment, any analysis of the long-range shielding effect of the carbonyl group must include the electrostatic effect (equation 2–2–5) on the local diamagnetic term, σ_d^L. Early attempts[1851,1208,1324,1413] to determine the anisotropy of the carbonyl group from chemical shift data, which ignored this term, are invalid. The only systematic attempt to separate σ_d^L and the long-range shielding effect is that reported by Zürcher.[2644] He has analysed chemical shift data for eight ketosteroids and has found that these data can only be accounted for if both the electrostatic and long-range shielding effects are taken into account. He obtains $\Delta\chi_x = \chi_x - \chi_z = 25\cdot7$ and $\Delta\chi_y = \chi_y - \chi_z = 12\cdot2$. These values cannot be directly compared with those of Pople since the former have not been broken down into the atomic contributions.

Figure 2–2–12 depicts the long-range shielding effect of the carbonyl group and is based on Pople's values for the principal susceptibilities of the two constituent atoms. We will now consider qualitative evidence which supports Pople's predictions. We note that the shielding in the plane of the bond is negative except for a small region near the x-axis at the "carbon end" of the bond and that, above and below the plane of the bond, in the vicinity of the z-axis the shielding is positive.

The protons of aldehydes lie in a region of intense deshielding and there is no doubt that their high δ-values (*ca.* 10 ppm) are partly due to this effect for, as Pople[1987] has pointed out, such a large deshielding effect could not be due solely to electron withdrawal by the $-I$ and $-M$ effects of the carbonyl group.

Evidence for positive shielding in the region of space near the x-axis at the "carbon end" of the bond is provided by the chemical shift of the protons of

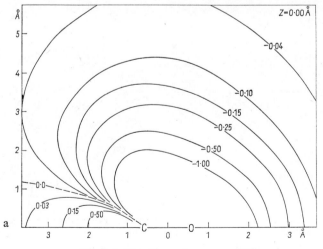

Fɪɢ. 2–2–12 a. (Legend, see page 90.)

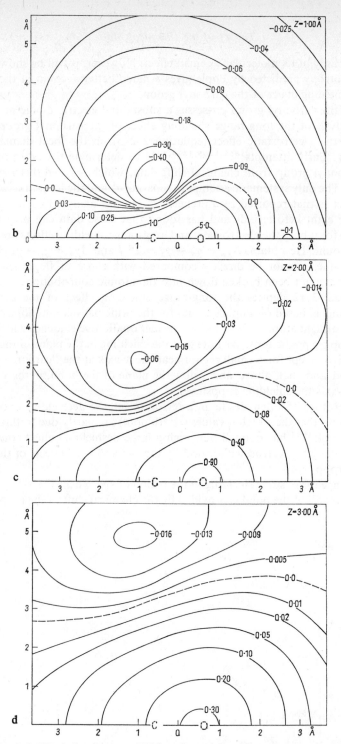

FIG. 2–2–12 b–d. Long-range shielding due to the carbonyl group, based on Pople's values[1997] for the principal susceptibilities of the two constituent atoms (Table 2–2–4). The figures on the contours represent shielding in ppm, negative values denoting deshielding. The four diagrams are sections in the X–Y plane with the Z coordinate equal to 0, 1, 2 and 3 Å.

the β-methylene group in cyclobutanone. These protons have exactly the same shift as the protons of cyclobutane.[2565] However, the effect of the electric dipole moment of the carbonyl group would be to deshield these protons by *ca.* 0·5 ppm, so that the long-range shielding term must be positive and of this order of magnitude.

There are many examples of the deshielding of protons which lie in the X–Z plane and close to the oxygen atom. For instance, it is well established that *cis*-β-protons are less shielded than *trans*-β-protons in α,β-unsaturated ketones,[1413,1259,1435] acids and esters,[1841,1259,1435,1983,1476] amides,[1476] and acid chlorides.[1476] Similarly, the protons of *cis*-β-methyl groups are less shielded than *trans*-β-methyl groups in aldehydes,[412] ketones,[1259] and acids and their derivatives.[1259] The work of Kossanyi,[1435] in particular, demonstrates that this effect is associated with an S-*cis* conformation of the carbonyl group. Martin, Defay and Greets-Evrard[1696] have demonstrated the existence of large deshielding effects in compounds of the types (XXXI), (XXXII) and (XXXIII) (the values appended to the formulae are the estimated deshielding effects in ppm of the carbonyl groups; see Chapter 3–6, Table 3–6–3

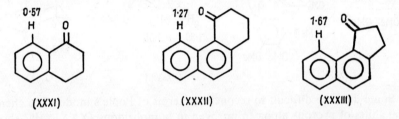

(XXXI)   (XXXII)   (XXXIII)

for additional data). Furthermore, it has been shown by Williams, Bhacca and Djerassi[2581] that 1β-protons in 11-keto-5α-steroids absorb near δ = 2·5 which corresponds to a deshielding by the carbonyl group of the order of 1 ppm. These protons, which are equatorial, are in the plane of the double bond and only about 2 Å from the oxygen atom. There is no doubt that the shifts described above are in part due to the electrostatic effect of the carbonyl dipole but they are too large to be ascribed solely to this cause.

There are several examples of shielding of protons situated above a carbonyl group. Crombie and Lown[562] have noted a large shift (1·55 ppm) for the 1-proton between the epimeric methyl ethers of rotenolone A (XXXIV) and B (XXXV). Models indicate that in the *trans*-isomer (XXXV) the 1-proton lies close to the oxygen atom of the 12-keto-group and in the plane of the carbon–oxygen double bond, whereas in the *cis* isomer (XXXIV) it is above the plane of the carbonyl group and near its z-axis. The magnitude of the

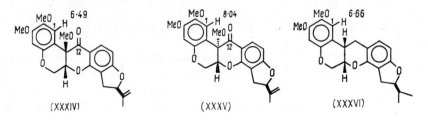

(XXXIV)   (XXXV)   (XXXVI)

shift suggests deshielding in (XXXV) and shielding in (XXXIV) although no really satisfactory model compounds have been examined to show that there is a definite *shielding* by the carbonyl group in the latter. The nearest model is 12-desoxy-6′,7′-dihydrorotenone (XXXVI) in which the 1-proton absorbs at 6·66. Chapman, Smith and King[468] have provided convincing evidence that α-lumicolchicine has the structure (XXXVII). They note that the aliphatic methoxyl groups, each of which lies above the plane of a carbonyl group, absorb at $\delta = 3·02$, compared with 3·3, the usual value for the protons of an aliphatic methoxyl group. Bergquist and Norin,[225] and Tori[2420] have deduced that the conformation of β-thujone is (XXXVIII) and conclude that the 6α-proton of the cyclopropane methylene group is shielded by the carbonyl group. Lehn and Vystrcil[1529,1536] quote data for triterpenes which indicate that protons above the plane of the double bond are shielded.

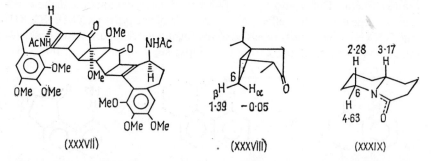

(XXXVII)   (XXXVIII)   (XXXIX)

Some data are difficult to reconcile in terms of Pople's model. The chemical shifts of protons alpha to nitrogen in quinolizidone (XXXIX)[282] show considerable variation and, in particular, shielding of the 6-axial proton is not predicted. Of course, the assumption that the anisotropy of the carbonyl group in amides is similar to that in aldehydes and ketones may be incorrect.

Perhaps the most important failure of Pople's model is its inability to account for the deshielding[417,244] of the C-19 protons of steroids (angular methyl group) by keto groups in the 1-, 3- and 7-positions. These protons lie fairly close to the *x*-axis at the "carbon end" of the carbonyl group and should be slightly shielded. A similar argument holds for the β-protons in cyclopentanone which are significantly less shielded than those of cyclopentane.[2565] It is unlikely that these discrepancies can be accounted for solely in terms of the electrostatic effect of the carbonyl group. It is more likely that the region of shielding near the *x*-axis is less extensive than indicated in Fig. 2–2–12, which would correspond to a rather larger anisotropy at the oxygen atom than given in Table 2–2–4.

(viii) SHIELDING BY THE CARBON–CARBON TRIPLE BOND

The diamagnetic anisotropy of the carbon–carbon triple bond can reasonably be assumed to arise largely from the π-electron system and the nature of the effect then follows from the axial symmetry of the π-electron distribution.[1987] Intuitively (p. 73), it is expected that circulations of the π-electrons will occur readily when the applied magnetic field is parallel to the molecular axis whereas they will be inhibited if the field is perpendicular to this axis. There will be excess diamagnetism along the molecular axis and $\chi_{\parallel} - \chi_{\perp} < 0$ since diamagnetism corresponds to a negative value of χ. The associated shielding field predicted by the point dipole approximation is shown in Fig. 2–2–13. Pople[1997] has calculated $\Delta\chi = -32$, and Cabaret, Didry and

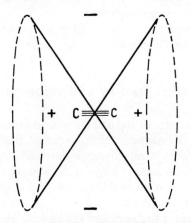

Fig. 2–2–13. The long-range shielding effect of the carbon–carbon triple bond.

Guy have estimated by a direct calculation, which does not involve the point dipole approximation, that the π-electron system contributes $+1\cdot92$ ppm to the shielding of the protons of acetylene.[417]

Confirmation of the predicted long-range shielding effect of the acetylene bond has been provided by Zeil and his coworkers[1145,2637] who have studied the chemical shifts in a series of mono-alkyl acetylenes and ethynyl halides. They estimated the effective electronegativity of the ethynyl group to be $3\cdot1$ and used this value to compute the chemical shifts of α- and β-protons. Differences between the calculated and observed values were then equated to the long-range shielding effect of the triple bond. By using equation (2–2–7) they obtained $\Delta\chi = -56$. Close agreement with Pople's value is not expected because of the inadequacies of the point dipole approximation which for the distances involved would lead to a higher value of $\Delta\chi$.

Reddy and Goldstein[2073] quote $\Delta\chi = -27$, derived from equation (2–2–7) by using ^{13}C—H coupling constant data to determine the local diamagnetic shielding of the protons of methylacetylene (p. 77).

(ix) Shielding by the Nitrile Group

The diamagnetic anisotropy of the nitrile group is expected to take the same form as that of the ethynyl group although the magnitude of $\Delta\chi$ could be different. Zeil and Buchert[2637] and Reddy and Goldstein[2073] give values of -34 and -27, respectively, which are identical with the respective values obtained for acetylene by these workers. In the former case no correction was made for the direct electrostatic effect of the appreciable dipole of the nitrile group.

Zürcher[2644] has analysed (p. 68) the shifts of the 19-protons in five cyanosteroids and has found that they can be rationalized solely in terms of the electrostatic effect. Similarly Cross and Harrison[568] have recognised that the long-range shielding effect of a 5α-cyano-substituent leads to shifts of the 19-proton signals which are in the wrong direction, and that the electrostatic effect is probably more important.

Clearly, further investigations are needed in order to establish the magnitude of $\Delta\chi$ for the nitrile group.

(x) Shielding by the nitro group

No theoretical estimates of the diamagnetic anisotropy of the nitrogroup have been reported. Yamaguchi[2622] has studied the long-range shielding effect of the nitro group in nitrobenzene derivatives, as a function of the angle between the group and the plane of the benzene ring. He has restricted the study to the effect on *ortho* methyl groups. The electrostatic contribution was determined by Buckingham's method[388] and the mesomeric effect by comparison with data for the *p*-methyl substituent. He finds that the nitrogroup deshields a proton in its plane and shields a proton situated above it. The anisotropy estimated by equation (2–2–6) is $\chi_x = -18$, $\chi_y = -3\cdot3$, and $\chi_z = -23$ where x is in the direction of the C—N bond and y and z are in- and out-of-plane, respectively. Wells[2524] has obtained a similar result for a series of methyl derivatives of α- and β-nitronaphthalene. Tori and Kuriyama[2433] reach the same conclusion from a study of the shifts in 6α- and 6β-nitrosteroids. These results explain the observations that aromatic protons are strongly deshielded by an *o*-nitro group and that the methyl group of *cis*-1-nitropropene is deshielded by 0·27 ppm relative to that of the *trans*-isomer.[187]

(xi) Shielding by aromatic rings

It has long been known that aromatic compounds are characterized by abnormally high diamagnetic susceptibilities and that this is essentially due to an excess of diamagnetism in the direction perpendicular to the plane of the ring(s). The diamagnetic anisotropy is believed to be due largely, but not entirely, to the π-electron system which, in monocyclic systems at least, provides a closed path suitable for the circulation of the π-electrons when the applied field is perpendicular to the plane of the ring. Aromatic molecules therefore provide examples of interatomic diamagnetic shielding (p. 63).

Considerable attention has been given to the direct calculation of the diamagnetic anisotropies and long-range shielding effects in aromatic molecules. The first attempt to calculate the shielding effect arising from π-electron systems of aromatic molecules was made by Pople[1995, 232] who extended

Pauling's[1956] classical model for the diamagnetism of aromatic molecules. In this treatment, the π-electron system of benzene is regarded as a circular loop with radius equal to that of the ring and the current induced by the applied field is calculated by Larmor's theorem. The secondary magnetic field associated with this ring current is then simulated by an equivalent point magnetic dipole situated at the centre of the ring from which the shielding

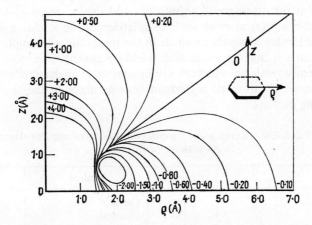

FIG. 2–2–14. Long-range shielding associated with the benzene ring. The co-ordinate axes are defined in the annexed diagram. (After Johnson and Bovey[1284].)

effect can be calculated by equation (2–2–7). This rather poor approximation has been superseded by a more refined treatment in which the physical size of the current loop is taken into account. Waugh and Fessenden[2507] and Johnson and Bovey[1284] have calculated the shielding effect for benzene by assuming a current ring situated 0·7 Å above and below the plane of the ring. Unlike the point dipole approximation, this model correctly predicts a positive shielding effect at all points inside the ring. The model is a useful one for semiquantitative estimations of the shielding effect of the benzene ring, and a contour diagram of a cross section of the benzene molecule is shown in Fig. 2–2–14. In 1937, London[1580] developed a molecular orbital method for estimating the ring current and this method has been extended to the calculation of chemical shifts by McWeeney[1645,1646] and Hall and Hardisson.[1081] Both the classical and the quantum mechanical methods have been used to calculate the chemical shifts of the ring protons in polycyclic aromatic hydrocarbons,[232,1991,1081,1739,1646,1299,1740] non-benzenoid hydrocarbons[3,2494,2211] and heterocyclic aromatic molecules.[3,22,1083,1082]

The values calculated for the diamagnetic anisotropy of benzene, using various theoretical methods, are given in Table 2–2–5 together with experimental values obtained from the Cotton–Mouton effect and from studies of single crystals. It is seen that the calculations underestimate the anisotropy and while this may be due to errors inherent in the theoretical methods, it has been pointed out[237,1824] that the interatomic diamagnetic contribution constitutes only about 70 per cent of the measured anisotropy, the remainder arising from local paramagnetic contributions of the type postulated for ethylene (p. 83), as well as contributions from the σ-framework. Indeed, it has been suggested by Musher[1824] that the concept of a ring current is an unsatisfactory one and that the diamagnetic anisotropy of benzene and particularly polycyclic aromatic hydrocarbons is better treated as the sum of local contributions. So far, Musher[1824] has only produced an empirical treatment for diamagnetic anisotropy, and the extension of his hypothesis to either empirical or *ab initio* calculations of the shielding effects of aromatic molecules is eagerly awaited. In the meantime, the simple concept of a ring current and its associated shielding contribution provides the organic chemist with an excellent semiquantitative understanding of the contribution which aromatic π-electron systems make to the shielding of neighbouring protons.

TABLE 2–2–5. CALCULATED AND EXPERIMENTAL VALUES OF THE DIAMAGNETIC ANISOTROPY OF BENZENE

Workers	Reference	Method	$-\Delta\chi$
Pauling, L.	1956	Classical	81·8
Dailey, B.P.	590	Classical	67·8
London, F.	1580	MO	54·5
Fujii, S. and Shida, S.	928	ASMO	54·4
Itoh, T., Ohno, K. and Yoshizumi, H.	1250	ASMO	44·2
Caralp, L. and Hoarau, J.	427	ASMO	54·7, 51·8
Davies, D.W.	604	SCFMO	54·5
Hall, G.G. and Hardisson, A.	1081	SCFMO	54·5
Dailey, B.P.	590	SCFMO + σ-bond correction	89·6
Pople, J.A.	2000	MO + local paramagnetic correction	83·3
Hoarau, J., Lumbrosa, N. and Pacault, A.	1177	Single crystal measurements	98·5
LeFevre, R.J.W., Williams, P.H. and Eckert, J.M.	1527	Cotton–Mouton effect	96·4

The form of the shielding effect indicated in Fig. 2–2–14 has been abundantly confirmed by chemical shift data.

The δ-value (7·25) of the protons of benzene itself is evidence of a considerable deshielding in the plane of, and exterior to, the ring. The theories also predict that, in polycyclic systems, each ring has its own ring current so that their shielding contributions are approximately additive. This means that in such systems the proton nearest to the largest number of rings will have the highest δ-value (i.e. most deshielded); a conclusion which is borne out by the shifts reported for the benzenoid hydrocarbons (see Chapter 3–6B). It is evident from Fig. 2–2–14 that a proton inside an aromatic ring should be heavily shielded and this is indeed the case. In [18]annulene (XL), bis-dehydro[14]annulene (XLI) and related systems,[1258,2333,2332,946] the inner protons are at extraordinarily high fields. It should be noted that in these molecules the outer protons are more deshielded than those of benzene itself. This follows from the fact that the shielding effect of the ring current is *proportional to the area of the ring*. A similar pattern of chemical shifts has been observed for porphin (XLII) and other porphyrins.[208,3,15a,446] It has been predicted[1581] that, in very large annulenes (> [24]), the ring current

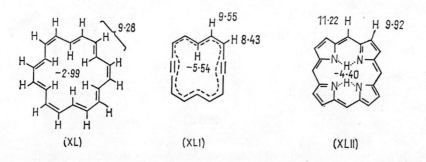

(XL)   (XLI)   (XLII)

will no longer exist and their spectra will resemble linear polyenes. The molecular orbital theory indicates[1644] that the proton spectrum of the cyclic hexaphenyl (XLIII) should resemble biphenyl rather than an annulene and this appears to be the case.[2345]

There are also a number of examples which illustrate that the central region above and below the plane of an aromatic ring system is strongly shielded. The novel dihydropyrene derivative (XLIV)[276] and the methylene bridged [10]annulene (XLV)[2483] show this effect to a marked degree. This

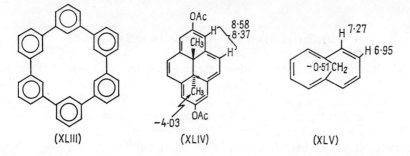

(XLIII)   (XLIV)   (XLV)

situation has also been encountered in the antibiotic rifomycin-B (XLVI) and its derivatives in which several methyl groups of the macrocyclic ring lie over the plane of the aromatic system,[2013] and give rise to signals near $\delta = 0.0$. Other examples include 1,4-decamethylenebenzene[2507] (XLVII) and the interesting series (XLVIII) which permits a comparison of the relative effects of the benzene, furan and thiophene rings.[619]

The references in the footnote† are a selection of additional examples in which long-range shielding effects of aromatic nuclei have been observed.

† References 57, 106, 328, 357, 432, 447, 448, 555, 556, 567, 582, 585, 948, 1027, 1057, 1138, 1219, 1247, 1483, 1569, 1582, 1693, 1781, 2144, 2289, 2354, 2484, 2485, 2596.

The chemical shifts of aromatic protons in general will be discussed in Chapter 3–6. In many instances unequal π-electron charge densities play an important role in determining δ-values.

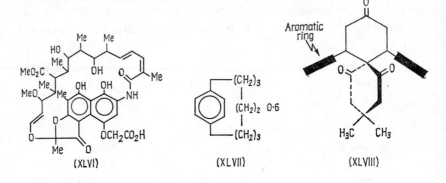

It has been suggested by Elvidge and Jackman[765] that, if the contribution of the ring current to the shielding of ring protons and ring methyl substituents can be determined, it can be used as a measure of the aromaticity of the system. This thesis has received support[619,606,760] and criticism.[22,1824] Certainly, chemical shifts have proved useful in demonstrating the striking difference between annulenes which obey Hückel's rule ($4n + 2\pi$-electrons) and those that do not ($4n$ π-electrons). The term "aromaticity" is of doubtful utility, however it is defined, and there certainly will be no general relation between the magnitude of the π-electron contribution to the chemical shift and other properties such as reactivity towards electrophiles.

(xii) SHIELDING BY THREE-MEMBERED RINGS

Cyclopropane has a molar diamagnetic susceptibility of $-65\cdot7$[1480] while the value predicted from Pascal's numbers is *ca.* -60. This exaltation of diamagnetism has been attributed to a ring current in the three-membered ring[1480] and the consequential long-range shielding effect has been considered by a number of workers.[1946,411,2425,857] Attempts have been made to account for the magnitude of the shielding field due to the ring current in terms of Pople's[1995,232] equivalent dipole model[411,857,2425] and by means of the Waugh and Fessenden[2507] treatment.[1946] It has also been suggested[2425] that the shielding field associated with the cyclopropane ring arises from the anisotropy of the three carbon–carbon single bonds which may have an anomalously high value (-33) in this system.

The qualitative prediction of the ring current model, *viz.* that the shielding is positive above the ring, is well supported by the available data. The highly shielded nature of the cyclopropane protons ($\delta = 0\cdot22$)[2565] is itself consistent with this prediction.[1946,411] Further evidence is provided by comparison of the shifts associated with the systems listed in Table 2–2–6. Tadanier and Cole[2375] have observed that the usual order of shifts for axial and equatorial protons ($\delta_{ax} - \delta_{eq} = 0.5$ ppm) is reversed for protons at the 6-position in the $3\alpha,5\alpha$-cyclosteroids (XLIX) and (L).

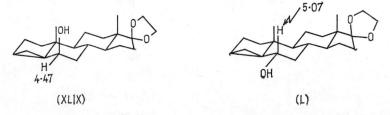

(XLIX)   (L)

It is interesting that the protons of the bicyclobutane ring in compounds (LI),[501] (LII)[505] and (LIII)[505] do not exhibit the characteristic upfield shift of cyclopropane itself. It is evident from ^{13}C—H coupling constants that the

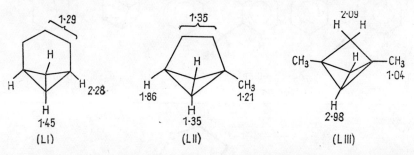

(LI)   (LII)   (LIII)

C—H bonds have considerably increased *s*-character in these molecules and this results in an appreciable decrease in local diamagnetic shielding. In this connection, it should be noted that the protons of cyclopropane are presumably more shielded by the ring anisotropy than indicated by their chemical shift, since the C—H bonds in this molecule also have greater *s*-character than those of acyclic methylene groups.

Comparison of ethylene oxide, ethyleneimine, and ethylene sulphide with the corresponding six-membered rings (Table 2–2–7) indicates that these systems probably have anisotropies similar to cyclopropane with the possible exception of ethylene sulphide. Jefferies, Rosich and White[1272] have reported one example in which intense shielding above the plane of the epoxide ring is apparent. They examined derivatives of $15\beta,16\beta$-oxidobeyerol (LIV)

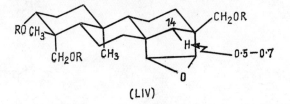

(LIV)

in which a proton assigned to the 14α-position is at an abnormally high field. The shifts of the 19-protons in steroidal epoxides and episulphides (Table 2–2–8)[2428] are less readily interpreted. In general, the results given in Table 2–2–8 are in accord with the qualitative predictions of the ring current model and suggest that the effect is largest for the episulphide ring.

TABLE 2–2–6. LONG-RANGE SHIELDING BY THE CYCLOPROPANE RING

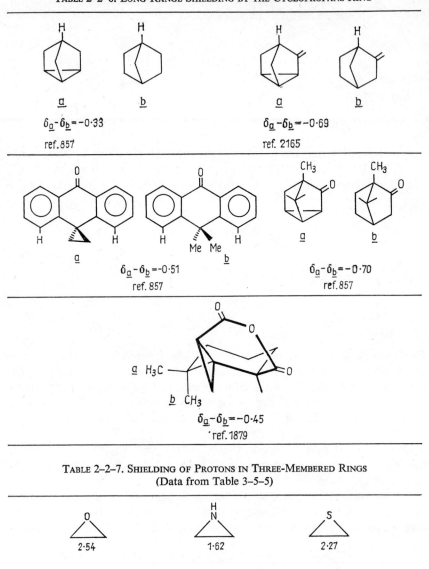

TABLE 2–2–7. SHIELDING OF PROTONS IN THREE-MEMBERED RINGS
(Data from Table 3–5–5)

However, the magnitudes of the shifts are quite small and other factors, such as distortion of the pure chair conformations and electrostatic effects may be comparable with the expected long range shielding. In this connection, Laszlo[1503] has pointed out that measurements of chemical shifts of

TABLE 2–2–8. 19-PROTON SHIFTS DUE TO THREE-MEMBERED RINGS IN STEROIDS[a]

Position	X=CH$_2$	X=O	X=S	Sign[b]
2α, 3α		0·01	−0·02	—
2β, 3β		−0·07	−0·13	+
3β, 4β			−0·20	+
5α, 6α		−0·26	−0·38	—
5β, 6β	−0·22[c]	−0·20	−0·35	—
8α, 9α		−0·10		—
9α, 11α		−0·20		—
9β, 11β		−0·22		—
11α, 12α		−0·07	−0·09	—
11β, 12β		−0·17	−0·15	—

[a] Taken from the paper of Tori, Komeno and Nakagawa[2428]. A positive sign represents shielding.
[b] The sign given in this column is that predicted from a consideration of Dreiding models.
[c] Ref. 2375.

epoxides are subject to significant solvent effects even when chloroform is the solvent.

Tori and his coworkers[2427,2426] have examined the effect of both an epoxide and an aziridine ring at the 2,3-position in norbornane, norbornene, benzonorbornene and benzobicyclo[2.2.2]octene systems and have concluded that both types of three-membered rings shield protons above their plane. However, they point out that other factors could also be operative in these systems.

Uebel and Martin[2453] have reported a remarkable differential shielding of the 2(4)-axial and 2(4)-equatorial protons in 8-thiabicyclo[3.2.1]octane-3-spiro-3′-diazirine (LV) which would indicate that the diazirine ring strongly shields a proton above its plane. The chemical shifts of the protons of cyclopropene (LVI)[2565] are consistent with an appreciable anisotropy of the type associated with a ring current.

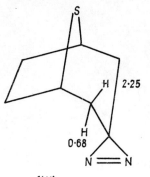

(LV)

(LVI)

D. CONTACT SHIFTS

The theory of proton chemical shifts outlined in the preceding sections is valid for molecules which do not possess unpaired electrons. The presence of unpaired electrons can make an extremely large contribution to the observed chemical shift. We have already seen that two nuclei can spin–spin couple (p. 13). In the same way, a proton can couple with an electron which has a spin number of one half. This is, in fact, the origin of the isotropic hyperfine splitting observed in electron spin resonance spectra. If, however, the spin of the coupled electron is changing rapidly, the proton experiences the average of the two electron spin states and, because of the slight Boltzmann excess of the lower spin state, this average is not zero. Since the magnetic moment of the electron is large, the Boltzmann average of the contribution of the two spin states to the field experienced by the proton can correspond to substantial chemical shifts, sometimes of the order of 100 ppm. This type of chemical shift is called a *contact shift* since in order for it to occur, the unpaired electron has to have a density at the proton so that coupling by the Fermi contact mechanism can occur (p. 116).

The observation of contact shifts is rarely possible, since, unless the elec-

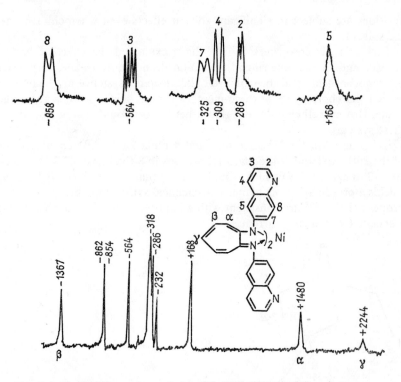

FIG. 2–2–15. 60 MHz spectrum of nickel(II) *N,N′*-di(6-quinolyl)-aminotroponeiminate. The chemical shifts are in Hz from TMS. (After Eaton, Josey, Phillips and Benson[721].)

tron relaxation time is extremely short, severe paramagnetic broadening of the proton spectrum occurs (p. 9). For instance, the relaxation times for organic free radicals are comparatively long and high resolution spectra of such species are generally not observed in solution. More favourable relaxation times operate in the case of certain paramagnetic transition metal ions and the proton spectra of organic ligands can be observed. In addition, in certain cases rapid ligand exchange can achieve the same result. A further mechanism for narrowing the proton resonance lines exists in complexes where there is a rapid intramolecular interconversion between high and low spin states of the metal ion.

While the above phenomena are of considerable interest to the inorganic chemist, some of the results also have direct bearing on the electronic structures of the organic ligands. Furthermore, because of the large chemical shifts, the proton spectra are usually first order (p. 124) and yield values of proton–proton coupling constants directly (Chapter 2–3 C). Figure 2–2–15 shows the spectrum of a tetrahedral nickel (II) complex obtained by Eaton, Josey, Phillips and Benson.[721]

An excellent review of this field has been published by Eaton and Phillips.[727]

E. THE HYDROGEN BOND

Protons, such as hydroxylic protons, which can form hydrogen bonds either inter- or intra-molecularly, usually have chemical shifts which depend on the strength and concentration of the hydrogen-bonded complex and on the nature of the donor, as evidenced by the following examples. Gaseous water absorbs 4·58 ppm to higher field than the liquid.[2212] The hydroxylic proton of phenol at infinite dilution in carbon tetrachloride has $\delta = 4\cdot25$ and reaches a limiting value of 6·75 at higher concentrations.[2006] In 2,5-dihydroxypropiophenone (0·36 M; deuterochloroform) the strongly intramolecularly bonded 2-proton absorbs at 12·02 compared with the 5-proton at 5·8.[1930] Finally, the δ-value of phenol at infinite dilution in benzene is *ca.* 1·5 ppm to *higher* field than in carbon tetrachloride.[2006]

Schneider, Bernstein and Pople[2212] have considered that the shifts associated with hydrogen bonding arise from three terms:

(i) The electrostatic effect of the donor atom or group.
(ii) The diamagnetic anisotropy of the donor atom or group.
(iii) Changes in the long-range shielding of the acceptor group.

Of these three effects, the first is dominant in the case of strong hydrogen bonds. The donor site generally consists of an atom bearing a partial negative charge (e.g. the oxygen atom of a carbonyl group). Calculations based on equation (2–2–5) give a value of the order of -3 ppm for a linear hydrogen bond to a carbonyl group.

The second term may have either sign. In the case of the carbonyl group its diamagnetic anisotropy is such (p. 88) that it will cause an additional deshielding probably of a large magnitude. Both the first and second term therefore contribute to the high δ-values of systems such as *o*-hydroxy-phenones.

In the case of weak hydrogen bonding the second term may become dominant. This is evidently the case in the phenol–benzene system, in which it is envisaged that hydrogen bonding involves the π-electrons so that the proton is situated in a region of intense shielding above the plane of the benzene ring.

The third term has little experimental foundation. It can only operate in cases of intermolecular hydrogen bonding. In the case of a linear molecule, H—X, hydrogen bonding to X with another molecule could reduce the axial symmetry of H—X and hence the long-range shielding of X (cf. the shielding in acetylene; p. 93). It has been suggested that this term is important in explaining the gas-to-solution shifts for acetylene and hydrogen iodide.[2212]

F. SOLVENT EFFECTS

We have seen in Part 1 that the vast majority of shielding values for protons in organic molecules are obtained from measurements in solutions and are expressed relative to the absorption line of an internal reference. We know, in fact, that such shielding values are sensitive to concentration and to the nature of the solvent. We therefore need to consider the contribution which the solvent can make to the shielding of a proton in a solute molecule, in order to assess how well shielding values, determined from measurements in solution, reflect the shielding for the corresponding molecules in a perfect gas. Furthermore, an understanding of solvent effects can be of considerable use in the elucidation of the structure and stereochemistry of organic compounds by n.m.r. spectroscopy.

The shielding contribution (σ_s; equation (2–2–2)) of the solvent to a proton in a solute molecule is conveniently discussed in terms of five separate effects as expressed in equation (2–2–8);

$$\sigma_s = \sigma_B + \sigma_w + \sigma_a + \sigma_E + \sigma_c \qquad (2\text{–}2\text{–}8)$$

where the first four terms on the right hand side of the equation are associated, respectively, with the effects of bulk magnetic susceptibility, van der Waal interactions, diamagnetic anisotropy of the solvent, and electric polarization and polarizability of the solvent. The fifth term is designed to include weak interactions, such as charge transfer and hydrogen bonding, which lead to some form of complex in which the solute and solvent molecules are specifically oriented with respect to each other.

The first term, σ_B, arises if the geometry of the sample differs from that of the hypothetical spherical cavity occupied by a molecule within the sample. As pointed out in Chapter 1–2D, the usual geometry of the sample corresponds to an infinite cylinder and the correction, σ_B, is then equal to $2\pi\varkappa/3$

where $\varkappa$ is the bulk susceptibility of the sample.[651] Bothner-By and Glick[303] have shown that this term is insufficient to account for the effect of solvents so that when it is removed by the use of spherical samples (p. 45) or, as is commonly the practice, by employing an internal chemical shift reference (e.g. tetramethylsilane), the remaining four terms in equation (2–2–8) must still be considered.

The second term σ_w is proportional to the time average of the *square* of the electric field provided by neighbouring solvent molecules. This will have a non-zero value even for non-polar solvent molecules. Bothner-By[301] was the first to consider the influence of van der Waal interactions in relation to solvent effects and he showed that the magnitude of σ_w was of the order (*ca.* 0·1 ppm) of the shift from gas to liquid (corrected for σ_B) observed for non-polar, magnetically isotropic molecules. Similar conclusions have been reached from studies of the pressure dependence of the chemical shift in pure gases,[1006,2066] binary gas mixtures,[2066] and from gas-to-solution shifts.[396,1585,1224] In all cases σ_w will be negative. Its direct calculation for complex molecules is difficult because it is not possible to define the spatial relation between solvent molecules and a particular proton in the solute molecule. Table 2–2–9 shows the solvent shifts for methane in a variety of solvents. The first eleven entries exhibit a linear relation with the heats of vaporization of the solvents, which is assumed to be a measure of the van der Waal interactions.[396] The halo-compounds (entries 12–26) show a similar relation but are displaced by *ca.* 0·17 ppm. There seems little doubt that the observed shifts in both cases are due to σ_w, and that this term can be of the order of 0·1–0·7 ppm. However, the data presented in Table 2–2–10 indicate that although σ_w varies from solvent to solvent its value for non-polar solutes in any given solvent is reasonably constant ($\pm 0·05$ ppm). Indeed, the deviations may well arise from errors in the corrections for σ_B. We may conclude that the contribution σ_w is not likely to introduce significant errors if spectra are calibrated against an internal reference such as tetramethylsilane.

The term σ_a, which arises from the diamagnetic anisotropy of the solvent, was first recognised by Bothner-By and Glick.[303,304] They showed that chemical shifts determined in aromatic solvents correspond to increased shielding relative to that of the isolated molecule. They interpreted this result, which holds for highly symmetrical non-polar molecules such as methane (cf. items 33–36 in Table 2–2–9), as a consequence of the disc-like shapes of aromatic molecules. The shielding regions (cf. Fig. 2–2–14) of these molecules are more exposed than the deshielding regions which are partially occupied by the ring protons or substituents. Thus, on purely statistical grounds the solute molecules will be shielded. The argument has been extended to rod-like molecules such as acetylene.[396] In this case the exposed region is transverse to the molecular axis and the diamagnetic anisotropy of the solvent molecule will cause a down-field shift relative to the gas phase, a prediction which is borne out by the entries 27–32 in Table 2–2–9. In the absence of interactions which specifically orient the solvent molecules with respect to the solute molecules, the above considerations are unlikely to result in differential solvent shifts between protons of the same or different

TABLE 2–2–9. VALUES of σ_s (PPM) FOR METHANE IN VARIOUS SOLVENTS[396]

(The gas phase shift is taken as zero and values are corrected for σ_B)

Solvent		Solvent	
1. Neopentane	−0·13	20. Carbon tetrachloride	−0·42
2. Cyclopentane	−0·16	21. Carbon tetrabromide	−0·57
3. n-Hexane	−0·20	22. CH_2ClBr	−0·41
4. Cyclohexane	−0·23	23. $CHClBr_2$	−0·70
5. *trans*-But-2-ene	−0·15	24. CCl_3Br	−0·44
6. *cis*-But-2-ene	−0·17	25. Silicon tetrachloride	−0·20
7. Acetone	−0·18	26. Stannic chloride	−0·31
8. Diethyl ether	−0·19	27. Propyne	−0·26
9. Ethyl acetate	−0·23	28. Acetonitrile	−0·46
10. Ethyl nitrate	−0·23	29. But-2-yne	−0·41
11. Triethylamine	−0·21	30. Carbon disulphide	−0·57
12. Bromine	−0·38	31. Diacetylene	−0·63
13. Methyl bromide	−0·38	32. Dicyanoacetylene	−0·75
14. Methyl iodide	−0·46	33. Benzene	+0·18
15. Methylene chloride	−0·35	34. Toluene	+0·13
16. Methylene bromide	−0·46	35. Chlorobenzene	+0·11
17. Methylene iodide	−0·71	36. Nitrobenzene	+0·27
18. Chloroform	−0·39	37. Nitromethane	−0·03
19. Bromoform	−0·68	38. Nitroethane	−0·11

TABLE 2–2–10. VALUES OF σ_w (PPM) FOR THE PROTONS OF VARIOUS NON-POLAR SOLUTES IN A SELECTION OF SOLVENTS

Solute \ Solvent	Cyclo-hexane	Cyclo-pentane	Neo-pentane	Acetone	Dioxan	Nitro-methane	Methy-lene chloride	Carbon tetra-chloride
Methane[a,b]	−0·24 −0·23			−0·17 −0·18	−0·35	−0·20	−0·35	−0·40 −0·42
Ethane[c,a]	−0·20 −0·15	−0·16		−0·08 −0·11	−0·23	−0·13	−0·30	−0·31
Neopentane[c]	−0·20	−0·18	−0·18	−0·10			−0·30	−0·33
Cyclopropane[a]	−0·23			−0·21	−0·31	−0·24		−0·38
Ethylene[c,a]	−0·22 −0·20	−0·20	−0·20	−0·11 −0·22	−0·34	−0·27		−0·37
Tetramethylsilane[c]	−0·25			−0·15			−0·32	−0·36
Cyclopentane[c]	−0·17	−0·15	−0·15	−0·08			−0·30	−0·29

[a] ref. 1585;　[b] ref. 396;　[c] ref. 301.

solutes.† In other words, the use of an internal reference should still give the same relative chemical shifts as observed in the gas phase. However, there are many examples (see below) in which substantial changes in relative chemical shifts are observed in aromatic and other solvents and these occur because of a variety of interactions such as electric dipole–dipole, electric dipole–quadrupole, charge transfer and hydrogen bond associations which result in specific solvent–solute orientations. In summary then, σ_a should not contribute to shielding determined with the use of an internal reference, although the diamagnetic anisotropy of the solvent will often make a contribution through the term σ_c.

In 1958, Stephen[2356] predicted that the shielding of a proton in a polar molecule would be influenced by the electric field associated with the polarization of the medium by the solute. This prediction was subsequently confirmed and the effect has been further analysed.[388,660,396] A quantitative treatment of electrostatic effects in pure gases and binary gas mixtures has been carried out by Raynes, Buckingham and Bernstein[2066] who have used a semi-empirical, statistical mechanical model involving two-body interactions. The results of this treatment are important since they show that, not only dipole–dipole and dipole–induced dipole interactions, but also dipole–quadrupole interactions are important, and that these interactions are frequently much larger than σ_w. This method is not readily applicable to the liquid state and has to be replaced by less satisfactory models. Stephen[2356] and Buckingham[388] have used the Onsager[1912] model to estimate the *reaction field*; i.e. the field induced in the medium by the solute dipole or quadrupole. In this model, the solute is represented by a sphere of radius r, containing a point electric dipole (or quadrupole) at its centre. The solvent is regarded as a continuous medium with a dielectric constant ε. The reaction field, R, which is parallel to the electric dipole of the solute, is then given by equation (2–2–9);

$$R = \frac{2(\varepsilon - 1)(n^2 - 1)}{3(2\varepsilon + n^2)} \cdot \frac{\mu}{\alpha} \tag{2-2-9}$$

where μ is the dipole moment of the solute, $\alpha = (n^2 + 1)\, r^3/(n^2 + 2)$ and n is the refractive index of the pure solute in the liquid or solid state. From equation (2–2–5), neglecting the second term, the component of the field R, and hence the effect on the shielding along the X—H bond for a proton in a given solute molecule, is proportional to $(\varepsilon - 1)(\varepsilon - n^2/2)$. The field R induced by a quadrupole is proportional to $(\varepsilon - 1)/(\varepsilon + 0.67)$,[388] so that the term σ_E will be recognizable by an approximately linear dependence on $(\varepsilon - 1)/(\varepsilon + 1)$ for a range of solvents. Buckingham has pointed out that the Onsager model is likely to be a rather poor approximation. More sophisticated models, which take into account molecular shape, have been in-

---

† Shape factors could be important in determining the orientation of non-associative collisions, and could therefore lead to a dependence of σ_a on the molecular shape of the solute. However, the protons of most organic molecules are peripherally located and it is doubtful if such effects are significant.

vestigated by Diehl and Freeman[660] and it is possible that these give better results.[660,1225]

It is clear that the term σ_E can lead to variations in chemical shifts determined in different solvents, even if an internal reference is employed. Errors introduced in this way will be minimized by the use of solvents of low polarity and polarizability (i.e. low dielectric constant). The sign of σ_E will depend on the direction of the solute dipole moment with respect to the C—H bond of the proton under consideration. The approximate dependence of the absolute magnitude of σ_E on $(\varepsilon - 1)/(\varepsilon + 1)$ thus offers a means of making chemical shift assignments and of distinguishing between two structural or stereochemical possibilities, if these differ in the relation of the molecular dipole to a given C—H bond. We will now consider the experimental evidence for the existence of σ_E in order to obtain some idea of its magnitude.

In his original paper, Buckingham[388] predicted the effect of the reaction field on the chemical shift of the o-, m- and p-protons of monosubstituted benzenes. For a dipole of the direction

$$Ph \rightleftharpoons X$$

the theory predicts that increasing the dielectric constant of the medium should increase the shielding of the o-proton and decrease that of both the m- and p-protons. Changing the direction of the dipole reverses this trend. A detailed analysis of solvent effects in substituted benzenes has been made by Diehl.[655,656] He has shown that solvent effects are additive and can be expressed by equation (2–2–10).

$$L_{\alpha,\text{total}} = \left(\sum L'_{n,X,\alpha} \right) + L_{H,\alpha} \qquad (2\text{–}2\text{–}10)$$

Here, α designates the solvent. $L_{H,\alpha}$ is the shift for benzene (5 mole %) in the solvent α relative to benzene (5 mole %) in n-hexane ($\varepsilon = 1.9$). X is the substituent and n its orientation (o, m, or p) to the proton under consideration. Values of $L'_{n,X,\alpha}$ for $\alpha =$ acetone and benzene are given in Table 2–2–11. The data for acetone ($\varepsilon = 20.7$) shows some agreement with Buckingham's predictions. For instance, the trends for NH_2, Cl and NO_2 are in fair agreement.

TABLE 2–2–11. ADDITIVE SOLVENT EFFECTS (PPM) FOR AROMATIC PROTONS[655]

X	Acetone ($L_H = -0.13$)			Benzene ($L_H = 0.38$)		
	L'_o	L'_m	L'_p	L'_o	L'_m	L'_p
NH_2	−0·05	0·10		0·07	−0·10	
OH	0·02	0·03		0·16	0·10	
OCH_3	0·01	−0·01	0·04	−0·17	−0·05	−0·07
F	−0·05	−0·07		0·06	0·23	
CH_3	0·02	0·01	0·04	−0·02	−0·05	
Cl	−0·02	−0·10	−0·13	0·09	0·26	0·26
Br	−0·01	−0·10	−0·10	0·09	0·27	0·27
CN	−0·09	−0·10		0·37	0·53	
I	0·00	−0·09		0·09	0·26	
NO_2	0·11	−0·13		0·26	0·61	

Both Buckingham[388] and Diehl[655] have pointed out that the Onsager model is a poor approximation for aromatic solutes.

Buckingham[388] predicted that since *cis*-dihaloethylenes are dipolar and the *trans*-isomers are non-polar, the former should give rise to a larger reaction field. This prediction has been confirmed by Hruska, Bock and Schaefer[1225] who have examined the solvent shifts (relative to internal cyclohexane) for *cis*- and *trans*-dichloro- and dibromo-ethylene in media of varying dielectric constant (Table 2–2–12). Although the shifts are quite small, the greater sensitivity of the *cis*-isomers is clearly demonstrated.

TABLE 2–2–12. VARIATION OF SOLVENT SHIFTS (PPM) WITH DIELECTRIC CONSTANT FOR *cis*- AND *trans*-DICHLORO- AND DIBROMO-ETHYLENE IN AQUEOUS DIOXAN SOLUTIONS[1225]

ε	*cis* $C_2H_2Cl_2$	*trans* $C_2H_2Cl_2$	*cis* $C_2H_2Br_2$	*trans* $C_2H_2Br_2$
2·21[a]	(0·000)	(0·000)	(0·000)	(0·000)
2·82	0·017	0·012	0·020	0·010
3·18	0·020	0·012	0·025	0·012
3·76	0·028	0·013	0·032	0·015
5·00			0·040	0·022
5·83	0·047	0·022	0·050	0·025
7·24	0·053	0·020	0·055	0·023
9·56	0·058	0·025	0·058	0·023

[a] Pure dioxan.

The magnitude of the reaction field shift for some organic solvents can be gauged from Table 2–2–13 which summarizes data obtained by Lumbroso, Wu and Dailey.[1585] Here the observed gas-to-solution shifts, corrected for σ_B and σ_w, are assumed to be due to the reaction field. The values are in qualitative agreement with those calculated by Buckingham's method. There have also been several investigations which demonstrate an approximately linear dependence of solvent shifts on $(\varepsilon - 1)/(\varepsilon + 1)$ or related expressions.[1502,396,1225,2660,656]

We now turn to a consideration of σ_c, the last term in equation (2–2–8). We have seen that the first three terms are eliminated by the use of a suitable internal reference and that σ_E will generally be less than 0·2 ppm. Solvent shifts associated with the specific solvent effect, σ_c, can however assume much larger values, sometimes of the order of 1 ppm. These effects are particularly important in the application of proton spectroscopy to organic chemistry because they can sometimes be used to remove overlap of absorption lines, and because they can yield both structural and stereochemical information. The most important type of specific solvent effect is that associated with aromatic solvents. Also of interest are effects which arise when solute and solvent can complex through hydrogen bonding. Frequently, the interactions between solute and solvent which give rise to σ_c are referred to as *collision complex* formation.

TABLE 2-2-13. OBSERVED AND CALCULATED VALUES OF σ_E (PPM) FOR SOME SIMPLE SOLUTES IN SOLVENTS OF VARYING DIELECTRIC CONSTANT[1585]

Solute	Cyclohexane ($\varepsilon = 2{\cdot}02$)		Dioxan ($\varepsilon = 2{\cdot}21$)		CCl$_4$ ($\varepsilon = 2{\cdot}23$)		Acetone ($\varepsilon = 20{\cdot}7$)		CH$_3$NO$_2$ ($\varepsilon = 35{\cdot}9$)	
	CH$_3$	CH$_2$	CH$_3$	CH$_2$	CH$_3$	CH$_2$	CH$_3$	CH$_2$	CH$_3$	CH$_2$
CH$_3$CN (obs)	$-0{\cdot}06$		$-0{\cdot}16$		$-0{\cdot}20$		$-0{\cdot}34$		$-0{\cdot}30$	
(calc)	$-0{\cdot}06$		$-0{\cdot}06$		$-0{\cdot}06$		$-0{\cdot}17$		$-0{\cdot}18$	
CH$_3$CH$_2$Br (obs)	$-0{\cdot}02$	$-0{\cdot}04$	$-0{\cdot}06$	$-0{\cdot}21$	$-0{\cdot}08$	$-0{\cdot}11$	$-0{\cdot}09$	$-0{\cdot}29$	$-0{\cdot}10$	$-0{\cdot}28$
(calc)	$-0{\cdot}01$	$-0{\cdot}02$	$-0{\cdot}09$	$-0{\cdot}02$	$-0{\cdot}01$	$-0{\cdot}02$	$-0{\cdot}02$	$-0{\cdot}06$	$-0{\cdot}03$	$-0{\cdot}06$
CH$_3$CH$_2$I (obs)	$0{\cdot}01$	$0{\cdot}02$	$-0{\cdot}07$	$-0{\cdot}16$	$-0{\cdot}07$	$-0{\cdot}04$	$-0{\cdot}17$	$-0{\cdot}29$	$-0{\cdot}13$	$-0{\cdot}27$
(calc)	$-0{\cdot}01$	$-0{\cdot}02$	$-0{\cdot}01$	$-0{\cdot}02$	$-0{\cdot}01$	$-0{\cdot}02$	$-0{\cdot}02$	$-0{\cdot}06$	$-0{\cdot}02$	$-0{\cdot}06$
CH$_3$CH$_2$CN (obs)	$0{\cdot}01$	$-0{\cdot}12$	$-0{\cdot}01$	$-0{\cdot}26$	$-0{\cdot}10$	$-0{\cdot}23$	$-0{\cdot}05$	$-0{\cdot}37$	$-0{\cdot}05$	$-0{\cdot}32$
(calc)	$-0{\cdot}02$	$-0{\cdot}04$	$-0{\cdot}02$	$-0{\cdot}05$	$-0{\cdot}01$	$-0{\cdot}05$	$-0{\cdot}06$	$-0{\cdot}12$	$-0{\cdot}06$	$-0{\cdot}13$
(CH$_3$CH$_2$)$_2$O (obs)	$0{\cdot}03$	$0{\cdot}07$	$0{\cdot}02$	$0{\cdot}03$	$-0{\cdot}01$	$0{\cdot}02$	$0{\cdot}11$	$0{\cdot}20$	$-0{\cdot}01$	$-0{\cdot}04$
(calc)	$0{\cdot}00$	$0{\cdot}01$	$0{\cdot}00$	$0{\cdot}02$	$0{\cdot}00$	$0{\cdot}02$	$-0{\cdot}01$	$0{\cdot}04$	$-0{\cdot}01$	$0{\cdot}04$
(CH$_3$CH$_2$)$_3$N (obs)	$0{\cdot}05$	$0{\cdot}08$	$0{\cdot}04$	$0{\cdot}05$			$0{\cdot}04$	$0{\cdot}04$	$0{\cdot}04$	$0{\cdot}02$
(calc)	$0{\cdot}00$	$0{\cdot}01$	$0{\cdot}00$	$0{\cdot}01$			$0{\cdot}00$	$0{\cdot}02$	$0{\cdot}00$	$0{\cdot}02$

Shifts due to specific solvation by benzene were early recognized by several groups of workers.[2188,2187,304,2081] Although external references were used, substantial variations in relative chemical shifts within the same molecule were observed. In particular, Schaefer and Schneider[2187,2188] observed that the relative chemical shifts of protons of aromatic molecules changed substantially in going from hexane to benzene as solvent and they attributed the effect to a very weak hydrogen bonding of the aromatic protons to the π-system of benzene. Abraham[2] studied acetonitrile in a series of solvents with methane as an internal reference and showed that the solvent effects correlated quite well with dielectric constants, with the exception of benzene which appeared to produce an upfield shift of about 1 ppm. Similar results were obtained by Buckingham, Schaefer and Schneider.[397]

Hatton and Richards[1092,1127] have studied the relative chemical shifts of the protons of the non-equivalent alkyl groups of N,N-dimethyl and N,N-diethyl formamide, and N,N-dimethylacetamide in a wide range of solvents. Only minor variations (<0.2 ppm) were observed for a series of non-aromatic solvents, whereas aromatic solvents produced considerable changes (0.2–1.7 ppm). A detailed consideration of these shifts and of the diamagnetic anisotropy of the aromatic nucleus led to the conclusion that, in the collision complex, the amide molecule is situated above the ring with the partially positive nitrogen atom close to the ring and the partially negative oxygen away from it as in (LVII).[1127] The effect is reduced by all substituents on the

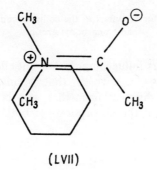

(LVII)

benzene ring, particularly by electron-withdrawing groups. The naphthalene nucleus produces a greater effect than the benzene ring. These results suggest the formation of a charge transfer complex in which the configuration and free energy of the complex is a function of the substituents. Similar complexes with solvent benzene appear to occur with acetonitrile,[1129] N-alkyl-formamides,[1498] aldehydes,[1408] α,β-unsaturated ketones[1128,2416] and a variety of aromatic compounds.[1815,2117,2522,656] Diehl[656] has shown that, in aromatic systems, such shifts are an additive function of the substituents. Bertelli and Golino[236] have established that the interaction of dimethyl sulphoxide with a series of polycyclic aromatic hydrocarbons is a function of the delocalization energies of the π-systems. This result probably reflects both the ease of complex formation and the magnitude of the diamagnetic anisotropy of the aromatic compound. Brown and Stark[367] have observed a correlation between the dipole moment of solutes and solvent shifts in benzene relative to carbon tetrachloride. Ledaal[1522] has examined the solvent shifts for twenty-five simple solutes in benzene vs. carbon tetrachloride and the solvent shifts of the acetonitrile protons in twenty aromatic solvents. A

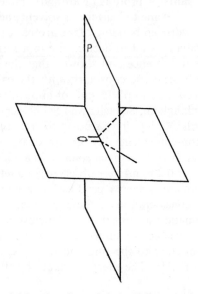

FIG. 2–2–16. A model for the effect of the collision complex between benzene and a carbonyl group.

study of the steric factors influencing the solvent effect of benzene has been made by Anderson.[71] It is noteworthy that the interaction of amides with pyridine appears to be different from that with benzene since the former causes downfield shifts of the absorptions of both groups of *N*-alkyl protons.[1128,1785]

The interaction of benzene or toluene with the carbonyl group is particularly important as it leads to solvent shifts ($\Delta = \delta_{CDCl_3} - \delta_{Ar}$) which can be useful in structural and stereochemical analysis. These effects have been extensively studied in the steroid field by Bhacca and Williams,[242,298,2582,1511,2578,2577] and in monocyclic, bicyclic and tricyclic ketones by Connolly and McCrindle,[526] Henrick and Jefferies,[1153] and Tori.[2420] Their results are rationalized in terms of the model shown in Fig. 2–2–16. The reference plane "P" divides the effect of the solvent into two regions. Protons to the left of the plane have negative Δ-values while those to the right have positive values. The value of Δ will be small in the region of the plane, and, of course, also in regions remote from the carbonyl group. This model immediately gives rise to certain important generalizations which hold for steroidal and similarly constituted ketones. For instance, it follows that axial protons or methyl groups α to the carbonyl group will have large positive Δ-values compared with *ca.* zero for their equatorial counterparts. Similarly, the model qualitatively predicts the observed consequence of the keto group in various positions of 5α-androstane on Δ for the 18- and 19-protons (i.e. the two angular methyl groups).[2578] Hashimoto and Tsuda[1119] have pointed out that, in addition to Δ, evidence of specific solvation is provided by the non-linearity of δ as a function of the mole fraction of benzene in benzene–chloroform mixtures.

Specific solvation effects involving non-aromatic systems have been observed.[1503] A remarkable interaction between halide ions and aryl tetra-*O*-acetyl-β-D-glucopyranosides (LVIII), leading to specific deshielding of the order of 1 ppm of the 1-, 3- and 5-protons, has been reported by Lemieux, Martin and Hayami.[1544]

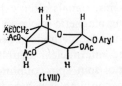

(LVIII)

A further discussion of the solvent effects in connection with stereochemical problems is given in Chapter 3–8 J.

THEORY OF SPIN–SPIN MULTIPLICITY

As ALREADY stated in Chapter 1–1 D and illustrated with spectra of ethanol (Figs. 1–1–7 and 1–1–10) n.m.r. signals often exhibit a fine structure due to the presence of neighbouring magnetic nuclei in the molecules examined.

In this chapter we shall consider methods of obtaining the magnitude (and sometimes the sign) of the internuclear coupling constants from n.m.r. spectra and the mechanism of spin–spin coupling.

The subject of spin–spin coupling has been dealt with in considerable detail by Pople, Schneider and Bernstein,[2004] Gutowsky,[1061] Conroy[531] and Bishop.[254] A number of review articles deal with spin–spin coupling in general,[162,1803,1933,2470,1013] while others are concerned with specialized details, such as the relation between spin–spin coupling and electronegativity,[1723] vicinal coupling,[1501] multiple resonance[147,1108] and long-range coupling.[2358] Some qualitative aspects of spectral analysis are covered in a recent book[241] on a level suitable for organic chemists, while two books[2564,2121] and a review article[540] are solely concerned with the mathematical aspects. Practically all important general references in this area can be found in excellent "condensed" reviews by Foster[864,865] and Lustig and Monitz[1588] which follow one by Reilly[2086] and in a more detailed review by Grant.[1013]

Obtaining structural chemical information from the phenomenon of spin–spin coupling involves three separate steps:

(a) the fine structure must be recognized as due to spin–spin interactions;
(b) the multiplet, or group of multiplets, must be "analysed", ideally to give all the principal parameters, i.e. the chemical shifts of all the nuclei involved and the magnitudes and signs of all the coupling constants;
(c) the information thus obtained may then be used to postulate the presence of particular arrangements of magnetic nuclei in the substance.

The last point will be dealt with in Part 4.

In practice, portions of spectra often cannot be examined in detail because of overlap, i.e. the lack of discrete resonances, and the analysis step is rarely complete. Nevertheless, examination of spin–spin splitting patterns nearly always gives important information and further, this type of information (i.e. the relative arrangements of magnetic nuclei) is usually less readily obtainable from any other chemical or physical measurements.

A. RECOGNITION OF SPIN–SPIN MULTIPLETS

Before proceeding to the examination of the spin–spin coupling pheno-menon we shall concern ourselves with the *recognition* of multiplets, i.e. with the differentiation between multiplicity due to spin–spin coupling and that due to chemical shift differences only.

In practice, multiplicity due to spin–spin coupling is usually instantly recognizable because it gives rise to characteristic patterns. However, even when this is not the case, a number of methods are available for showing whether a signal is split because of the presence of another magnetic nucleus (or group of nuclei) or whether the components are due to nuclei present in different environments. First of all, in the n.m.r. spectra of pure substances, the relative intensities of individual lines due to groups of nuclei in different chemical environments are always simple ratios. Thus an isolated signal which is not equivalent to an integral number of protons must be part of a spin–spin multiplet. However, this method may not be applicable to some complicated spectra (e.g. steroids) where a comparison would have to be made between the intensities of a very weak signal (due to, say, an olefinic proton or two such protons) and a large, unresolved signal due to the remain-ing protons and where it would therefore be very difficult to determine the exact ratio of intensities. Similarly, in some very simple spectra, consisting of only very few lines, all lines may be of the same intensity, or assignments based on intensities alone may be ambiguous. In such cases, where the prob-lem could amount to differentiating between a doublet due to coupling to a nucleus *not being observed*, e.g. ^{19}F in a proton spectrum, and two singlets of equal intensity, advantage could be taken of the fact that while chemical shifts are field dependent, coupling constants are not (see below). A decision could therefore be made by taking the spectrum at two spectrometer frequencies. However, facilities for performing this experiment are usually not available in a routine investigation, and with *second order* (see below) spectra the result would not be simple. Chemical shifts are generally more sensitive to solvent changes than coupling constants, so that determination of the spec-trum of a substance in more than one solvent could clarify the problem. It must, however, be noted (Part 4) that under certain circumstances, *coupling constants may be solvent dependent*. Finally, spin decoupling (Section C) could be used.

B. THE MECHANISM OF SPIN–SPIN COUPLING

So far (Chapter 1–1 D) we have merely stated that the field experienced by any group of nuclei can be influenced by the presence of other magnetic nuclei which may be in their various spin states, and that as a result, the signals in the n.m.r. spectrum due to these nuclei will be split.

Now, we have already considered spin–spin interactions (p. 8) and have noted that, while local fields are possible in solids, in liquids they are almost completely averaged to zero by rapid molecular motions. In order to explain the present observations (e.g. Fig. 1–1–7) we must have an interaction mechanism which does not depend on the direct transmission of the magnetic field through space. An explanation, which has been advanced by Ramsey,[2045] is that the bonding electrons are responsible for communicating the spin state of a nucleus to a neighbouring nucleus. Let us consider a hypothetical single covalent bond between two atoms, A and B ($I_A = I_B = \frac{1}{2}$) and let us denote high and low energy spin states, whether they be of nuclei or electrons, by α and β, respectively. It is known, from the way in which the two electrons of the covalent bond correlate, that at any instant there is a high probability of finding one electron in the immediate vicinity of nucleus A and the other in the neighbourhood of nucleus B. It is also known that nuclear and electron spins tend to pair, i.e. when they are closely associated they tend to have opposite spins. Thus, if nucleus B has an α spin, the spin of the electron in its vicinity will most frequently be β. Therefore, by the Pauli principle, the electron in the neighbourhood of A must most frequently have an α spin. Because of this, the energy of the β spin state of A is lowered and that of the α state is raised. In this way the spin of nucleus B influences the transition energies of nucleus A as is shown diagramatically in Fig. 2–3–1. This mechanism is referred to as the Fermi contact mechanism.

Another mechanism, put forward by Gutowsky and his co-workers,[1071] is that orbital motions (i.e. induced electron circulations) may so shield the direct interaction between nuclei that the local fields constituting this interaction might not be averaged to zero by random molecular motions. Direct calculations have shown that, for protons, Ramsey's mechanism is the predominant one. Pople[1992] has examined the contribution of orbital coupling (Gutowsky's mechanism) between fluorine atoms and has concluded that if highly polar (in the electrical sense) bonds are involved the orbital coupling is significant.

It is evident from the above mechanisms that *electron coupled spin–spin interactions are independent of the strength of the applied field.*[1071]

Clearly, if one of the nuclei (say B) had $I = n$ (and hence $2n + 1$ energy levels) the signal due to nucleus A would be split not into *two* but into $2n + 1$ lines. The effectiveness of the spin–spin interaction is measured by the energy difference between the two transitions $A\beta$ to $A\alpha$ with B in the α state, and $A\beta$ to $A\alpha$ with B in the β state. This energy difference can be expressed in Hertz (H_3), which has the dimensions of energy, and is then called the *coupling constant* between A and B. The coupling constant is usually denoted by the symbol J and is often directly obtainable from certain line separations in the experimental spectrum. A convenient way of denoting the coupling constant between A and B is by writing $J_{A,B}$ or simply, J_{AB}. Now, because the effectiveness of the electron coupled spin–spin interactions is independent of the strength of the applied field, it follows by definition that the coupling constant, J, is also field independent. However, because it does not always depend *only* on the magnitude of J, the *fine structure* of a multiplet may alter with field in some cases (see below).

If the energy of interaction between the nuclei A and B is more favourable for the antiparallel arrangement of spins ($\alpha\beta$ or $\beta\alpha$) than for the parallel arrangement of spins ($\alpha\alpha$ or $\beta\beta$), then the coupling constant, J_{AB}, is said to be *positive in sign*. In fact, J_{AB} in Fig. 2–3–1 is positive, because the energy for the transition $B(\beta)\,A(\beta) \to B(\beta)\,A(\alpha)$ is seen to be smaller than for the transition $B(\alpha)\,A(\beta) \to B(\alpha)\,A(\alpha)$. It follows, that if Ramsey's mechanism[2045]

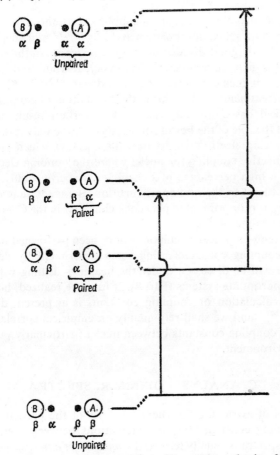

FIG. 2–3–1. Diagrammatic representation of Ramsey's mechanism for electron coupled spin–spin interactions.

is correct, the coupling constants between directly bonded nuclei should always be positive in sign. Now, although a number of methods exist for the determination of *relative* signs of coupling constants in systems where more than one is possible (see below), until recently no means of verifying the absolute signs was available, because the appearance of the n.m.r. spectrum is unaffected by a complete reversal of all signs. However, it has been predicted[394,391] that if a system whose n.m.r. spectrum is being determined is simultaneously subjected to a strong electric field, a partial alignment of the electric dipoles with this field occurs and direct magnetic dipole–dipole interactions are no longer averaged to zero. This additional magnetic field contribution may either add to or subtract from the splitting due to spin–spin coupling, in a manner which is fully calculable and which depends on, *inter alia*, the sign of J. An experiment of this type has been performed on *p*-nitrotoluene and it was found that the coupling constant between the *ortho* protons was *positive in sign*.[392] Because a number of relevant *relative* sign determinations was already available, this experiment also established that the sign of the direct coupling $J_{13_{C,H}}$ is also positive in accord with theory. A possibility exists that absolute signs of coupling constants may also be determined from certain relaxation phenomena[1600,2260,72] and spectra of liquid crystals.[2326,2172] A good discussion of the significance of the signs of coupling constants is given in a paper by Gutowsky, Karplus and Grant.[1069]

A number of refinements have been introduced[1613,1994,2001] into Ramsey's original treatment[2045] permitting the calculation of coupling constants on a theoretical basis. Possibly the most important result has been the vindication of the role of the Fermi contact term[45] (the only term in the wave equation involving electron density near the nucleus), which should in turn cause more effective coupling for nuclei where the bonding electrons are in orbitals with a high percentage of *s*-character. Experimentally, it has been shown that, at least to the first approximation, the magnitude of the direct coupling $J_{13_{C,H}}$ is proportional to per cent *s*-character in the C—H bond.[1312, 2269,1807] (see Chapter 4–5A).

Numerous semi-empirical calculations have been performed in an attempt to calculate coupling constants using some experimentally derived parameters. These will be discussed under the headings dealing with spin–spin coupling in appropriate systems (Part 4). It must be realized, however, that the *accurate* calculation of coupling constants is at present difficult,[2002, 1994,162,1354,2001] and we shall rely mainly on empirical correlations of the magnitude of coupling constants between nuclei (particularly protons) with molecular environment.

C. ANALYSIS OF N.M.R. SPECTRA

The process of extracting the chemical shifts of the participating nuclei and the coupling constants for the interactions between them, from the experimental spectra is usually termed the *analysis of n.m.r. spectra*. Because the process is entirely mathematical, and is based ultimately on quantum-mechanical calculations, organic chemists have either tended to ignore the important information which can be thus gained, or in the other extreme, have used certain simplified procedures (first order analysis, see below) in cases where they are not justified. In this chapter we shall *not* attempt to present the exact theory in detail, but shall concentrate instead on demonstrating the uses and limitations of the first-order theory. We shall confine our discussion to protons and occasionally mention other nuclei of spin $\frac{1}{2}$, because nearly all the nuclei of interest to organic chemists (^1H, ^{19}F, ^{13}C, ^{31}P) are of this type. Further, in case of nuclei with $I > \frac{1}{2}$ (e.g. deuterium, $I = 1$) the coupling, while causing the appearance of more lines, is usually simple to unravel because it is due to nuclei of different isotopes and first-order theory often applies.

(i) SOME DEFINITIONS

The term *equivalent* may take on two meanings in connection with n.m.r. spectroscopy. Two or more nuclei are said to be *chemically equivalent* when they are in identical chemical environments and hence experience the same shielding, i.e. they will have the same chemical shift.

On the other hand, a group of two or more nuclei are said to be *magnetically equivalent* when they not only experience the same shielding but, further, *the set of coupling constants to all other nuclei is identical for each member of the group.*

As an illustration, consider the two protons in 1,1-difluoroethylene (I), which are obviously chemically equivalent but which can experience two types of interaction with the fluorine nuclei, namely a *cis* and *trans* coupling which are generally different. The two fluorine nuclei are also chemically equivalent but neither of the pairs are magnetically equivalent.

On the other hand, the three protons of the methyl group in propene (II) form a group of magnetically equivalent nuclei (rapid rotation about the bond joining the methyl group to the rest of the molecule being assumed) because, while they are coupled with *different* coupling constants to each of the three olefinic protons the three methyl protons participate equally in each interaction.

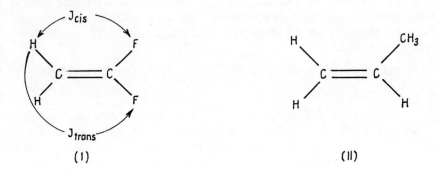

(I)   (II)

The combination of factors which determine the chemical shift of a nucleus (Chapter 2–2) is sometimes such that two nuclei (or groups of nuclei) which are in *different structural* environments experience the same *total* shielding and hence resonate at the same frequency. Such nuclei are said to be *accidentally equivalent* to distinguish them from nuclei which are equivalent by symmetry. A group of accidentally equivalent nuclei behaves, in all respects, like a group of chemically equivalent nuclei which sometimes leads to puzzling complications in the spectra. Fortunately, because of low symmetry, compounds whose spectra exhibit accidental equivalence would not generally preserve it in different solvents.

An example of this behaviour was encountered in an investigation[361] where the structure of an unexpected transformation product was postulated to be (IIIa) or (IIIb) on the basis of other evidence.

However, the n.m.r. spectrum of the material in deuterochloroform showed a sharp singlet of six proton intensity at $\delta = 2\cdot74$ whereas in either of the structures (III), the two methyl groups should have absorbed at different frequencies. Accidental equivalence being suspected, the spectrum was also obtained in pyridine solution where two singlets of three proton intensity each, appeared at $2\cdot68$ and $2\cdot58$, respectively.

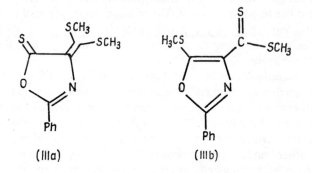

(IIIa)   (IIIb)

In the present text, the term *equivalent* will henceforth refer to *magnetic equivalence* while chemically equivalent nuclei will be termed *isochronous*.[1,1933] Accidental equivalence will always be referred to as such.

The term *spin system* is simply defined as a group of magnetic nuclei which interact (the terms *interact* and *spin–spin couple* are used interchangeably) among each other but do not interact with any nuclei outside the spin system. It is not necessary for *all* nuclei within the spin system to be coupled to *all* the other nuclei.

$$CH_3\!-\!CO\!-\!O\!-\!CH_2\!-\!CH_3 \qquad\qquad H_3C\!-\!CH_2\!-\!O\!-\!CH_2\!-\!CH_2\!-\!CH_3$$
$$\text{(IV)} \qquad\qquad\qquad\qquad \text{(V)}$$

In a large number of cases, the spin system embraces a whole molecule; for instance 1,1-difluoroethylene (I) and propene (II) are spin systems of four and six nuclei respectively. However, the presence of *insulating groups* sometimes causes the appearance of two, or more, independent spin systems within one molecule. Thus ethyl acetate (IV) contains *two* spin systems, *viz.* the five-spin ethyl group and the three-spin methyl group because the coupling across the carboxyl group is generally not observable. Similarly, ethyl n-propyl ether (V) contains two spin systems, the five-spin ethyl group and the seven-spin n-propyl group. The latter system illustrates the case where not all the nuclei are coupled to all the other nuclei, because the coupling between the terminal methyl and the methylene adjacent to the oxygen is negligible. Nevertheless, the seven spins must be considered as one spin-system, and an attempt to analyse the spectrum due to the n-propyl part of ethyl n-propyl ether (V) as *two* spin systems (five-spin system consisting of the terminal methyl and the central methylene and a four-spin system consisting of the methylene adjacent to the oxygen and the central methylene) may lead to serious errors. The only cases where such a procedure is permissible is where only very weak coupling is involved across the point at which we introduce the simplifying *break*, i.e. in cases where a situation almost equivalent to the presence of an insulating group exists.

We shall next consider the definition of *strongly* and *weakly* coupled nuclei. This concerns the ratio of the chemical shift differences between the interacting nuclei to the coupling constants between them, which is the *critical parameter determining the appearance of the spectrum*. For the purpose of analysing spin multiplets, chemical shifts are best expressed in terms of frequency, i.e. in Hz and hence are dependent on the operating frequency of the spectrometer. The difference between the chemical shifts of any two nuclei, or groups of isochronous nuclei, is then symbolized as Δv and expressed in Hz. Thus one may speak about the chemical shift difference between nuclei A and B as Δv_{AB}, and the ratio referred to above becomes $\Delta v_{AB}/J_{AB}$. When this ratio exceeds about 10, the nuclei can usually be safely defined as *weakly coupled*, otherwise they may be *strongly coupled*. It is obvious that the dividing line is quite arbitrary and that there is, in fact, a gradual transition between the properties of strongly coupled and weakly coupled sets of nuclei. If a *borderline* strongly coupled set is analysed as a weakly coupled set, generally only minor numerical errors will result. However, attempts to consider a strongly coupled set as a weakly coupled one may lead to complete misinterpretation (see below). It should be quite obvious that *any spin system may have within it sets of both strongly and weakly coupled nuclei*.

We may now introduce the commonly accepted *nomenclature used for describing spin systems*. The system used here is based on that due to Pople;[2004] it has been discussed in some detail at a qualitative level by Musher and Corey,[1830] Anet,[91] Parello[1933] and, on a more fundamental level, in numerous publications dealing with specific spin systems (see references in Table 2–3–1, p. 138); the work of Abraham and Bernstein[8] is a particularly good example of this sort of discussion and is relatively easy to follow.

A set of n equivalent nuclei of type A is denoted by A_n. If a set of isochronous nuclei contains sub-sets of magnetically equivalent nuclei, these may be denoted by primes. Thus a set of four isochronous nuclei, which comprises two sub-sets of magnetically equivalent nuclei may be symbolized by A_2A_2'.

We shall reserve the groups of letters A, B, C, D, E, ... X, Y, Z and M, N ... for sets of strongly coupled nuclei; the coupling between nuclei symbolized by *different* groups of these letters is weak.

Clearly, a completely correct description of a spin system in terms of these symbols would require an *a priori* knowledge of chemical shifts and coupling constants which is the object of the whole analysis. This paradox is, fortunately, only apparent because even the approximate knowledge of magnitudes of coupling constants and chemical shifts usually enables us to assign symbols to spin systems without any difficulty.

As we have not yet discussed the characteristic values of coupling constants (see Part 4), we shall simply assume that most coupling constants have magnitudes of the order of 10 Hz and try to assign symbols to the examples of spin systems encountered above.

As the molecule of 1,1-difluoroethylene (I) comprises two sets of different nuclei which resonate *millions* of Hz apart, and as we have already discussed the type of equivalence encountered there, it follows that this spin system is of the AA'XX' type.[552]

Our knowledge of chemical shifts of methyl and olefinic protons (Fig.1–1–5) enables us to propose that propene (II) be treated as an $ABCX_3$ system. It would take a detailed knowledge of the relations within the vinyl group and probably an inspection of the spectrum to decide whether further simplification (possibly to an $ABMX_3$) is feasible.

Our knowledge of chemical shifts and the fact that rotation about C—C bonds at ordinary temperatures is rapid on the n.m.r. time-scale (Chapter 2–1), leading to magnetic equivalence in the protons concerned, enables us to state immediately that ethyl acetate (IV) will contain two independent spin systems, a trivial A_3 for the methyl group and an A_2X_3 system for the ethyl group.

Similarly, ethyl n-propyl ether (V) contains an A_2X_3 system due to the ethyl group and either an $A_2X_2M_3$ or an $A_2X_2Y_3$ system for the n-propyl group, depending on the exact magnitude of $\Delta v/J$ for the methyl group and the central methylene group. Other examples of assigning symbols to spin-systems will be encountered below.

A further refinement, explicitly discussed by Musher and Corey[1830] is the writing of spin systems in a manner which enables us to visualize the coupling constants more easily. For instance, 2,3-dibromobutane (VI) is correctly described as an $X_3X_3'AA'$ system, but it may also be written as X_3–A–A'–X_3' which helps us to visualize that the methyl groups are *not* equally coupled to each of the methine protons (coupling between A and X_3 or A' and X_3' is across three bonds while coupling between A and X_3' or A' and X_3 is across *four* bonds and much smaller).

$$\begin{array}{cccc} X_3 & A & A' & X_3' \\ H_3C\!-\!CH\!-\!&\!CH\!-\!CH_3 \\ & | & | \\ & Br & Br \end{array}$$

(VI)

The above definitions may appear to be artificial and, historically, they were not introduced until *after* certain problems associated with spectral analysis had become apparent, but, because of the importance of clearly understood definitions, we have thought it profitable to present a number of them in one section.

(ii) THE EXACT ANALYSIS OF N.M.R. SPECTRA

The energy levels (and hence the line positions in the n.m.r. spectrum) and transition probabilities (and hence the relative intensities of the spectral lines) of the transitions associated with the spin–spin interactions can be calculated by quantum mechanical methods[2045,1071,1617,1079,1078,1070] requiring the solution of the *spin Hamiltonian* which bears a formal resemblance to similar quantum-mechanical calculations involving electronic energy levels, with which some organic chemists have had experience.

It is a major triumph of quantum mechanics that *exact* correspondence between experiment and prediction has been achieved and it is widely believed that the quantum-mechanical calculations relating to the problem of spin–spin interactions are of interest, not only in connection with the problem at hand, but also as offering a "back-door" entry into a major field in theoretical chemistry.

The solution of the quantum mechanical problem yields line positions and intensities corresponding to a set of all coupling constants and chemical shifts for a given spin-system. Thus for the general three-spin system, ABC, six† parameters are involved, namely v_A, v_B, v_C, J_{AB}, J_{AC} and J_{BC}. In certain, rather simple spin systems, general algebraic equations relating these quantities have been derived.

In this way a theoretical spectrum may be obtained for any set of parameters. The values of these parameters may be then altered until the theoretical spectrum matches the experimental one. Clearly, this procedure is ideally suited for the utilization of high-speed computers[2555] and iterative programs have been developed[2372,440] which converge to the correct solution. This procedure is rapidly gaining in popularity and its use will undoubtedly increase still further. However, it requires the assignment of the correct transitions to the experimental lines, which is not without difficulty in complex cases.

In those cases for which algebraic relations are available, e.g. AB, ABX and A_2X_2, a complete analysis, or the extraction of important parameters, can be performed analytically, i.e. by operating on directly measurable spacings. Details of these procedures are given in all general treatments (see below) and will be touched upon in the section on first-order theory.

The majority of authors dealing with analysis of n.m.r. spectra confine themselves to discussing one or two systems and we list a number of such contributions in Table 2–3–1 (p. 138) after discussing the first-order theory in detail. Extensive *general* treatments of analysis of n.m.r. spectra are given by Pople, Schneider and Bernstein,[2004] Corio,[540] Roberts,[2121] Wiberg and Nist,[2564] Conroy,[531] Gutowsky[1061] and Primas.[2017] More recent general methods or aspects of treatments have been described by Banwell and Primas,[160] Whitman,[2556] Arata, Shimizu and Fujiwara,[118] Ferguson and Marquardt,[810] Grimley[1026] and Gioumoisis and Swalen.[486] A somewhat different approach utilizes variants of a perturbation theory[127,1628a,451,1144] including iterative perturbation methods[1186,2090,1012] and second-order perturbations.[1200] Anderson and McConnell[83] discuss analysis by the method of second moments, while Shimizu and Fujiwara describe a method which utilizes a harmonic oscillator model and completely avoids quantum-mechanical treatment.[2261] Under certain conditions, more complex spectra may be considered as built up from simpler systems (subspectra); examples of this type of analysis are given by Pople and Schaefer,[2003] Diehl and

† Actually only two chemical shift parameters are needed as the third one is often considered as the arbitrary origin of the spectrum.

Pople,[662] Corio[541] (who treats the general case A_nBX_m), by Hoffman and his collaborators[972,969,970] (who e.g. consider the analysis of some A_2B_2 systems as a superimposition of two AB systems) and by Diehl.[658]

The above could be construed as signifying that *any* conglomeration of spins can be analysed quantitatively to yield all the desired parameters. In *practice*, because of difficulties in the mathematical procedure, full analysis can only be conveniently performed for systems of less than about 10 spins even when drastic simplifications due to symmetry are possible and for only 5–7 spins where no such simplifications exist.

In many spectra, the resonances overlap to such an extent, that the assignment of lines to specific transitions, or even the identification of spin-systems is quite impossible. Extreme examples of this type of spectrum are those of steroids and triterpenes where the majority of protons give rise to a broad unstructured absorption. Another type of limitation occurs in the case of near degenerate spectra: for instance the aromatic protons of toluene, which must form an AA′BB′C spin system comprising well over 100 lines, give rise to a slightly broadened singlet at 60 MHz† due to the extreme similarity of all chemical shifts. Quite clearly, fitting an "envelope" of this kind to a computed spectrum would lead to an ambiguous result. Other examples of degenerate spectra will be encountered below.

(iii) THE FIRST-ORDER THEORY

In spite of the existence of highly reliable quantitative methods for the analysis of n.m.r. spectra the vast majority of interpretations of multiplets have been carried out by means of the first-order theory described below, and there is every reason to believe that this will always be the practice adopted by organic chemists.

As might be expected, first-order theory applies only to *first-order spectra* which result from certain combination of parameters v and J, namely those in which *all the ratios $\Delta v/J$ are large*. Within the framework of the definitions already given, it can be alternatively stated that first-order spectra are given by weakly coupled spin systems.‡ This can be easily demonstrated by comparing series of computed spectra[540,2564] and some examples of such series will be given below for certain systems. It may thus be stated that "it follows from the mathematics" that certain combinations of parameters will lead to the simplified, first-order spectra. This, however, is not very satisfying intellectually, and the majority of lecturers and reviewers have in fact adopted a different approach by considering simple, "common-sense" energy diagrams. We have illustrated this approach in Chapter 1–1D but it must be stated at the outset, that this approach does not predict all the features of first-order theory and that it is not capable of being extended to second-order (i.e. general) spectra, except by means of relatively sophisticated mathematical treatments.

---

† At 200 MHz this spectrum is sufficiently resolved to permit the assignment of the aromatic protons.[326]

‡ It is interesting that certain *very* strongly coupled spectra are also simplified.[2508]

The set of *rules governing first-order analysis* may be stated as follows:

(1) First-order analysis may only be applied to spin systems where *all* the $\Delta v/J$ ratios are large. Thus, by definition, the AMX spin system gives a first-order spectrum while the ABX system does not. It must be noted, that spin systems where chemical equivalence *only* is present (e.g. AA'X where $J_{AA'} \neq 0$) will in general,[8] not give true first order spectra.

(2) The nuclei of an equivalent group (e.g., the three protons of the methyl group in ethanol) do not interact with each other in such a way as to cause observable multiplicity. This failure of equivalent nuclei to produce mutual splitting is not because they do not interact with one another, but, because for the combination of parameters which give rise to first-order spectra, the corresponding transitions are forbidden by the appropriate selection rules.[1079] However, spin systems composed *entirely* of equivalent nuclei (A_n) will always give rise to singlets.†

(3) When a nucleus interacts with a group of n equivalent nuclei of spin I it will give rise to a signal of $(2nI + 1)$ lines; for the case of protons ($I = \frac{1}{2}$) this of course reduces to $(n + 1)$. If there are more than two interacting groups (A_n, M_m, X_p, ...) all of spin $\frac{1}{2}$, the multiplicity of the signal due to A will be given by $(m + 1) \cdot (p + 1)$ ... i.e. the part of the spectrum due to the nuclei A has a form of a multiplet of submultiplets. *Note that the number n does not enter into the expression.*

(4) The spacings within each multiplet due to a particular interaction are equal to the coupling constant (J) for this interaction. In the case of more than two groups of interacting nuclei, the submultiplet spacings are pertinent (see the AMX example below). These spacings are, of course, completely independent of the operating frequency (see above).

(5) The intensities of a multiplet are symmetric about the mid-point of the band, which corresponds to the origin of the multiplet and is equal to the chemical shift. When the multiplicity is produced by a group of equivalent nuclei with $I = \frac{1}{2}$, the relative intensities of the lines within the multiplet are given by the coefficients in the binomial expansion, i.e., doublet: 1:1; triplet: 1:2:1; quartet: 1:3:3:1; quintet: 1:4:6:4:1; sextet: 1:5:10:10:5:1; septet: 1:6:15:20:15:6:1; octet: 1:7:21:35:35:21:7:1; nine lines: 1:8:28:56:70:56:28:8:1; ten lines: 1:9:36:84:126:126:84:36:9:1. These relations are almost never *exactly* obeyed in practice (see examples below).

Before proceeding to some examples of first-order spectra, it must be stated that, theoretically, *pure* first-order spectra *do not exist* because $\Delta v/J$ is never *infinitely* large, but in many cases first-order analysis leads only to minor errors.‡

The 100 MHz spectrum of ethanol (Fig. 1–1–7), with the coupling due to the OH group removed by rapid exchange is, of course, the archtype of the

† See sub-sections (vii) and (viii) for methods of analysis of such systems.
‡ It is not strictly correct to consider "the centre of gravity" of a multiplet as the *exact* value of the chemical shift, but experimental methods (double integration) exist[2543] for the determination of the true "centroids". For most practical purposes the differences are not important.

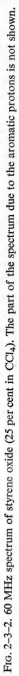

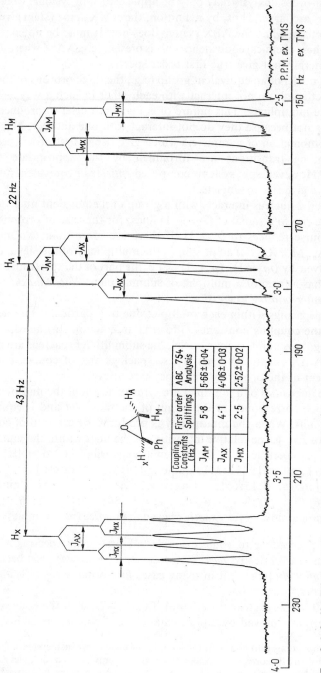

Fig. 2–3–2. 60 MHz spectrum of styrene oxide (25 per cent in CCl₄). The part of the spectrum due to the aromatic protons is not shown.

first-order spectra, with the spin system which we have defined as A_2X_3. It can be seen that the multiplicity is as predicted (the A part is split into $3 + 1$ lines and hence gives rise to a quartet while the X part is split into $2 + 1$ lines, i.e. it is a triplet). The spacings within each multiplet are equal and the spacings within the two multiplets are the same, as expected, because only one coupling constant is involved. However, it can be seen at a glance that the intensities do *not* follow the rule (5) exactly and in particular the *inner lines in each multiplet (i.e those closer to the other multiplet) are more intense than the outer ones*. This type of distortion is quite characteristic, and is, in general the most noticeable deviation from first-order rules.

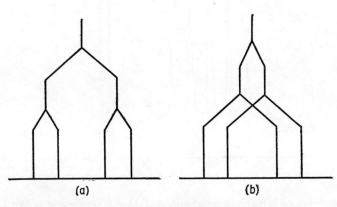

(a)   (b)

FIG. 2–3–3. Equivalent, alternative ways of presenting splitting.

The 60 MHz spectrum of styrene oxide (Fig. 2–3–2) is a good example of a first-order AMX spectrum. As expected 12 lines are observed in all, each proton signal being split into a doublet of doublets. Figure 2–3–2 also illustrates the common method of presenting this type of spectra, although it is not usual to include the distances between the origins of the multiplets, which are given here so that the various $\Delta v/J$ ratios can be estimated by the reader. The sloping of intensities away from the resonance of the proton involved in the particular interaction is noteworthy.

A minor source of confusion arises from the fact that the splitting diagrams of first-order spectra are often drawn in two apparently different ways. Thus any doublet-of-doublets (e.g. the X part of an AMX spectrum) can be represented as shown either in Fig. 2–3–3a or 2–3–3b. A moment's reflection should convince the reader that the two presentations are equivalent. Similarly, there is obviously no difference between the terms "doublet of triplets" and "triplet of doublets".

The last example of a first-order spectrum shown in this section is that of isopropyl chloride (Fig. 2–3–4) which approximates to an AX_6 system and consists, as expected, of a septet and a doublet. The point of interest is the appearance of the septet due to the methine proton. It is quite obvious that in many cases the outer lines would not be observable and it is therefore important to obtain particularly high-gain spectra if a multiplet of this sort is suspected, and care must be taken to avoid confusion between low intensity components and spinning side bands. The universal type of intensity distortions (towards the other multiplet) is again observed.

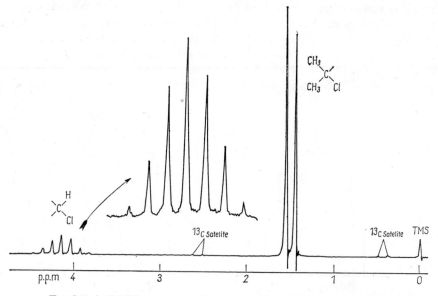

Fig. 2–3–4. 60 MHz spectrum of isopropyl chloride (25 per cent in CCl₄).

(iv) Recognition of second-order spectra

In general, second-order spectra (i.e. those given by strongly coupled spin systems) differ from first-order spectra in a number of obvious ways, so that mistakes due to straightforward applications of first-order analysis to obvious second-order spectra are rare.

Second-order spectra do not usually show regular splitting patterns or intensity ratios, and the total number of lines may be larger than predicted on the basis of first-order theory, although some of the lines may overlap. More subtle departures from first-order spectra occur however, in some cases, and the recognition of these is important and is discussed below, in particular in connection with the ABX and AA′XX′ patterns.

(v) Discussion of individual spin systems

We shall discuss now a number of commonly encountered spin systems in essentially qualitative terms with particular attention to the limitations of first-order analysis in each case. Each of these spin systems has been discussed in detail and in quantitative terms in a number of general references[1061,531, 540,1933,2564,2121,2004,2470] as well as in the pertinent original papers. A by no means exhaustive list of original references dealing with other systems is tabulated (Table 2–3–1). We shall only deal with protons, but clearly the theory applies to any nuclei with $I = \frac{1}{2}$.

Before we proceed, it is important to note the exact significance of the nomenclature used for naming the spin systems. As already pointed out, ideal first-order spectra do not exist and the transition between the near-ideal first-order spectra and second-order spectra is gradual. The criterion used by most authors in classifying spin systems as weakly or strongly coupled (e.g. A_nB_m or A_nX_m) is based on whether or not any *further* increase in the difference in the chemical shifts of the nuclei or groups of nuclei *materially* alters the appearance of the spectrum.

a. The two-spin system (AB)

In this spin system, the coupling constant can be extracted for all values of the parameters by direct measurement of the experimental spectrum, and the origin of each multiplet, and hence the chemical shift, may be determined by a simple calculation. The AB spectrum always consists of four lines, two for the A part of the spectrum and two for the B part, and is often referred to as "an AB quartet". The separation between the two lines in each pair is exactly equal and is also exactly equal to the magnitude of J_{AB}. In other words, labelling the lines as shown in Fig .2–3–5 we can write equation (2–3–1), which gives the absolute value, but not the sign, of J_{AB}

$$(3 - 4) = (1 - 2) = J_{AB} \qquad (2\text{–}3\text{–}1)$$

The difference between the chemical shifts of protons A and B (ν_{AB}) is given by equation (2–3–2)

$$\nu_{AB} = \sqrt{(1 - 4)(2 - 3)} \qquad (2\text{–}3\text{–}2)$$

and the actual values of ν_A and ν_B can then be obtained by adding and subtracting $\nu_{AB}/2$ from the centre of the AB quartet. The relative line intensities are given by equation (2–3–3).

$$\frac{\text{Intensity of outer lines}}{\text{Intensity of inner lines}} = \frac{(2\text{–}3)}{(1\text{–}4)} \qquad (2\text{–}3\text{–}3)$$

Clearly, this relationship could be used to calculate the chemical shift of the B proton where the A lines are visible but the B lines are obscured by other resonances. It can be seen from the above equations that the AB quartet is *perfectly symmetrical*,† and in fact, deviations from symmetry in this (and other related spin systems such as A_2B_2, AA'BB' etc.) have been used to postulate[2142,1870] the existence of some interaction with nuclei *outside* the AB system which is not of equal strength for the nuclei A and B. This interaction could be a small spin–spin coupling[2142] or a broadening due to quadrupole relaxation. Figure 2–3–5 shows a number of calculated AB spectra with different $\Delta\nu/J$ ratios (in this case ν_{AB}/J_{AB}). Two points of interest

---

† Assuming relaxation times and saturation factors to be essentially the same, as appears to be the case at r.f. power levels usually employed.

emerge: firstly it can be seen that, except for the hypothetical, pure AX case and for the trivial A_2 case, the characteristic pattern of intensities for the AB quartet is weak–strong–strong–weak. Secondly, this series of spectra illustrates graphically how the outer transitions due to mutual splitting of two nearly equivalent nuclei become less probable when the nuclei approach equivalence, finally becoming forbidden in the limiting A_2 case. However, it must *not* be assumed that interactions between equivalent nuclei do not influence the appearance of more complex spectra.

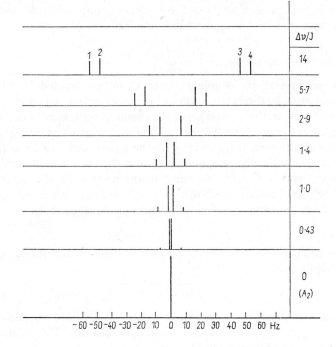

FIG. 2–3–5. Calculated AB spectra. J_{AB} was set at 7 Hz and ν_{AB} was varied to the ratios shown.

b. The three-spin systems: The AB_2 to AX_2 cases

First-order rules allow us to predict the appearance of an AX_2 spectrum. The A part will give rise to a symmetrical triplet with spacings equal to J_{AX} and the X part to a symmetrical doublet with spacings also equal to J_{AX}. However, as can be seen from the set of calculated spectra shown in Fig. 2–3–6,† for the values of $\Delta\nu/J$ smaller than approximately 5, this simple five-line pattern can no longer be recognized, and a complex AB_2 spectrum of up to 9 lines results. One of these lines is a weak combination line due to a transition involving two nuclei simultaneously and is often not observable. The mathematical analysis of the AB_2 system is straightforward:[2121] the

† The numbering of lines in Fig. 2–3–6 and some subsequent Figs. refers to assignments of transitions in the theoretical analyses of the spectra and may not be sequential.

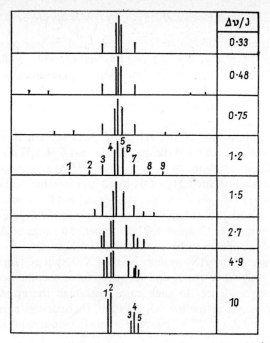

FIG. 2–3–6. Calculated AB$_2$ spectra. (After Anderson[2470].)

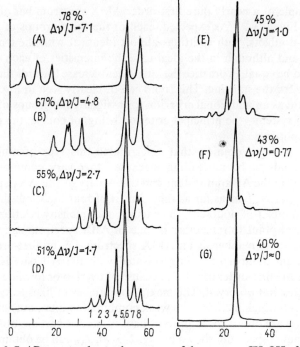

FIG. 2–3–7. AB$_2$ spectra due to the protons of the group —CH$_2$OH of benzyl alcohol. Percentages denote concentrations of benzyl alcohol in acetone, the scale is in Hz at 60 MHz (and is arbitrary) and the gain is not uniform throughout the series. (After McGreer and Mocek[1625].)

origin of the multiplet due to A (i.e. the chemical shift of A) is given by the position of line 3 (see Fig. 2–3–7D for an actual example) while the chemical shift of B is the mid-point between lines 5 and 7. Graphical methods[2121] or tables[2004] may then be used to obtain J_{AB}.

We shall refer to rules which give spectral parameters directly from line positions in second-order spectra as *interval rules*.[822,290]

The actual appearance of the AB_2 and AX_2 spectra is well illustrated (Fig. 2–3–7) by a series of 60 MHz spectra of the —CH_2OH portion of benzyl alcohol at different concentrations in acetone.[1625] At all concentrations there is rapid rotation about the CH_2—OH bond and intermolecular exchange of the OH protons is slow on the n.m.r. time scale, but the *chemical shift* of the OH proton is considerably altered due to changes in the degree of association (hydrogen bonding, see Chapter 2–2 E) and hence a range of $\Delta\nu/J$ values can be obtained. J_{AB} appears to be unaltered at 5·6 Hz.

A special case of the AX_2 system is the AXX' spin system, which can in practice arise only when the X and X' nuclei have become isochronous due to accidental equivalence. In such cases, although the appearance of the spectrum is the same as for the AX_2 system, the spacings correspond to an *average* of J_{AX} and $J_{AX'}$.[2108,8] The same is true for the general AA'B case.[2108]

c. The three-spin systems: The AMX, ABX *and* ABC *cases*

An example of a nearly pure first-order AMX spectrum has already been discussed (Fig. 2–3–2). As expected, each of the three groups of signals is a doublet-of-doublets, with splittings almost identical with the true coupling constants, and although in the ideal case the intensities of each of the four lines should be equal, in practice the intensities increase towards the resonance responsible for the splitting. This trend can be clearly seen in Fig. 2–3–2 and can be useful as an additional criterion for assigning the groups of multiplets to one spin system—the principal criterion being, of course, the presence of matching splittings.

It is the authors' opinion that referring to this type of multiplet as a "quartet" tends to be misleading, because other very common four line patterns (e.g. in the AB spin system and the A part of the AX_3 spin system) are also "quartets" in as far as they consist of four lines, but differ greatly from the "doublet of doublets" in spacings and intensity relations.

The crucial point to remember is that an ABX spectrum cannot be analysed in the same manner as an AMX spectrum, i.e. by first-order theory. The maximum number of lines associated with the AB part is 8, i.e. the same as for the first-order case but, because of overlap or low intensity, some lines are often not observed. The maximum number of lines associated with the X part is 6 which includes two weak combination lines, but sometimes as few as 3 lines are observed. If all eight lines of the AB part are observed (e.g. in the computed spectrum shown in Fig. 2–3–8), J_{AB} can be obtained directly from the spectrum.† However, neither J_{AX} nor J_{BX} can be obtained directly,

† In Fig. 2–3–8, $J_{AB} = (1 - 3) = (2 - 4) = (5 - 7) = (6—8)$. However, the correct numbering of the lines in other ABX spectra is not necessarily the same.

e.g. by measuring the separations of lines 1 and 2, 3 and 4 etc. Similarly, neither J_{AX} nor J_{BX} can be obtained directly from line separations, in the X part of the spectrum. The separation between the lines 9 and 12 (i.e. the two *strong* outer lines) does, however, correspond exactly to the sum $J_{AX} + J_{BX}$ and this interval rule is often useful. The complete analysis of the

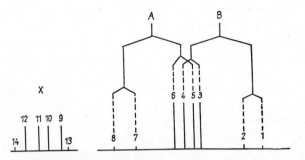
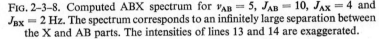

FIG. 2–3–8. Computed ABX spectrum for $\nu_{AB} = 5$, $J_{AB} = 10$, $J_{AX} = 4$ and $J_{BX} = 2$ Hz. The spectrum corresponds to an infinitely large separation between the X and AB parts. The intensities of lines 13 and 14 are exaggerated.

ABX spectrum is not difficult and has been fully described in an easily followed form.[2121] It is important to realize that the ABX system can still be solved analytically, i.e. by operating on the experimentally determined parameters. A particularly convenient method of handling ABX spectra is by the technique of subspectral analysis.[2003]

The "dividing line" between the ABX and AMX cases shows different sensitivity to the various parameters, but as a rule of thumb, it may be assumed that if the smallest chemical shift difference in an ABX system (by definition, ν_{AB}) is at least twice as large as the largest coupling constant, the result of interpreting the ABX spectrum as an AMX spectrum will lead only to minor errors. The most obvious of these will usually be a convergence of the values of J_{AX} and J_{BX} obtained directly from the splittings of the X part, towards a mean of the true J_{AX} and J_{BX}. In cases of more strongly coupled systems far more serious errors may result, leading to a complete misinterpretation of the spectrum.

One of the most common errors involves systems where A and B are strongly coupled and J_{AX} is negligible. In such spectra, the X part still gives rise to four or six lines and it may be said that the protons A and X are *virtually coupled*. This type of spectrum is fully discussed by Musher and Corey,[1830] who also consider analogous complications in other spin systems, and will be dealt with in sub-section (xii).

A very common difficulty connected with the analysis of ABX spectra is that often only the X part is observed while the AB part is concealed beneath other resonances. The problem is then to decide whether the observed X resonance is a part of an AMX, an A_2X or an ABX system, i.e. whether it is legitimate to obtain J_{AX} and J_{BX} from the X part of the spectrum by direct measurement.

This problem is by no means trivial and it cannot be solved by direct calculation. One method is to obtain the spectrum at two frequencies.[184] If the appearance of the multiplet due to the X proton remains unaltered, the splittings correspond to the coupling constants, and hence the proton X is a part of an AMX or A_2X system. It may also be possible to solve the problem by use of double irradiation techniques (see below). The problem of analysing this type of spectrum is commonly encountered with steroids and has been discussed at some length by Williams and Bhacca.[241]

The general three spin system, ABC, is considerably more complicated than those discussed above and no useful information can be obtained from ABC spectra by attempts at first-order analysis. In fact, computing and matching techniques are usually applied.[2372,440] The ABC spectrum consists of up to 15 lines (including three combination lines) and has no characteristic appearance.

d. The multispin systems related to ABX: The ABX_n system

The first-order AMX_n system is of course trivial, the A and M parts each giving rise to $2(n + 1)$ lines and the X part to a doublet-of-doublets.

The ABX_n system is related to the AMX_n system in a manner exactly analogous to the AMX and ABX cases. Both of these systems have been discussed at some length, using the ABX_3 case as an example and a number of theoretical spectra calculated for the various parameter combinations have been published.[1439,813]

The spectra of *cis-* and *trans*-isoeugenol[2144] shown in Fig. 2–3–9 are an example of a nearly first-order spectrum and a completely distorted spectrum, respectively. It is instructive to note that the analysis of the more strongly coupled spectrum gives the relative signs, as well as the magnitudes of some of the coupling constants. Figure 2–3–9 b also shows that two coupling parameters can be derived from inspection in the strongly coupled system (interval rules[813]). The error involved in analysing the weakly coupled case (Fig. 2–3–9 a) as an AMX_3 system is negligible.

e. The symmetrical four-spin systems: A_2X_2, A_2B_2, AA'BB' and AA'XX'

Using the definitions given above, it is quite obvious that the appearance of an A_2B_2 spectrum will be governed by *four* parameters (ν_{AB}, J_{AB}, J_{AA} and J_{BB}), while the appearance of an AA'BB' spectrum will be governed by *five* parameters ($\nu_{AB} = \nu_{AB'} = \nu_{A'B'}$, $J_{AA'}$, $J_{BB'}$, $J_{AB} = J_{A'B'}$, $J_{A'B} = J_{AB'}$). In the literature some discussions of the general A_2B_2 (and A_2X_2) systems apply in fact, to the AA'BB' (and AA'XX') cases. The notation used in these papers shows the existence of *two* different coupling constants between the different types of protons as J_{AB} and J'_{AB} or J_{AX} and J'_{AX}, respectively.

An example of a common type of A_2B_2 (or A_2X_2) system is the methylene protons in freely rotating molecules† of the type $X—CH_2—CH_2—Y$, where

_____

† I.e. where only the average $J_{vicinal}$ is pertinent (cf. Chapter 4–2).

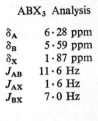

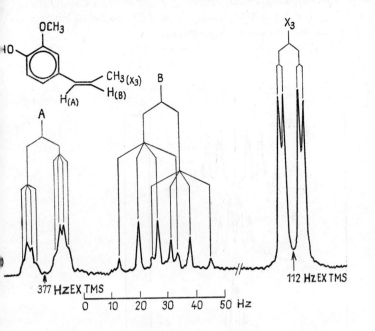

ABX$_3$ Analysis

δ_A	6·28 ppm
δ_B	5·59 ppm
δ_X	1·87 ppm
J_{AB}	11·6 Hz
J_{AX}	1·6 Hz
J_{BX}	7·0 Hz

377 Hz EX TMS

112 Hz EX TMS

| 0 | 10 | 20 | 30 | 40 | 50 Hz |

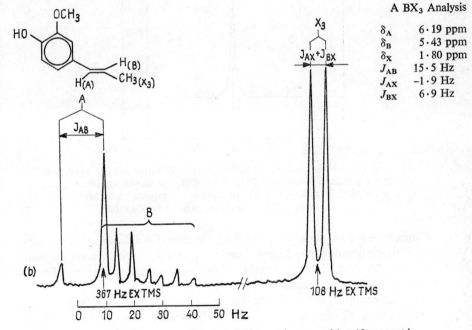

A BX$_3$ Analysis

δ_A	6·19 ppm
δ_B	5·43 ppm
δ_X	1·80 ppm
J_{AB}	15·5 Hz
J_{AX}	–1·9 Hz
J_{BX}	6·9 Hz

$J_{AX}+J_{BX}$

367 Hz EX TMS

108 Hz EX TMS

(b)

| 0 | 10 | 20 | 30 | 40 | 50 Hz |

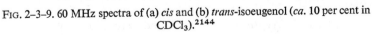

Fig. 2–3–9. 60 MHz spectra of (a) *cis* and (b) *trans*-isoeugenol (*ca.* 10 per cent in CDCl$_3$).[2144]

X and Y denote different magnetically inactive groups, while common examples of the AA′BB′ (or AA′XX′) systems are *para* disubstituted benzenes, furan, pyrrole, some disubstituted cyclopropanes, etc.

All the above spin systems give rise to *symmetrical* spectra, the A and B (or A and X) parts being mirror-images of each other. The maximum number

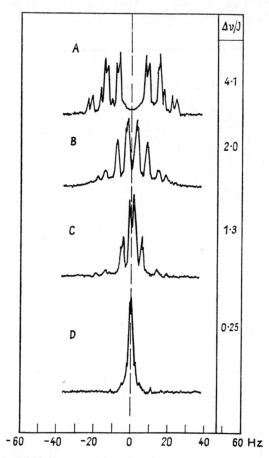

Fig. 2–3–10. 60 MHz spectra of benzyl acetone (only resonances due to the two methylene groups of Ph—CH_2—CH_2—CO—CH_3 are shown). (a) 10 mole per cent in benzene; (b) neat; (c) 10 mole per cent in carbon disulphide; (d) 10 mole per cent in acetone. The scale is arbitrary. (After Danyluk[592].)

of lines is 24 (12 for each half) and the minimum (for a finite value of coupling between protons of different chemical shifts) is 6 (two triplets). In addition, the AA′ and XX′ parts are themselves symmetrical multiplets. This is not true for AA′BB′ spectra.

It is important to note that the simple six-line spectrum will be given not only by the A_2X_2 case, as predicted from first-order rules, but also by some AA′XX′ cases (notably the spectrum of furan as a neat liquid[8]). Such spectra have been termed *deceptively simple*[8,657] and will be discussed in section (xii).

The AA′XX′ spectrum may be treated by an analytical approach, certain sums and difference of coupling constants being given directly from the spacings of lines in the experimental spectra.[2121] The AA′BB′ case requires analysis by computing and matching procedures.

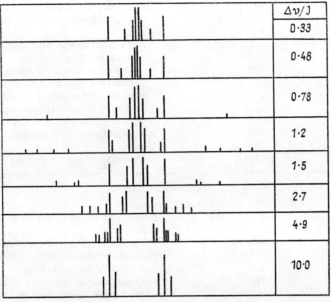

	$\Delta v/J$
	0·33
	0·48
	0·78
	1·2
	1·5
	2·7
	4·9
	10·0

FIG. 2–3–11. Calculated A_2B_2 spectra. (After Anderson[2470].)

An example of the changes of appearance of an A_2B_2 spectrum with the change of $\Delta v/J$ is given in Fig. 2–3–10 where the chemical shifts of the two pairs of methylene protons are altered by means of solvent effects, while J_{AB} is only slightly altered.[592] A series of calculated A_2B_2 spectra,[2470] including an almost pure A_2X_2 case is shown in Fig. 2–3–11. The existence of weak satellite lines is well shown in some of the examples, and these should always be searched for if a spectrum of this sort is to be analysed.

f. More complex systems

While first-order spectra of spin-systems containing quite large numbers of nuclei can be analysed by inspection, even a qualitative description of second-order spectra of any complexity is beyond the scope of this book. In Table 2–3–1 the authors have attempted to summarize some of the more accessible and better known discussions dealing with spectral analysis. It should be remembered that, while such papers are sometimes of considerable mathematical sophistication, often useful quantitative results can be obtained by inspection (e.g. interval rules) and, in nearly all cases, examples are given which should enable the reader to identify the type of spectrum he has encountered.

Simpler systems are also, of course, discussed at length in the general references given above.

TABLE 2–3–1. REFERENCE LIST FOR ANALYSIS OF INDIVIDUAL SPIN SYSTEMS

THREE-SPIN SYSTEMS

SYSTEM	REFERENCES AND REMARKS	
ABC	442, 814	: General discussion and examples
	2536	: Use of double resonance for analysis
	441, 450, 449, 2003, 1281	: Some aspects of the ABC case, including ambiguities, simplifications etc.
	890	: Application to obtaining relative signs of geminal and vicinal coupling constants
	2088, 2089	: Comparison of results obtained by ABX and other approximations with full analysis. Examples
	444	: Analysis of some vinyl spectra
AA′B	2260, 8	: General
AB_2	2508	: Degenerate, very strongly coupled case

FOUR-SPIN SYSTEMS

SYSTEM	REFERENCES AND REMARKS	
A_2B_2 and AA′BB′	658, 1015, 673, 672	: General
	8, 1014, 1174	: Problem of chemical and magnetic equivalence
	970, 969	: Perturbation using $(AB)_2$ approximation
	2508	: Degenerate, very strongly coupled case
	4	: Analysis of dioxolan spectra
	1015	: Detailed analyses including effects of relative signs. Application of furan, thiophene and disubstituted benzenes
	5, 6, 2069	: Ambiguities in analyses of furan spectra and related cases
	1839	: p-disubstituted benzenes
AB_3	2508	: Degenerate, very strongly coupled case
ABX_2, AB_2X, ABC_2	10	: m-dinitrobenzene: AB_2X at 60 MHz, AB_2C at 30 MHz
	1440	: Influence of field on ABX_2
	2541, 516, 1795	: Disubstituted propenes
ABXY	2116, 8	: General
	1444, 1442	: 2- and 3-substituted pyridines. Comparison of first order, ABXY approximation and ABCD analysis
	2056	: 2,4-dinitrofluorobenzene
ABCX	161	: Vinyl fluoride and 3-methylbut-1-ene
ABCD	2091	: Glycidaldehyde
	193	: Use of nomograms
General four-spin systems		: 2536, 2058

FIVE-SPIN SYSTEMS

SYSTEM	REFERENCES AND REMARKS	
$A_2A_2'X$	1243, 2181	: General and cyclopropylamine
AB_2X_2	2210	: General
A_2B_2X	658, 2109, 972, 120	: Includes $(AB)_2X$ approximation
AB_2C_2	234	: General
ABC_3 and ABX_3	813	: General
	1439	: Field dependence and relative signs
	2542	: Analysis of propenes

TABLE 2–3–1 *(cont.)*

FIVE-SPIN SYSTEMS *(cont.)*

A_2B_3	2508	: Degenerate, very strongly coupled case
	1623	: Exact solution for ethyl group
AB_4	272	: General
	458, 460	: Glycerides

SIX-SPIN SYSTEMS

A_3X_3	2597	: General
A_2X_4	1597, 1599, 1760	: General
A_2B_3X and A_2B_3C	1850, 1947	: Ethyl compounds, e.g. $(CH_3CH_2)_3P$
A_3BCD	820	: 1,2-dibromopropane
A_2B_2CX	930	: monofluorobenzene
$AA'A''XX'X''$	1305	: *sym*-trifluorobenzene
$AA'A''A'''XX'$, $AA'A''XX'X''$	658	: Subspectral analysis

SPIN SYSTEMS OF MORE THAN SIX SPINS

A_6B and A_6BX	2053	: General *iso*propyl case including Me_2CHF etc.
A_3B_2X	987	: General
A_6B_2 and A_9B	810	: General
ABM_3X_3	107	: 2,3-dimethylindolines
$A_3XX'A_3'$	2257	: Propane by means of ^{13}C satellites
$A_4B_2C_2$	1795	: Norbornadiene
$ABMNX_3Y_3$	933	: 2,4-disubstituted pentanes
$A_3(X_3)_3$	31	: Mesitylene
$A_2X_2Y_6$ and A_2XYZ_6	1642	: General
AB_4 (+ other nuclei)	272	: Where AB_4 is the group $—SF_5$

GENERAL TREATMENTS

AB_n	2506	
$X_nAA'X_n'$	1111, 1599	
A_mB_n	2506	
AB_2X_n	662	
$AB_2M_nX_m$	662	
ABM_nX_m	2003	: Solution by building up from simpler spectra
A_nBX_m	541	: Reducible to ABX, ABX_2, A_2BX etc.

(vi) FIELD VARIATION AS AN AID IN THE ANALYSIS OF SPECTRA

As coupling constants are field invariant, while chemical shifts are field dependent, it follows that given a strong enough field, all spectra would behave as first-order spectra. The simplification of spin–spin splitting patterns is one of the most important reasons for the development of spectrometers operating at higher magnetic fields (Chapter 1–2).

Figure 1–2–2 (p. 24) provides an excellent example where a reasonable first-order analysis can be performed for a 100 MHz spectrum but not for that obtained at 60 MHz.

Under some circumstances, however, it is more profitable to analyse spectra obtained at *lower* fields because there is inherently more information in second-order spectra, although the information is harder to obtain. The most obvious information which is not available from the analysis of first-order (or near first-order) spectra is the magnitudes of coupling constants between equivalent protons (c.f. definition of first-order spectra above). Further, the relative signs of coupling constants can generally be obtained by calculation only in the case of strongly coupled systems[909,2324,1067,1113,2542,44,1848,312,2128] and on occasions, exceptionally low fields have been used[907,156,10] to obtain such data.

Finally, full analysis at two fields provides an excellent check for the assignments,[814] a result which is also sometimes obtainable by analysis of spectra obtained in two solvents (see below). Some further examples of field dependence of n.m.r. spectra are given in the literature.[1439,1440,1813]

(vii) ANALYSIS OF PROTON SPECTRA WITH THE AID OF ^{13}C SATELLITE BANDS

Any very strong signal in the proton n.m.r. spectrum due to protons bound to carbon atoms is usually flanked by a number of small satellite signals, i.e. small bands which are symmetrically distributed about the central resonance. Often most of these are spinning side bands (Chapter 1–2B) but at least one pair, located at a distance of $\pm(50–150)$ Hz from the central peak is due to spin–spin splitting by the ^{13}C $(I = \frac{1}{2})$ present at its natural abundance (1·1 per cent) and bound directly to the proton whose resonance is being observed (see Fig. 2–3–4).

These bands, usually called ^{13}C *satellites*, can be conveniently distinguished from spinning side bands because their position does not alter when the rate of spinning of the sample is changed.

In some cases, it is also possible to observe a pair, or several pairs of satellites located more closely to the central resonance. These are due to a smaller coupling between protons and non-adjacent ^{13}C nuclei (for values of ^{13}C,H coupling constants see Chapter 4–5A). The ^{13}C satellite bands located close to the central resonance are more difficult to observe because they are usually obscured by the strong central signal and may appear only as shoulders, but can be observed by means of special techniques.[904,903,78]

Due to the influence (spin–spin coupling) of a different isotopic species, the protons concerned become *isotopically non-equivalent* and will be split by other protons in the molecule which are otherwise equivalent. Another way of looking at this is to say that an effective chemical shift,[1601] equal to $\frac{1}{2}J_{13C,H}$ has been introduced between the otherwise equivalent nuclei. Thus the ^{13}C satellite bands can give information not only about coupling between protons and ^{13}C, but also about interproton coupling constants.

As an example, consider the schematic spectrum of acetylene[1601] at natural isotopic abundance (Fig. 2–3–12). The central peak (which will contain approximately 99 per cent of the total signal) is due to acetylene species which contains no ^{13}C; the species $H—^{13}C≡C—H$ gives rise to the four satellite doublets whose total intensity is approximately 1 per cent of the signal. The inner pair is due to the protons attached to ^{12}C but split by the neighbouring ^{13}C and the outer pair is due to protons directly attached to ^{13}C. Each signal is further split into a doublet by the proton on the other carbon atom and all three coupling constants can then be obtained by direct measurement. The

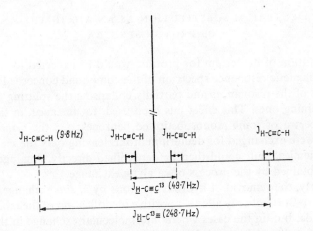

FIG. 2–3–12. Schematic spectrum of acetylene at natural isotopic abundance. Not to scale. (After Lynden-Bell and Sheppard[1601].)

actual constants quoted in Fig. 2–3–12 were obtained from an isotopically enriched sample.[1601,1008] It can be seen that we have obtained a "hidden" coupling constant ($J_{H,H}$) which would not have caused any splitting in an isotopically pure ^{12}C species (A_2 case). Good examples of analyses of this type are given by Muller.[1805]

Clearly with more complex molecules the satellite pattern would not be as simple, but it is always relatively easy to analyse because, at least, the further removed satellites have favourable $\Delta\nu/J$ ratios due to the large coupling between directly bonded ^{13}C and ^{1}H.

The principles of this type of analysis were first discussed by Sheppard[2257,517] and have since been applied to the analysis of the spectra of acetylene, ethylene and ethane,[1601,2070,1008] some ABC spin systems in vinyl derivatives,[1729,1179,901,1598,689] methyl acetylene,[2273] cyclopropane derivatives,[1946] cyclic olefins,[1507] thiophenes,[1713] methyl acroleins,[689] 1,2-disubstituted ethylenes,[1805] cyclic anhydrides,[777,1244] substituted benzenes,[1238] three-membered heterocycles,[1796] furan[2069] and many others. The last example[2069] contains a good discussion of certain inherent advantages of spectral analysis with the aid of ^{13}C satellites.

Quite clearly, an analogous analysis can be performed by observing ^{13}C satellites in the ^{19}F spectra of fluoro compounds[1114] and similar satellites due to ^{119}Sn, ^{29}Si, ^{199}Hg, etc. can also be observed in proton spectra.[1412,1852] Reference to results obtained in this manner is made in other sections of this book. With the aid of devices for improving the signal-to-noise ratio[1730,618] such as the CAT (Chapter 1–2C) satellite signals of sufficient intensity can be obtained to permit the analysis of relatively complex substances. A good example of such an analysis is reported by Laszlo and Schleyer[1509] who analysed the spectra of some norbornene derivatives with the aid of ^{13}C satellite patterns.

(viii) DEUTERIUM SUBSTITUTION AS AN AID IN THE ANALYSIS OF PROTON SPECTRA

Substitution of deuterium for protium would be expected to simplify the proton magnetic resonance spectrum of the compound concerned by removing some of the resonances and partially collapsing the splitting patterns in the remaining ones. This effect has been used, for instance, in the analysis of the spectra of some monosubstituted benzenes;[1493] the *ortho* and *para* protons were exchanged for deuterium atoms leaving only the *meta* protons whose chemical shift could then be determined directly. Analogous results can be obtained by the process of catalytic exchange.[1620]

Similarly, the removal of hydroxyl protons by *in situ* exchange with D_2O (Chapter 1–2G) not only allows a positive identification of the labile protons to be made, but, in the cases where intermolecular exchange in the original substance is slow (Chapter 4–2Axi), also causes a collapse of splitting in the signals due to the neighbouring protons. An example of such behaviour is shown in Fig. 2–3–13. In n-butanol, the intermolecular exchange rate of the hydroxyl protons is evidently slow under the conditions of the determination of the spectrum and the hydroxyl proton gives rise to an approximate triplet. The methylene protons on the carbon atom α- to the hydroxyl group appear as an approximate quartet, which can possibly be interpreted by first-order rules, as being due to roughly equal coupling to the methylene protons on the β-carbon atom and to the hydroxyl proton. Removal of the hydroxylic proton by exchange with deuterium (Fig. 2–3–13) causes the disappearance of the signal assigned to it and also a simplification of the signal assigned to the α-methylene to an approximate triplet.

In practice, the systems where simplification by deuteration is convenient are limited to those containing labile protons, e.g. those on carbon atoms α- to carbonyl groups, but in some cases, relatively elaborate syntheses of partially deuterated compounds have been carried out.[95,1543,2016,1811,104]

Besides these obvious effects, deuterium substitution can be used as a more subtle aid in the analysis of proton spectra. Deuterium is magnetic ($I = 1$) and there is a simple relation between the coupling constants $J_{H,H}$ and $J_{H,D}$ for nuclei in identical environments. This is the ratio of the gyromagnetic constants and is numerically equal to:

$$J_{H,H} = J_{H,D} \times 6 \cdot 514$$

Thus coupling to deuterium is considerably weaker, and is often further obscured by a broadening (p. 58) due to the quadrupole moment of deuterium,[1562,661] but $J_{H,D}$ can usually be obtained. As the spin number of deuterium is 1, one deuterium nucleus will split a group of equivalent protons into a triplet whose components are of equal intensity, while two equivalent deuterium nuclei will split a group of equivalent protons into a quintet whose components are of relative intensity $1:2:3:2:1$. The spectra will tend to be first-order in character because the resonance frequency difference between protium and deuterium is for all practical purposes infinitely large compared with $J_{H,D}$.

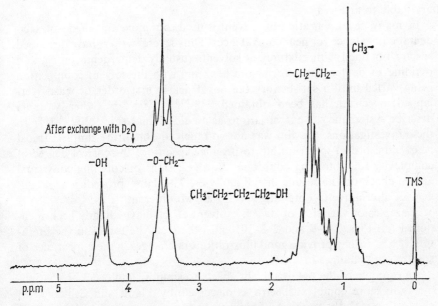

Fig. 2–3–13. 60 MHz spectrum of n-butanol (25 per cent in CCl_4).

Thus, not only will substitution of deuterium for protium result in some simplification of the proton spectrum but, at least in principle, all the coupling constants present in the original system can be derived by an examination of the partially deuterated sample. This procedure has been used, *inter alia*, for the analyses of the ABC spectrum of acrylonitrile[995] and 1,2-diphenylpropionic acid.[889]

Partial deuteration can also be used to obtain "hidden" coupling constants (e.g. $J_{A,A}$ in the A_n system). Thus, the partial deuteration of an isolated methyl group (A_3) will result in the species —CH_2D and —CHD_2 (or a mixture) which will give rise to a triplet or a quintet, respectively, from the spacings of which $J_{H,H}$ can be obtained by multiplying by the factor 6.514. Many geminal coupling constants[233,1620] have been obtained in this manner and the method has also been applied to the analysis of the spectrum of ethylene.[2070] Deuterium spectra[661] have also been obtained.

(ix) SOLVENT EFFECTS AS AN AID IN THE ANALYSIS OF SPECTRA

In general, with the exception of flexible systems, chemical shifts are more solvent dependent than coupling constants (however see Part 4 for examples of solvent dependence of coupling constants) and hence by obtaining the spectrum of the same substance in two or more solvents, the spin system (or systems) present may change some of the $\Delta v/J$ ratios and hence become more amenable to analysis.

In many cases, variation of solvent is used to remove the effects of accidental equivalence (or near equivalence). Thus several writers have described spectra obtained with mixtures of solvents (usually deuterochloroform and pyridine or deuterochloroform and benzene) with the solvent composition being varied until a satisfactory (i.e. most nearly first order or least overlapped) spectrum has been obtained.[2310,2298,1472,118,13] Other workers describe systematic solvent variations as an aid in analyses.[2191,2183,2542] In these investigations, the aim was not so much to obtain the simplest (most nearly first order) spectrum but to produce spectra of the correct degree of complexity for obtaining additional data (e.g. signs of coupling constants) and to check on the results of full analyses. This latter procedure is a convenient alternative to obtaining spectra at two different fields.

Examples of variation of $\Delta v/J$ by solvent effects have already been mentioned (cf. Figs. 2–3–7 and 2–3–10) and others can be found in the literature.[543,1242] They provide good illustrations of changes in the appearance of spectra with changes in $\Delta v/J$.

An example of the use of solvent effects to simplify and check the assignments in more complex spectra is discussed by Flautt and Erman[831] in connection with the problem of identifying the X part of an ABX or AMX system. This paper is well worth reading as it pertains to an actual problem in organic chemistry and is non-mathematical.

(x) CONTACT SHIFTS AS AN AID IN THE ANALYSIS OF SPECTRA

Contact shifts are very large chemical shifts which can be observed in the spectra of certain paramagnetic materials (Chapter 2–2D). Because the total spread of the resonances is very large (of the order of 10,000 Hz at 60 MHz for protons) these spectra are of an almost pure first-order type and all coupling constants may be obtained by inspection.[724] While, naturally, only very few substances exhibit this phenomenon, it could be utilized to obtain the coupling constants in practically any aromatic or heterocyclic system by preparing suitably substituted ligands for chelation with Ni^{II}, Co^{II} or other ions. This approach could yield an independent set of values for coupling constants in systems which are rather difficult to analyse, but undoubtedly this procedure is essentially of academic interest.

(xi) ANALYSIS OF SPECTRA BY SPIN DECOUPLING

As will be shown in Chapter 2–4, spin decoupling is one of the results obtainable by the technique of multiple irradiation. For the present purposes, it suffices to state that spin decoupling results in an effective removal of the irradiated proton from the spin system concerned. One could imagine that multiple spin decoupling would be capable of reducing any spin system to a case which would become trivially easy to analyse. In fact, there are certain limitations which will be mentioned here. There is no doubt, however, that spin decoupling, and in particular multiple, frequency-sweep decoupling at 100 MHz (or higher) operating frequency is a powerful method for the analysis of complex spectra.

Independent of the type of instrumentation, spin decoupling is inherently incapable of significantly simplifying symmetrical spectra, e.g. of the AA'XX' type. Irradiation of the X part would result in a singlet appearing at the origin of the A part thus giving no direct information about the coupling constants. However, other types of double irradiation experiments could be more fruitful although far more difficult to interpret (Chapter 2–4).

Further, there are practical limitations, in particular with the field-sweep type of equipment, to the minimum frequency difference between the r.f. field used for spin decoupling (H_2) and that used for observation of the spectrum (H_1). Thus, unfortunately, it is precisely the more strongly coupled, and hence difficult to analyse, spectra which are the most difficult to simplify by spin decoupling.

Thus for the ABX case, it may be possible to observe the AB part of the multiplet while irradiating the X part and so obtain J_{AB} and ν_{AB} by a method even simpler than the analysis of the ABX system, but it may be more difficult to observe the X part while irradiating either the A or the B part or to observe A while irradiating B or B while irradiating A.

The actual procedure to be used in simplifying complex spectra may be quite involved. A good example of analysis of complex (ABA'B'XX'MM' and ABCDXYM) spectra by a combination of double and triple spin-decoupling followed by analysis is described by Rao and Baldeschwieler,[2057] while a method for analysing three-spin systems with the aid of field-sweep double resonance is given by Whipple.[2536] An instructive example of the application of spin-decoupling for the simplification of an n.m.r. spectrum in a favourable case (complex spectrum but large $\Delta\nu/J$ ratios) has been described by Shoolery.[2271] The normal spectrum of 1,2,5,6-tetrahydropyridine is shown in Fig. 2–3–14. It is apparent that the 2- and 5-protons are involved in extensive spin–spin coupling with other protons, as well as with each other, and that direct measurement of $J_{2,5}$ from this spectrum is not feasible. This parameter, however, is readily obtained from the spectrum (d) by strong irradiation at 575 Hz (the approximate resonance frequencies of the 3- and 4-protons). Under these conditions, the 2-proton appears as a triplet, the spacings (3 Hz) of which are equal to $J_{2,5}$. The inset (f) illustrates a triple irradiation experiment, in which simultaneous strong irradiation at

575 Hz *and* 333 Hz (2-proton) reduces the complex multiplet at 207 Hz (5-proton) to a triplet with spacings identical with those of the triplet at 295 Hz (6-proton). Finally, double irradiation (insets a and b) leads to the determination and assignments of the chemical shifts of the strongly coupled 3- and 4-protons.

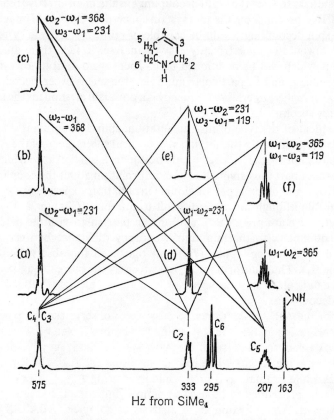

FIG. 2–3–14. 100 MHz spectrum of 1,2,5,6-tetrahydropyridine. (After Shoolery[2271].)

It will be noted that certain of the frequency separations between the irradiating and observing fields do not correspond *exactly* to the chemical shift differences between the observed and irradiated protons. Optimum decoupling is observed with a frequency separation slightly less than the chemical shift difference. This effect is real and can be calculated (see Chapter 2–4).

Other methods related to multiple irradiation, such as transitory selective irradiation,[851,1190,1191] and the observation of multiple quantum transitions[2628,2574,1823,1826,81,1631,31] can be utilized for the analysis of spectra but their theoretical complexity puts their consideration outside the scope of this book.

(xii) VIRTUAL COUPLING AND DECEPTIVELY SIMPLE SPECTRA

Virtual coupling and deceptive simplicity are not physical phenomena at all; they are simply terms which were introduced[1830,8] to describe types of second-order spectra which, because of their appearance, are liable to be misinterpreted when an attempt is made to apply first-order rules to them. Such spectra arise when certain combinations of spectral parameters (chemical shifts and coupling constants) occur, and the limiting values for the combinations of parameters associated with some of the spectra of this type have been worked out for a number of important cases.[1830,91,657,8]

It will be shown below that the errors which result are directly attributable to a careless interpretation of the limitations of first-order analysis which, in turn, is due to a lack of understanding of the basic definitions used to describe spin systems.

a. Virtual coupling

As mentioned above, the X part of an ABX spectrum may consist of more than two lines even if H_X is coupled *only* to one of the AB nuclei. Thus the signal assigned to H_X suffers a complication (which could be mistaken for evidence for additional spin–spin coupling) due to the presence of a proton *to which it is not actually coupled*, but to which it may be considered as being "virtually" coupled.[1830]

The numerical combinations of spectral parameters which give rise to this effect in the ABX system have been defined by Musher and Corey;[1830] qualitatively, it is simple to remember that the signal due to H_X departs most from the expected first-order (i.e. AMX) case when H_A and H_B are most strongly coupled.

An example of virtual coupling in a linear ABX system is shown in Fig. 2–3–15 which depicts a spectrum of an aldehydic component extracted from the green-shield bug.[982] It can be seen that although the aldehydic proton gives rise to four lines, analysis shows it to be significantly coupled to only one of the vinylic protons. However, as for all ABX cases, the total distance between the strong outer lines of the X part of the spectrum is equal to $J_{AX} + J_{BX}$. Other examples may be found in the literature.[539,1276,1139,1547]

Provided that the whole spectrum can be seen, there is little difficulty in either analysing the system[1830,982] or in removing the ambiguity by obtaining the spectrum at a higher field[2280] or through solvent effects.[1547] Difficulties arise, however, when only the X part is visible and in such cases *the possibility of virtual coupling must always be considered.*

Analogous complications also occur in other spin systems[1830] and often involve primary or secondary methyl groups.[91] The mathematical treatment of such cases is somewhat more complex[1830,91] and, in any case, the lines are usually insufficiently resolved for the purpose of analysis. The same general principle applies, i.e. that a signal due to a proton, or a group of equivalent protons X_n, may be complicated by the presence of protons A_m,

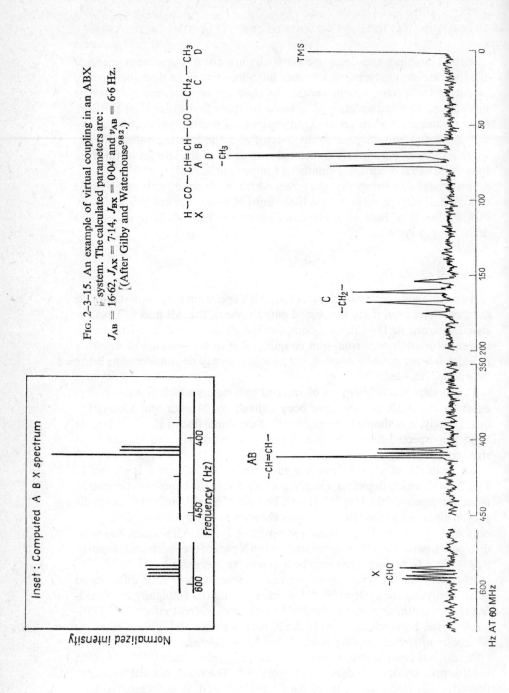

FIG. 2-3-15. An example of virtual coupling in an ABX system. The calculated parameters are:
$J_{AB} = 16·62$, $J_{AX} = 7·14$, $J_{BX} = 0·04$ and $\nu_{AB} = 6·6$ Hz.
(After Gilby and Waterhouse[982].)

to which X_n is *not coupled* but which are *strongly* coupled to other protons B_r to which X_n *is* coupled, however weakly.

An example of virtual coupling involving a terminal methyl group is shown in Fig. 2–3–16a. It can be seen that increasing $\Delta v/J$ by obtaining the spectrum at a higher frequency, Fig. 2–3–16b, removes the ambiguity.[2472] The 100 MHz spectrum shows the methyl group as a slightly distorted doublet, as expected for a secondary methyl from first-order considerations,

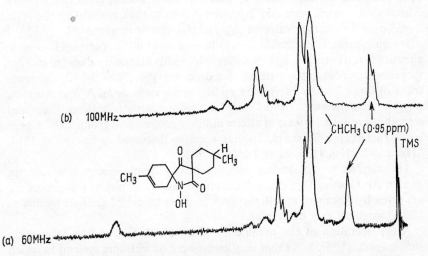

(b) 100MHz

CHCH₃ (0·95 ppm)

TMS

CH₃

(a) 60MHz

FIG. 2–3–16. An example of virtual coupling involving a secondary methyl group.[2472]

but in the 60 MHz spectrum only an unresolved broad absorption is present. The total width of this band is, however, the same as that of the doublet. This example is related to the ABX_3 case in which the width of the X_3 part of the spectrum corresponds to $J_{AX} + J_{BX}$. It is important to realize that at *both* frequencies the methyl group itself is well removed from the methine signal. It is the small chemical shift and appreciable coupling between the methine proton and *its* neighbours (at 60 MHz) which is responsible for the appearance of extra lines between the essential doublet of the methyl group, this giving rise to an unresolved signal. For further examples of spectra of this type see references.[2303,955,1820,1969]

Combination of reduced chemical shift differences between the methyl group and the adjoining methylene group and virtual coupling to further removed methylene groups, gives a characteristic appearance to signals due to terminal *primary* methyl groups in relatively long-chain compounds (e.g. Fig. 2–3–13). It can be seen that the basic, first-order triplet is scarcely recognizable, but the signal is sufficiently characteristic to be used for identification purposes.

It is evident from the above that virtual coupling could be mistaken for actual coupling if an attempt was made to interpret, by first-order rules, a spectrum of a spin system which is only *partially* weakly coupled.

b. Deceptively simple spectra

Certain combinations of parameters, notably in the ABX and AA'XX' cases,[8] often lead to spectra which contain fewer lines than expected, and such spectra have been termed "deceptively simple",[8,657] because they resemble certain first-order spectra. In fact, deceptively simple spectra are partially degenerate and contain less information than the apparently more complicated cases. For example, when an ABX system gives rise to only 5 lines (3 for the X part and 2 for the AB part, thus resembling the A_2X case)[8] only the average value of J_{AX} and J_{BX} can be extracted.[8]

Deceptive simplicity should always be suspected if the spectrum is of such a nature that its analysis by first-order rules yields identical values for two (or more) sets of coupling constants of a different type. Thus the six-line spectrum of furan,[8] when interpreted by first-order rules as an A_2X_2 spectrum, suggests that the vicinal and "*meta*" coupling constants between the α and β protons are equal, a state of affairs unlikely on chemical grounds (cf. Chapter 4–3). Deceptively simple spectra are also discussed in references 1014, 2184, 2190, 2180, 1950, 92 and 2508.

The simplest way of dealing with deceptively simple spectra is to attempt to alter $\Delta v/J$ through specific solvent effects, and examples of this procedure are given by Abraham and Bernstein.[8] In other cases[2069] analysis by means of ^{13}C satellites may be helpful.

A consideration of the parameter values which give rise to deceptively simple spectra[8,657] shows that they are associated with spin systems in which groups of nuclei are isochronous, or nearly so, but not magnetically equivalent. Thus whenever chemical equivalence is present without magnetic equivalence, first-order analysis may not be applicable.

THEORY AND APPLICATIONS OF MULTIPLE IRRADIATION

A. THEORY

We have seen (Chapter 2-1) that for spin–spin coupling to give rise to observable multiplicity of an absorption band, the interacting nuclei must spend a time in any one spin state which is long compared with the reciprocal of the coupling constant between them. In the absence of chemical exchange or a highly efficient spin–lattice relaxation mechanism (e.g. quadrupole relaxation), this condition is invariably fulfilled. *Spin-decoupling*, a type of *multiple irradiation*, is an artificial way of reversing this condition so that the coupling between nuclei is removed. This is achieved by strongly irradiating a nucleus at its resonance frequency while observing another to which it is coupled. The effect of the strong radiofrequency field is to stimulate transitions and so reduce the life-time in any one spin-state. This process is sometimes referred to as "stirring" the nuclei and the strong radiofrequency field is the "stirring" field. We will now introduce a set of symbols, adopted by Baldeschwieler and Randall,[147] for the parameters involved in multiple resonance experiments.

In the applied field, H_0, the ith nucleus will have a resonance frequency ν_{0i} given by equation (2–4–1).

$$\nu_{0i} = -\gamma_i H_0 (1 - \sigma_i)/2\pi \qquad (2\text{–}4\text{–}1)$$

The weak radiofrequency field used for *observing* the absorption spectrum has a strength H_1 and rotates with a frequency equal to $\omega_1/2\pi$. It is important to realize that the field strength of H_1 can be expressed in units of frequency by equation (2–4–2).

$$\nu_{1i} = -\gamma_i H_1 (1 - \sigma_i)/2\pi \qquad (2\text{–}4\text{–}2)$$

Similarly, a strong stirring field has strength H_2 (gauss) or ν_{2i} (Hz) and rotates with a frequency equal to $\omega_2/2\pi$. In a typical double resonance experiment of an AX spin system in which A is observed and X is strongly irradiated we will have an observing field of frequency $\omega_1/2\pi$ and strength ν_{1A}, and a stirring field of frequency $\omega_2/2\pi$ and strength H_2 ($\equiv \nu_{2X}$). The chemical shift between the two nuclei is $\Delta\nu_{AX} = (\nu_{0X} - \nu_{0A})$ and the coupling constant is J_{AX}.

We can now discuss the conditions for a successful spin-decoupling experiment with an AX spin system. There are basically two ways in which the experiment can be performed, *viz.* by using a field sweep or a frequency sweep (p. 26).

In the *frequency sweep mode*, the frequency of the strong radiofrequency field (H_2) is adjusted to coincide with ν_{0X}, so that the X nucleus is stirred by a field of strength H_2 ($\equiv \nu_{2X}$). If H_2 is sufficiently large, the spectrum of the A nucleus appears as a singlet. Clearly, if $H_2 = 0$ the normal spectrum is observed and we may ask how large must H_2 be before the A spectrum collapses to a narrow singlet. The answer is that ν_{2X} ($\equiv H_2$) must be approximately equal to $2J_{AX}$. In practice, the spectrum can be observed with varying values of H_2 until complete decoupling occurs. We will consider later the effect of using smaller values of H_2.

An alternative method is to use the *field sweep mode*. In this case, the frequency $\omega_2/2\pi$ of the strong field is adjusted so that the *difference* in frequency between the fields H_1 and H_2 is equal to the difference in chemical shift (in Hz) between the nuclei A and X. This means that the two nuclei will resonate at the same value of the applied magnetic field and that A is observed while X is stirred so that decoupling may result. Of course, in both techniques it is necessary to employ methods of detection which suppress the unwanted signal from the X nucleus, but this is readily achieved. The field sweep method does not require a frequency lock system and a number of methods for performing such experiments have been described.[912,76,910,898,1321,756,2451,1876,1251] The strength of the field required to effect complete decoupling in the field sweep mode is $\nu_{2X} > J_{AX}/2$.

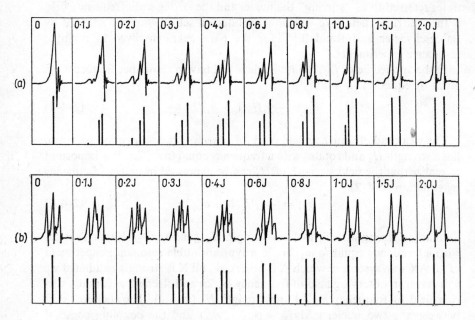

FIG. 2–4–1. Theoretical and experimental spectra of the A part of an AX system (dichloroacetaldehyde) for various values of $(\nu_{0X} - \omega_2/2\pi)$ with H_2 held constant at $\nu_{2X} = J_{AX}/2\pi$: (a) field sweep mode; (b) frequency sweep mode. (After Freeman and Whiffen[912].)

Quantum mechanical methods have been successfully used to predict the phenomenon of spin-decoupling.[147,269,174,2264,80,778] In particular, the theory predicts rather complicated spectra for situations in which H_2 is smaller than the value required for complete decoupling and when $\omega_2/2\pi$ is not *precisely* equal to ν_{0i} for the nucleus which is being irradiated. Indeed, with low values of H_2, the observed spectra are remarkably sensitive to the difference $(\nu_{0i} - \omega_2/2\pi)$, and the behaviour in a field sweep experiment is different from that observed with a frequency sweep. This is well illustrated by the spectra in Fig. 2–4–1.[912] The theory can be extended to more complex spin systems.[903a]

The double irradiation technique is not confined to nuclei of the same isotope. When it is, it is called *homonuclear* spin-decoupling. The analogous technique for nuclei of different isotopes is termed *heteronuclear* spin-decoupling. Heteronuclear spin-decouplers employ either two separate oscillators, or techniques which involve the synthesis of one of the frequencies from the other.[986,2047,138] This technique permits the indirect determination of accurate chemical shifts of other isotopes from the more readily observed proton spectra.[148]

It is possible to decouple simultaneously more than one nucleus by multiple irradiation experiments[2270] and an example of this technique has already been discussed (Fig. 2–3–14).

B. APPLICATIONS

Spin decoupling techniques play a very important role in the analysis of complex multiplets and in the general interpretation of spectra. The uses of the technique are:

  (i) Simplification of complex spin–spin multiplets.
  (ii) Determination of relative signs of coupling constants.
  (iii) Highly accurate determination of chemical shifts.
  (iv) Location of "hidden" absorptions.
  (v) Interrelation of two or more coupled groups of nuclei.
  (vi) Investigation of moderately rapid chemical exchange reactions.

(i) SIMPLIFICATION OF COMPLEX SPIN–SPIN MULTIPLETS

Some aspects of this type of application have already been considered in Chapter 2–3 C (xi). An example is provided by the fungal metabolite monascin which has the structure (I). It was of interest to determine the configuration

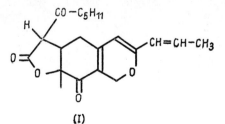

(I)

of the disubstituted double bond which constitutes a rather tightly coupled ABX_3 spin system. A decoupling experiment[1253] simplified the AB region and gave $J_{AB} = 15$ Hz, so that the double bond has the *trans*-configuration.

Experiments of this type are best carried out by the frequency sweep method, although they are feasible by the field sweep technique provided the multiplet under investigation is not too wide. It should be noticed that both methods are inherently incapable of completely decoupling tightly coupled nuclei. For example, an ABCD spectrum can not be simplified in this way.

(ii) Determination of relative signs of spin–spin coupling constants

As pointed out above (p. 117), relative signs of coupling constants are not available from purely first-order spectra. Fortunately, in these cases multiple irradiation can often be used.

Consider the case of an AMX system in which all coupling constants have the same sign, arbitrarily taken as positive in Fig. 2–4–2. If, while the spectrum is being observed, the A nucleus is irradiated with a field of frequency $\omega_2/2\pi$ equal

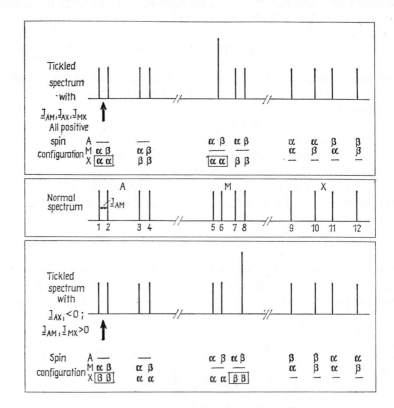

Fig. 2–4–2. Determination of the relative signs of coupling constants in an AMX system by double irradiation. Field sweep experiment.

to the mean of the frequency of lines 1 and 2 and with a value of v_{2A} (H_2) slightly greater than J_{AM}, only those systems in which the X nucleus is in the α spin state will be affected and, in these, A will be decoupled from M. This means that lines 5 and 6 in the M multiplet will collapse to a singlet and 7 and 8 will be unchanged (Fig. 2–4–2a). If, however, the sign of J_{AX} is reversed with respect to J_{MX}, lines 7 and 8 will collapse and 5 and 6 will remain. This type of analysis can be extended to other first-order and some second-order spectra. There are a number of examples of the determination of relative signs of coupling constants.† The field sweep method is well suited for this type of application.

(iii) HIGHLY ACCURATE DETERMINATION OF CHEMICAL SHIFTS

It is apparent from Fig. 2–4–1 that, for low values of H_2, the pattern of the partially decoupled spectrum is very sensitive to the value of $\omega_2/2\pi$. It is therefore comparatively easy to determine the chemical shift with an accuracy of $\pm 0.01J$ by interpolation. This procedure is particularly valuable when chemical shifts of nuclei other than protons are being determined indirectly from proton spectra by heteronuclear spin–decoupling.

The frequency difference $(\omega_1 - \omega_2)/2\pi$ is related to the chemical shift Δv between the two coupled nuclei by equation (2–4–3).

$$\left(\frac{\omega_1 - \omega_2}{2\pi}\right)^2 = (\Delta v)^2 + v_{2i}^2 \tag{2–4–3}$$

Since, for effective decoupling, $v_{2i} > J$ is sometimes used, small errors may be introduced in homonuclear decoupling experiments. The necessary correction implied in equation (2–4–3) will seldom exceed 3 Hz[2451] provided v_{2i} is maintained at a value *just sufficient* to effect complete decoupling, otherwise somewhat larger discrepancies arise (e.g. Fig. 2–3–14).

Baldeschwieler[145] has pointed out that in systems of the type $A_nB_mX_o$ in which A and B are so strongly coupled as to give rise to a spectrum which is indistinguishable from that expected for exact equivalence, the small AB chemical shift can be determined from the appearance of the AB spectrum when X is irradiated at low values of H_2.

(iv) LOCATION OF "HIDDEN" ABSORPTIONS

A common situation in dealing with the spectra of complex organic molecules is that in which some well defined multiplets are observed but others overlap extensively to give more-or-less featureless envelopes. In these cases, it is possible to locate some absorptions within the envelopes by decoupling

† References 96, 175, 442, 509, 512, 593, 754, 755, 756, 784, 785, 786, 845, 847, 851, 905, 906, 907, 911, 940, 971, 1189, 1229, 1239, 1241, 1281, 1515, 1664, 1665, 1666, 1667, 1826, 1837, 1838, 1947, 1948, 2263, 2353, 2405 and 2407.

experiments in which the well-defined multiplets are studied. An example of this technique is provided by the study of the bitter principle clerodin (II).[184] The spectrum is shown in Fig. 2–4–3.

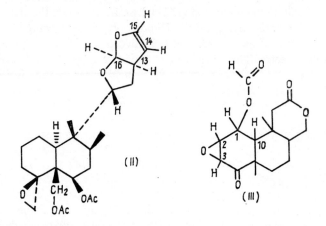

In connection with the elucidation of the structure of this compound, it was important to determine the chemical shift of the proton which was responsible for the splitting of the resonance at $\delta = 5.91$ (16-proton). This was accomplished by observing this doublet and varying $\omega_2/2\pi$ until it collapsed to a singlet. In this way, the 13-proton was located at 3·3. The same type of experiment showed that this proton absorbed at 1·9 in 14,15-dihydroclerodin, conclusively proving it to be allylic to the 14,15-double bond.

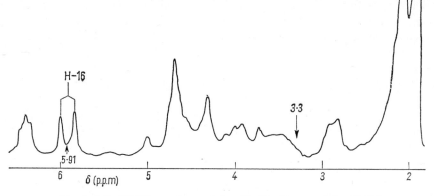

FIG. 2–4–3. 60 MHz spectrum of clerodin (*ca.* 10 per cent in CDCl$_3$).

The great utility of spin-decoupling techniques for locating "hidden" absorptions cannot be too strongly stressed as it greatly increases the information available from the spectra of complex molecules.

(v) INTERRELATION OF TWO OR MORE COUPLED GROUPS OF NUCLEI

If its is believed that two multiplets in a spectrum arise from mutually coupled groups of nuclei, *unequivocal confirmation* can be provided by a spin–decoupling experiment. Frequently this process can be extended to other groups. For instance the sequence of protons at positions 1, 2, 3 and 10 in the degradation product (III) of the bitter principle palmarin was established by a series of spin-decoupling experiments starting with the formyl proton which is long-range coupled to the 1-proton.[142] Another example which illustrates the way in which extensive interrelations can be made is the proof of structure of the macrocyclic antibiotic radicicol.[1608,1762] A selection of references to other examples is given in the footnote.†

It is noteworthy that spin-decoupling can be used even if the coupling is so small as to cause slight line broadening rather than resolvable fine structure. In this case decoupling is manifest as a decrease in line width.‡ This method should be most useful in studying the orientation of methoxyl groups in aromatic systems since these are weakly coupled to ring protons, presumably in the *ortho* relation (cf. Chapter 4–4).

(vi) INVESTIGATION OF MODERATELY RAPID CHEMICAL EXCHANGE REACTIONS

Forsen and Hoffman[850] have shown how double resonance experiments can be used for this purpose.

---

† References 12, 13, 97, 421, 469, 644, 677, 738, 900, 1087, 1292, 1749, 1914, and 2057.
‡ The decrease in line width is, of course, accompanied by an increase in peak height. However, the latter effect is not in itself a criterion for decoupling as it could be a consequence of a nuclear Overhauser effect[102a] which depends on proximity of protons and *not* on spin–spin coupling. The nuclear Overhauser effect results in an enhancement of intensity which can be detected by integration.

PART 3

APPLICATIONS OF THE CHEMICAL SHIFT†

CHAPTER 3–1

GENERAL CONSIDERATIONS

FROM the discussion of the factors influencing the chemical shifts of protons, (Chapter 2–2) it is clear that chemical shifts may give information about electronic configuration in molecules, in particular about electron densities at various points and about magnetic susceptibilities of electrons in bonds. However, applications of this type are of secondary importance to organic chemists compared to the fact that the phenomenon of the chemical shifts permits us to treat hydrogen as a functional group and thus identify and quantitatively estimate hydrogen atoms present in different environments.

The principal application to be discussed in this section is thus essentially a form of functional group analysis; we are concerned with deriving structural and stereochemical information from proton chemical shifts. Now, it has been already shown (Chapter 2–2) that an accurate a priori calculation of the chemical shifts of protons in defined environments is not feasible at present, so that the approach adopted here will necessarily be empirical. However, the major influences on chemical shifts are sufficiently well understood to enable us to predict the directions and approximate magnitudes of chemical shift changes due to various factors, and thus usefully extend the range of situations to which our models may be applied.

Before proceeding to a systematic discussion of the chemical shifts of protons in various molecular environments, it is important to enumerate certain limitations of correlations based on chemical shifts. While most of the points below are either trivially obvious or follow in a perfectly straightforward manner from the theory of chemical shifts, it is the authors' experience that even the more elementary of the precautions presented here are often not followed in practice.

(i) Correlations based on *small* chemical shift effects (say changes of less than 0·2 ppm) should be avoided, unless very stringent precautions are taken. In the vast majority of routine spectra reported in the literature, little effort was made to obtain chemical shifts extrapolated to infinite dilution in carbon

---

† Extensive compilations of n.m.r. spectral data appear, *inter alia*, in Varian catalogs[239] and Sadtler n.m.r. spectra, while references to spectra published up to 1963 have been collected by Hershenson[1156] in the form of a chemical compound index.

tetrachloride or even to report concentrations. In many cases, chemical shifts obtained in carbon tetrachloride, carbon disulphide, cyclohexane and deutero-chloroform have to be compared, even though solvent effects involved are not always negligible[1503] (Chapter 2–2F). Further, even with spectrometers equipped with some field-frequency locks, the sweep-width calibration may vary with time by as much as 2 per cent. Besides these experimental factors (which are outside the research worker's control when literature data have to be used), it should be clear than even if all the possible influences on chemical shifts of analogous protons in two substances are correctly considered, subtle and little understood effects could still remain.

(ii) Steric relationships are very important in determining chemical shifts, and hence if a model is sought for a particular structure, the pertinent proton (or group) must not only be surrounded by identical groups *but they must be in the same steric relations.* The existence of long-range effects on shielding makes it practically impossible to obtain a *perfect* model substance and hence a very close correlation between chemical shifts. In particular it is very dangerous to apply chemical shift data from acyclic model compounds to cyclic compounds.

(iii) While a number of successful additivity rules for predicting chemical shifts in certain environments are very useful, in general, additivity relationships are at best approximate, partially because, in many environments, the introduction of further groups may cause the alteration of some existing steric relations, and because of other factors. The operation of the latter is illustrated by the chemical shifts of halomethanes shown in Table 3–1–1. It can be seen that for example, the substitution by a chlorine atom in methane ($\delta = 0.23$)[2403] leads to a deshielding of the remaining protons by 2·82 ppm, the substitution by another chlorine to a shift of 2·28 ppm and the substitution of the third proton is associated with a further shift of only 1·91 ppm.

TABLE 3–1–1. CHEMICAL SHIFTS IN HALOGENATED METHANES

Substance \ X	F	Cl	Br	I
CH_3X	4·26[a], 4·10[b]	3·05[a]	2·68[a]	2·16[a]
CH_2X_2	5·45[b]	5·33[a]	4·94[a]	3·90[a]
CHX_3	6·25[b] 6·49[c]	7·24[a]	6·82[a]	4·91[c]

[a] Ref. 2403.
[b] Ref. 881.
[c] Results from authors' laboratories, for approx. 2 per cent solutions in CCl_4.

The above limitations affect the presentation of the material in the remainder of this Part. While we shall attempt to list as many references as possible to significant collections of data, we shall limit ourselves to tabular presentation of the chemical shifts of only those substances which we feel to be useful models for a reasonably wide range of compounds. In these tables, *references to papers containing data for derivatives of the parent compounds are frequently given.* Further, unless otherwise stated the data refer to "dilute" (arbitrarily taken as less than 10 per cent) solutions in carbon tetrachloride, carbon disulphide, cyclohexane and deuterochloroform. Similarly, in keeping with the now widely adopted referencing system, chemical shifts will be quoted in parts per million (ppm) from tetramethylsilane as internal reference[2403] unless otherwise stated, i.e. in δ-values.

Chemical shifts characteristic of protons in various environments can be very conveniently represented in the form of bar-graphs and several compilations of this type are available.† The bar-graph shown in Fig. 3–1–1 is of the

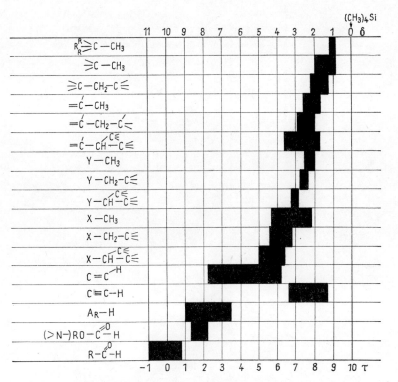

Fɪɢ. 3 1–1. Approximate ranges of proton chemical shifts (R = H or alkyl; Y = —SR, —NR₂; X = —OR, —NHCOR, —O—COR, halogen).

same type, but is simplified by omission of data for labile protons such as —NH₂ and —OH which are, in practice, readily identifiable by *in situ* exchange with D₂O and are discussed in Chapter 2–7. Because effects due to strong long-range influences or unusual combinations of factors were not taken into consideration in compiling Fig. 3–1–1 the ranges shown are only

† References 321, 666, 716, 1602, 1767, 1768, 1883, 2028, and 2286.

indicative and of limited value in making firm spectral assignments. They probably merely represent a logical starting point in assessing a spectrum of an unknown compound.

Although several classification schemes[1883,239] have been proposed (it has even been suggested[321] that automatic data processing methods could be applied to obtaining structural information from n.m.r. spectra), the authors feel that most organic chemists prefer to base n.m.r. spectral assignments on key model compounds followed by a consideration of additional effects in the light of what is known about the factors influencing the shielding of protons.

PROTONS BONDED TO NON-CYCLIC
sp^3 CARBON ATOMS

THE chemical shifts of protons bonded to non-cyclic sp^3 carbon atoms are expected to be less dependent on long-range effects than those of protons in compounds of fixed configuration, because any long-range influences tend to be attenuated by averaging through rapid conformational changes. Nevertheless, stereochemistry plays a part in determining the chemical shifts of protons in acyclic compounds by influencing rotamer distribution and this has to be considered in any *thorough* analysis.

To illustrate the magnitude of this type of effect, consider the chemical shifts of pairs of isomeric 2,3-disubstituted butanes[312] shown in Table 3–2–1. It is evident that differences in chemical shifts between *meso* and *dl* stereoisomers, which can be best attributed to conformational effects, may be quite appreciable and can be used for making configurational or conformational assignments.†

TABLE 3–2–1. CHEMICAL SHIFTS IN 2,3-DISUBSTITUTED BUTANES, $H_3C—\overset{\displaystyle H}{\underset{\displaystyle X}{C}}—\overset{\displaystyle H}{\underset{\displaystyle X}{C}}—CH_3$.

The values are taken from reference 312 and are for approximately 10 per cent solutions in carbon disulphide

X	δ CH$_3$ (ppm)			δ CH (ppm)		
	meso	*dl*	$\varDelta\delta$	*meso*	*dl*	$\varDelta\delta$
Br	1·852	1·735	+0·117	4·070	4·380	−0·310
Cl	1·600	1·527	+0·073	3·912	4·132	−0·220
OAc	1·132	1·115	+0·017	4·827	4·806	+0·021
Ph	0·965	1·218	−0·253	2·808	2·842	−0·034

The chemical shifts of protons in methyl, ethyl, n-propyl, isopropyl and t-butyl groups attached to common functional groups are given in Table 3–2–2. These data serve not only as an obvious source of model chemical shifts but also enable us to draw important general conclusions. It can be seen at a glance that while the chemical shifts of protons bound to carbon atoms α

† References 37, 169, 961, 1402, 1711, 2049, 2071, 2634.

TABLE 3–2–2. CHEMICAL SHIFTS IN COMMON ALKYL DERIVATIVES RX

R X	Methyl CH$_3$	Ethyl CH$_2$	Ethyl CH$_3$	n-Propyl αCH$_2$	n-Propyl βCH$_2$	n-Propyl CH$_3$	iso-Propyl CH	iso-Propyl CH$_3$	t-Butyl CH$_3$
H	0·23	0·86	0·86	0·91	1·33	0·91	1·33	0·91	0·89
—CH=CH$_2$	1·71[c]	2·00[c]	1·00[c]				1·73[e]		1·02[d]
—C≡CH	1·80[a,l]	2·16[d,l]	1·15[d,l]	2·10[l]	1·50[l]	0·97[l]	2·59[m]	1·15[m]	1·22[l]
—Ph	2·35	2·63	1·21	2·59	1·65	0·95	2·89	1·25	1·32
—F	4·27	4·36[g]	1·24[g]						
—Cl	3·06	3·47	1·33	3·47	1·81	1·06	4·14	1·55	1·60
—Br	2·69	3·37	1·66	3·35	1·89	1·06	4·21	1·73	1·76
—I	2·16	3·16	1·88	3·16	1·88	1·03	4·24	1·89	1·95[d]
—OH	3·39	3·59	1·18	3·49	1·53	0·93	3·94	1·16	1·22
—O—	3·24	3·37	1·15	3·27	1·55	0·93	3·55	1·08	1·24
—OPh	3·73	3·98[d]	1·38[d]	3·86[d]	1·70[d]	1·05[d]	4·51[d]	1·31[d]	
—OCOCH$_3$	3·67	4·05	1·21	3·98[d]	1·56[d]	0·97[d]	4·94[d]	1·22[d]	1·45
—OCOPh	3·88[d]	4·37[d]	1·38[d]	4·25[d]	1·76[d]	1·07[d]	5·22[d]	1·37[d]	1·58[d]
—O—SO$_2$—C$_6$H$_4$—pCH$_3$	3·70[d]	4·07[d]	1·30[d]	3·94[d]	1·60[d]	0·95[d]	4·70[d]	1·25[d]	
—C(=O)H	2·18	2·46	1·13	2·35	1·65	0·98	2·39	1·13	1·07[d]
—COCH$_3$	2·09	2·47[a]	1·05[a]	2·32[a]	1·56[a]	0·93[d]	2·54[d]	1·08[d]	1·12[d]
—COPh	2·55	2·92[a]	1·18[a]	2·86[a]	1·72[a]	1·02[d]	3·58[d]	1·22[a]	
—COOH	2·08	2·36	1·16	2·31	1·68	1·00	2·56	1·21	1·23
—COOCH$_3$	2·01	2·28[d]	1·12[d]	2·22[d]	1·65[d]	0·98[d]	2·48[d]	1·15[d]	1·16[d]
—CONH$_2$	2·02[d]	2·23[a]	1·13[a]	2·19[d]	1·68[d]	0·99[d]	2·44[d]	1·18[d]	1·22[d]
—NH$_2$	2·47	2·74	1·10	2·61	1·43	0·93	3·07	1·03	1·15[a]
—NHCOCH$_3$	2·71[d]	3·21[d]	1·12[d]	3·18	1·55	0·96	4·01[d]	1·13[d]	1·43[d]
—SH	2·00[b]	2·44	1·31	2·46[d]	1·57[d]	1·02[d]	3·16[d]	1·34[a]	
—S—	2·09	2·49	1·25	2·43	1·59	0·98	2·93	1·25	1·32[d]
—S—S—	2·30[b]	2·67	1·35	2·63	1·71				
—C≡N	1·98	2·35	1·31	2·29	1·71	1·03	2·67	1·35	1·37[d]
—N≡C	2·85[h]			3·30[h,i,j]	1·71	1·11	4·83[h,l]	1·45[h,l]	1·44[h]
—NO$_2$	4·29	4·37	1·58	4·28	2·01	1·03	4·44	1·53	

Unless otherwise stated the results are from references 452 and 2403. [a] ref. 239, [b] ref. 1971, [c] ref. 311, [d] results from authors' laboratories for ~7% solutions in CCl$_4$ except for amides which are for ~7 per cent solutions in CDCl$_3$. The chemical shifts were obtained by approximate calculations only.

to a substituent vary over a range 3–4 ppm with a common range of substituents, protons bound to carbon atoms β to the same substituents vary only over 0·7–1 ppm, while the chemical shifts of the protons on carbon atoms γ to the substituent in propyl compounds range over only 0·3 ppm. Thus, in acyclic systems, in the absence of factors strongly favouring particular conformers, the influence of a substituent on protons attached to a γ-carbon atom (i.e. H—C—C—C—X) (and by implication also on further removed protons) is too small to be of value in making structural assignments unless a careful study is made of groups of closely related compounds.

It can also be seen that the methyl protons in MeX are invariably less deshielded than the methylene protons in R—CH₂—X which are, in turn, generally less deshielded than the methine protons in Me₂CH—X. On the other hand, the differences in the chemical shifts of the methylene protons on the α-carbon atoms of ethyl and n-propyl derivatives are generally negligible.

The influence of substituents on the chemical shifts of protons on β-carbon atoms (i.e. H—C—C—X) are relatively small, and as they are probably strongly influenced by long-range effects of X (and hence indirectly by conformation factors) assignments based on β-effects must be made with caution. It is generally only possible to ascertain whether an electronegative substituent is present or absent on a carbon atom β to the methyl, methylene or methine group in question, but it is usually not possible to obtain any definite information about its nature.

The presence of *two* electronegative substituents (or groups with appreciable long-range effects due to e.g. diamagnetic anisotropy) on a β-carbon atom(s) may cause quite pronounced downfield shifts, but generalizations are difficult because it is evident from the miscellaneous data collected in Table 3–2–3, that the effects are only very approximately additive and hence probably strongly dependent on stereochemistry. Only one attempt to derive general additivity relationships for β-protons has been reported.[2019]

Methyl (and other) groups bonded to certain conformationally rigid structures (e.g. steroids) may be subject to pronounced long-range shielding influences. However, although methyl groups are acyclic, these effects will be best discussed separately under the heading of stereochemistry and the chemical shift (Chapter 3–8).

We shall now consider the influence of the principal functional groups on the chemical shifts of protons bonded to the α-carbon atom and discuss the influence of further removed structures only where systematic studies are available. We shall consider the chemical shifts of methyl, methylene and methine groups together, i.e. the heading H—C—X implies CH₃—X, CH₂—X and CH—X groups.

In practice, signals due to methyl, methylene and methine protons are easily distinguished from one another (if no severe overlap is present) because of differences in integrated intensities and splitting patterns. In particular, there are obvious limitations to the number of vicinal spin–spin interactions with a methyl group.

Signals due to protons of methyl groups have come to assume a special importance in practical n.m.r. spectroscopy because they are strong, they

TABLE 3–2–3. EFFECT OF MULTIPLE β-SUBSTITUTION ON CHEMICAL SHIFTS.[a]
Unless otherwise stated the values are drawn from references 239 and 2403

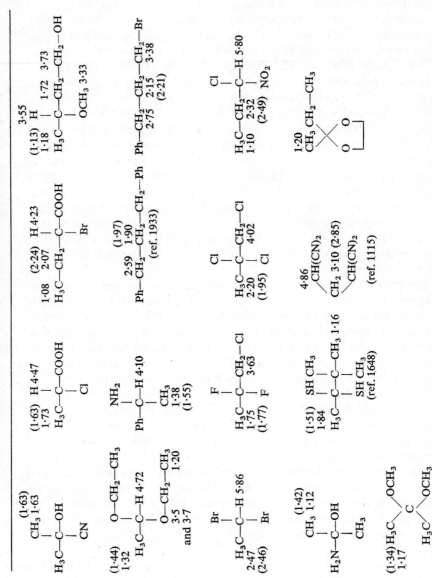

[a] Values in parentheses have been calculated from the appropriate data in Table 3–2–2 on an assumption of com-

usually show first-order splitting patterns (see however Chapter 2–3 for exceptions), and they can be influenced by only one group in the α-position. Furthermore, because of the unique three-fold axis of symmetry and very low rotational barriers, the three protons of the methyl group are always magnetically equivalent.†

In contrast, chemical shift information on methine groups in acyclic compounds is least reliable because the presence of methine protons implies a relatively high degree of substitution and hence possible complications due to the existence of preferred conformations.

These considerations not only make the signals due to methyl groups relatively easy to identify, but also lead to rather narrow ranges of characteristic chemical shifts and to recognition of subtle influences due to long-range effects.

$$\text{A. } \overset{\alpha}{\text{H—C}}\text{—}\overset{\beta}{\text{C}}$$

This group comprises protons bound to sp^3 hybridized non-cyclic carbon atoms which bear no substituents other than carbon or hydrogen. We shall exclude the case where the β-atom is a C=O or related groups (cf. Section B). The most important influence on the chemical shifts is the hybridization of the β carbon atom(s) and it is convenient to consider compounds with sp^3, sp^2 (non-aromatic) and sp^2 (aromatic) β-carbon atoms in turn. Data for compounds with sp hybridized β-carbon atoms will be presented together with those for protons bonded directly to acetylenic carbon atoms.

Unless drastically simplified by symmetry, the spectra of aliphatic hydrocarbons are not easily analysed and relatively few accurate data are available. The chemical shifts of all protons in methane, ethane and propane are given in Table 3–2–2 and additional data are given in Table 3–2–4. It can be seen that the chemical shifts of both methyl and methylene groups fall within very narrow ranges in the absence of β-substituents. The data for methine protons in hydrocarbons are difficult to obtain, but it is estimated[186] that their chemical shifts are in the range 1·5–1·8 ppm. In spite of the apparent insensitivity of their chemical shifts to environment, a careful consideration of a large number of data together with the characteristic shapes assignable to the methyl signals makes it possible to obtain a considerable amount of structural data even from the spectra of paraffinic hydrocarbons.[186] In closely related series of compounds, such as isobutyrophenone derivatives[1395] and ring-substituted ethylbenzenes,[2590] the chemical shifts of terminal methyl groups, while occurring over a narrow range, may be correlated with substitution.

Slomp and MacKellar[2299] have presented some correlations involving the methylene and non-angular methyl protons in steroids, but unlike the much used data relating to the angular methyl groups, (cf. Chapter 3–8) these correlations have not been exploited.

† This would not be the case if the internal rotation about the bond of the methyl carbon to the rest of the molecule could be slowed down. However, at present this has not been accomplished.

TABLE 3–2–4. CHEMICAL SHIFTS OF PROTONS ATTACHED TO sp^3 HYBRIDIZED CARBON
ATOMS NOT BEARING SUBSTITUENTS OTHER THAN CARBON OR HYDROGEN

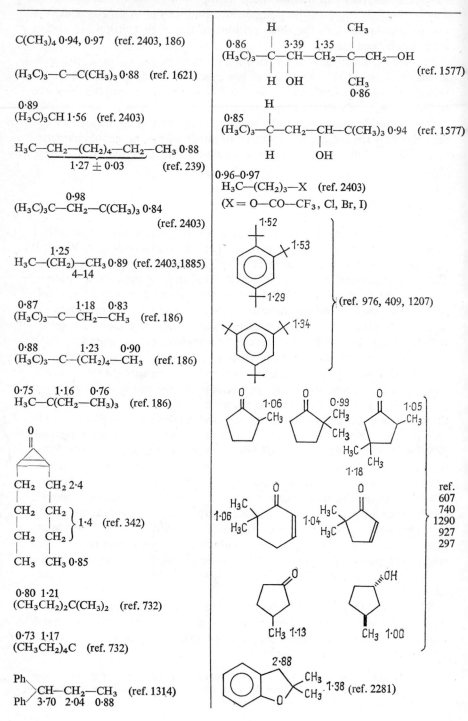

C(CH₃)₄ 0·94, 0·97   (ref. 2403, 186)

(H₃C)₃—C—C(CH₃)₃ 0·88   (ref. 1621)

0·89
(H₃C)₃CH 1·56   (ref. 2403)

H₃C—CH₂—(CH₂)₄—CH₂—CH₃ 0·88
 1·27 ± 0·03   (ref. 239)

0·98
(H₃C)₃C—CH₂—C(CH₃)₃ 0·84
                    (ref. 2403)

1·25
H₃C—(CH₂)—CH₃ 0·89 (ref. 2403,1885)
 4–14

0·87   1·18 0·83
(H₃C)₃—C—CH₂—CH₃   (ref. 186)

0·88   1·23   0·90
(H₃C)₃—C—(CH₂)₄—CH₃   (ref. 186)

0·75   1·16   0·76
H₃C—C(CH₂—CH₃)₃   (ref. 186)

CH₂ CH₂ 2·4
CH₂ CH₂ } 1·4   (ref. 342)
CH₂ CH₂
CH₃ CH₃ 0·85

0·80 1·21
(CH₃CH₂)₂C(CH₃)₂   (ref. 732)

0·73 1·17
(CH₃CH₂)₄C   (ref. 732)

Ph
 >CH—CH₂—CH₃   (ref. 1314)
Ph   3·70 2·04 0·88

H   CH₃
0·86 | 3·39 1·35 |
(H₃C)₃—C—CH—CH₂—C—CH₂—OH
        | | |
        H   OH   CH₃
                      0·86   (ref. 1577)

H
0·85 |
(H₃C)₃—C—CH₂—CH—C(CH₃)₃ 0·94   (ref. 1577)
       | |
       H   OH

0·96–0·97
H₃C—(CH₂)₃—X   (ref. 2403)
(X = O—CO—CF₃, Cl, Br, I)

1·52
1·53
1·29
1·34
                     } (ref. 976, 409, 1207)

1·06 CH₃
0·99 CH₃ CH₃
1·05 CH₃
H₃C CH₃
1·18

1·06 H₃C H₃C
1·04 H₃C H₃C

ref.
607
740
1290
927
297

CH₃ 1·13
OH CH₃ 1·00

2·88
CH₃ 1·38 (ref. 2281)
CH₃

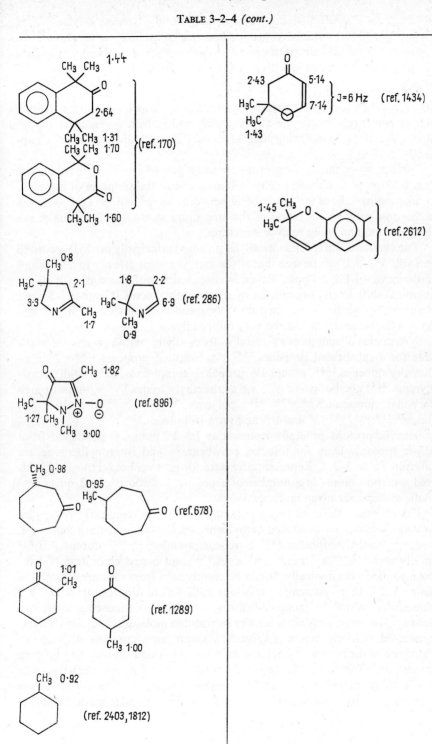

Listed in Table 3–2–4 are some examples of crowded tertiary butyl groups[976,980,409] and methyl groups attached to cyclic structures. It can be seen that while the actual range of chemical shifts is only approximately 0·6 ppm, they are quite sensitive to small changes in the environment.

The chemical shifts of allylic protons ($\overset{\alpha}{H}$—$\overset{\beta}{C}$—$\overset{\gamma}{C}$=C) are expected to be strongly dependent on the nature of γ-substituents which can adopt a definite (*cis* or *trans*) relation to the allylic group, and in fact, correlations of this nature are useful in assigning stereochemistry about double bonds (cf. Chapter 3–8).

Further, when the allylic protons are not a part of a methyl group, important differences in chemical shifts will arise due to the adoption of preferred conformations about the single bond between the sp^3 and sp^2 carbon atoms with consequent modification of the long-range shielding effects due to the double bond and more remote structures.

The chemical shifts for some allylic protons (principally methyl) are listed in Table 3–2–5. It can be seen that allylic methyl groups give rise to resonances in the range of 1·5–2·3 ppm. Taken in conjunction with the sensitivity of the chemical shift to the orientation of remote substituents (Chapter 3–8) and the characteristic fine structure due to allylic and homoallylic coupling (Chapter 4–4C), these signals can be very informative.

Systematic collections of chemical shifts of allylic methyl groups are available for α-substituted propenes,[2542] β-substituted propenes,[1259,1358] α,α'-dimethylstilbenes,[1247] group IV propenyl compounds,[2244] β,β-dimethyl-styrenes,[2144] alkyl-ethylenes,[313] α,β,β-trimethylstyrenes,[1870] α,β-unsaturated carbonyl compounds,[887,891,1841]† polyenes,[165,610,163,600] long-chain olefins,[188,189,190,1693,119] and tri- and tetra-isobutenes.[876]

Benzylic protons generally resonate at 2·3–2·9 ppm, i.e. downfield from allylic protons. Data for toluene, ethylbenzene and isopropylbenzene are given in Table 3–2–2. Representative data for protons bound to sp^3-hybridized α-carbon atoms in a number of aromatic, heterocyclic and quasi aromatic systems are given in Table 3–2–6.

The chemical shifts of benzylic (and even further removed) protons in-among others, p-substituted ethylbenzenes,[2590] 4,6-disubstituted-o-cresols,[1427] methylthiophenes,[1030] nitro compounds,[2524,2622] durenes,[663,399] mesitylenes,[663,399] xylenes,[886] toluenes,[1845] and benzyl chlorides[1246] have been studied systematically. It can be readily seen from examples quoted in Table 3–2–7 that systematic variations with substitution occur, which are principally related to group anisotropy and factors influencing π-electron density. However, with more heavily substituted molecules additional effects connected with orientation of substituents with respect to each other and to the plane of the aromatic ring, and even possible ring buckling may become important.‡ The data in Table 3–2–7 were chosen to illustrate the magnitude of the effects rather than to enable the reader to calculate the shifts due to substituents. In some series of compounds, in particular methylnaphtha-

---

† See also references quoted in Table 3–3–3.

‡ References 49, 567, 663, 886, 976, 980, 2524, 2622, 2629.

TABLE 3–2–5. CHEMICAL SHIFTS OF ALLYLIC PROTONS
(Unless otherwise stated the values are taken from reference 1259)

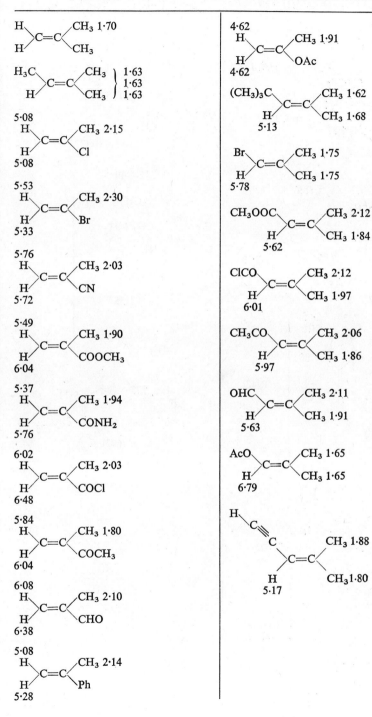

TABLE 3–2–5 *(cont.)*

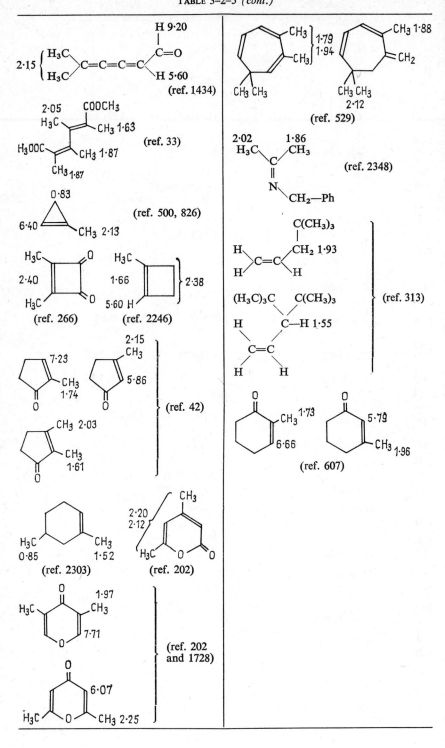

H 9·20

2·15 {H₃C / H₃C C=C=C=C / C=O / H 5·60 (ref. 1434)

2·05 COOCH₃ H₃C CH₃ 1·63 H₃OOC CH₃ 1·87 CH₃ 1·87 (ref. 33)

0·83 6·40 CH₃ 2·13 (ref. 500, 826)

H₃C O 2·40 H₃C O (ref. 266)

H₃C 1·66 5·60 H 2·38 (ref. 2246)

7·23 CH₃ 1·74 O

2·15 CH₃ 5·86 O

CH₃ 2·03 CH₃ 1·61 O

(ref. 42)

H₃C CH₃ 0·85 1·52 (ref. 2303)

2·20 2·12 CH₃ H₃C O O (ref. 202)

O H₃C 1·97 CH₃ 7·71 O

O 6·07 H₃C O CH₃ 2·25

(ref. 202 and 1728)

CH₃ 1·79 CH₃ 1·94 CH₃ CH₃

CH₃ 1·88 CH₂ CH₃ CH₃ 2·12

(ref. 529)

2·02 1·86 H₃C CH₃ C N CH₂—Ph (ref. 2348)

C(CH₃)₃ H CH₂ 1·93 H C=C H H (H₃C)₃C C(CH₃)₃ H C—H 1·55 C=C H H

(ref. 313)

O CH₃ 1·73 6·66

O 5·79 CH₃ 1·96

(ref. 607)

TABLE 3–2–6. CHEMICAL SHIFTS OF BENZYLIC METHYL GROUPS

Compound	References
CH_3 2·65 (naphthalene); CH_3 2·46 (naphthalene)	2629, 1835, 916
Pyrrole: 5·89, 5·72, 6·36, CH_3 2·16, N–H 7·2; Pyrrole: 5·85, 2·05 CH_3, 6·42, 6·28; Pyrrole: 5·82, 1·96 CH_3, 6·28, CH_3 2·02, N–H 7·1	1169, 2292, 1038
Furan: CH_3 2·17; Furan: CH_3 1·94	26, 2127, 997
Thiophene: CH_3 2·41; Thiophene: CH_3 2·21	1030
Imidazole: CH_3 2·27, H–N; Imidazole: *H–N, CH_3 2·42	2074
Pyrazole: 2·05 H_3C, 7·15, 7·15, N–H*; Pyrazole: 5·75, CH_3 2·79, 2·79 H_3C, N–H*	518
Thiazole: 2·47 CH_3; Thiazole: H_3C 2·74; Oxazole: CH_3 2·18	26, 1076
Indole: CH_3 2·30, 6·80; Indole: 6·13, CH_3 3·20	508

TABLE 3–2–6 *(cont.)*

Compound	References
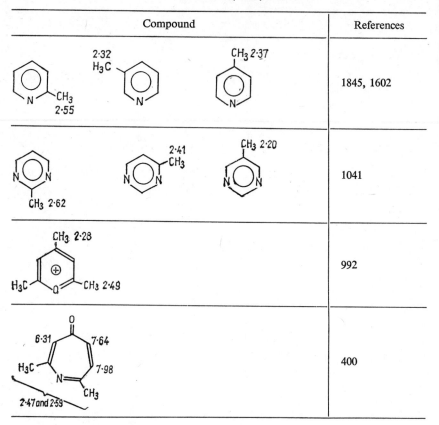	1845, 1602
	1041
	992
	400

* Rapid exchange between starred protons leads to equivalence of certain positions.

lenes,[1835,2629] halogenated toluenes,[2355] veratrole derivatives,[16] substituted *o*-nitrotoluenes[2622] and some substituted alkyl benzenes,[976,980] sufficient data are available to enable correlations to be made, and some of the papers[976,980,567,2629] also contain considerable discussion of the factors influencing the chemical shifts of benzylic methyl groups. Data are also available on the chemical shifts of methyl groups attached to some polynuclear hydrocarbons,[567,916] and chemical shifts of methyl groups attached to various heterocyclic systems may be found in references listed in connection with chemical shifts (Chapter 3–6) or coupling constants (Chapter 4–3) in aromatic and heterocyclic systems.

B. H—C—X

This group comprises protons bonded to sp^3 hybridized carbon atoms carrying the usual range of functional groups. It can be seen that the division between these compounds and those listed under section A above is quite

TABLE 3–2–7. INFLUENCE OF SUBSTITUENTS ON THE CHEMICAL SHIFTS OF BENZYLIC PROTONS.
Unless otherwise stated the results are from ref. 2403 or from authors' laboratories

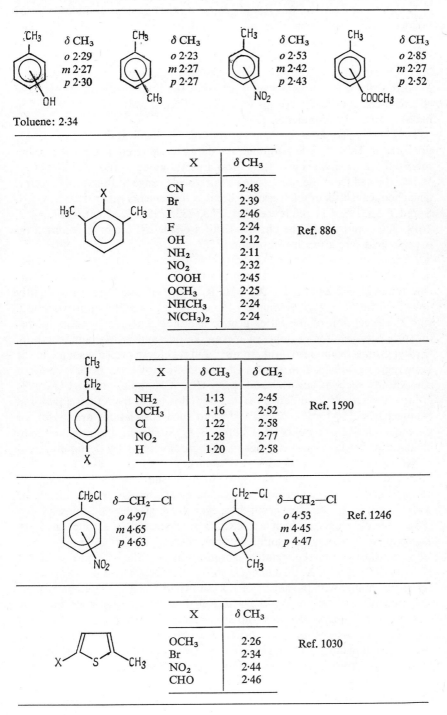

Toluene: 2·34

X	δ CH₃
CN	2·48
Br	2·39
I	2·46
F	2·24
OH	2·12
NH₂	2·11
NO₂	2·32
COOH	2·45
OCH₃	2·25
NHCH₃	2·24
N(CH₃)₂	2·24

Ref. 886

X	δ CH₃	δ CH₂
NH₂	1·13	2·45
OCH₃	1·16	2·52
Cl	1·22	2·58
NO₂	1·28	2·77
H	1·20	2·58

Ref. 1590

X	δ CH₃
OCH₃	2·26
Br	2·34
NO₂	2·44
CHO	2·46

Ref. 1030

arbitrary and was made mainly to conform to the traditional meaning of the term "functional group".

The effect of common functional groups on the chemical shifts of protons bonded to the same carbon atom can easily be seen from the data listed in Table 3–2–2. In this section, we shall consider shielding by some less common functional groups, systematic studies which enable us to evaluate shielding contributions due to subtle changes in molecular environment, and, most important from the practical point of view, the inter-relations between the effects of certain specific groups which have been shown to be very useful in making spectral assignments.

The effects of some less common groups on the chemical shifts of H—C—X are listed in Table 3–2–8. Although the data are scattered, and most of them refer only to methyl groups, they form a useful guide to the shielding effect to be expected from the presence of a fairly wide range of functional groups. The chemical shifts of alkyl groups bonded to elements other than C, O, N, S and P are listed in references 491, 588, 733, 735, 832, 1000, 1137, 1412, 1619, 2205, and 2304. The chemical shifts of methyl esters of common inorganic acids are given by Hammond.[1099]

An inspection of Table 3–2–2 shows that the chemical shifts of protons α to a carbonyl group in the fragment H—C—CO—X occur over a very narrow range (2·0–2·2 ppm) for X = H, R, OH, OR and NH_2 and slightly downfield for X = aryl (aromatic ketones). Thus, to a first approximation, with the exception of the latter group, it is not generally possible to distinguish between, for example, a methyl ketone and an acetamido group. Methyl groups in enolized β-diketones[50] and β-ketoesters also resonate in the same region (Table 3–2–8, cf. also Chapter 5–3). However, in some series of compounds such as sugars, axial —O—CO—CH_3 and —NH—CO—CH_3 groups resonate approximately 0·1–0·2 ppm downfield from their equatorial counterparts[1565,141,1566,863,2331,1554] and such correlations are useful for stereochemical assignments (cf. Chapter 3–8). Similarly, it is possible[154] to distinguish between —NH—CO—CH_3 groups ($\delta = 1·9$–1·96 ppm) and —$N(COCH_3)_2$ groups ($\delta = 2·31$–2·37 ppm).

Careful studies[1327] show that the chemical shifts of methyl groups in acetates and methylene groups in succinates exhibit systematic variations with the nature of the ester group but, even when groups with known long-range shielding effects, such as phenyl, are included, the total range of chemical shifts is only 0·24 and 0·38 ppm, respectively.

In contrast with the relative insensitivity of the chemical shift of H—C—CO—X with X, the chemical shifts of H—C—O—X vary in a characteristic manner with the nature of X as shown in Table 3–2–9. In particular, acylation of alcohols causes a characteristic downfield shift of approximately 0·5 ppm for primary alcohols and approximately 1·1 ppm for secondary alcohols, and this phenomenon, often called "the acylation shift", is of considerable importance since it can often be used to identify the absorptions of the protons α to primary or secondary hydroxyl groups.

The differences in the magnitude of the acylation shift between primary and secondary alcohols have been shown[575] to be due to steric factors viz.

TABLE 3–2–8. CHEMICAL SHIFTS OF H—C—X WHERE X IS A LESS COMMON FUNCTIONAL GROUP
(Unless otherwise stated the values are taken from refs. 239 and 2403)

(A) Point of attachment of X is carbon or oxygen

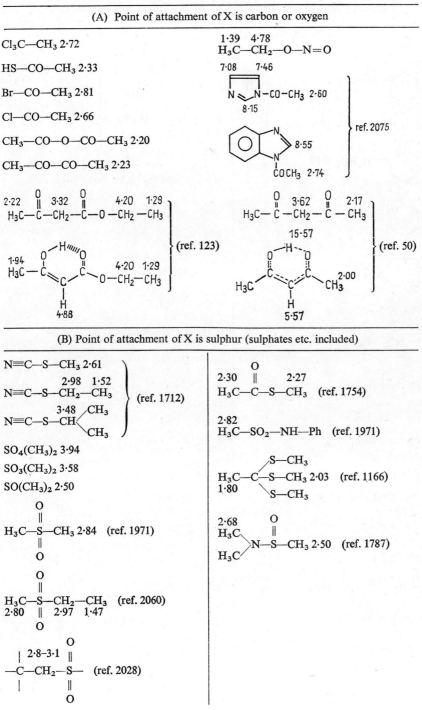

Cl₃C—CH₃ 2·72

HS—CO—CH₃ 2·33

Br—CO—CH₃ 2·81

Cl—CO—CH₃ 2·66

CH₃—CO—O—CO—CH₃ 2·20

CH₃—CO—CO—CH₃ 2·23

1·39 4·78
H₃C—CH₂—O—N=O

7·08   7·46
N≡N—CO—CH₃ 2·60
8·15

8·55
CO CH₃ 2·74   } ref. 2075

2·22 3·32 4·20 1·29
H₃C—C—CH₂—C—O—CH₂—CH₃   } (ref. 123)

1·94
H₃C—C 4·20 1·29
O—CH₂—CH₃
H 4·88

O 3·62 O 2·17
H₃C—C—CH₂—C—CH₃   } (ref. 50)
15·57

H₃C 2·00
CH₃
H 5·57

(B) Point of attachment of X is sulphur (sulphates etc. included)

N≡C—S—CH₃ 2·61

2·98 1·52
N≡C—S—CH₂—CH₃   } (ref. 1712)

3·48 CH₃
N≡C—S—CH
CH₃

SO₄(CH₃)₂ 3·94

SO₃(CH₃)₂ 3·58

SO(CH₃)₂ 2·50

O
‖
H₃C—S—CH₃ 2·84 (ref. 1971)
‖
O

O
‖
H₃C—S—CH₂—CH₃ (ref. 2060)
2·80 ‖ 2·97 1·47
O

O
| 2·8–3·1 ‖
—C—CH₂—S— (ref. 2028)
| ‖
O

O
2·30 ‖ 2·27
H₃C—C—S—CH₃ (ref. 1754)

2·82
H₃C—SO₂—NH—Ph (ref. 1971)

S—CH₃
H₃C—C—S—CH₃ 2·03 (ref. 1166)
1·80 S—CH₃

2·68 O
H₃C ‖
N—S—CH₃ 2·50 (ref. 1787)
H₃C

TABLE 3–2–8 *(cont.)*

(C) Point of attachment of X is nitrogen (isocyanates included)

$H_3C(CH_2)_{(3-17)}NHCH_3$ 2·42 (ref. 1602)	$S=C=N-CH_3$ 3·37

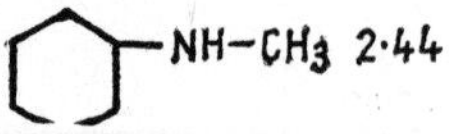

—NH—CH₃ 2·44 (ref. 1602)

$(CH_3CH_2)_2NCH_3$ 2·19 (ref. 1602)

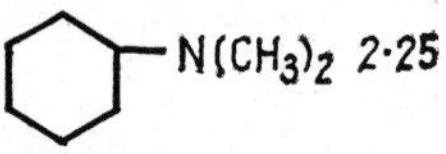

—N(CH₃)₂ 2·25 (ref. 1602)

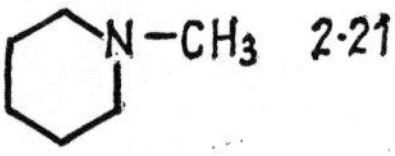

N—CH₃ 2·21 (ref. 1602)

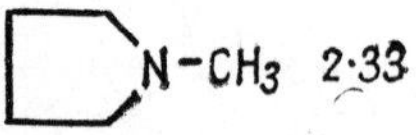

N—CH₃ 2·33 (ref. 1602)

Ph—NH—CH₃ 2·66 (ref. 1602)

p—NO₂—C₆H₄—N(CH₃)₂ 3·09 (ref. 1602)

Ph—N(CH₃)₂ 2·85 (ref. 1602)

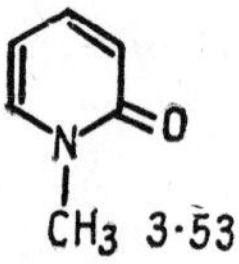

CH₃ 3·06
|
Ph—N—NH₂ (ref. 1602)

|
CH₃ 3·53 (ref. 1602)

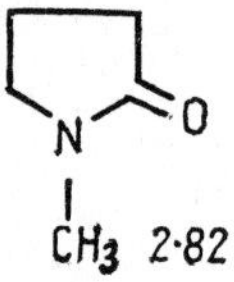

|
CH₃ 2·82 (ref. 1602)

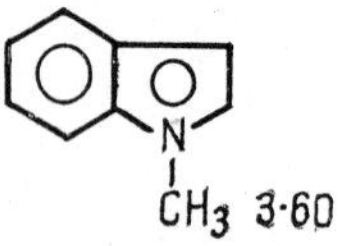

|
CH₃ 3·60 (ref. 1961)

$S=C=N-CH_2-CH_3$
3·64 1·40

$S=C=N-\overset{\displaystyle CH_3}{\underset{\displaystyle CH_3}{C}}-H$ 3·98

(ref. 1712)

3·37 1·20
O=C=N—CH₂—CH₃ (ref. 1044)

4·54 1·45
N≡C—O—CH₂—CH₃ (ref. 1044)

4·16
H₃C
    \
        3·16
   ⊕N=N—CH₃ (ref. 1431, 897)
    |
    O⊖

1·48   ⊕   1·28
(CH₃)₃C—N=N—C(CH₃)₃ (ref. 897)
        |
        O⊖

   O
   ‖   CH₃ 2·96
H₃C—C—N⟨   (ref. 1602)
          CH₃ 2·85

1·14 3·33
H₃C—CH₂\
            NH (ref. 267)
3·68 H₃CO—C/
          ‖
          O

1·26 3·27
H₃C—CH₂\
          NH
          | (ref. 267)
         /C=O
CH₃—CH₂—NH/

2·78 5·16   4·14 1·23
H₃C—NH—C—O—CH₂—CH₃
         ‖
         O

H₃C—C—N—CH₃ 3·18
      ‖ |
      O CH₂OH

(ref. 484)

TABLE 3–2–8 *(cont.)*

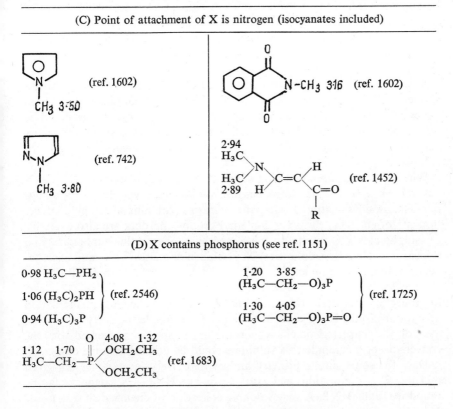

(C) Point of attachment of X is nitrogen (isocyanates included)

(D) X contains phosphorus (see ref. 1151)

0·98 H₃C—PH₂
1·06 (H₃C)₂PH } (ref. 2546)
0·94 (H₃C)₃P

1·20   3·85
(H₃C—CH₂—O)₃P

1·30   4·05
(H₃C—CH₂—O)₃P=O } (ref. 1725)

1·12 1·70 O 4·08 1·32
H₃C—CH₂—P⟨ OCH₂CH₃
          OCH₂CH₃   (ref. 1683)

TABLE 3–2–9. CHEMICAL SHIFTS IN H—C—O—X
(Data based on Table 3–2–2 and results from authors' laboratories)

X	CH₃—O—X	—CH₂—O—X	⟩CH—O—X
H	3·39	3·6	3·9
alkyl	3·3	3·4	3·6
Ph	3·73	3·9	4·5
CO—alkyl	3·7	4·0	5·0
CO—Ph	3·88	4·3	5·1
CO—CF₃	3·96	4·3	5·3

differences in the average orientation of the ester carbonyl group. While the shift accompanying etherification is much smaller, in some molecular environments the general upfield tendency is sufficiently large[1854] to be of some utility. The differences in the magnitudes of the acylation shift between primary (0·4–0·6 ppm) and secondary (1·0–1·15 ppm) protons is paralleled by similar phenomena accompanying benzoylation and can also be seen in the series of alkyl phenyl ethers and alkyl phenyl ketones listed in Table 3–2–2. Undoubtedly, all of these involve conformational preferences which could be very markedly affected by molecular environment.

A large number of data is available for the methoxyl groups which commonly resonate in the range $\delta = 3\cdot2$–$3\cdot3$ for aliphatic ethers, $3\cdot5$–$4\cdot1$ for enol and aryl ethers, and $3\cdot5$–$3\cdot9$ for methyl esters. While the total range is small, careful studies show a marked dependence of resonance frequency in substituted anisoles and phenetoles,[1141,49,2645,2122,1469] and in methyl cinnamates[1357] on substitution, and correlations based on these studies could be useful for some groups of compounds. On the other hand small differences between the axial and equatorial methoxyl groups in sugars[141,993] are not reliable.

When X is a group containing sulphur, the chemical shifts of H—C—X occur over a very wide range because of the wide variety of groupings possible. Besides scattered data, the literature contains a valuable collection[1971] of chemical shifts for *S*-methyl groups and data are also available for sulphinates[2603], sulphides, sulphones, sulphonates, thiosulphonates and thiocarboxylates[2028], sulphones[2060,209], thiosulphonates[52,1712] and miscellaneous sulphur compounds[1397,1918,1166,2028]. Data for simple alkyl mercaptans, sulphides, disulphides and tosylates are given in Table 3–2–2 and for other sulphur derivatives in Table 3–2–8. While chemical manipulation of sulphur groups is not quite as trivial as for example the acetylation of an alcohol, the characteristic changes in the n.m.r. spectra associated with the conversion of a sulphide to a sulphone could be utilized in some cases.

Data for some simple primary amines and *N*-alkylacetamides are given in Table 3–2–2 and additional examples in the H—C—N category may be found in Table 3–2–8. A very valuable collection of chemical shift data for N—Me groups is given by Ma and Warnhoff.[1602] Other collections contain data on amines,[84,1555,1223] α,β-unsaturated aminocarbonyl compounds,[1452] *N*-nitrosourethanes,[1797] hydroxymethylamides,[484] dialkylaminophosphorus derivatives,[548] akylsulphinamides,[1788] cyanates and isocyanates,[1044] *N*-methylindoles,[1961] ureas and urethanes[267] and nitroalkanes.[1200]

As can be seen from the examples in Tables 3–2–2 and 3–2–8 methyl groups (and also methylene and methine groups, provided no significant conformational factors occur) in primary, secondary and tertiary aliphatic amines resonate over a relatively narrow range[1602] and those in arylamines occur slightly downfield. Acetylation causes a significant (generally more than 0·5 ppm) downfield shift reminiscent of the acylation shift in alcohols. Very important for diagnostic purposes[1602,84] is the pronounced (generally 0·5–1·0 ppm) downfield shift on changing the solvent from CDCl$_3$ to CF$_3$COOH, i.e., on conversion of basic nitrogen to the positively charged species. In practice, it is often possible to observe this effect by simply adding a few drops of CF$_3$COOH to the chloroform solution, as trifluoroacetates of many nitrogenous bases are soluble in chloroform. In this way it is possible to estimate the number of H—C—N protons and often also to remove overlap. It is important to realise that only *pronounced* downfield shifts can be considered diagnostic in this respect, because smaller effects can be caused by the interaction of trifluoroacetic acid with other functional groups[742] and, for this reason, the addition of approximately equimolar amounts of trifluoroacetic acid to deuterochloroform solutions may be preferable to the use of pure trifluoroacetic acid.† Of additional diagnostic utility is the fact that proton exchange in NH groups in trifluoroacetic acid solution is generally slow and additional fine structure due to the coupling in H—C—N̅—H can often be observed.[84]

Some chemical shift data for compounds containing phosphorus are listed in Table 3–2–8. Many other data are available in the literature.[1216,1725, 602,2282,1721,407,1683]

Specific solvation effects, (Chapter 2–2 F) have been little investigated as an aid in the interpretation of spectra of acyclic compounds, although they have some utility in assignment of aromatic resonances and in compounds with defined configuration (cf. Chapters 3–6 and 3–8). This does not, however, indicate that the magnitude of such shifts is small. With methoxy,[1141] methyl[1845] and acetyl[252] groups attached to aromatic nuclei, solvent shifts or degree of protonation in strong acids can be correlated with substituents. With amines and ethers[374,553] complex formation with Lewis acids leads to effects interpretable in terms of effective changes in the electronegativity of the heteroatoms. Similarly, with many compounds containing hydroxyl, carbonyl etc., groups, protonation in strong acids leads to some well-defined shifts[742] which, while not as pronounced as those observed with amines,[84,1602] could be used for assignment of resonances and simplification of spectra.

C. EFFECT OF MULTIPLE SUBSTITUTION AT THE α-CARBON ATOM

Although we would not expect the shielding effects of functional groups to be strictly additive, an empirical procedure ("Shoolery's rules") has been developed[2268] for predicting the chemical shifts of methylene protons in compounds of the type Y—CH_2—X and methine protons in compounds of

the type $$Y—\overset{\displaystyle X}{\underset{\displaystyle |}{CH}}—Z.$$ The procedure consists of adding the values of the "effective shielding coefficients" for the groups X, Y and Z to the chemical shift of methane (0·23 ppm). The effective shielding coefficient of X is derived empirically by averaging the chemical shifts caused by successive substitution of methane by X and a number of such coefficients are listed in Table 3–2–10.

Thus, for example, the chemical shift of the methylene group in Br—CH_2—Cl would be calculated as $0·23 + 2·53 + 2·33 = 5·09$ ppm as compared to the actual value of 5·16 ppm. In general, the calculated values for X—CH_2—Y are accurate to better than 0·1 ppm and the discrepancy

---

† Pure trifluoroacetic acid rapidly converts many alcohols to their trifluoroacetates, a transformation which can also cause shifts of the order of 0·5–1·4 ppm.

TABLE 3–2–10. SHOOLERY'S EFFECTIVE SHIELDING CONSTANTS

Group	Shielding constant (ppm)	Group	Shielding constant (ppm)
—CH$_3$	0·47[a]	—I	1·82[a]
—CF$_3$	1·14[b]	—Ph	1·85[b]
—C=C	1·32[a]	—S—C≡N	2·30[d]
R—C≡C—	1·44[a,c]	—Br	2·33[a]
—CO—OR	1·55[b]	—OR	2·36[a]
Ar—C≡C—	1·65[c]	—NO$_2$	2·46[e,f]
—NR$_2$	1·57[a]	—Cl	2·53[a]
—CO—NR$_2$	1·59[b]	—OH	2·56[b]
—SR	1·64[a]	—N=C=S	2·86[d]
R—C≡C—C≡C—	1·65[c]	—O—CO—R	3·13[b]
—CN	1·70[a]	—O—Ph	3·23[b]
—CO—R	1·70[a]		

[a] Ref. 2268; [b] Ref. 2286; [c] Ref. 2394; [d] Ref. 1712; [e] Ref. 1200; [f] Not considered reliable.

TABLE 3–2–11. EFFECT OF MULTIPLE SUBSTITUTION AT C$_\alpha$.

Numbers in parentheses are calculated from Shoolery's effective shielding constants (Table 3–2–10). Unless otherwise stated, experimental values are from references 239 and 2403

(7·31)
4·96 H—C(OEt)$_3$

     Ph
     |
(7·14)
6·61 H—C—Cl
     |
     Cl

     OCH$_3$
(6·50) |
4·82 H—C—COOCH$_3$
     |
     OCH$_3$

(7·61)
7·52 H—C(NO$_2$)$_3$ (ref. 1200)

(4·22)
4·50 H—C—Ph$_3$ (ref. 2319)

     CH$_3$
(3·93) |
4·06 H—C—Ph (ref. 1314)
     |
     Ph

     Ph
(6·50) |
6·66 H—C—O—COCH$_3$ (ref. 1217)
     |
     COPh

(4·46)
4·07 ClCH$_2$CN

(3·19)
3·56 F$_3$C—CH$_2$—I

(5·15)
6·10 CH$_2$(NO$_2$)$_2$ (ref. 1200)

(6·07)
6·08 Ph$_2$CH$_2$ (ref. 2319)

(4·95)
5·56 CH$_2$(OCH$_3$)$_2$ (ref. 881)

(3·63)
4·13 CH$_2$(CN)$_2$ (ref. 1115)

$$\overset{\displaystyle Y}{\underset{\displaystyle |}{}}$$

rarely exceeds 0·3 ppm. For X—CH—Z the agreement is very poor, discrepancies of up to 1 ppm occurring. Some results for both systems are listed in Table 3–2–11, the examples being chosen mainly to illustrate the limitations of additivity, especially for the methine protons.

Besides their predictive value, Shoolery's effective shielding constants are a useful approximate guide to the relative deshielding effects of common functional groups.

Similar additivity relationships can be calculated for successive substitution on elements other than carbon.[2152]

PROTONS BONDED TO NON-AROMATIC sp^2 CARBON ATOMS

THE chemical shifts of protons bonded to sp^2 hybridized non-aromatic carbon atoms are profoundly affected by the nature of the element to which the double bond is attached. It is thus convenient to discuss the case of H—C=C and H—C=X (where X is a heteroatom) separately.

A. OLEFINIC PROTONS H—C=C

A vast amount of data is available in the literature on the chemical shifts of olefinic† protons. Fortunately a very useful generalization and several important collections of data are available thus allowing us to obtain all the important correlations without the necessity of referring to scattered data.

Unlike protons bonded to sp^3 hybridized carbon atoms, olefinic protons occupy a better defined position in space with respect to both α and β-carbon atoms and hence the effect of substituents on their chemical shifts is expected to vary in a reasonably regular manner, since conformational effects are confined to the more remote groups. In fact, the examination of the chemical shifts of over 1000 olefinic protons has led Pascual, Meier and Simon[1941] to postulate a set of additivity rules summarized in Table 3–3–1. In the majority of olefins, the substituent shielding coefficients Z, listed in Table 3–3–1, when added to the chemical shift of ethylene[2068] (5·28 ppm) in the manner shown in equation (3–3–1), give values which are within 0·3 ppm of the experimentally determined data.

$$\delta_{C=C-H} = 5{\cdot}28 + \sum Z \qquad (3\text{–}3\text{–}1)$$

Some examples of ethylenes with other substituents and of values significantly outside these limits are listed in Table 3–3–2. The exceptional chemical shifts can be rationalized in terms of phenomena such as extended conjugation, competitive electron withdrawal, steric inhibition of resonance,[799] incorporation into strained rings, long-range effects due to unusual conformations or inductive phenomena etc., all of which might be expected to affect chemical shifts (Chapter 2–2). Other examples are listed in reference 2223.

A by no means comprehensive list of references to significant collections of data for chemical shifts of olefinic protons is given in Table 3–3–3 for the purpose of obtaining more accurate correlations, and determining the in-

---

† The term "vinylic" is often used instead of olefinic, although it should, strictly speaking, be reserved for mono-substituted ethylenes.

The *Z*-factors for conjugated substituents are used when *either the substituent or the double bond* are further conjugated with *other* substituents. The *Z*-factors for "alkyl-ring" signify that the alkyl groups together with the double bond form a ring. $\delta_{CH_2=CH_2} = 5.28$ ppm

(After Pascual, Meier and Simon[1941])

Substituent	(ppm)		
	Z_{gem}	Z_{cis}	Z_{trans}
—H	0	0	0
—Alkyl	0·44	−0·26	−0·29
—Alkyl-Ring	0·71	−0·33	−0·30
—CH_2O—, —CH_2I	0·67	−0·02	−0·07
—CH_2S—	0·53	−0·15	−0·15
—CH_2Cl, —CH_2Br	0·72	0·12	0·07
—CH_2N<	0·66	−0·05	−0·23
—C≡C—	0·50	0·35	0·10
—C≡N	0·23	0·78	0·58
—C=C	0·98	−0·04	−0·21
—C=C conj.	1·26	0·08	−0·01
—C=O	1·10	1·13	0·81
—C=O conj.	1·06	1·01	0·95
—COOH	1·00	1·35	0·74
—COOH conj.	0·69	0·97	0·39
—COOR	0·84	1·15	0·56
—COOR conj.	0·68	1·02	0·33
H / —C=O	1·03	0·97	1·21
N< / —C=O	1·37	0·93	0·35
Cl / —C=O	1·10	1·41	0·99
—OR, R: aliph.	1·18	−1·06	−1·28
—OR, R: conj.	1·14	−0·65	−1·05
—OCOR	2·09	−0·40	−0·67
Aromatic	1·35	0·37	−0·10
—Br	1·04	0·40	0·55
—Cl	1·00	0·19	0·03
—F	1·03	−0·89	−1·19
R / —N< R: aliph. \R	0·69	−1·19	−1·31
R / —N< R: conj. \R	2·30	−0·73	−0·81
—SR	1·00	−0·24	−0·04
—SO_2—	1·58	1·15	0·95

TABLE 3–3–2. CHEMICAL SHIFTS OF SOME OLEFINIC PROTONS

Values in parentheses were calculated by means of equation (3–3–1)

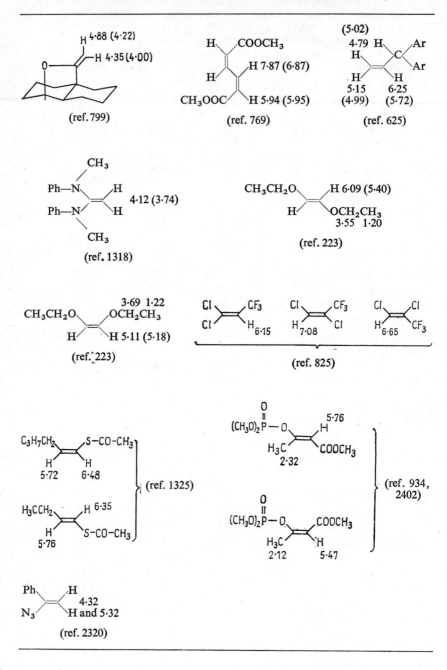

fluence of less common substituents. Although the shielding constants Z in Table 3–3–1 were derived from, *inter alia*, some cyclic olefins, the relation clearly breaks down for cyclopropene[2565,826] and cyclobutene[296] and for this reason collections of data for the chemical shifts in cycloalkenes and hetero-cycloalkenes are given in Table 3–3–4 together with references to selected compounds.

TABLE 3–3–3. REFERENCE LIST FOR CHEMICAL SHIFTS OF OLEFINIC PROTONS

Butadiene derivatives and polyenes:	163, 164, 165, 248, 255, 306, 307, 429, 610, 761, 769, 936, 1178, 1222, 1436, 2573, 2619
Vinyl derivatives:	159, 380, 2182, 2067
Conjugated acetylenes:	280, 1173, 1261
Vinyl ethers and thioethers:	223, 279, 280, 799, 1180, 1477, 1944, 2168, 2496
Enol acetates:	1218, 2168
α,β-unsaturated carbonyl derivatives:	215, 685, 689, 761, 770, 931, 1259, 1260, 1435 1476, 1689, 1840, 1841, 2061, 2446, 2573, 2601
C-7 alkenes:	119
tri- and tetra-*iso*butenes:	876
β-amino-α,β-unsaturated carbonyl derivatives:	1452, 1690
Nitroalkenes:	187, 1429
Group IV derivatives:	1464, 1558, 2244
Hydroxymethylene compounds:	639, 949, 953
Styrene and stilbene derivatives:	168, 228, 585, 586, 974, 1201, 1247, 1254, 1357, 1358, 1359, 1370, 1870, 1874, 1944, 2144, 2339, 2447, 2450, 2571, 2572, 2607
Miscellaneous substituted ethylenes;	119, 144, 311, 313, 649, 876, 995, 1259, 1260, 1358, 1476, 1510, 1867, 1918, 2135, 2541, 2542

Similarly, as can be seen from some data in Table 3–3–2, the additivity relationship[1941] for olefinic protons is particularly prone to inaccuracies in polyenes and here also references listed in Table 3–3–3 would have to be consulted before any far-reaching conclusions could be drawn from the chemical shifts in this group of compounds.

In spite of its approximate nature, the additivity relationship summarized

TABLE 3–3–4. CHEMICAL SHIFTS IN CYCLIC OLEFINS

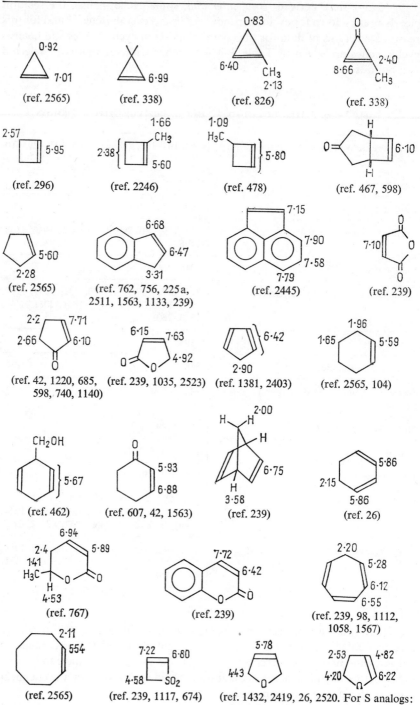

TABLE 3–3–4 *(cont.)*

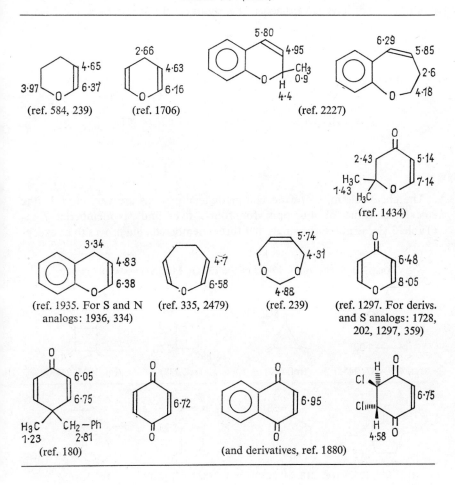

(ref. 584, 239)	(ref. 1706)	(ref. 2227)	(ref. 1434)

(ref. 1935. For S and N analogs: 1936, 334) (ref. 335, 2479) (ref. 239) (ref. 1297. For derivs. and S analogs: 1728, 202, 1297, 359)

(ref. 180) (and derivatives, ref. 1880)

in Table 3–3–1 and equation (3–3–1) is very valuable. Not only does it allow us to predict the chemical shifts of olefinic protons in the vast majority of molecular environments with remarkably high accuracy, but the list of the shielding constants in Table 3–3–1 contains within it many useful correlations which have been independently arrived at as a result of specific investigations. Thus, it can be seen that the downfield shift associated with geminal methyl substitution and the upfield shift associated with vicinal methyl substitution[995,2068,1841,311] is embodied in the Z-factors for the alkyl group. Similarly, the relative magnitudes of downfield shifts associated with *cis*- and *trans*-substitution by carbonyl groups[1259,1260] appear in the Z-factors for the various —CO—X groups listed in Table 3–3–1. However, for the purpose of making stereochemical deductions (cf. Chapter 3–8), a more subtle analysis may be necessary. Thus, the chemical shifts of compounds (I) and (II),[1867] while in good agreement with the values in parentheses obtained by means of Z-factors and equation (3–3–1), yield *opposite* relative magnitudes and thus might have led to incorrect assignment if not examined more fully.

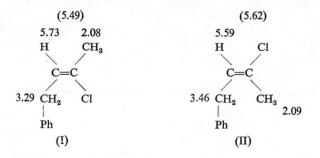

The chemical shifts of exocyclic methylene protons are very close to the calculated value of 4·65 ppm for four-, five- and six-membered rings (Table 3–3–5); not surprisingly the three-membered ring proves to be excep-

TABLE 3–3–5. CHEMICAL SHIFTS IN EXOCYCLIC METHYLENES AND ALLENES

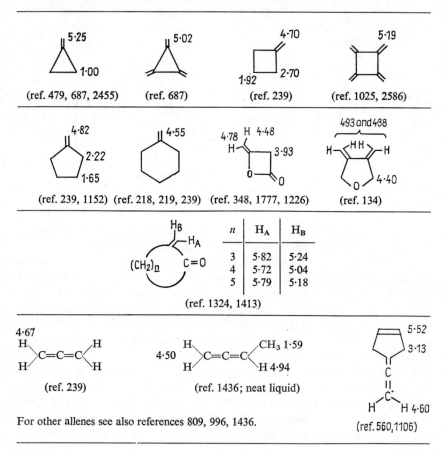

For other allenes see also references 809, 996, 1436.

TABLE 3–3–6. CHEMICAL SHIFTS OF FORMYL PROTONS

Compound	Reference	Compound	Reference
$\overset{O}{\overset{\|\|}{\text{R—C—H}}}$ 9·26–10·11 (R = akyl or alkenyl)	689, 1410, 2266	$\underset{\text{NOH}}{\overset{\text{Ar}}{\underset{\|\|}{\text{C}}}}\diagup\text{H}$ 7·2–8·6	1134, 1175, 1963, 2012
$\overset{O}{\overset{\|\|}{\text{Ar—C—H}}}$ *meta* and *para* subst.: 9·65–10·20 *ortho* subst.: 10·20–10·50	1408, 2005	$\underset{\text{N—NH—Ar}}{\overset{\text{R}}{\underset{\|\|}{\text{C}}}}\diagup\text{H}$ 6·1–7·7	1340, 1347, 2251
$\underset{\text{Br}}{\overset{\text{O Br O}}{\overset{\|\|\ \|\ \|\|}{\text{H—C—C—C—H}}}}$ 9·72	2444	$\underset{\overset{\text{N}}{\diagup}\overset{\text{X}}{\underset{\text{C—NH}_2}{}}}{\overset{\text{R}}{\underset{\|\|}{\text{C}}}}\diagup\text{H}$ 6·5–7·5 X = O,S	2007
$\underset{\text{Me}}{\overset{\text{Me}}{}}\text{C=C=C=C}\overset{\text{C=O} \quad 9·20\ \text{H}}{\underset{\text{H 5·60}}{}}$	1434	$\overset{O}{\overset{\|\|}{\text{MeO—C—H}}}$ 8·03	740, 1445
$\underset{\text{Me}}{\overset{\text{Me}}{}}\text{N—CH=CH—C}\overset{\text{H 8·96}}{\underset{\text{O}}{}}$	1690, 1452	$\overset{O}{\overset{\|\|}{\text{Me}_2\text{N—C—H}}}$ 7·84	740, 2007
C≡C–$\overset{O}{\overset{\|\|}{\text{C}}}$–H 9·38	1410	Ph—CO—NH—NH—$\overset{O}{\overset{\|\| \quad 8·03}{\text{C}}}$—H	2007
O=C—H 10·38 10·14 H C=O O=C—H 11·51	1408	$\underset{\text{Me}}{\overset{\text{Ph}}{}}\text{N—C—H}$ 8·33	
10·71 H $\overset{O}{\underset{O}{}}$ H	2005	$\underset{\text{H}}{\overset{\text{Ph}}{}}\text{N—C—H}$ 8·17–8·68	2032
		Me—Me 0·9 $\underset{\text{N}}{}$=6·9 Me—Me 1·2 $\overset{\oplus}{\underset{\overset{\text{N}}{\underset{\ominus\text{O}}{}}}{}}$=6·6 (1·2 2·2 2·0 2·4)	286
$\underset{\text{NOH}}{\overset{\text{R}}{\underset{\|\|}{\text{C}}}}\diagup\text{H}$ 6·8–7·9	371, 584, 1240	8·35 H $\overset{\text{CHMe}_2}{\underset{\overset{\text{N}}{\underset{\oplus}{}}}{}}$	1553, 1551

tional. There are also some differences in the effect exerted by carbonyl sub-stitution in rings of different size and in the chemical shifts of radialenes. Together with the characteristic values of the geminal and allylic coupling constants (Chapters 4–2 and 4–4), the chemical shifts assembled in Table 3–3–5 make these signals very informative. The chemical shifts of some allenic pro-tons are also included in Table 3–3–5.

The chemical shifts of β-protons (and alkyl groups, cf. Chapter 3–2) in α,β-unsaturated carbonyl compounds are always downfield from the corre-sponding α-substituents (see also Chapter 3–8). It is interesting that in cyclic systems this difference decreases from approximately 1·6 ppm in cyclopen-tenone and approximately 1·0 ppm in cyclohexenone (Table 3–3–4) to approx-imately half of this value in cycloheptenone and even less in the eight- and nine-membered rings, mainly due to a decrease in the deshielding of the β-protons.[1140] However, as the exact conformations of the larger rings are not known, this correlation should be used with caution.

The chemical shifts of olefinic protons in some propenes,[1358] styrenes[1358] and cinnamic acid derivatives[1201,1370,1357] have been related to Hammett constants and the "*ortho*" or "*cis*" shift due to halogens has been the sub-ject[1227] of an exhaustive investigation.

Characteristic changes in chemical shifts of substituted ethylenes have been reported to take place upon complex formation with silver fluoroborate.[2029]

B. FORMYL PROTONS H—C=X

The great majority of compounds containing protons in this environment are aldehydes and their derivatives, and substances related to formic acid. Formyl protons invariably occur at very low fields and the rather small vicinal coupling constants generally associated with them[1445,931,1337,1342] (Chapters 3–2 and 3–4) afford an additional aid to their indentification.

Table 3–3–6 shows some representative ranges as well as data for some less common types of formyl protons. The ranges for derivatives of aldehydes include both *syn* and *anti* forms, but these can be further assigned (cf. Chap-ter 3–8).

The chemical shifts of aldehydic protons occur in a relatively narrow range. They are influenced by alkylation on the α-carbon atom and α,β-unsaturation in the case of aliphatic aldehydes,[1410] and they correlate with Hammett para-meters in the case of *meta* and *para* substituted benzaldehydes,[1408] but these correlations must be applied with caution. In some groups of natural pro-ducts of well defined stereochemistry[1410,1401,1152,457] useful correlations with stereochemistry have been made (cf. Chapter 3–8). Other data on aro-matic and heterocyclic aldehydes have been reported[1446] and effects due to intramolecular hydrogen bonding[2005] and association with solvents[1409] have been investigated.

PROTONS BONDED TO *sp* CARBON ATOMS

A NUMBER of systematic n.m.r. investigations of acetylenic compounds are available in the literature and some typical data are given in Table 3–4–1. Acetylenic protons resonate at frequencies which are also characteristic of other types of protons and thus chemical shift data by themselves do not

TABLE 3–4–1. CHEMICAL SHIFTS IN SOME ACETYLENIC COMPOUNDS

H—C≡C—H	1·80 (ref. 1455)
R—C≡C—H	1·73–1·88 (ref. 1455, 1309, 560)
Ar—C≡C—H	2·71–3·37 (ref. 532, 1309)
C=C—C≡C—H	2·60–3·10 (ref. 1309, 1455, 1173, 1261)
X—C—C≡C—H $\overset{\|}{O}$	2·13–3·28 (ref. 1311)
C≡C—C≡C—H	1·75–2·42 (ref. 1311, 640, 285)
H_3C—C≡C—C≡C—C≡C—H	1·87 (ref. 1311)
X—CH_2—C≡C—H (X = halogen, $\overset{R}{\underset{R'}{>}}$C—C≡C—H $\underset{OH}{\|}$ —S—, >N—, —O—)	2·0–2·4 (ref. 1309, 2287, 1455, 2539) 2·20–2·27 (ref. 1309)
RO—C≡C—H	approx. 1·3 (ref. 2288, 1309)
X—C≡C—H	(X is not carbon) 1·3–4·5 (ref. 2288, 2287, 733, 1970)
$\overset{1·12–1·13}{}\quad\overset{2·17–2·34}{}$ H_3C—CH_2—C≡C—X	(X=Cl, Br, I) (ref. 2637)
$\overset{1·10}{}\;\overset{2·16}{}\qquad\overset{3·27}{}$ H_3C—CH_2—C≡C—CH_2—COOH	(ref. 589)
$\overset{1·2}{}\;\;\overset{2·3}{}\qquad\qquad\overset{1·95}{}$ H_3C—CH_2—C≡C—C≡C—H	(ref. 640)
$\overset{1·08}{}\;\;\overset{2·06}{}\qquad\overset{1·71}{}$ H_3C—CH_2—C≡C—CH_3	(ref. 331)
$\overset{3·1–3·6}{}$ R—C≡C—CH_2—C≡C—R^1 $\overset{1·8–2·2}{}$ R—C≡C—C≡C—CH_2—R^1	(ref. 2394)

afford a very reliable diagnostic feature for the presence of acetylenic hydrogen. Fortunately, the multiplicity of such resonances is fairly characteristic, only splitting due to long-range coupling being possible (cf. Chapter 4–4), and thus the signals are in practice easy to identify.

A number of references to acetylenic compounds are listed in Chapter 4–4. Some data listed in Table 3–4–1 also give representative examples of the chemical shifts of protons on the α- and β-carbon atoms and several of the references (in particular, 2287, 1311, 1309, 2637, 532, 1145, 2539, 2394, 2287 and 1455) contain a large number of data. The chemical shifts of non-acetylenic protons in some simple alkyl acetylenes are listed in Table 3–2–2.

The chemical shifts of acetylenic protons are seen to vary in a fairly characteristic fashion with substitution,[1309,1311,532] and electronegativity effects are also transmitted through the triple bond.[1145,2637] Acetylenic protons have been found to exhibit marked solvent shifts,[1455,1311,532] which are related to their acidity and to specific associations involving the π-electron cloud.

PROTONS BONDED TO *sp³* CARBON ATOMS IN NON-AROMATIC CYCLIC STRUCTURES

THE presence of cyclic structures implies certain limitations on conformational mobility (and hence long-range shielding effects different from those in acyclic compounds) and, most obviously in the case of strained rings, possible partial rehybridization with the attendant chemical shift changes. Further, even in the absence of substituents, the components of the rings themselves may exert characteristic shielding influences on each other (cf. Chapter 3–8).

The variety of cyclic structures is quite astronomical and it is very important to realize the limitations to the applicability of data obtained from model compounds. In particular, to be a useful model for cyclic structures, a compound must be of known conformation. As an example, in the absence of supporting data, the chemical shift of the methine proton in cyclohexanol could not be used as a reliable model for the chemical shift of the methine proton at the carbon bearing the hydroxyl group in a rigid six-membered ring, even if it was possible to ignore long-range influences due to substituents. This is because the former is an average of an unknown percentage of forms with axial and equatorial hydroxyl groups while in the latter the conformation of the hydroxyl group and hence the methine proton is fixed. Thus the "frozen" (cf. Chapter 3–8 and Part 5) spectrum of cyclohexanol, which would yield the chemical shifts of the pure axial and pure equatorial methine protons, would be a more useful model than the averaged spectrum. With smaller rings, conformational differences are not quite so marked and in any case, data for conformationally pure compounds are practically non-existent.

The series of dialkyl ketones[385] shown in Table 3–5–1 illustrates the large differences in chemical shifts which may occur when the same substituent is

TABLE 3–5–1. CHEMICAL SHIFTS OF METHINE PROTONS IN SOME DIALKYL KETONES
(after Bryce-Smith and Gilbert[385])

Ketone	δ of methine protons
di-cyclopropylketone	1·97
di-cyclobutylketone	2·10
di-cyclopentylketone	2·30
di-cyclohexylketone	2·83
di-isobutylketone	2·16
di-isopropylketone	2·67

introduced into different cyclic structures. It can be seen that, even if the datum for dicyclopropyl ketone is ignored, the methine protons in dicyclo-alkyl ketones derived from cyclobutane, cyclopentane and cyclohexane reso-nate over a range of 0·7 ppm and the shifts are not simply related to the average chemical shifts in the parent cycloalkanes (cf. Table 3–5–2).

TABLE 3–5–2. CHEMICAL SHIFTS IN SOME CYCLOALKANES

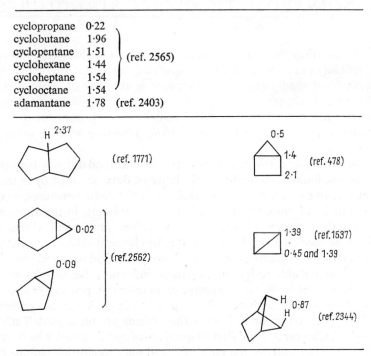

cyclopropane	0·22	
cyclobutane	1·96	
cyclopentane	1·51	(ref. 2565)
cyclohexane	1·44	
cycloheptane	1·54	
cyclooctane	1·54	
adamantane	1·78	(ref. 2403)

The data presented in Tables 3–5–2 to 3–5–5 are principally for unsub-stituted compounds. Clearly, the principles discussed in Chapter 2–2 and data from Chapter 3–8 could be used to predict, at least qualitatively, the changes in chemical shifts associated with the introduction of substituents, provided that the conformational changes associated with the introduction of the substituents themselves are taken into consideration.

It can be seen that most of the data refer either to essentially rigid systems or to averaged chemical shifts in flexible rings. The latter are of very limited value in predicting the chemical shifts of their derivatives, but some of the references quoted also give results for substituted compounds and the im-portant problem of substituted cyclohexanes will be considered separately in Chapter 3–8 G.

Chemical shift data for cyclic systems containing double bonds are given in Table 3–3–4.

As already discussed in Chapter 2–2, and exemplified in Tables 3–3–4, 3–3–5, 3–5–2 and 3–5–5, the chemical shifts of the ring protons in cyclo-

propanes and heterocyclopropanes are abnormal. Further, such protons
exhibit a special dependence of chemical shifts on substitution, because these
always occupy well-defined positions in space (cf. Chapter 3–8), and because
of the possibility of transmission of electron withdrawal by, for example,
carbonyl groups, through conjugative effects. Some data on the chemical
shifts of ring protons in cyclopropanes and substituted cyclopropanes are
given in Chapter 3–8.

TABLE 3–5–3. CHEMICAL SHIFTS IN SOME HYDROAROMATIC COMPOUNDS
(Unless otherwise stated the data are taken from references 239 and 2403)

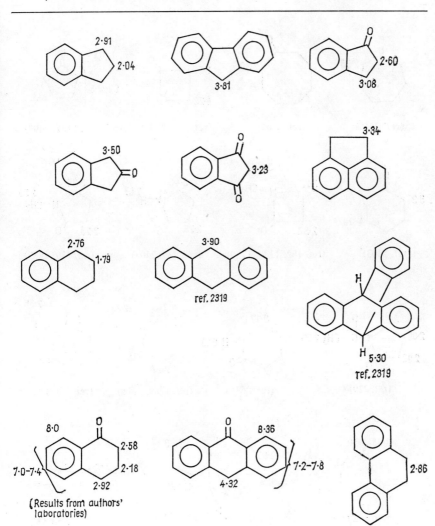

TABLE 3–5–4. CHEMICAL SHIFTS IN SOME CYCLIC KETONES, LACTONES AND LACTAMS

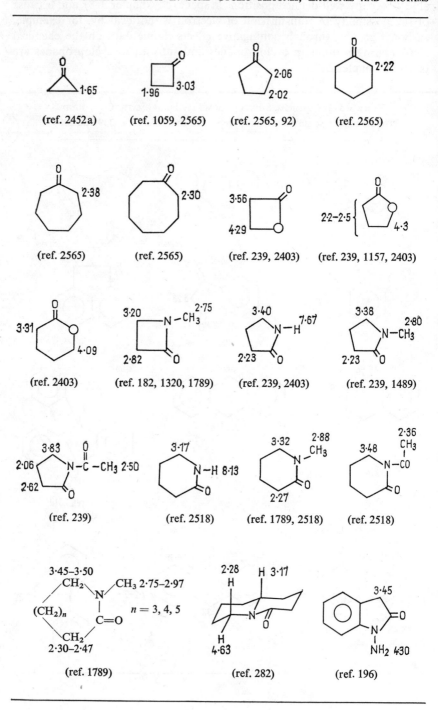

(ref. 2452a)   (ref. 1059, 2565)   (ref. 2565, 92)   (ref. 2565)

(ref. 2565)   (ref. 2565)   (ref. 239, 2403)   (ref. 239, 1157, 2403)

(ref. 2403)   (ref. 182, 1320, 1789)   (ref. 239, 2403)   (ref. 239, 1489)

(ref. 239)   (ref. 2518)   (ref. 1789, 2518)   (ref. 2518)

(ref. 1789)   (ref. 282)   (ref. 196)

TABLE 3–5–5. MISCELLANEOUS NON-AROMATIC HETEROCYCLIC COMPOUNDS

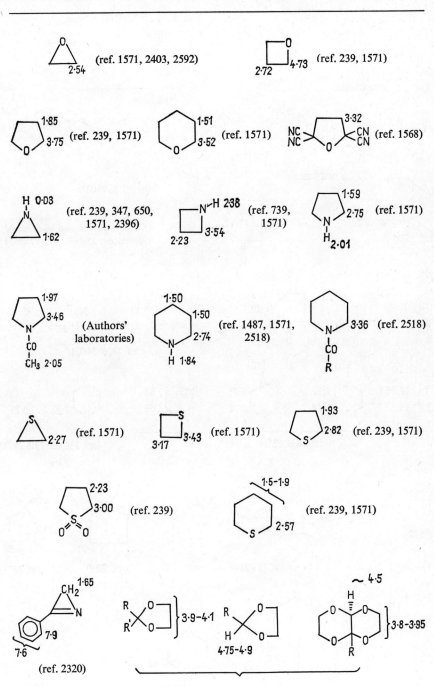

(Values for steroid derivatives, ref. 438
and 439. See also ref. 41 and 113)

Table 3–5–5 *(cont.)*

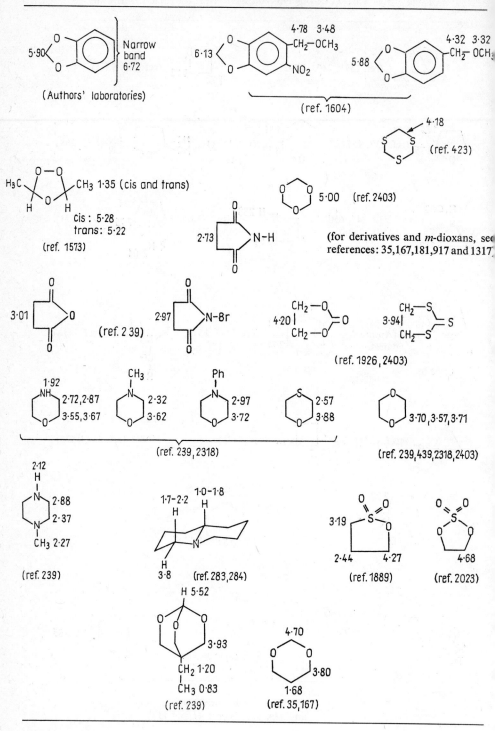

PROTONS BONDED TO AROMATIC AND HETEROCYCLIC CARBON ATOMS

FROM the theory of chemical shifts (Chapter 2–2), it is clear that the chemical shifts of aromatic and heterocyclic protons will depend principally on the magnetic anisotropy of the ring ("ring-current effect"), the π-electron charge density on the carbon atom to which the proton is attached, and on long-range effects associated with substituents and/or heteroatoms. The separation of these effects is difficult, in particular as substituents will almost always affect π-electron densities and exert long-range shielding, especially if situated *ortho* to the proton concerned. Further, if more than one substituent is present, they may interact both electronically and sterically thus making the substituent effects non-additive and, in some cases, may even cause ring buckling.†

There is also a practical difficulty in obtaining accurate chemical shift data for aromatic compounds. On one hand lightly substituted species may give rise to complicated multiplets which require good quality spectra for analysis and on the other hand the chemical shifts in aromatic compounds tend to vary significantly with concentration thus requiring spectra of very dilute solutions.

Fortunately, while the above considerations cause some difficulty in obtaining and interpreting high quality data for purposes such as correlation of chemical shifts with π-electron densities,[2616] valuable structural information can be relatively easily obtained from even superficial studies of aromatic resonances, especially as the magnitudes of coupling constants in aromatic systems (cf. Chapter 4–3) are fairly characteristic. In particular, while data for most heterocyclic systems are incomplete, it appears that substituent effects on chemical shifts are *qualitatively* similar to those in benzene derivatives, and this fact, in conjunction with the chemical shift in the parent heterocyclic systems and coupling data makes it possible to assign resonances with reasonable confidence.

A. BENZENE DERIVATIVES

A great deal of systematic work has been carried out regarding the influence of substituents on the chemical shifts of aromatic protons in benzene derivatives.‡ In spite of some minor quantitative discrepancies, which are caused

† Cf. however references 2086, 976, 975.
‡ References 135, 399, 542, 654, 663, 846, 886, 991, 1357, 1427, 1494, 1684, 1685, 2102, 2160, 2192, 2222, 2262, 2306, 2341, 2367, 2370, 2524, 2616 and 2624.

partly by solvent effects and partly by different methods of analysing data, it appears that the introduction of substituents into the benzene nucleus results in characteristic chemical shifts at the *ortho, meta* and *para* positions and that the magnitude of the latter two is related approximately to Hammett's substituent constants. Further, as long as no two substituents are *ortho* to one another, i.e. in *meta* and *para* disubstituted benzenes and 1,3,5-trisubstituted benzenes, the substituent effects are approximately (± 0.1 ppm) additive and adjusted parameters[1685] give even better agreement.

The substituent constants listed in Table 3–6–1 have been collected from the above sources, values for dilute solutions in carbon tetrachloride or cyclohexane obtained by complete analyses being quoted whenever available. Besides their obvious, although limited, utility in predicting the chemical shifts in some di- and tri-substituted benzenes, these data give a *qualitative*

TABLE 3–6–1. THE EFFECT OF SUBSTITUENTS ON THE CHEMICAL SHIFT (7·27 PPM) OF BENZENE. NEGATIVE SIGN DENOTES DOWNFIELD SHIFT

Substituent	Shift relative to benzene (ppm)		
	ortho	*meta*	*para*
NO_2	-0.95^a	-0.17^g	-0.33^a
CHO	-0.58^a	-0.21^a	-0.27^a
COCl	-0.83^b	-0.16^b	-0.3^e
COOH	-0.8^d	-0.14^g	-0.2^e
$COOCH_3$	-0.74^c	-0.07^g	-0.20^c
$COCH_3$	-0.64^b	-0.09^b	-0.3^e
CN	-0.27^b	-0.11^g	-0.3^e
Ph	$-0.18^{f,h}$	$0.00^{f,h}$	0.08^f
CCl_3	-0.8^e	-0.2^e	-0.2^e
$CHCl_2$	-0.1^d	-0.06^g	-0.1^e
CH_2Cl	0.0^e	-0.01^g	0.0^e
CH_3	0.17^h	0.09^h	0.18^h
CH_2CH_3	0.15^h	0.06^h	0.18^h
$CH(CH_3)_2$	0.14^h	0.09^h	0.18^h
$C(CH_3)_3$	-0.01^h	0.10^h	0.24^h
CH_2OH	0.1^e	0.1^e	0.1^e
CH_2NH_2	0.0^e	0.0^e	0.0^e
F	0.30^a	0.02^g	0.22^a
Cl	-0.02^a	0.06^g	0.04^a
Br	-0.22^a	0.13^g	0.03^a
I	-0.40^a	0.26^g	0.03^a
OCH_3	0.43^a	0.09^g	0.37^a
$OCOCH_3$	0.21^d	0.02^d	—
OH	0.50^c	0.14^g	0.4^e
$O-SO_2-p-C_6H_4-Me$	0.26^d	0.05^d	—
NH_2	0.75^a	0.24^g	0.63^a
SCH_3	0.03^d	0.0^d	—
$N(CH_3)_2$	0.60^a	0.10^a	0.62^a

[a] Ref. 2341; [b] Ref. 1685; [c] Ref. 654; [d] Ref. 2306; [e] Ref. 542—some values are approximate only; [f] Ref. 1731; [g] Ref. 1493, 1494; [h] Ref. 326.

insight into the direction and approximate magnitude of substituent effects in all aromatic and heterocyclic systems.

The anomalies associated with substituent effects on the chemical shifts of ring protons when two substituents are *ortho* to one another, are illustrated by the chemical shifts in compounds (I–V), where the values refer to chemical shifts relative to benzene (negative sign denotes downfield shift) and the numbers in parentheses were calculated from data similar to those in Table 3–6–1.[654]

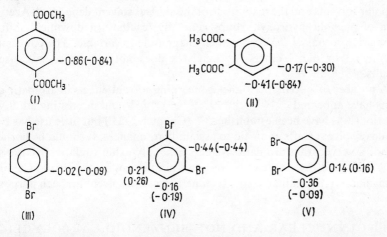

Clearly, although various additive parameters for particular substitution patterns are useful,[654,1685,1427,886,399,663] the prediction of chemical shifts in heavily substituted benzene derivatives with multiple buttressing effects and electronic interactions would be highly complicated.

Substituent constants for a wider range of substituents than those quoted in Table 3–6–1 have been given.[2306]

Besides attempts to obtain general relations between substitution patterns in benzene derivatives and the chemical shifts of ring protons, a number of detailed studies of more limited groups of compounds are available, among them xylenes,[886] halobenzenes,[2355] phenols,[2223,1949,356,1164] methoxybenzenes,[2645,1141,1622,846] durene and mesitylene derivatives,[399,663] polyalkylated benzenes,[976] 4,6-disubstituted *o*-cresols,[1427] methoxyaniline derivatives,[2646] benzylic carbanions,[2160] amines[2102,2035,2370,2033] and nitrocompounds.[2524,2033] Aromatic proton chemical shifts are useful for defining the oxygenation patterns in a variety of plant phenolics.[1252a]

Of some interest for the purpose of assigning aromatic resonances are the well-defined changes associated with the conversion of aniline derivatives to their salts,[2370,2102,1602] which in practice can be achieved by addition of trifluoracetic acid. The shifts vary somewhat with substitution[2102] but with the exception of diphenylamine derivatives, a downfield shift of approximately 1 ppm for the *ortho* and *para* protons and of approximately 0·5 ppm for the *meta* protons can be expected by changing the solvent from carbon tetrachloride to trifluoroacetic acid.

Acetylation of aniline derivatives[362] generally causes downfield shifts of the order of 0·8, 0·3 and 0·5 ppm for the *ortho*, *meta* and *para* protons, respectively, but when the carbonyl group of the amide takes up a preferred conformation due to hydrogen bonding between an *ortho* substituent and the amide proton, deshielding of the remaining *ortho* proton by approximately 2 ppm results.

The conversion of phenols to phenolate ions causes appreciable upfield shifts in the *ortho* and *para* positions and smaller upfield shifts in the *meta* positions.[1164,356] Unfortunately, spectra of this sort can only be observed in polar solvents and the magnitude of the shifts is solvent dependent. Acetylation of phenolic hydroxyl groups generally results in a downfield shift of aromatic protons,[2119,1164] but their magnitude is variable. The conversion to methyl or picryl ethers, or tosylates does not appear to give useful results.[1164,2199]

A number of systematic studies concerning solvent effects in aromatic systems have appeared[1245,655,1409,656,1815,2188,2187] and interesting additivity relationships have been established[656] (Chapter 2–2F) but little use has been made of solvent shifts as an aid to assigning resonances. It should be borne in mind, however, that aromatic systems generally exhibit relatively large solvent shifts and that this can be very useful in removing accidental chemical equivalence (cf. Chapter 2–3). Acetone is a useful solvent for such purposes.

B. POLYNUCLEAR AND NON-BENZENOID CARBOCYCLIC AROMATIC COMPOUNDS

Chemical shift data for the more common types in this class are listed in Table 3–6–2. Data on other polynuclear aromatic hydrocarbons may be found in references 1299, 1694, 1699, 1699a, 1741 and 193 and spectra of some substituted tropolones have been reported.[2368,253] As discussed previously (Chapter 2–2), the ring-current effect is approximately additive[1299] and thus, for example, in naphthalene the α-protons resonate downfield from the β-protons. However, some protons for instance H-4 and H-5 in phenanthrene, are abnormally deshielded[1299] (presumably by van der Waal effects p. 71) and this makes it comparatively easy to assign resonances in angularly condensed polycyclic compounds.[1694]

Systematic studies of substituent effects have been carried out on naphthalenes (cf. references in Table 3–6–2), phenanthrenes,[1700] benzo(C)phenanthrenes,[1697] pyrenes[1701] and triphenylenes.[1695] In general, while the substituent effects are not equal to those in benzene, they operate in the same direction and are of the same order of magnitude.

Two novel features can however be observed. Firstly, substituents have generally very pronounced effects on the chemical shifts[700] of *peri* protons (and other *peri* substituents in general) and secondly, even when no obvious steric reasons can be advanced, the two *ortho* positions are not equally affected. Both phenomena can be seen in the effects of α- and β-nitro groups in naphthalenes (when the latter do not suffer from any specific steric interactions such as occur when *ortho* or *peri* groups are introduced)

TABLE 3–6–2. CHEMICAL SHIFTS IN NON-BENZENOID AROMATIC CARBOCYCLIC COMPOUNDS

Compound	References
7·81 7·46	213, 644, 700, 760, 1299, 1367, 1633, 2522, 2524, 2525
8·26 7·82 7·51 7·33 7·13	595, 1744, 1745, 2192
8·31 7·91 7·39	1299, 1699a
7·82 7·88 8·12 8·93 7·71	198, 1299, 1699a, 1700
7·99 8·16 8·06	1299, 1699a, 1701
7·95 8·65 7·5–7·9 8·7	1699a, 1742
8·56 7·61	1299, 1695, 1699a

TABLE 3–6–2 *(cont.)*

Compound	References
6·60 6·47	1367, 1702
Coronene: 8·84	1299, 1699a
1·74 H 7·02 H H H 8·10 7·56	2332
6·11 and 6·44 =CH₂ 5·78	999, 1405, 1746, 1861, 2317
4·15 Fe	1556, 1833, 2041, 2117, 2118
7·27 6·95 CH₂ −0·51	1057. For many related ex- amples see: 1027, 1654, 1655, 2481, 2483, 2485, 2487

shown in structures (VI) and (VII). The values given are chemical shifts with respect to naphthalene; the negative sign denotes that the shifts are towards lower fields.[2524]

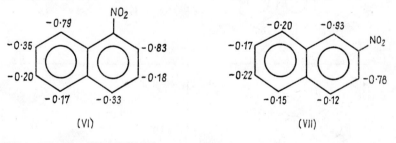

(VI)   (VII)

The *peri effect*[700,1696,1633] leads to deshielding *with all types of sub-stituents* but is particularly well exemplified in a series of polycyclic ke-tones[1696] shown in Table 3–6–3. Other "proximity" interactions are also possible in polynuclear compounds[1138,328,860] and appear to cause pronounced deshielding in all cases so far reported (cf. references above and Chapter 2–2B(v)).

TABLE 3–6–3. EFFECT OF CARBONYL GROUPS ON RING PROTONS IN SOME POLYNUCLEAR AROMATIC SYSTEMS
The figures in brackets are shifts in ppm relative to the parent system. Negative sign denotes a downfield shift. All values are from ref. 1696

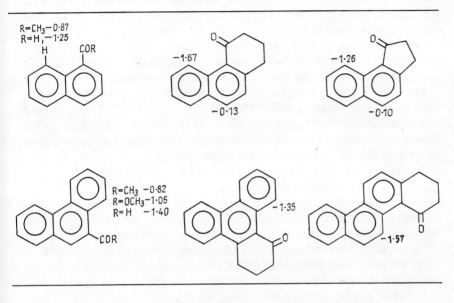

C. HETEROCYCLIC COMPOUNDS

The presence of a nitrogen or oxygen atom causes a pronounced downfield shift in ring protons, especially in the α-position, by comparison with the corresponding carbocyclic analog. These relatively large "built-in" chemical shift differences between ring-protons, combined with the fact that the presence of the hetero-atom implies an insulating group (Chapter 2–3), makes the spectra of most heteroaromatic compounds relatively easy to analyse.

Table 3–6–4 gives the chemical shifts of the ring protons in the parent compounds of some common heterocyclic systems and lists references to substituted derivatives. Unfortunately, many heterocyclic derivatives are insoluble in solvents commonly used for n.m.r. spectra and, at the same time, their chemical shifts are quite susceptible to solvent and concentration changes, thus often making direct comparison between various classes impossible.

Although the number of systematic investigations of substituent effects on the chemical shifts of heterocyclic protons is limited, many data have been collected on furans,[1043,1940] thiophenes,[1037] pyrimidines,[1041] pyridines,[2617] quinazolines,[1369] nitroquinolines,[258] pyridazines,[2435] imidazo[1,2a]pyridines,[1951] pyrazoles,[819] purines[507] and pyrroles.[9,1038] Other references to data on substituted heterocyclic systems are listed in Table 3–6–4.

Tables 3–6–5, 3–6–6 and 3–6–7 summarize some of the data for the most thoroughly investigated systems. It can be seen that, while substituent effects tend to follow the order exhibited in benzene derivatives (Table 3–6–1), the magnitude of the effects varies over a large range. It is possible to rationalize most of these variations, but as already stated in Chapter 2–2, and in a number of references,[1041,1043,1369,2617] the effects are complex and it is unlikely that accurate predictions of chemical shifts in substituted heteroaromatic compounds can be made at the present stage of development, especially as data for suitable solvent systems are difficult to obtain. Tables 3–6–5, 3–6–6 and 3–6–7 together with Table 3–6–1 should however enable the reader to gain a qualitative picture of the effects of common substituents.

The effect of methyl substitution on the chemical shifts of ring protons has been specifically investigated[2074,997,2435,1080,1030,1657,1713,2075,198] and can be used[2617,1881,198] to assign chemical shifts when a series of methyl substituted compounds is available. A methyl substituent invariably causes an up-field shift (0·1–0·5 ppm) of the *ortho* proton.

By analogy with polycyclic hydrocarbons (Table 3–6–2), one would expect specific shielding effects in polynuclear heterocyclic systems, not only from *peri*-substituents or groups otherwise brought into proximity, but also through long-range effects associated with the heteroatoms themselves. Some examples of this type are given in Table 3–6–8. It can be seen that some of the chemical shifts thus induced[669] are very large. Similarly, large downfield shifts are associated with the presence of *peri* carbonyl groups, as for example in 4-quinolone derivatives.[2122,1002] These are, of course, analogous to the effect already noted in α-tetralone (p. 91).

The marked solvent sensitivity of the chemical shifts of ring protons in heterocyclic compounds has already been alluded to and has been commented upon by many authors (cf. references quoted in Table 3–6–4); these effects are particularly pronounced with some indoles[2096,1271] where shifts of the order of 0·8 ppm have been observed between neat liquids and 30 per cent solutions in carbon tetrachloride.[2096] In general, protons in different positions will show marked divergence in the magnitudes of the solvent shifts and this can be very useful in making assignments or removing accidental equivalence, for example in substituted quinolines.[1080]

With nitrogen-containing heterocyclic systems, well defined changes occur on protonation by, for instance, trifluoroacetic acid. However, even when only protonation on nitrogen occurs, as with pyridines[2192,2307,1595,2262,80] or quinoline,[1448] the chemical shifts change in a rather complex manner with concentration. Generally, γ-protons appear to undergo the largest, and α-protons the smallest downfield shifts (p. 82) and this could be useful in checking assignments or removing degeneracy.

Useful data on protonation of pyrazoles,[819] imidazoles,[1669] thiazoles[492] and azaindolizines[1953] are also available (cf. Chapter 3–9A(vi) and Chapter 5–3).

TABLE 3–6–4. CHEMICAL SHIFTS IN HETEROAROMATIC COMPOUNDS[a]

Compound	References
(furan) 6·30, 7·40; O	997, 1015, 1043, 1520, 1940, 2069, 2316
(thiophene) 7·04, 7·19; S	997, 1015, 1030, 1031, 1037, 1040, 1043, 1146, 1193, 1195, 1196, 1197, 1520, 1713, 2334, 2386, 2387
(pyrrole) 6·05, 6·62; N–H 7·70	9, 11, 239, 997, 1038, 1039, 1146, 1169, 1910, 2292
(selenophene) 7·12, 7·70; Se	1146
(pyrazole) 7·55, 6·25 b, 7·55, N–N–H 13·7; (N-methylpyrazole) 7·30, 6·14, 7·36, N–N–CH$_3$ 3·81	239, 518, 742, 819, 1077, 2399, 2584
(imidazole) 7·14 b, HN–N, 7·70; (N-methylimidazole) 7·05, 6·86, H$_3$C–N–N, 7·40	213, 239, 1307, 1669, 2074, 2075
(thiazole) 7·41 S, 7·98, 8·88, N	69, 213, 492, 1076, 2395
(oxadiazole) 8·19, N–O–N; (thiadiazole) 8·58, N–S–N	1907
(benzofuran) 7·49, 6·66, 7·13, 7·52, 7·19, 7·42, O	259, 563, 762, 763, 1752
(indole) 6·34, 6·54, N–H 7·00	28, 239, 259, 508, 763, 1142, 1271, 1678, 2096

TABLE 3-6-4 *(cont.)*

Compound	References
	763, 1703, 1752, 2383, 2384, 2385
	257
	256, 2075
	256
	262
	262, 1951, 1952
	1953
	507, 1268, 1716, 1718, 2075

TABLE 3–6–4 *(cont.)*

Compound	References
7·59 7·86 7·19 7·50 (structure with N)	277, 1255
7·46 7·06 8·50 (pyridine structure)	27, 194, 279, 960, 979, 981, 1197, 1442, 1444, 1450, 1595, 1815, 1837, 2192, 2307, 2617
7·68 9·17 (pyridazine structure)	533, 616, 766, 768, 960, 1815, 2435
9·15 7·09 8·60 (pyrimidine structure)	616, 766, 960, 1036, 1041, 1815, 1887, 2074, 2435, 2477
8·24 8·98 7·34 8·43 (pyrimidine N-oxide structure)	1887
8·5 (pyrazine structure)	766, 960, 1815, 2435
9·18 (triazine structure)	616, 766
7·26 6·15 6·57 7·31 (2-pyridone structure, N–H, =O)	379, 765
7·68 8·00 7·43 7·26 7·61 8·81 8·05 (quinoline structure)	89, 258, 669, 960, 1080, 1448, 1947, 2233

TABLE 3–6–4 *(cont.)*

Compound	References
	2437
	261, 669, 960, 2435
	2437
	260, 960, 2435
	260, 960, 1717
	260, 376
	260, 960, 1369, 1717
	669, 1423

TABLE 3–6–5. PROTON CHEMICAL SHIFTS IN 4-SUBSTITUTED PYRIDINES[2617]
The chemical shifts are differences from pyridine. All spectra are for dilute solutions in dioxan. Negative signs denote downfield shifts. Pyridine: $H\alpha$: 8·50 ppm, $H\beta$: 7·06 ppm

4-substituent	Chemical shifts (ppm)	
	H-3	H-2
NH_2	0·808	0·510
OMe	0·427	0·207
Me	0·158	0·175
Cl	—0·137	0·051
Br	—0·280	0·135
CN	—0·428	—0·216
CHO	—0·468	—0·271
COMe	—0·493	—0·192
NO_2	—0·815	—0·332

TABLE 3–6–6. SUBSTITUENT EFFECTS IN QUINAZOLINES[1369]
The chemical shifts (in ppm) are differences from quinazoline. Negative signs denote downfield shifts. All spectra were taken in dimethyl sulphoxide

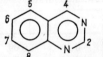

Substituent position	Proton position	Substituent				
		OMe	OH	Me	Cl	NO_2
5	6	0·69	0·80	—	—	—0·73
6	5	0·80	0·93	—	—0·05	—0·94
7	8	0·80	0·92	0·33	0·07	—0·65
8	7	0·62	0·79	0·31	—0·14	—0·49
6	7	0·54	0·53	—	0·12	—0·53
7	6	0·52	0·57	0·28	0·08	—0·69
5	7	0·20	0·31	—	—	—0·11
6	8	0·26	0·22	—	0·06	—0·08
7	5	0·21	0·26	0·22	0·04	—0·23
8	6	0·13	0·30	0·24	0·04	—0·11

Chemical shifts in the parent compound: H2: 9·22, H4: 9·28, H5: 7·84, H6: 7·57, H7: 7·83, H8: 8·00.

Footnotes to Table 3–6–4

[a] Data on other systems includes: pyrimidones;[1036,1424,2630,1041,2477,195] chromones and thia- and benzo-analogs;[1700a,1714] aza-indolizines;[1657,262] pteridines;[1717] *N*-oxides of miscellaneous heterocyclic bases;[2437,1887,2436,1886] miscellaneous aza-derivatives of polynuclear aromatic compounds;[431,669,1758] dibenzofuran and carbazole;[259,59] furano-quinolines;[2122] oxadiazoles;[1799] isothiazoles;[2347a,74] oxazoles;[1076] isoxazoles;[2627,2435] thienopyrroles and thienothiophenes;[1662,1911,2448,965,2149,1205] triazoles;[1596] tetra-zoles;[1673,1779] sydnones;[2361,1519] 1,2-dithiole-3-ones and 1,2-dithiole-3-thiones.[362]

[b] Rapid proton exchange makes the spectrum symmetric (i.e. the two nitrogen atoms are equivalent on the n.m.r. time scale).

[c] In dimethyl sulphoxide.

2-substituted series

Substituent	2-substituted furan			2-substituted thiophene		
	H-3	H-4	H-5	H-3	H-4	H-5
CN	−0·63	−0·13	−0·11	−0·47	0·00	−0·28
CHO	−0·80	−0·19	−0·22	−0·65	−0·10	−0·45
COCH$_3$	−0·73	−0·10	−0·06	−0·57	0·00	−0·28
COOCH$_3$	−0·76	−0·07	−0·07	−0·70	0·05	−0·20
SCN	−0·47	−0·13	−0·21	−0·30	0·05	−0·28
I	−0·19	0·06	−0·10	−0·13	0·33	−0·01
Br	0·08	0·01	0·03	0·05	0·27	0·11
SCH$_3$	0·04	−0·02	−0·02	0·03	0·18	0·05
CH$_3$	0·43	0·14	0·18	0·37	0·24	0·28
OCH$_3$	1·27	0·16	0·58	0·94	0·43	0·82

3-substituted series

Substituent	3-substituted furan			3-substituted thiophene		
	H-2	H-4	H-5	H-2	H-4	H-5
CN	−0·55	−0·29	−0·09	−0·63	−0·20	−0·15
CHO	−0·59	−0·44	−0·03	−0·79	−0·45	−0·03
COCH$_3$	−0·56	−0·43	0·02	−0·68	−0·47	0·02
COOCH$_3$	−0·55	−0·40	0·03	−0·78	−0·47	0·05
SCN	−0·30	−0·26	−0·14	−0·25	−0·05	−0·05
I	0·02	−0·12	0·11	−0·06	0·00	0·19
SCH$_3$	0·08	−0·02	0·05	0·33	0·10	0·03
CH$_3$	0·24	0·17	0·14	0·45	0·22	0·14
OCH$_3$	0·36	0·21	0·27	1·10	0·38	0·20

[a]Chemical shifts in the parent compounds: Furan: Hα: 7·40; Hβ: 6·30 ppm; Thiophene: Hα: 7·19, Hβ: 7·04 ppm.

TABLE 3–6–8. CHEMICAL SHIFTS IN SOME POLYCYCLIC AZA-DERIVATIVES[669]

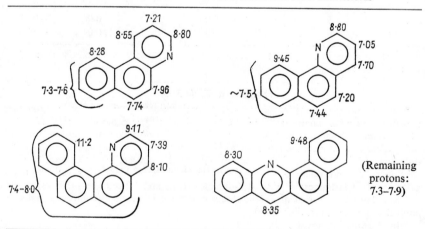

PROTONS BONDED TO ELEMENTS
OTHER THAN CARBON

PROTONS attached to elements other than carbon fall into two broad classes, namely the familiar OH, NH and SH protons in alcohols, phenols, enols, carboxylic acids, amines, amides, thiols etc., and a large variety of protons attached to other elements in compounds which are generally not considered "organic", for example organometallics, pi-complexes, phosphine and arsine derivatives etc. We shall only consider the former class; references to papers containing chemical shift data for the latter types can be found in Chapter 4–5.

A. —OH, —NH AND —SH GROUPS

The chemical shifts of these protons are, to a greater or lesser extent, influenced by the related phenomena of intermolecular exchange and hydrogen bonding. Exchange which is "fast" on the n.m.r. time-scale (Chapter 2–1 and Chapter 5–1) leads to the averaging of the chemical shifts of, for instance, two types of —OH protons and the degree and type of hydrogen bonding has a very strong influence on the chemical shifts (Chapter 2–2 E).

The chemical shifts of such protons are generally very sensitive to concentration which affects intermolecular hydrogen bonding, i.e. the position of equilibrium between the various polymeric hydrogen bonded species and the monomeric species which is the sole species at infinite dilution.

For the same reasons, the chemical shifts of such protons are particularly sensitive to solvent type and to temperature,† and the degree of sensitivity to all these factors will itself very with the type of group and molecular environment. It follows that reliable information on the chemical shifts of this group of protons is not routinely available.

However, exchange with D_2O and the characteristic *appearance* of signals due to —OH, —NH[2408] and —SH protons (Chapters 4–2 A (ix) and 5–1 B) makes these groups very easily identifiable by n.m.r. Further, as will be shown below, more detailed studies of the dependence of chemical shifts on molecular environment and concentration yield results of interest to organic chemists.

Table 3–7–1 lists the chemical shift ranges for several commonly encountered types of protons bound to elements other than carbon. The results refer to "ordinary experimental conditions", i.e. moderate (5–10 per cent)

† With most instruments variation of sample temperature can be used to displace such resonances as an alternative to deuterium exchange.

concentrations in chloroform or carbon tetrachloride and have been drawn from previous compilations[1768,2286,716,666,321,239] and from data obtained in the authors' laboratories. They refer principally to small molecular species, do not cover cases of strong intramolecular H-bonding unless stated and, as is clear from the factors discussed above, they are very approximate. The values for amide protons, which are less labile than other types listed in Table 3–7–1, are somewhat more characteristic but this is offset by the fact that the presence of the amido group is often associated with very limited solubility in suitable solvents and complications may also arise due to restricted rotation about the amide bond (Chapter 5–1) and extreme broadening of signals.[2408]

TABLE 3–7–1. CHEMICAL SHIFTS OF PROTONS ATTACHED TO ELEMENTS OTHER THAN CARBON IN COMMON FUNCTIONAL GROUPS

	δ range (ppm)
Alcohols[a]	0·5–5·5
Phenols (Intramolecular hydrogen bonding)	10·5–16
Other phenols	4–8
Enols (Intramolecular hydrogen bonding)	15–19
Carboxylic acids	10–13
Oximes[b]	7·4–10·2
Aliphatic thiols (R—SH)	0·9–2·5
Aromatic thiols (Ar—SH)	3–4
Sulphonic acids	11–12
Amines[c] (R—NH_2, R_2NH)	0·4–3·5
Amines[c] (Ar—NH_2, Ar—NH—R, Ar—NH—Ar)	2·9–4·8
Amides (R—CO—NH_2, Ar—CO—NH_2)	5–6·5
Amides (R—CO—NH—R', Ar—CO—NH—R)	6–8·2
Amides (R—CO—NH—Ar, Ar—CO—NH—Ar')	7·8–9·4

[a] With species of small molecular weight, where intermolecular association is not hindered, the hydroxylic protons generally resonate in the region of $\delta = 3$–5·5 ppm unless very high dilutions in an inert solvent, such as carbon tetrachloride, are used. However, with many large molecules, e.g. steroids, the hydroxylic protons often resonate near $\delta = 8$ ppm even at relatively "high" concentrations, partially because the *molar* concentration is low and partially due to steric effects.

[b] For further data see Chapter 3–8.

[c] These resonances exhibit pronounced downfield shifts in trifluoroacetic acid.

The chemical shifts of NH protons in some miscellaneous compounds are listed in Table 3–7–2.

Of the large number of detailed studies† concerned with the effect of molecular environment, solvent and concentration on the chemical shifts of alcoholic protons, only a limited number have obvious applications in organic

† See e.g. references 1570, 36, 1724, 812, 1686, 528, 205, 612, 1891, 1727, 1692 and references quoted in Chapter 5–1.

TABLE 3–7–2. CHEMICAL SHIFTS OF —NH PROTONS IN MISCELLANEOUS COMPOUNDS [a,b,c]

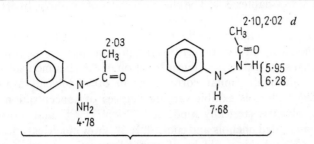

ref. 196

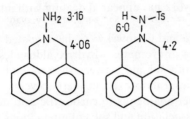

ref. 433

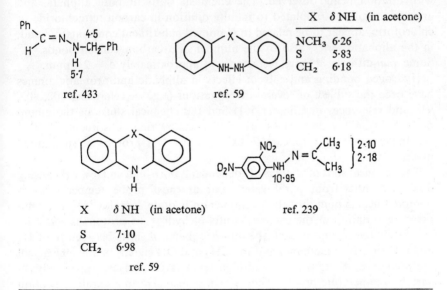

ref. 433   ref. 59

X	δ NH (in acetone)
S	7·10
CH₂	6·98

ref. 59   ref. 239

[a] For pyrrole, indole and other aromatic NH resonances see Table 3–6–4.
[b] For cyclic amides see Table 3–5–4.
[c] For cyclic amines (especially ethyleneimine) see Table 3–5–5.
[d] The doubling of the amide resonances is due to restricted rotation about the NH—CO bond. The signals coalesce at elevated temperatures.

chemistry. Most important are correlations between the steric environment of the hydroxyl groups and their chemical shifts, the discussion of which is deferred to Chapter 3–8. The chemical shifts of alcoholic hydroxyl protons in dimethyl sulphoxide occur over a relatively narrow range (4–6·5 ppm)[464] but accurate correlations with structure have not yet been developed.

The chemical shifts of the hydroxylic protons in *para* substituted phenols in dimethyl sulphoxide are a linear function of the Hammett substituent constant[1921] although similar correlations in other solvents are more complex.[1949,511] Phenols exhibit varying degrees of concentration dependence in inert solvents: sterically hindered[2620,2621,2330,47] and intramolecularly hydrogen-bonded phenols and enols[2005,166] show much less effect than other phenols[1949] and this could be of some diagnostic utility. Good correlations exist between the infrared stretching frequencies and chemical shifts of the hydroxyl protons in a number of phenols and enols.[844,1135,1010]

Much interesting work has appeared dealing with intramolecularly hydrogen-bonded phenols.[844,2077,166,837,2005,47,2080,1930]

In particular, Porte and Gutowsky[2005] have related the chemical shifts of the hydroxyl protons in aromatic *o*-hydroxyaldehydes and methyl ketones with the bond order of the bond between the *ortho* substituents. Some *meta*-substituted phenols appear to exist in inert solvents in the form of stable dimers[106] in a manner analogous to carboxylic acid dimers.

The effect of self-association and solvent interactions on the chemical shifts of —SH protons has been studied,[839,1715,2145,519] the usual upfield shifts with dilution being observed. The chemical shifts of both aliphatic and aromatic thiols, extrapolated to infinite dilution in carbon tetrachloride or chloroform, appear to be related to Hammett substituent constants[1672] and, in the aliphatic series, also to the number of α-carbon–carbon bonds.[1672] Some *gem*-dithiols[1648] give resonances at approximately $\delta = 2\cdot4$ ppm.

Hydrogen bonding and solvent effects in aliphatic and aromatic amines have been the subject of several investigations (e.g. see references 802, 803, 978 and references in Chapter 5–1) and the chemical shifts of the amino protons in aniline derivatives are influenced by ring substituents.[717,2631] A similar correlation appears to exist for the NH protons in substituted uracils.[1424]

The chemical shift of water is pH dependent, both high and low pH causing downfield shifts from pure water, and at most probe temperatures is observed near 5 ppm, but the phenomenon is quite complex.[1457] The temperature variation of the chemical shifts of hydroxylic protons is generally accurately linear, and in fact, the internal chemical shifts between the CH_3 and OH signals in methanol and the CH_2 and OH signals in ethylene glycol are used to calibrate the low and high temperature ranges, respectively, in variable temperature probes. These spacings also give an accurate indication of the normal probe temperature whenever this is required, and avoids the risk of breakage of the probe insert which almost invariably follows attempts to insert a thermometer into the probe.

The intermolecular interactions of hydrazine, which has some potential as an n.m.r. solvent, have been reported.[36]

STEREOCHEMISTRY AND THE CHEMICAL SHIFT

A NUMBER of useful correlations of chemical shifts with stereochemical features are based on intermolecular interactions (Section J) and often important stereochemical assignments can be made very elegantly on the basis of symmetry of some features of the n.m.r. spectrum (Section A). However, the vast majority of information relating chemical shifts and stereochemistry is based on long-range shielding effects (Chapter 2–2).

It is important to reiterate in the present context that the chemical shift of any proton, or group of equivalent protons, depends on the *sum* of all remote influences (thus making the effect of a particular group difficult to estimate) and on the relative position in space of the groups concerned, whether determined by configurational or conformational factors, which are generally not easily separated. Further, it will be recalled (Chapter 2–2) that a group may influence the chemical shift of a proton via a variety of mechanisms, which are, to a greater or lesser extent, dependent on stereochemical factors and which need not necessarily all act in the same direction, much as the inductive and resonance effects due to a particular group often oppose each other.

It is the presence of these severe limitations which prompts us to discuss some of the actually achieved applications of chemical shifts to stereochemistry, rather than to enumerate and illustrate the sort of information which one would logically expect to be forthcoming on the basis of the theory of chemical shifts. While the material in this Chapter will be arranged into sections dealing with various types of structures, another important division between the numerous correlations should become apparent. It will be seen that these are either of a fairly general type, based on relatively well established and prominent long-range effects, for example those due to carbonyl groups and aromatic rings, or upon smaller effects which are typically, rather than exceptionally, the result of a number of minor influences, for instance the differences between the chemical shifts of axial and equatorial groups attached to six-membered rings. It is very important to remember that the latter are generally applicable only to rigidly defined groups of compounds.

In the great majority of applications discussed in this Chapter, the chemical shift differences between stereoisomers are insufficient to permit the assignment of stereochemistry on the basis of the n.m.r. spectrum of a single isomer, although it is often possible to distinguish between a pair of isomers.

It should be clear from the above, that the most reliable stereochemical deductions can be made in systems where the relative positions of the groups causing and experiencing long-range shielding effects are accurately known, *and* when the shielding effect is large and well understood. Even where the "framework" is rigid, as in ethylene derivatives, uncertainties about the average conformation are very likely to occur with all groups other than the methyl group, the t-butyl group, trihalomethyl groups, the halogens, the cyano group, the ethynyl group, the proton and trimethylsilyl, thus rendering prediction of long-range effects difficult. It is unfortunate that this is especially true in the case of highly substituted compounds, where chemical shifts, rather than coupling constants, are the principal n.m.r. data available. It might thus be said that conformational uncertainty is the curse of chemical shift data. Conversely, it follows that when a long-range shielding effect is well understood, chemical shifts can provide information about average conformation.

A. SYMMETRY ARGUMENTS

In the present context a symmetry argument implies simply the presence or otherwise of chemically or magnetically equivalent protons or groups of protons (cf. Chapter 2–3 for definitions of equivalence). It follows that applications of symmetry arguments are not confined to stereochemical problems. Further, it should be apparent that symmetry arguments are extremely simple and powerful and do not require a knowledge of n.m.r. parameters. However, a good understanding of the effects of chemical (as against magnetic) equivalence on the splitting patterns, the recognition of spin multiplets due to symmetric groups and the realization that fast exchange processes may result in a spuriously symmetrical spectrum, are essential.

The following examples are chosen to illustrate the variety of such applications and are placed in this Chapter because the majority of them have been concerned with stereochemical problems.

(i) STRUCTURAL APPLICATIONS

In many cases, otherwise difficult problems can be unequivocally solved simply by observation of asymmetry in the n.m.r. spectrum. Thus the thiosulphonate structure (I) was proved correct and the disulphoxide structure (II) incorrect, by observation of two sets of signals from otherwise identical R groups;[52,2141] the 7-norbornadienyl carbonium ion[2362] cannot have the otherwise attractive symmetrical† structure (III) because the n.m.r. spectrum showed two groups of olefinic protons; the polymer obtained by oxidation of 2,6-xylenol must be composed solely of the regular unit (IV) because its n.m.r. spectrum shows only two signals, one for the benzylic methyl groups and one for the aromatic protons.[416]

Similarly, a tetra-t-butylbenzene must have the symmetrical structure (V),[126] because there are only two sharp lines in its spectrum and supporting evidence

† Other examples of the application of symmetry considerations to the determination of the structure of carbonium ions are given in Chapter 3–9.

for the structure (VI) consists of a slight non-equivalence[665] between the carbomethoxy methyl signals.

When searching for non-equivalence which is likely to be small, as with (V) and (VI), it is essential to obtain the spectrum in another solvent if a coincidence of signals is observed, in order to reduce the possibility of accidental equivalence (Chapter 2–3).

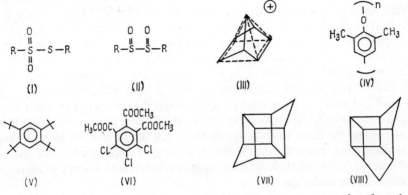

Perhaps the most commonly encountered symmetry argument involves the recognition of the symmetrical AA'BB' splitting patterns characteristic of all *para*-disubstituted benzenes, 4-substituted pyridines, symmetrical *ortho*-disubstituted benzenes etc. In particular, the patterns associated with *para*-disubstituted benzenes, and equivalent systems, are very easy to recognize as they resemble simple AB quartets due to the relatively large magnitudes of *ortho* coupling constants (Chapters 2–3 and 4–3).

Symmetry arguments have often been applied in the case of cage structures. Thus, evidence for structure (VII) rather than (VIII) was obtained[1605] from the observation of three signals of equal intensity in the n.m.r. spectrum.

We have alluded above to the possibility of errors arising from the observation of an averaged "pseudo-symmetrical" spectrum resulting from rapid interchange of the environments of protons, and examples of situations where this occurs are mentioned in Chapter 5–3. Often, a rapid interchange must be invoked, to explain a symmetrical spectrum: e.g. the formazan (IX) must undergo the rapid tautomerism as shown[2413] to give equivalent sets of signals for two of the aromatic rings, and (X) must undergo rapid conformational inversion to give the observed symmetrical AA'XX' pattern for the ring protons and a single signal for the t-butyl groups.[2026]

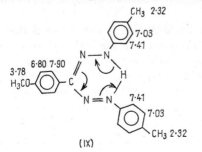

(ii) STEREOCHEMICAL APPLICATIONS

The type of spectra expected from acyclic systems of varying degrees of symmetry under conditions of slow and rapid conformational interchange are discussed in Chapter 5–2. It is of considerable interest that n.m.r. allows us to distinguish clearly between *meso* and *dl* forms, in e.g. 2,4-disubstituted pentanes and 2,5-disubstituted hexanes,[1315,1642,1641,1786] because only the former give rise to spectra attributable to symmetrical spin systems.

Numerous examples of applications of symmetry considerations to the assignment of stereochemistry in cyclic compounds have been reported. For example, the mere observation of *two* separate signals for the methyl groups in (XI)[498] (absolute stereochemistry not implied) is sufficient to assign the *trans*-1,2-dimethyl stereochemistry shown, and the isomers (XII) and (XIII) can be firmly assigned because the former gives rise to a symmetrical AA'BB' pattern and the latter to an unsymmetrical AA'BC pattern.[681] In each case, the underlying reason for the chemical equivalence of two protons, or groups of protons, is due to their identical spatial relation with respect to *all* other groups in the molecule.

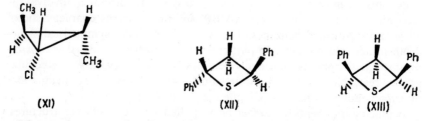

(XI)   (XII)   (XIII)

Many other examples of arguments of this type applied to four-membered rings,[2560,109,110,2200] five-membered rings, [573] fused ring systems,[859,2161,2300] and metal chelates[796,797] can be found in the literature, and several examples appear in Table 3–8–7 (p. 235).

B. GEOMETRICAL ISOMERISM IN ETHYLENE DERIVATIVES

Structures containing double bonds (cf. also section C) are clearly likely to have protons or groups in a definite juxtaposition and we have already referred to this in discussing the chemical shifts of allylic and olefinic protons (Chapters 3–2 and 3–3).

From the results listed in Table 3–3–1, it can be seen that in simple ethylene derivatives olefinic β-protons *cis* to the carbonyl group in ketones, carboxylic acids and carboxylic acid derivatives resonate 0·3−0·9 ppm downfield from the corresponding *trans* β-protons. The influence of the aldehydic groups is not so clear cut.[1259,1941,689] Some examples of actual chemical shifts are given in Table 3–2–5 and the relationship has been firmly established in a number of comprehensive investigations[2563,1260,1259,1435,1136,2601,1476,944] (cf. also references listed in Table 3–3–3). Methyl groups attached to the β-carbon atom *cis* to the carbonyl function also resonate downfield from their *trans* counterparts (cf. Tables 3–2–5 and 3–3–5) and examples of this type for some cyclic systems are given in Table 3–8–1.

TABLE 3–8–1. CHEMICAL SHIFTS OF OLEFINIC PROTONS AND ALLYLIC
METHYL GROUPS IN SOME ISOMERIC CYCLIC SYSTEMS

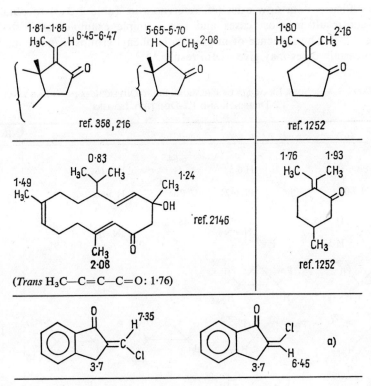

ref. 358, 216

ref. 1252

ref. 2146

(*Trans* H_3C—C=C—C=O: 1·76)

ref. 1252

a)

a G. P. Newsoroff and S. Sternhell, unpublished results.

Although few systematic investigations on the relation between conformational effects and the relative shielding of β-*cis* and β-*trans* protons and methyl groups have so far appeared,[1413,1435,179,1324] it is clear that the average conformation of the carbonyl group is of importance. The authors are not, however, aware of any reversal of the general relations stated above, with the exception of some aldehydes.[1941,689]

The case of the relative shielding of *cis* and *trans* olefinic protons and alkyl groups by aromatic rings is complex, due to the variety of average conformations assumed by the strongly anisotropic aromatic rings with respect to the double bond in different molecular environments. The average conformation is affected by the effective size of the aromatic ring (i.e. the number of rings and substituent type, particularly *ortho* substitution[2144,1870]) and the bulk of the α- and β-*cis* substituents, as well as by the tendency of conjugated systems to assume coplanarity.

The effect of the average conformation of α-, β-*cis*- and β-*trans*-phenyl groups on the chemical shifts of olefinic protons and methyl groups can be estimated by reference to the contour map, Fig. 2–2–14 (p. 95). Clearly, for any pair of isomers of styrene or stilbene derivatives, the relative shielding effects at any particular position will be quite sensitive to the nature of substituents and thus the problem becomes essentially conformational. However, sufficient data now exist (cf. references in Table 3–3–3) to obtain a clear-cut result in most cases and some simple examples are listed in Table 3–8–2. The influence of more remote phenyl groups is much smaller, but in certain cases may give useful results.[1693]

TABLE 3–8–2. SOME EXAMPLES OF CHEMICAL SHIFT DIFFERENCES IN *cis* AND *trans* 1,2-DIPHENYL AND 1,2-DIMETHYL ISOMERS

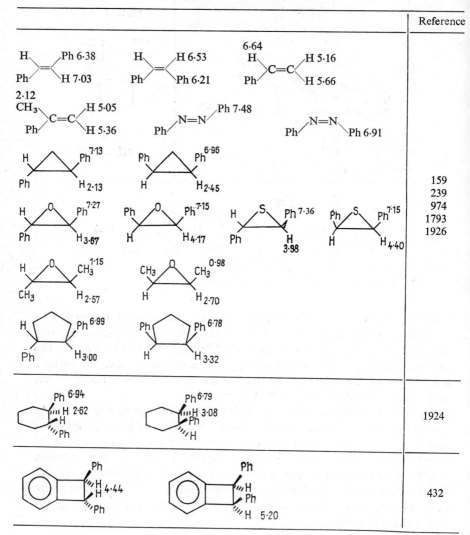

	Reference
	159 239 974 1793 1926
	1924
	432

It can be seen from Table 3–3–1 that most groups besides carbonyl deriva-tives and aromatic rings cause a differential shielding between *cis-β* and *trans-β* protons and this is also generally found for *cis-β* and *trans-β* alkyl groups.[1259,1260] As the actual differences between the chemical shifts are generally below 0·2 ppm, it is obvious that minor variations due to con-formational effects or remote influences could lead to wrong assignments and an example of a situation of this type has been quoted in Chapter 3–3 (p. 189). However, when a series of closely related compounds is carefully examined, very useful information may be obtained even when the actual magnitude of the effect is small.

Thus, the nitrile group has been found[995,1476] to deshield both *β-cis* pro-tons and *β-cis* methyl groups relative to their *trans* counterparts. Similarly, many useful results[476,2093,876,190,119,1222] have been obtained from the differences[188,189] between the chemical shifts of methyl groups in (XIV) and (XV), although the latter generally resonate less than 0·1 ppm downfield from the former.

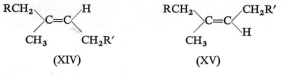

(XIV)   (XV)

In butadiene derivatives, and polyenes generally, (cf. references in Table 3–3–3), often quite pronounced effects can be discerned. In particular, "in-chain" methyl groups in carotenoids and similar compounds resonate upfield from the "end-of-chain" methyl groups, and *cis* and *trans* configurations about the double bonds may be assigned.[610,165,248,163,164] Similarly, in the side chain (XVI), the configuration may be assigned[429] from the chemical shift of protons in the 1,3-relation to alkyl or hydrogen, and even more pro-nounced shifts occur when carbonyl groups are present (cf. Tables 3–3–2 and 3–2–5 and references in Table 3–3–3).

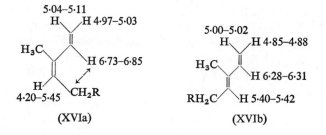

(XVIa)   (XVIb)

It will be recognized that the 1,3 relation possible in butadienes, and poly-enes in general, (cf. arrow in structure XVIa) is analogous to the *peri*-effect considered in connection with the chemical shifts in aromatic and hetero-cyclic compounds (Chapter 3–6). However, the influence of the double bond itself and of all substituents, both direct and *via* changes induced in average conformations, would have to be considered separately in each case.

C. GEOMETRICAL ISOMERISM IN SYSTEMS
CONTAINING C=N AND N=N BONDS

The systems considered in this section, which of course include cases of *syn–anti* isomerism, are in general simpler from the n.m.r. point of view than ethylene derivatives because of the smaller number and range of substituents, and n.m.r. spectroscopy has been widely used for the assignment of stereochemistry in this field.

In simple derivatives of aliphatic aldehydes,[1341] such as hydrazones, alkyl and aryl hydrazones, oximes,[239,1586] oxime sulphonates etc., the formyl proton *syn* to the group Z in the generalized structure (XVII) always resonates downfield from the corresponding *anti* proton, the separation being 0·25–1·0 ppm. However H_α and H_β protons *syn* to Z in (XVII) and in aliphatic ketone derivatives (XVIII) may be either deshielded or shielded with respect to the corresponding *anti* protons[1341] due to conformational factors[1342] which may place them in different relative positions with respect to Z. A general method for assigning resonances due to H_α and H_β, based on solvent shifts[1341] and coupling constants[1342] (in aldehyde derivatives), is available.

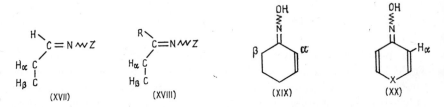

The assignment of resonances in oximes of cyclohexanone and its derivatives is uncertain,[2156,2157,2158,2440] but in the unsaturated systems (XIX) and (XX) (X = carbon or a heteroatom) the protons, H_α, *syn* to the oxime hydroxyl always appear downfield from their *anti* counterparts by approximately 0·5–1·0 ppm[220,1881,714,2627,1732,2301] and a smaller shift in the same direction has been observed for H_β.[2301]

Several collections of data for oximes of aromatic aldehydes are available;[1175,1963,2012,559] the formyl protons *syn* to the oxime hydroxyl always appear significantly downfield from their *anti* counterparts, and the chemical shifts of both show dependence on Hammett parameters.[1963]

An extensive and detailed study of n.m.r. spectra of carbonyl derivatives and related compounds has been carried out by Karabatsos and his co-workers[1336,2251,1341,1342,1340,1347,1343] and data on semicarbazones,[584, 1336,1347] aryl hydrazones,[584,1336,2251,1884,1347,1340] carbethoxyhydrazones of α-diketones,[2138] alkyl hydrazones[1344] and thiosemicarbazones[1347] may be found in the literature.

Few data for ketimines[1488] and Schiff's bases[2348] have been reported. It appears that in the latter[2348] *syn* alkyl groups resonate upfield from their *anti* counterparts.

Although the *syn* and *anti* isomers of *N*-nitroso amines cannot be obtained as separate compounds, the rotational barrier about the N—N bond is generally sufficiently high (cf. Chapter 5–1) for observation of separate n.m.r. spectra.[1343,1112,1463,2207,2369,355] It is interesting that the —NO group *shields* protons and alkyl groups which are co-planar and *syn* to it, with respect to their *anti* counterparts, i.e. it has the opposite effect to the N—OH group in oximes. The same relation can be found in azoxy compounds,[897] anions derived from quinone monoximes[1881] and possibly with *p*-nitrosodimethylaniline.[1643]

Other systems for which data are available include alkyl nitrites,[355] amidoximes,[2456] nitronic esters,[1429] methazonic acid,[371] azo and azoxy compounds[897,2509,1431] and azines.[124] Some references containing chemical shift data for systems with partial double-bond character (and hence a type of *syn–anti* isomerism) are given in Chapter 5–1.

D. CYCLOPROPANES AND HETEROCYCLOPROPANES

A large number of chemical shift data for cyclopropanes,† epoxides,[1534, 2592,2428,974,1973,291] and other heterocyclopropanes[2428,1985,1571,650,347] are available in the literature (cf. also references listed in Chapters 4–1 and 4–2 in connection with geminal and vicinal coupling constants in three-membered rings).

As substituents in three-membered rings should occupy well-defined relative positions in space, one would expect regularities in the variation of chemical shifts with stereochemistry of substituents, and this has in fact been noted by several authors. However, as symmetry considerations and the very characteristic values of coupling constants in three-membered rings (Part 4) are usually sufficient to establish the configuration of the substituents, chemical shift data have rarely[1958,2488,2241] been used in the solution of structural problems involving stereochemistry in three-membered rings.

Perhaps the most firmly established result is that an aromatic ring attached to a three-membered ring will shield the proton (or proton-containing group) *cis* to it with respect to its *trans* counterpart[503,1462,506,1924,585,1926] and this important relation also applies to four-, five- and six-membered rings[1926] (cf. below and Table 3–8–2), indicating that, in all cases, the aryl ring tends to assume an average conformation which places the eclipsed proton (or substituent) in its shielding zone. This may be contrasted (cf. Section B above), with the situation in styrene and stilbene derivatives where the analogous (*cis*) proton (or group) may be either within the shielding or deshielding zone.

When a carbomethoxy group is attached to a cyclopropane ring, the introduction of a *cis* phenyl ring shifts the methoxy resonance upfield by 0·1–0·3 ppm.[1462] The average conformation of the phenyl ring with respect to the cyclopropane ring also affects the chemical shift of the aromatic protons.[503] Further examples of shielding by a *cis* phenyl group are given in Tables 3–8–2 and 3–8–3.

† References 503, 506, 585, 1009, 1239, 1243, 1462, 1924, 1946, 1958, 1974, 2241, 2243, 2267, 2488, 2519, 2535, 2568 and 2592.

Applications of the Chemical Shift

TABLE 3–8–3. CHEMICAL SHIFT DATA FOR SOME SUBSTITUTED CYCLOPROPANES AND HETEROCYCLOPROPANES

The chemical shifts of the parent compounds are as follows: cyclopropane: 0·22, ethylene oxide: 2·54, ethylene sulphide: 2·27, ethyleneimine: 1·62 ppm (cf. Chapter 3–4 for source of these data)

Compounds					Reference

X	H_A	H_B	H_C		Reference
— COOH	1·06	0·97	1·58		2567
— Br	0·88	1·00	2·84		2519
0					1243
$-C(=O)$—⟨O⟩—OCH_3	1·13	0·91	2·55		239
— NH_2	~0·3	~0·4	2·30		
— Ph	0·68	0·74	1·80		

X	H_A	H_B	H_C		Reference
—OAc	1·58	1·84	4·28		
—OCH_3	1·52	1·67	3·62		2592
—Br	1·58	2·08	3·45		
—Ph	1·73	1·85	2·80		
—COOH	2·02	1·87	2·52		
—$Si(CH_3)_3$	1·11	1·43	0·58		

Structures		Reference
CH_3 1·03 / H 1·77 / 0·68 H / CH_3 1·17 / $COCH_3$ 2·12 / $H_{1·17}$	CH_3 / H 2·6 / 0·38 H / CH_3 / Cl / H 0·51	1241 1974

Structures			Reference
1·10 H / Ph / 1·47 H H / $COOCH_3$ / H	1·33 H / Ph / 1·66 H H / CN / H	0·60 H / CH_3 / 1·16 H H / $COOCH_3$ / H	1958

X	H_A	H_B	H_C			Reference
—$COCH_3$	2·84	2·96	3·28	($COCH_3$ = 1·96)		1973
—CH_3	2·23	2·60	2·80	(CH_3 = 1·32)		291
—OAc	2·58	2·76	5·34			2592
—Cl	2·75	2·83	4·90			
—Ph	2·52	2·82	3·62			
—CHO	3·10	3·17	3·35			
—COOH	2·93	2·99	3·48			
—CN	3·00	3·12	3·50			

TABLE 3–8–3 *(cont.)*

Compounds	Reference
1·35 CH₃ H 3·13 — COCH₃ 1·94 H 2·97 O / 1·34 CH₃ H 3·15 1·20 CH₃ COCH₃ 2·06 O	1973
Ph H 1·48 H 2·66 N H 1·86 N–H 1·14 Ph CH₃ 0·66 H 2·63 N CH₃ 1·02 N–H Ph CH₃ 0·53 H N H N–H 2·14 J=5·4 Hz Ph H H N CH₃ 0·83 N–H 2·14 J=5·4 Hz	347
2·02 H H 2·83 H 2·39 S CH₃ 2·46 H H 3·54 H 2·30 S Ph	1571 1666

Some chemical shifts for ring protons and methyl groups in cyclopropanes and heterocyclopropanes are collected in Table 3–8–3. It can be easily seen that correlations of obvious utility could be drawn from these data and the many further data contained in references cited above. In particular, it appears that *generally* all substituents tend to cause protons *cis* to them to appear at higher fields than those *trans* to them.[2592] However, caution is strongly indicated in utilizing this correlation, because several exceptions can be seen among the data listed in Table 3–8–3. While those involving the acetyl[1974] and carboxyl[2592] groups can be easily rationalized by arguments involving changes in the preferred conformations (which have been explored in some detail by several authors[1974,2535]), the effect of chlorine in 1-chloro-2,2-dimethylcyclopropane[1241] (Table 3–8–3), when compared with other data (e.g. the effect of chlorine in monochloroethylene oxide,[2592] Table 3–8–3), appears to be anomalous. Unfortunately, comparatively few data for monosubstituted cyclopropanes appear in the literature, presumably due to the tediousness of analysing their AA′BB′X spectra.

A valuable collection of data on steroidal epoxides and episulphides is available.[2428]

E. MULTI-RING STRUCTURES

As with structures containing double bonds or three-membered rings, the presence of fused rings often implies relative rigidity of the molecular framework and hence some likelihood of obtaining useful stereochemical information from chemical shifts. In this section we shall confine our attention to bicyclo[2.2.1]heptanes, bicyclo[2.2.2]octanes, their unsaturated analogs and to other, less common systems. The most commonly occurring of the rigid multi-ring systems, viz. the steroid nucleus and related compounds will be considered separately in section (H).

The extraction of stereochemical assignments from chemical shift data in systems such as bicyclo[2.2.1]heptanes is more difficult than with, for example, ethylene derivatives, because their cage-like structure brings more groups within distances at which long-range shielding effects are expected to be significant.

The n.m.r. spectra of substituted norbornanes and related systems have been thoroughly investigated and a list of key references is given on pp. 288 and 335. Some data (Table 2–2–3) have already been considered in connec-

TABLE 3–8–4. CHEMICAL SHIFT DATA FOR SOME NORBORNANES, NORBORNENES AND NORBORNADIENES

R	H_a	H_d	H_e	H_f	
OAc	2·84	3·12	5·17	2·10	
NO$_2$	3·00	3·53	4·94	2·14	
Cl	2·90	3·10	4·47	2·27	
Br	2·82	3·10	4·24	2·27	ref. 613,
SO$_2$—CH=CH$_2$	2·99	3·22	3·52	2·07	1508, 2614
Ph	2·90	2·95	3·32	2·14	
COCH$_3$	2·82	3·12	2·87	—	
CHO	2·91	3·20	2·92	1·87	
COOH	—	3·19	2·99	2·06	
COOEt	2·86	3·16	2·86	1·84	
CN	2·99	3·20	2·79	2·12	

R	H_A	H_B	H_C	
CN	2·17	2·70	3·40	
COOH	2·44	2·72	3·62	
Ph	2·38	2·86	2·87	ref. 2588
Cl	2·22	3·08	4·72	
OH	1·90	2·78	4·64	
OAc	1·90	2·95	5·50	

tion with the diamagnetic anisotropy of the carbon–carbon double bond. An inspection of data shown in Table 3–8–4 reveals regularities which could be used to obtain stereochemical results, and in fact, a number of such studies, principally concerned with *endo-exo* isomerism in norbornanes and norbornenes and with *syn-anti* isomerism at C–7 in norbornenes, have appeared in the literature. It will also be shown (cf. Chapters 4–2 and 4–4) that both vicinal and long-range coupling constants are very useful in assigning the stereochemistry in these systems.

Fraser has shown[888] that the hydrogenation of the 2,3-double bond in norbornenes produces fairly small (0·1–0·2 ppm) shifts in *exo* and *endo* protons and methyl groups at C–5 and C–6, whose magnitude vary with their configuration due to their different relative positions with respect to the double bond in the unsaturated series. This correlation must, however, be used with caution because (cf. Table 3–8–5) both *endo* and *exo* protons at C–5 may undergo substantial upfield shifts with some substituents[2614] on hydrogenation of the double bond.

TABLE 3–8–5. EFFECT OF HYDROGENATION ON THE CHEMICAL SHIFTS OF *exo* AND *endo* PROTONS IN NORBORNENES (*cf.* ref. 2614)

Structure \ X	(structure 1)	(structure 2)	(structure 3)	(structure 4)
—OH	4·46	4·17	3·82	3·66
—OAc	5·19	4·87	4·57	4·54
—Brosylate	5·14	4·83	4·49	4·45

The chemical shifts of protons and functional groups at C–7 in norbornenes[2323,1199,2366,2422,613] are not always simply related to their stereochemistry with respect to the double bond, but appear to be a complex function of the overall substitution pattern.[2366,2422] Nevertheless, useful correlations can be obtained with the numerous model compounds which have been investigated (cf. Table 3–8–4 and refs. above).

Oullette, Booth and Liptak[1922] have shown that, for a number of 2-hydroxynorbornanes and methylsubstituted 2-hydroxynorbornanes, the limiting slope of the chemical shift of the hydroxyl proton vs. concentration in carbon tetrachloride can give reliable information about the configuration (*endo* or *exo*) of the hydroxyl groups.

In 2-substituted norbornanes and many of their derivatives (cf. e.g. data in Tables 3–8–4 and 3–8–5) there is a general tendency for the *endo* protons to resonate upfield from the corresponding *exo* protons and this might be thought of as an analogy[831] to the well known generalization about the relative chemical shifts of axial and equatorial protons in six-membered rings

TABLE 3–8–6. CHEMICAL SHIFT DATA FOR MISCELLANEOUS MULTI-RING STRUCTURES

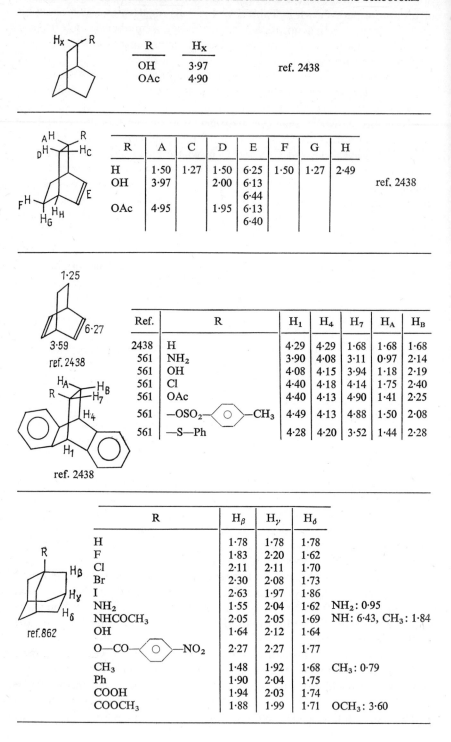

R	H_X
OH	3·97
OAc	4·90

ref. 2438

R	A	C	D	E	F	G	H
H	1·50	1·27	1·50	6·25	1·50	1·27	2·49
OH	3·97		2·00	6·13			
				6·44			
OAc	4·95		1·95	6·13			
				6·40			

ref. 2438

1·25

6·27

3·59

ref. 2438

Ref.	R	H_1	H_4	H_7	H_A	H_B
2438	H	4·29	4·29	1·68	1·68	1·68
561	NH_2	3·90	4·08	3·11	0·97	2·14
561	OH	4·08	4·15	3·94	1·18	2·19
561	Cl	4·40	4·18	4·14	1·75	2·40
561	OAc	4·40	4·13	4·90	1·41	2·25
561	$-OSO_2-\langle O \rangle-CH_3$	4·49	4·13	4·88	1·50	2·08
561	$-S-Ph$	4·28	4·20	3·52	1·44	2·28

ref. 2438

R	H_β	H_γ	H_δ	
H	1·78	1·78	1·78	
F	1·83	2·20	1·62	
Cl	2·11	2·11	1·70	
Br	2·30	2·08	1·73	
I	2·63	1·97	1·86	
NH_2	1·55	2·04	1·62	NH_2: 0·95
$NHCOCH_3$	2·05	2·05	1·69	NH: 6·43, CH_3: 1·84
OH	1·64	2·12	1·64	
$O-CO-\langle O \rangle-NO_2$	2·27	2·27	1·77	
CH_3	1·48	1·92	1·68	CH_3: 0·79
Ph	1·90	2·04	1·75	
COOH	1·94	2·03	1·74	
$COOCH_3$	1·88	1·99	1·71	OCH_3: 3·60

ref. 862

TABLE 3–8–6 *(cont.)*

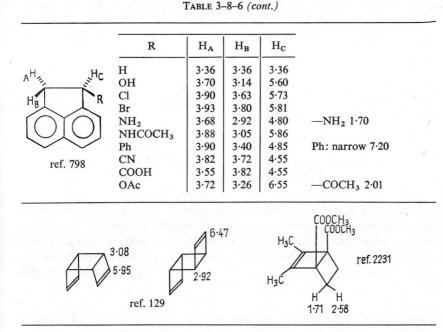

R	H_A	H_B	H_C	
H	3·36	3·36	3·36	
OH	3·70	3·14	5·60	
Cl	3·90	3·63	5·73	
Br	3·93	3·80	5·81	
NH_2	3·68	2·92	4·80	—NH_2 1·70
$NHCOCH_3$	3·88	3·05	5·86	
Ph	3·90	3·40	4·85	Ph: narrow 7·20
CN	3·82	3·72	4·55	
COOH	3·55	3·82	4·55	
OAc	3·72	3·26	6·55	—$COCH_3$ 2·01

ref. 798

ref. 129

ref. 2231

(Section G). However, the shielding in the bicyclic systems may be due to more complex factors[1828] and in any case, with more highly substituted compounds the influence of other substituents would have to be carefully considered.

Some of the data in Table 3–8–4 show that most substituents at C-2 cause the protons eclipsed by them to appear at higher fields[2366] than those where the dihedral angle is approximately 120°, in a manner analogous to the pattern in three-membered rings and other rigid bicyclic systems (cf. Table 3–8–6), and it thus appears that, at least with substituents where conformational factors are not important (e.g. halogen, cyano, etc.), this correlation could be useful.

In the case of *para*-substituted phenylhexachlorbicyclo[2.2.1]heptenes, the chemical shifts of both *α*- and *β*-protons show dependence upon the Hammett *σ*-function, similar to that in ethyl benzenes,[2590] but the shifts are the same for the *endo-β*- and *exo-β*-protons, i.e. the small long-range effect is apparently transmitted entirely through the bonds.

Some data for derivatives of camphor and norcamphor are given in the literature.[1737,538,831,526]

The bicyclo[2.2.2]octane system and its benzo-derivatives have received some attention.[2438,892,1386,561] As with norbornene derivatives, the *endo* protons at C-5 in bicyclo[2.2.2]oct-2-enes[892] appear to give resonances upfield from their *exo* counterparts, and it is possible to assign the configuration of some vicinal carbomethoxy groups from their relative chemical shifts.[1386] A systematic study in this series has been carried out by Tori and his collaborators,[2438] who have also used structures based on the bicyclo[2.2.2]octane skeleton[2425] to determine the long-range shielding effects due to various structural features. Some examples derived from the above references are given in Table 3–8–6, which also contains data for other fused-ring systems. The chemical shifts in the latter have not, in fact, been used to any extent for the purpose of assigning their stereochemistry and are included for a qualitative indication of the influence of β- and more remote substituents on chemical shifts. Further examples can be found, *inter alia* in references 859, 235, 447, 2214 and 2065.

F. *CIS-TRANS* STEREOCHEMISTRY IN FOUR-, FIVE- AND SIX-MEMBERED RINGS

Unlike the systems discussed in Sections B–E, and others with well-defined molecular frameworks (e.g. steroids, Section H), the relative position in space of any two protons, or groups, attached to the above ring systems is subject to additional uncertainty due to conformational flexibility of the rings. This is most serious in the case of saturated four- and five-membered rings, and when added to the effect due to the uncertain average conformation of many functional groups, makes stereochemical assignments, for example, *cis-trans* relationships, in four- and five-membered rings very hazardous. Nevertheless, as shown by the examples listed in Tables 3–8–2 and 3–8–7, regularities of obvious usefulness exist.

These will be most reliable in two cases. In the first case the effect of a particular group is very pronounced. For instance the phenyl ring causes resonances due to the *cis*-vicinal protons to appear upfield of those due to the *trans*-vicinal protons[2229,974,2200,432,1926] (cf. Table 3–8–2). The second case is that in which enough trigonal atoms are incorporated into the ring to make it effectively planar, thus leading to relationships similar to those found in three-membered rings and multi-ring structures.

This situation is closely parallel to the relationship between *cis*- and *trans*-vicinal coupling constants (cf. Chapter 4–2) in four- and five-membered rings, whose relative values cannot be considered a reliable indication of the configuration of substituents, unless the conformation of the molecule is known.

The case of six-membered rings is somewhat different. If we confine our attention to rings which exist in either the perfect chair conformation or which do not deviate from it sufficiently to affect significantly the relative positions of protons or groups in space, their mutual interactions should produce characteristic effects on their chemical shifts and thus lead to generalizations of considerable importance in making stereochemical assignments, at least with substituents whose effect cannot be altered by rotation about the bonds joining them to the rings.

In practice, however, the situation is quite complex. Firstly, the chemical shifts of protons and functional groups attached to six-membered rings in the chair conformation vary considerably with their own orientation (axial or equatorial) with respect to the ring (Section G), and there are a number of

TABLE 3–8–7. CHEMICAL SHIFT DATA FOR SOME SUBSTITUTED CYCLOBUTANE
AND CYCLOPENTANE DERIVATIVES

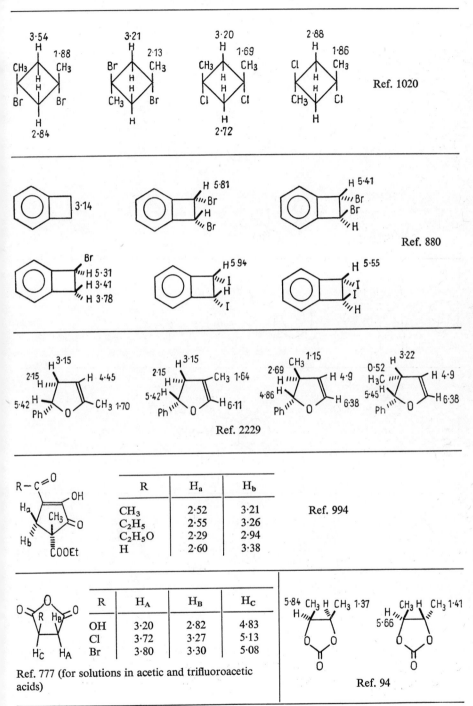

Ref. 1020

Ref. 880

Ref. 2229

R	H_a	H_b
CH_3	2·52	3·21
C_2H_5	2·55	3·26
C_2H_5O	2·29	2·94
H	2·60	3·38

Ref. 994

R	H_A	H_B	H_C
OH	3·20	2·82	4·83
Cl	3·72	3·27	5·13
Br	3·80	3·30	5·08

Ref. 777 (for solutions in acetic and trifluoroacetic acids)

Ref. 94

TABLE 3–8–7 *(cont.)*

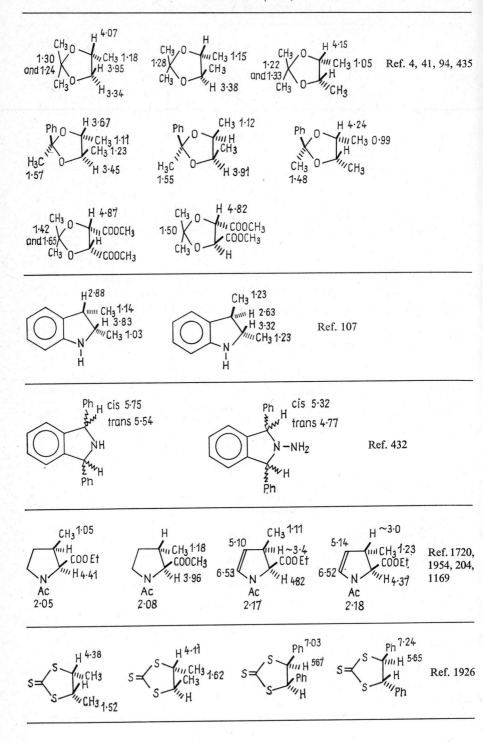

Ref. 4, 41, 94, 435

Ref. 107

Ref. 432

Ref. 1720,
1954, 204,
1169

Ref. 1926

possible relative orientations between substituents. Further, lightly substituted cyclohexane derivatives often consist of mixtures of rapidly interconverting chair conformers (Chapter 5–1) and their n.m.r. spectra are difficult to analyse because of the large number of overlapping resonances.

For these reasons, most of the systematic work in this area has in fact been carried out on steroids (cf. Section H), although many scattered data for simpler systems can be found in the literature (for references see pp. 290–365).

Bhacca and Williams[241] have summarized the effect of a number of substituents on ring protons. It appears that:

(i) The introduction of a vicinal axial substituent, such as OH, OAc, SH, generally causes the resonance of an equatorial proton to move upfield by 0·1–0·3 ppm and the resonance of an axial proton to move downfield by approximately 0·3 ppm.

(ii) The introduction of a vicinal equatorial substituent of the same type causes the resonance of either an axial or an equatorial proton to move upfield by up to 0·3 ppm, which is not surprising in view of the symmetrical disposition of an equatorial substituent with respect to the two vicinal protons (see also the data for 1-substituted adamantanes in Table 3–8–6).

(iii) A large variety of substituents cause pronounced (generally 0·3–0·5 ppm) downfield shifts when in 1,3-diaxial configurations with respect to protons.[241,1548,434,1543,748]

(iv) Other interactions, i.e. 1,3-diequatorial, 1,3-axial–equatorial and 1,4-interactions appear to be associated with smaller effects.

Unfortunately, exceptions exist to all the above relationships, in particular, not surprisingly in view of conformational uncertainty, the effect of the —OAc group appears to vary widely with environment.

The effect of alkyl substituents (in particular the methyl group[748]) on the chemical shifts of ring protons has been summarized by Booth.[292] The introduction of an equatorial methyl group causes significant (0·3–0·5 ppm) upfield shifts in the resonances of both axial and equatorial protons on the adjacent carbon atom, while the introduction of an axial methyl group causes a shielding of the adjacent equatorial proton (*ca.* 0·4 ppm) and a deshielding of the adjacent axial proton (*ca.* 0·2 ppm). 1,3-Diequatorial and 1,3-axial–equatorial methyl-hydrogen interactions result in very small (less than 0·1 ppm) shifts in the resonances of the ring protons, but 1,3-diaxial interactions cause significant downfield shifts (0·2–0·3 ppm). It must be remembered that other alkyl groups may not cause effects paralleling exactly those of the methyl group, and, in particular, the effects due to the t-butyl group may be quite marked and in the reverse direction.[292]

Collections of data for *m*-dioxans and 1,3,5-trioxans are available in the literature.[1671,181,2036,1317,35,622]

G. THE CHEMICAL SHIFTS OF AXIAL AND EQUATORIAL PROTONS AND GROUPS ATTACHED TO SIX-MEMBERED RINGS IN CHAIR CONFORMATION

In the absence of complicating factors, equatorial protons in the cyclo-hexane ring give rise to resonances downfield from their axial counterparts. Some examples of this are listed in Table 3–8–8 and others are given in references listed on pp. 290–365, while a very extensive list of references can be found in a review by Franklin and Feltkamp.[882]

The chemical shifts listed in Table 3–8–8 were obtained either from spectra taken at low temperatures, under which conditions conformational inversion is sufficiently slow to permit observation of the superimposed spectra due to the separate conformers, or by the method of "reasonable fixed models", in this case the 4-t-butyl derivatives. It can be seen that, whenever direct comparison is possible, the values obtained by means of the two methods are in reasonable agreement,[744,882] especially as the details of experimental conditions (e.g. concentration) were in general not identical. However, with some substituents the possibility of serious distortion cannot be excluded.[114]

The principal utility of data of this type has been in investigating conformational equilibria in monosubstituted cyclohexane derivatives and related systems[744,882,70] where the average chemical shift observed under the conditions of rapid conformational equilibrium can be simply related to the proportions of the individual conformers present (cf. Chapters 2–1 and 5–1). An analogous method employing averaged vicinal coupling constants is also often utilized (Chapter 4–2).

For routine stereochemical assignments, the correlation embodied in Table 3–8–8 should be used with caution, firstly because even the relatively minor influences of many common substituents on β- or γ-carbon atoms are often of the same order of magnitude as the differences between axial and equatorial shifts (cf. preceding section), and secondly because some substituents can exert *pronounced* long-range effects which are capable of *inverting* the usual order.

The most important example of such an inversion is the general observation of the resonances due to equatorial α-protons in α-halocyclohexanone derivatives at *higher* fields than those due to their axial counterparts[952,2521,1873,954] [cf. e.g. (XXI)[2521] and (XXII)[952]] although in some compounds[15] the differences between them are negligible.

The same inversion may be observed in the spiro-compound (XXIII),[2453] the pair of decalones (XXIV) and (XXV),[1221] in some steroidal ketones[238] and in other compounds.[1547,1557,2280,2375] The chemical shifts of both protons and acetyl groups in many flavan derivatives[494] are subject to so many relatively strong influences that little stereochemical information can be gained, although a large number of related compounds have been examined.[494,1710]

It is interesting that the axial and equatorial homoallylic protons in cyclo-hexene[104,947] which give rise to a resonance at 1·65 ppm in the rapidly

TABLE 3-8-8. CHEMICAL SHIFTS OF AXIAL AND EQUATORIAL PROTONS IN CYCLOHEXANE DERIVATIVES

X	^{a} X–H_{eq}	^{a} H_{ax}–X	Δ_{ax-eq}	X–H_{eq}	H_{ax}–X	Δ_{ax-eq}	References[b]
—D	1·60	1·12	0·48				323, 102, 1811, 1279, 1278
—Cl	4·40	3·68	0·72	4·34	3·63	0·71	1859, 744, 2082, 322
—Br	4·60	3·82	0·78	4·62	3·81	0·81	1859, 2082, 743
—I	4·72	3·98	0·73				1859, 229
—OH	3·87	3·27	0·60	3·93	3·37	0·56	1859, 746, 2220, 748, 1559
—OAc	4·80	4·49	0·31	4·98	4·46	0·52	2016, 46, 746
—NH$_2$				3·15	2·52	0·63	1484, 745, 295, 293, 294, 804
—NHCH$_3$				2·70	2·08	0·62	804
—SH				3·43	2·57	0·86	751
—Ph				2·98	2·47	0·51	955, 1232, 2415, 1233
—NO$_2$		4·23		4·43	4·23	0·20	1859, 1200, 2439, 805, 1235

^{a} From low temperature spectra.

[b] See also references listed in Table 1 of ref. 882. Data for several pairs of 4-t-butyl derivatives are listed in ref. 746.

TABLE 3–8–9. CHEMICAL SHIFTS AND CONFIGURATION OF FUNCTIONAL GROUPS
ATTACHED TO SIX-MEMBERED RINGS IN THE CHAIR CONFORMATION

Group	Relationship[a]	References	Exceptions (references)
—O—CO—CH$_3$	Equatorial upfield	1539, 1232, 1086, 1554, 2331, 1566, 141, 1565, 1564	155
—NH—CO—CH$_3$	Equatorial upfield	1566, 1565	
—OCH$_3$	Axial upfield	114, 141	993, 155
—CH$_3$	Equatorial upfield[b]	1289, 1290	
—OH	Axial upfield	2415, 2220, 464, 1920, 1922, 1965, 445, 1923	2220
—OSi(CH$_3$)$_3$	Equatorial upfield	1308	

[a] The differences are generally 0·1–0·2 ppm except in the case of hydroxyl groups where they may reach 0·5 ppm. The ranges generally overlap.
[b] Cyclohexanone derivatives.

TABLE 3–8–10. CHEMICAL SHIFTS OF AXIAL AND EQUATORIAL PROTONS IN SOME SATURATED
HETEROCYCLIC COMPOUNDS

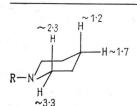

ref. 1487

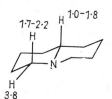

ref. 283, 284, 1093

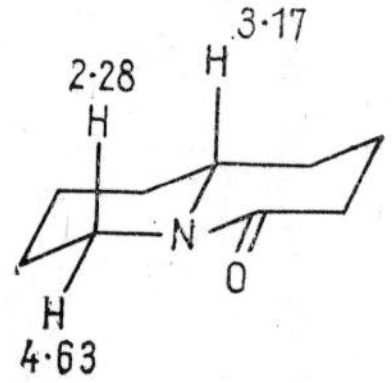

ref. 282

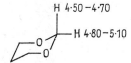

ref. 1538, 167, 35, 2036, 622

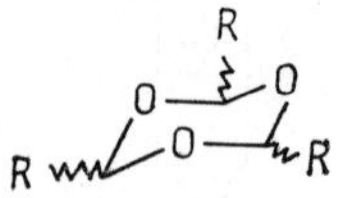

H_{eq}: 5·8
H_{ax}: 5·0–5·1

ref. 181, 1317

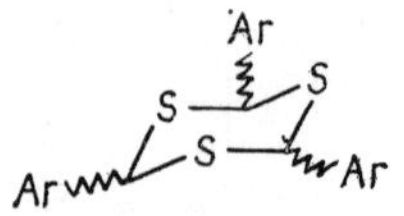

H_{eq}: 5·5–6·1
H_{ax}: 5·1–6·0

ref. 425

inverting mixture at room temperature, appear to be separated by approximately 0·4 ppm at low temperature, a value not greatly different to that in cyclohexane itself.

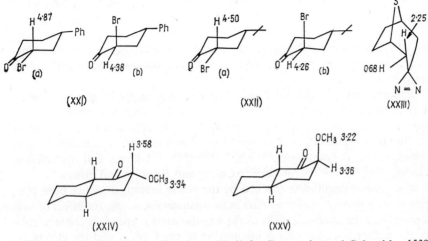

(XXI)   (XXII)   (XXIII)

(XXIV)   (XXV)

Since the early work by Lemieux, Kullnig, Bernstein and Schneider,[1539] many attempts have been made to correlate the chemical shifts of protons in the functional groups attached to six-membered rings (principally cyclohexane and pyranose monosaccharides) with their configuration, and a number of such correlations are summarized in Table 3-8-9. It can be seen that exceptions have been observed and hence assignments should be based on closely related compounds. In the case of the hydroxyl group, studies by Ouellette[1920,1923,1922] indicate that their chemical shifts at high dilution in carbon tetrachloride appear to be little affected by their environment.

In saturated six-membered heterocyclic compounds there is the additional possibility of different steric relations to the heteroatoms which may have long-range effects on chemical shifts (Chapter 2-2). A large amount of data are available for the pyranose carbohydrates,[1085] some of which are quoted in Table 3-8-9, and data for other systems are given in Table 3-8-10. It can be seen that, while the general relationship observed for cyclohexane derivatives appears to apply in most cases, in systems containing nitrogen large differences between the chemical shifts of axial and equatorial protons can be observed which should prove very valuable in stereochemical assignments.

H. CHEMICAL SHIFT—STEREOCHEMISTRY CORRELATIONS IN SOME COMPLEX NATURAL PRODUCTS

Some groups of natural products and their modified derivatives exhibit a great variety of functional groups in relatively fixed juxtaposition and their n.m.r. spectra have yielded correlations between stereochemistry and chemical shifts which have proved of value and even led to general conclusions about the nature of some long-range shielding effects (Chapter 2-2).

At the same time, it must be kept in mind that even the well-established correlations apply only if the steric relations in the model compound are preserved; this involves not only the conformational rigidity of the molecular framework, but also the rotation about single bonds joining substituents, not possessing cylindrical symmetry, to the framework. Thus, ironically, it is precisely in the case of the largest effects (for example, 1,3-diaxial interactions in six-membered rings when one of the groups has a pronounced long-range effect, as with phenyl, acetyl, nitro or carboalkoxy) that most caution must be exercised.

The amount of data correlating the chemical shifts of angular methyl groups in steroids with substitution patterns is very large.† The first compilation due to Shoolery and Rogers[2275] appeared in 1958 and was followed by the monumental work of Zürcher[2640,2642] and important contributions by Tori and his collaborators,[2428,1474,2429,2432,2421] Cross and his collaborators,[1419,566,1418,565] Smith,[2309] Cohen and Rock[514] and others.‡

The general conclusion reached by the above investigators has been that, *provided the steroid skeleton exists in its usual conformation*, the effect of substituents on the chemical shifts of the angular methyl groups is characteristic of the nature, location and configuration of the former, and the effects are *additive*.

The procedure used for calculating the chemical shifts of the angular methyl groups consists of adding characteristic "substituent shifts" to the chemical shifts of the angular methyl groups in the four androstanes listed in Table 3–8–11.

TABLE 3–8–11. CHEMICAL SHIFTS OF THE ANGULAR METHYL GROUPS
IN ANDROSTANES[241]

	C-19 CH$_3$	C-18 CH$_3$
5α, 14α-Androstane	0·792	0·692
5α, 14β-Androstane	0·767	0·992
5β, 14α-Androstane	0·925	0·692
5β, 14β-Androstane	0·900	0·992

Very large compilations of substituent shifts have been assembled by Bhacca and Williams, [241] Cohen and Rock[514] and Smith,[2309] and the selection given in Table 3–8–12 must be regarded as illustrative only.

An inspection of models of the steroid skeleton reveals that many positions are in an equivalent steric relation with respect to the C-19 protons, and it has been shown,[1659,241] that, provided the substituents are of such a nature that their effect does not depend on the average conformation about the single bond joining them to the steroid framework or if no particular hin-

† E.g. the compilation by Smith[2309] dealing solely with the effects of the hydroxyl, acetoxyl and keto groups, lists 98 references.

‡ For some significant compilations see references 227, 437, 483, 1263, 1416, 1428, 1473, 1478, 1659, 1791, 1792, 1794, 1977, 2510, and 2533.

TABLE 3–8–12. SUBSTITUENT SHIFTS FOR ANGULAR METHYL GROUPS IN STEROIDS[241]
Positive values indicate shifts towards lower field (ppm)

		Substituent	C-19 CH_3	C-18 CH_3
Ring A substituents	5α steroids	1-Oxo	0·375	0·017
		2-Oxo	−0·025	0·008
		3-Oxo	0·242	0·042
		4-Oxo	−0·033	0·017
		1α-OH	0·017	0·017
		1β-OH	0·050	0·008
		2β-OH	0·250	0·008
		2β-OAc	0·150	0
		2β-Br	0·233	0
		3α-OH	0	0·008
		3β-OH	0·033	0·008
		3α-OAc	0·025	0·017
		Δ^4-3-Oxo	0·417	0·075
		Δ^4	0·250	0·042
		Δ^5	0·233	0·042
		5α-OH	0·058	0·008
		5α-CH_3	0·150	0
		5α-Cl	0·250	−0·008
	5β steroids	1-Oxo	0·217	0
		3-Oxo	0·117	0·042
		3α-OH	0·008	0·008
		3α-OAc	0·025	0·008
		3β-OH	0·050	0·008
		3β-OAc	0·058	0·008
Ring B and C substituents. Results are applicable to 5α *and* 5β steroids unless otherwise stated		6-Oxo	−0·050	0·017
		7-Oxo	0·275	0·008
		11-Oxo	0·217	−0·033
		12-Oxo	0·100	0·375
		Δ^6	−0·025	0·050
		6β-OH in 5α-steroids	0·225	0·042
		6β-OH in 5β- and Δ^4-steroids	0·192	0·042
		6β-CH_3	0	0
		6β-Cl	0·008	0
		6β-Br	0·317	0·058
		8β-OH	0·183	0·183
		9α-OH	0·142	−0·067
		11α-OAc	0·092	0·058
		11β-OAc	0·067	0·117
Ring D substituents[a]	14α steroids	Δ^{14}	0·008	0·250
		15-Oxo	0·008	0·075
		17-Oxo	0·017	0·167
		14α-OH	0	0·117
		15α-OH	0·008	0·033
		15β-OH	0·033	0·267
		17α-OH	−0·008	−0·008
		17β-OH	0	0·033
		17β-$COCH_3$	−0·008	−0·083
		17β-COOH	−0·008	0·025
		17α-$COOCH_3$	−0·017	0·158
		17β-$COOCH_3$	−0·008	−0·050

TABLE 3–8–12 *(cont.)*

Substituent[a]			C-19 CH$_3$	C-18 CH$_3$
Ring D substituents[a]	14β steroids	15-Oxo	−0·042	0·192
		17-Oxo	0·017	0·083
		14β-OH	0·017	−0·025
		15α-OH	0	−0·033
		17β-OH	0·008	0·025
		17α-COOCH$_3$	−0·008	0·158
		17β-COOCH$_3$	−0·008	−0·067

[a] The effect of ring D substituents on both angular methyl groups appears to be independent of the configuration at C-5.

drance to rotation exists, the effect of substituents located in "equivalent" positions (e.g. 2β, 4β, 6β, 8β, and 11β in 5α steroids) is nearly the same.

The influence of the configuration at the ring junctions and of substituents on the chemical shifts of the angular methyl groups has also been studied in decalins,[2594,2593,599] yohimbine alkaloids,[2247] triterpenes† and other terpenoids.[1879,2529,1406,2635,1265,132] Some of the latter results dealing with derivatives of diterpenoid acids have already been considered in connection with the establishment of long-range shielding effects due to the double bond (Chapter 2–2).

Besides the characteristic shielding values for the angular methyl groups, *cis*- and *trans*-decalins, hydrindanes etc., also give rise to other stereochemically diagnostic features in their n.m.r. spectra,[1157,1771,289,493,2255] but, mainly because of the presence of complex and largely unresolvable resonances, spectra of this type are not easily interpreted.

In several series of diterpenes carrying a methyl and a —CH$_2$—OR, —CHO or —COOR substituent at C-4, the stereochemistry at this position may be deduced from the exact values of chemical shifts,[2529,1401,1152, 1856,457] e.g. the formyl protons of axial aldehydes at C-4 appear to resonate downfield from those in their equatorial counterparts.

Besides the much discussed substituent effects on the chemical shifts of the angular methyl groups, n.m.r. spectra of steroids offer many other examples of chemical shift data for fixed spatial relationships. It has been remarked[865] that "nowhere is there a richer lode of n.m.r. material buried than in the steroid literature" and, in fact, Bhacca and Williams[241] have presented the whole field of applications of n.m.r. spectroscopy in organic chemistry almost entirely from that point of view.

One interesting fixed relationship, somewhat resembling the *peri* effect (cf. Chapter 3–6), can be found between protons (or substituents) located at C-4 and C-6 in Δ⁴-3-oxo-steroids, where only slight[2432] distortion would be expected except in the case of 4,6α-disubstituted compounds.[2279] Some data illustrating the effect of this type of 1,3-interaction on the chemical shifts are given in Table 3–8–13, and the pronounced effect of several functional groups in the spatially close C-4 and C-6α juxtaposition can be readily observed.

† References 482, 515, 648, 816, 1236, 1264, 1479, 1516, 1517, 1529, 1531, 1532, 1533 and 1536.

TABLE 3–8–13. CHEMICAL SHIFTS IN Δ^4-3-OXOSTEROIDS[a]

C-4	C-6α	C-6β
5·55	<2·42	<2·42
5·65	CH_3	?
5·34	?	CH_3
6·19	OH	4·3
5·62	4·2	OH
5·71	OAc	5·4
5·81	5·3	OAc
6·22	Br	4·8
5·75	4·9	Br
6·05	F	5·10
5·89	5·05	F
5·53	NO_2	5·33
6·08	4·84	NO_2
OCH_3	3·07	?
OAc	2·72 (?)	?
Cl	3·26	?
Br	3·28	?
OH	3·04	?
CH_3 (1·72)	2·79	?
CH_3 (1·70)	5·33	Br

[a] See references 520, 2432, 1428, 2418, 2433, 759, 2609.

I. CONFORMATION

Studies of conformation by means of n.m.r. spectroscopy have dealt either with the evaluation of the position of mobile equilibrium and equilibrium kinetics between conformers, or with the actual conformation of a single species. In principle, chemical shift parameters can be used for the solution of either problem and we have already alluded to the use of averaged chemical shifts in the study of conformation of six-membered rings (cf. Section G above). Analogous investigations concerning, *inter alia*, the stereochemistry of ring A in steroids[241] have been performed.

When no model chemical shifts, i.e. those obtained from low-temperature spectra or model compounds, are available as is generally the case with acyclic compounds,[169,2535,647] the relation between chemical shifts and conformation becomes rather complex because it must involve:

(i) An estimate of the relative energy content of all conformers, and hence their relative populations, since this determines the average relative position in space of the groups whose long-range effects on chemical shifts are being considered.

(ii) An estimate of long-range effects on chemical shifts exerted by particular groups in particular combinations.

(iii) The average conformation *of* the groups exerting the long-range effect if they do not possess cylindrical symmetry—i.e. rotation about the bond(s) joining the group(s) to the acyclic structure.

In spite of these complications, arguments of this type may lead to convincing configurational and conformational assignments.[169]

It is clear that, if the long-range shielding effect of a particular group is believed to be well understood and vary in a predictable manner with its conformation towards a proton or a proton-containing group, the actual, or average, conformation of this group can be established from the chemical shift of the proton in question.

This approach has been utilized to some extent with the carbonyl,[1324,179, 1435,1413] and nitro,[2033,2433,2524,2622] groups and, above all, with aromatic rings.† Results of this type cannot be routinely applied to all new situations, but some data in Tables 3–8–2, 3–8–3, 3–8–4, 3–8–6 and 3–8–7 should give some qualitative indication of the magnitude of long-range effects due to "freely-rotating" phenyl groups in various molecular environments, while the contour chart (p. 95) can be used to obtain semi-quantitative estimates of this shielding.

J. STEREOCHEMISTRY AND THE SOLVENT SHIFT

As already discussed in Chapter 2–2F, the origin of *substantial* differences in chemical shifts measured from internal references observed when solvent, concentration or temperature are altered, must be sought in associative phenomena which affect the magnetic environment of the internal reference and the solute in a different manner.

In particular, the (often pronounced) changes in chemical shift commonly observed when the n.m.r. spectra of the same substance are recorded using the same internal reference but a different solvent, are referred to as *specific solvent shifts* or simply *solvent shifts* and arise principally‡ from the formation of *collision complexes* (p. 109) between the solvent and the solute (see however reference 656).

Two features of the collision complexes are of interest to us in the present context. Firstly, any collision complex†† must involve the formation and

† References 503, 582, 585, 619, 1219, 1483, 1687, 1731 and 1816.

‡ Differences in self-association of the solute and in association between the solute and the internal reference accompanying changes of solvent might have to be considered in some cases, but at moderate concentrations and with tetramethylsilane as the internal reference, solvent–solute interactions would be expected to account for the major part of the solvent shift. Further, quite pronounced solvent shifts may arise from merely geometrical considerations affecting the proximity between solvent and solute molecules on the one hand and the solvent and the internal reference on the other. However, distinctions between various types of weak interactions are generally difficult to make.

†† Intermolecular interactions, including solute–solvent association phenomena, may involve more than one collision complex for any two species. However, normally one would expect one type to be energetically favourable, and, in any case, the averaged effect on the chemical shifts of a multi-component system could be treated in the same manner as that in a two-component system.

breaking of a bond (which may be for example a charge-transfer or hydrogen bond), and hence there is an equilibrium between the complexes and the un-complexed species of the solute which should be affected by the formation constant of the complex, temperature and mass-action. In all cases so far reported, the reactions involved in these equilibria have been fast on the n.m.r. time-scale, thus resulting in the observation of averaged spectra.

Secondly, as bonding of any type is, to a greater or lesser extent, associated with a specific geometry, the collision complexes will, in general, have a "shape" and thus various portions of the solute molecules will bear different spatial relationships to the complexed solvent molecule(s). They will therefore experience, at least in principle, different magnetic environments associated with any long-range effects due to the solvent molecules.

It follows from the above, that solvent shifts might give information about the relative positions in space of the group(s) of protons experiencing them and the structural feature(s) responsible for the formation of the collision complex, and hence about some aspect of the stereochemistry of the solute. Before giving examples of applications of this type, let us consider the conditions necessary for the observation of solvent shifts from which useful stereochemical information might be derived.

Firstly, the solute must contain sites capable of specific bonding with the solvent, such as π-electron clouds, polar groups or groups capable of participating in the formation of hydrogen bonds.

Secondly, at least two solvents must be available which differ widely in their ability to form collision complexes. Typically, one of the pair is an "inert" solvent, i.e. one which is not expected to interact strongly with the solute. Carbon tetrachloride or cyclohexane are most suitable for this purpose, although deuterochloroform is most commonly used, in spite of the fact that its interactions with some solutes are by no means negligible.[1503]

Thirdly, the "active" solvent must not only be capable of specific bonding with a site in the solute molecule, but must also have strong, long-range, directed effects on chemical shifts. Aromatic solvents, generally benzene or pyridine, fulfil these conditions, and the majority of applications of solvent shifts to the solution of stereochemical problems have involved the pairs of deuterochloroform–benzene or deuterochloroform–pyridine.

Finally, a proton or group of protons in the solute must be in reasonable proximity and suitable spatial relation to the site responsible for the formation of the collision complex.

Information useful for the establishment of some stereochemical relationships can be obtained from solvent shifts associated with reaction field effects[1724,1502,1504,60] (p. 107), but the most valuable results have been obtained with solvent shifts associated with aromatic solvents.[1522]

It has been shown that solvent shifts between carbon tetrachloride and toluene[1511] and chloroform and pyridine[1107] in a number of molecules decrease with an increase of temperature, providing a convincing demonstration of the existence of collision complexes and even making it possible to evaluate the heat of formation of some of them.[1511]

It appears from the study of steroidal[2577,2578,2582,2575,241] and other[2578,]

[526,1289,1290,2420,298,1504] ketones, that deuterochloroform–benzene and deuterochloroform–pyridine solvent shifts, *generally* cause protons behind the orthogonal plane, P, to shift upfield in the aromatic solvent and protons in front of this plane to shift downfield,[2578,2577,526] as shown in Fig. 2–2–16 (p. 112). However, in view of the large number of model substances available (cf. references above), it is clearly profitable to consider solvent shifts in closely related compounds rather than to rely solely upon the above generalization. The solvent shifts caused by pyridine[2575] appear to be similar, but not quite identical, with those caused by benzene. The commonly encountered range of magnitudes of solvent shifts is 0–0·3 ppm.

An excellent discussion of the possible geometry of the ketone–benzene collision complexes and modifications of results due to the presence of other substituents has been given by Williams and Wilson.[2582] It is important to realize that the presence of polar substituents near the ketone function may drastically alter the magnitude of solvent shifts.

The use of solvents such as dimethylsulphoxide[1101] and dimethylformamide[571] in conjunction with, for example, pyridine[1102] has been investigated for a large range of steroids, and these combinations may be useful when solubility limitations preclude the use of carbon tetrachloride or deuterochloroform.

Clearly, once the solvent shifts associated with protons in a fixed relationship to the active centre (here the keto group) have been established, it becomes possible to use the magnitude of the solvent shift in flexible species for conformational investigations.[298]

Other compounds for which useful correlations have been established between the magnitude of the solvent shifts associated with aromatic solvents and the position of the protons involved with respect to functional groups are α,β-unsaturated ketones,[2416] δ-lactones,[668] acid anhydrides and lactones,[526a] quinones,[320] amides,[1128,1127] some cyclopropyl derivatives,[2242] spiro compounds[792] and *m*-dioxans.[71]

Much useful information can also be derived from several series of closely related compounds (e.g. gibberellin derivatives[1107] and kauranes[1153]) and aromatic solvent shifts can give precise information[1340,1141] regarding *syn–anti* isomerism in common derivatives of carbonyl compounds, the upfield shifts associated with α- and β-protons being characteristically different and also differing between the various classes of derivatives.

CARBONIUM IONS, CARBANIONS
AND RELATED SYSTEMS

MANY unstable entities, such as carbonium ions and carbanions which are postulated as short-lived intermediates in organic reactions, can often be stabilized for appreciable times by suitable choice of solvent and counterion. Proton magnetic resonance spectroscopy then offers a highly effective method for detecting these species and for establishing their chemical and electronic structures. The method is usually less ambiguous than other spectroscopic techniques because of the one-to-one relation between intensity and concentration.† In addition, there often exists the possibility of establishing the life-time of the ion when it is equilibrating between two or more structures. Finally, since these structures carry formal charges, equation (2–2–5) (p. 68) offers a means of estimating the extent to which the change is delocalized over various atomic centres in the system.

A major problem in this area is the measurement of meaningful δ-values. The types of solvent systems which have to be used are, of course, very limited and frequently tetramethylsilane is too insoluble for use as an internal reference. Unfortunately, some spectra in the literature have been referenced externally or by unstated means. In addition, ion pairing can be important[1426] and chemical shifts may be a function of temperature, solvent, counterion and concentration. However, the shifts caused by the presence of formal charges are usually large and the factors just considered will only be important in the consideration of electronic fine structure.

A. CARBONIUM IONS

Long-lived carbonium ions have been generated in a variety of solvents including strong acids ranging from aqueous mineral acids to oleums. Other solvents which have been used are trifluoroacetic acid, chlorosulphonic acid, fluorosulphonic acid, liquid anhydrous hydrofluoric acid and sulphur dioxide. The ions have been derived from alcohols, halo-compounds, olefinic and aromatic compounds, and carboxylic acids and their fluorides, by the action of protonic and Lewis acids. The most suitable internal reference for these systems appears to be the tetramethylammonium cation[631,794] prefer-

---

† For instance, the reported u.v. light absorption of t-alkyl carbonium ions[2134] has been shown to arise largely from cyclopentenyl carbonium ions formed under the conditions of preparation.[633,1898]

ably as its tetrafluoroborate or hexafluoroantimonate. δ-Values† ranging from 3·10 to 3·29 have been found for this ion.[631] Sometimes measurements must be made at sub-zero temperatures in order to avoid decomposition or rapid proton exchange. In this connection, concentrated sulphuric acid or oleums are rather unsatisfactory since, even at room temperature, they are sufficiently viscous to cause appreciable line broadening (p. 8).

(i) Alkyl and alicyclic carbonium ions

The first reported spectra of this class were of the t-butyl and isobutyl carboniums ions, by Olah, Baker and their co-workers[1906] who generated them from the alkyl fluorides in antimony pentafluoride. The same group[1898] subsequently reported data for isopropyl, t-butyl and t-amyl carbonium ions in the same solvent and noted that the same ions are also formed from the isomeric fluorides. For instance, the four isomeric butyl fluorides give the t-butyl carbonium ion. Similar observations were made independently by Brouwer and Mackor.[351] They used hydrofluoric acid as a solvent and generated the ions from alkyl bromides, alcohols and ethers with antimony pentafluoride. In this system, the equilibrium:

$$(CH_3)_2\overset{\oplus}{C} \cdot CH_2 \cdot CH_3 \rightleftharpoons CH_3 \cdot CH_2 \cdot \overset{\oplus}{C}(CH_3)_2$$

causes coalescence (p. 57) of the absorption bands at approximately room temperature and the activation energy for the process is of the order of 14 kcal/mole. Characteristic shifts are given in Table 3–9–1. We note that the shift of the 2-proton in the isopropyl cation is of the order predicted by equation (2–2–5).

Table 3–9–1. Chemical Shifts for Alkyl Carbonium Ions

$(CH_3)_2CH^{\oplus}$				$(CH_3)_3C^{\oplus}$	
5·06╪	13·5╪			4·34╪	ref. 1898
				3·85	ref. 351
ref. 1898					

$(CH_3)_2\overset{\oplus}{C}$—$CH_2$—$CH_3$

2·27╪	4·93╪	2·27╪	ref. 1898
3·8	4·2	1·65	ref. 351

CH₂—CH₂ 2·45
| |
CH₂   CH₂ 4·05
  \   /
    C⊕
    |
  CH₃ 3·8
  ref. 351

† In this section, chemical shifts will be based on $(CH_3)_4N^{\oplus}$ with $\delta = 3\cdot10$. Values based on unstated, or external, references are denoted by ╪.

The particularly interesting 1-adamantyl carbonium ion (I) has been investigated by Schleyer, Olah and their co-workers.[2201] This ion arises from both 1-adamantyl fluoride and 2-*exo*-chloro-*exo*-trimethylenenorbornane (II) in liquid sulphur dioxide–antimony pentafluoride mixtures. The chemical shift of the bridge-head protons is claimed to indicate a special type of delocalization of the positive charge in this ion, although calculations based on equation (2–2–5) suggest that the cause may be essentially electrostatic.[1252]

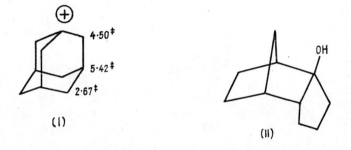

(I)

(II)

(ii) ARYL AND ARYLALKYL CARBONIUM IONS

In 1959, Moodie, Connor and Stewart[1773] reported the spectra of a number of triaryl carbonium ions prepared from the corresponding alcohols in trifluoroacetic acid—trifluoroacetic anhydride mixtures and observed complex multiplets, arising from the ring protons, at approximately 2·5 ppm to lower fields than benzene. Investigations of partially deuterated triphenylmethyl cations by Dehl, Vaughan and Berry[617] in the same year showed that on the n.m.r. time scale (cf. Chapter 2–1) the three rings are equivalent. Subsequently, a number of such ions have been examined and their spectra analysed in detail.[1895,793] In addition to the examples in Table 3–9–2 the spectra of polymethylbenzyl and related benzyl carbonium ions,[579] as well as the α,α-difluorobenzyl and α-fluorobenzhydryl carbonium ions,[1900] have been studied. It is interesting to note that a rapid exchange, resulting in coalescence of absorptions, occurs between triaryl carbonium ions and the corresponding halides.[895]

TABLE 3–9–2. PROTON CHEMICAL SHIFTS IN ARYL AND ARYLALKYL
CARBONIUM IONS IN CHLOROSULPHONIC ACID[793]

	o	m	p	α-H	α-CH$_3$
$(C_6H_5)_3\overset{\oplus}{C}$	7·69	7·87	8·24	—	—
$(C_6H_5)_2\overset{\oplus}{C}H$	8 46	7·98	8·38	9·81	—
$(C_6H_5)_2\overset{\oplus}{C}\cdot CH_3$	8·03	7·88	8·28	—	3·70
$C_6H_5\overset{\oplus}{C}(CH_3)_2$	8·80	7·97	8·45	—	3·57

(iii) ALKENYL AND CYCLOALKENYL CARBONIUM IONS

There have been extensive investigations, mainly by Deno and his group, of a number of alkenyl and cycloalkenyl carbonium ions. Much of the interest in this field arises from Deno's findings that many alkyl and alkenyl carbonium ions are eventually converted to derivatives of the cyclopentenyl carbonium ion.

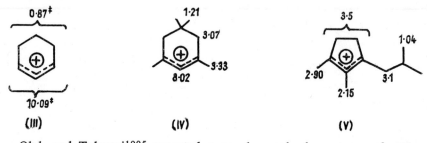

(III) (IV) (V)

Olah and Tolgyesi[1905] reported a poorly resolved spectrum of cyclohexenyl tetrafluoroborate (III) in nitromethane which is difficult to reconcile with the much more readily assigned spectrum of the 1,3,5,5-tetramethylcyclohexenyl carbonium ion (IV) reported by Deno, Richey, Hodge and Wisotsky.[632] Subsequently, data for a number of cyclopentenyl and cyclohexenyl carbonium ions were reported, of which (IV) and (V)[627] are representative.[2335,1904,2336,629,631,628,627]

The allyl, methylallyl, 1,1,3,3-tetramethylallyl and pentamethylallyl carbonium ions have also been investigated[1899,631] (Table 3–9–3).

TABLE 3–9–3. PROTON CHEMICAL SHIFTS FOR ALLYL CARBONIUM IONS

Substituent	1-H	1-Me	2-H	2-Me	3-H	3-Me
Nil[a]	8·97		9·64		8·97	
1-Me[a]		3·85			8·95	
1,1,3,3-Me₄[b]	7·70	{2·95[c] 2·97				{2·95[c] 2·97
Me₅[b]		2·95		2·12		

[a] Ref. 1899; Externally referenced.
[b] Ref. 631.
[c] These methyl groups are non-equivalent due to hindered rotation.

(iv) ALKYNYL CARBONIUM IONS

Several alkynyl carbonium ions have been investigated by Richey and his co-workers.[2112,2113] The appropriate data are given in Table 3–9–4. These results indicate substantial delocalization of the positive charge by the triple bond. The extent to which this occurs depends on the ability of the other substituents to stabilize the positive charge.

TABLE 3–9–4. CHEMICAL SHIFTS OF ETHYNYL AND PROPYNYL PROTONS IN ALKYNYL CARBONIUM IONS [2113]

R	R'	$HC\equiv C\cdot \overset{\oplus}{C}RR'$	$CH_3\cdot C\equiv C\cdot \overset{\oplus}{C}RR'$
$CH_3C\equiv C-$	$CH_3C\equiv C-$	—	2·76
Cyclopropyl	Cyclopropyl	5·68	2·43
Phenyl	Phenyl	5·70	2·60
Anisyl	Anisyl	6·96	2·88
Phenyl	Methyl	7·24	2·94

(v) CYCLOPROPYL CARBONIUM IONS

The tricyclopropylmethyl cation (VI) gives rise to only one absorption band, a singlet at 2·26[633,634] which, by comparison with the cyclopropyl-

(VI)   (VII)

ammonium ion, [634] corresponds to an appreciable fraction of positive charge at the β-positions. It is, of course, well known from solvolysis studies of cyclopropylmethyl derivatives that this type of ion has appreciable stability and several structures have been proposed.[634,1733,1982] Proton spectra of a number of other cyclopropyl carbonium ions have been reported[634,1982,2113] and have thrown further light on the structures of these interesting ions. Perhaps the most striking observation is that the two methyl groups in the cyclopropyldimethyl carbonium ion (VII) are non-equivalent. This requires that the mean configuration of the ion does not have a plane of symmetry. The structure (VIII) has accordingly been proposed. The configuration (IX) rather than (X) or (XI), is favoured for the dicyclopropylmethyl cation on the basis

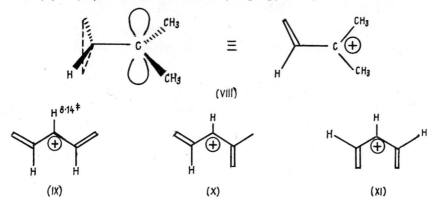

(VIII)

(IX)   (X)   (XI)

of the large vicinal coupling constant ($J = 13$ Hz). It is not obvious why the fifteen protons of the tricyclopropyl carbonium ion are equivalent. The proposed structure (XII) allows three sets of non-equivalent protons. Usually

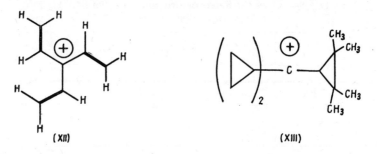

(XII)   (XIII)

the absorptions of the α-protons are displaced by 0·5–1·0 ppm to lower fields from those of the β-protons. It has been suggested that the long-range shielding of the cyclopropane rings (p. 98) increases the shielding of the α-protons by this amount. Even if this is so, the observed result still requires a second accidental degeneracy to explain the equivalence of the two types of β-protons. Furthermore, the published spectrum[634] of the 2,2,3,3-tetramethyl derivative (XIII) seems to suggest that all four methyl groups are also equivalent.

(vi) ARENONIUM IONS

The term benzenonium ion denotes the σ-complex form of the conjugate acid of benzene (XIV).[684] There is, of course, considerable interest in structures of this type as they are involved as intermediates in electrophilic aro-

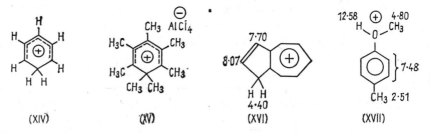

(XIV)   (XV)   (XVI)   (XVII)

matic substitution. The first direct observation of the proton spectra of cations of this type was made by Doering and his co-workers[684] and MacLean, van der Waals and Mackor.[1637] The first group established the structure of heptamethylbenzenonium aluminium tetrachloride (XV). The second group showed that a series of arenonium ions were formed from alkyl benzenes and benzenoid polycyclic hydrocarbons in liquid anhydrous hydrofluoric acid containing boron trifluoride or phosphorus pentafluoride. A number of arenonium ions have since been studied. These include benzenonium and methylbenzenonium ions,[251,1634,1896] (Table 3–9–5), hydroxy and methoxy derivatives,[352,1458,2221,353,251,249] polycyclic arenonium ions,[1634] and an extensive series of azulenonium ions.[1744,1745,1736,2225,595]

TABLE 3–9–5. PROTON CHEMICAL SHIFTS[a,b] FOR BENZENONIUM HEXAFLUORO-
ANTIMONATE AND ITS METHYL DERIVATIVES IN SULPHUR DIOXIDE
(Temp. −60° to 70°C)[1898]

Substituent	Ring Protons						Methyl Groups					
	1	2	3	4	5	6	1	2	3	4	5	6
Nil	3·6	8·8	7·2	7·5	7·2	8·8	—	—	—	—	—	—
2-Me	3·8						—	2·4	—	—	—	—
4-Me	3·8						—	—	—	2·6	—	—
3,4-Me₂	4·4	7·5	7·6	—	—	8·3	—	—	2·2	2·4	—	—
2,4-Me₂	4·6	—	7·1	—	7·7	8·4	—	2·7	—	2·9	—	—
2,5-Me₂	4·7	—	7·6	7·8	—	8·2	—	2·4	—	—	2·2	—
2,3,4-Me₃	4·6	—	—	—	7·6	8·3	—	2·2	2·0	2·6	—	—
2,4,5-Me₃	4·6	—	7·8	—	—	8·3	—	2·2	—	2·6	2·0	—
2,4,6-Me₃	4·6	—	7·7	—	7·7	—	—	2·8	—	2·9	—	2·8
2,3,4,6-Me₄	4·4	—	—	—	7·4	—	—	2·5	2·2	2·7	—	2·5
2,3,4,5-Me₄	4·4	—	—	—	—	7·9	—	2·4	2·2	2·7	2·2	—
2,3,5,6-Me₄	4·7	—	—	7·8	—	—	—	2·8	2·5	—	2·5	2·8
2,3,4,5,6-Me₅[c]	4·8	—	—	—	—	—	—	2·8	2·4	2·9	2·4	2·8
1,2,3,4,5,6-Me₆	4·2	—	—	—	—	—	1·9	2·8	2·6	3·0	2·6	2·8

[a] Relative to external tetramethylsilane.
[b] The assignments of the methyl protons are the same as in ref. 1898. The assignments of the ring protons are based on the published spectra. In some cases the assignments are unique or are based on multiplicity. In other cases, the assignments are equivocal and have been made so as to give a self-consistent set of chemical shifts.
[c] Data taken directly from the published spectrum.

There is no doubt that the ions formed from hydrocarbons are the thermo-dynamically most stable ones and frequently the structures, and hence the position of protonation, can be unequivocally determined from an analysis of the spectrum as in the case of the azulenonium ion (XVI).[1745] Similarly, it has been shown that pyrroles protonate preferentially at the 2-position[11,480] and indolizine at the 3-position.[885] It is also noteworthy that at low temperatures (⩽ −80°C) in HF−BF₃ mixtures, anisole and its alkyl derivatives [e.g. (XVII)] undergo *O*-protonation as well as *C*-protonation.[352] In the *C*-protonated form there is an appreciable barrier to rotation about the bond between the ring and the oxygen atom so that the protons on the two "sides" of the molecule are non-equivalent (cf. Fig. 3–9–1).[353]

By studying the spectra of the hexamethylbenzenonium ion as a function of temperature (Fig. 3–9–2), MacLean and Mackor[1628] have detected two exchange processes. At low temperatures, the ion is a σ-complex with a life-time which is long on the n.m.r. time scale. The spectrum of this species consists of three bands† (intensities 3:6:6) from the 4, 3 and 5, and 2 and 6 methyl groups, respectively, and a quartet and doublet from the CH—CH₃ system. At higher temperatures, the spectrum changes to a nineteen-line multiplet equivalent to one proton, and a doublet equivalent to eighteen protons. This shows that a rapid intramolecular hydrogen shift of the single

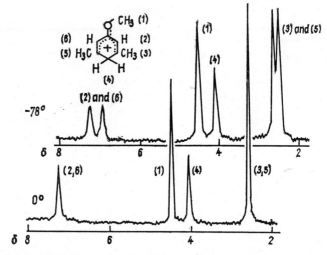

FIG. 3–9–1. 56·4 Mc/s spectra (in HF) of the cation of 3,5-dimethylanisole.
(After Brouwer, Mackor and MacLean[353].)

proton from one carbon atom to another takes place, so that it is equally coupled to the protons of all six methyl groups. At still higher temperatures, this spectrum collapses to the single line spectrum of hexamethylbenzene, due to rapid proton exchange with the solvent. Several detailed kinetic analyses of processes of these types have been reported.[1628,1636,249,251]

(vii) NON-CLASSICAL CARBONIUM IONS

Proton spectroscopy has played an important role in the controversial problem of distinguishing between non-classical or bridged carbonium ions (XVIII) and equilibrating open ions (XIX). Evidence for one or other of

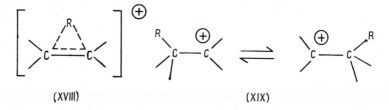

(XVIII)   (XIX)

† The two intense bands are singlets but the third, which is by symmetry assigned to the 4-methyl group, shows evidence of a curious long-range coupling (2 Hz) with the 1-proton.

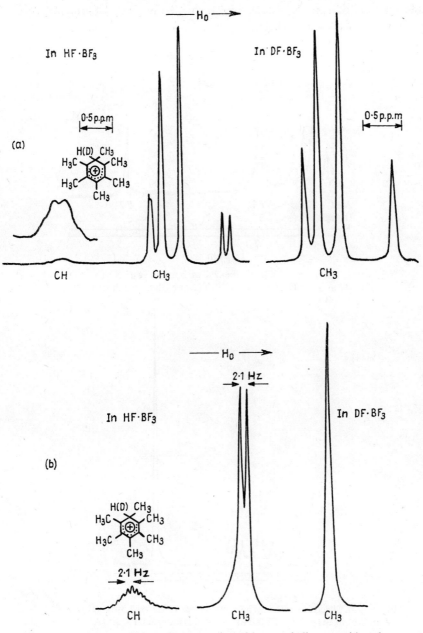

FIG. 3–9–2. Spectra of the proton complex of hexamethylbenzene (a) at low temperature (b) at high temperature. (After Mackor and MacLean[1628].)

these possibilities is obtained from considerations of chemical shifts or from the analysis of rate data accessible through the study of time-dependent phenomena (Chapter 2–1).

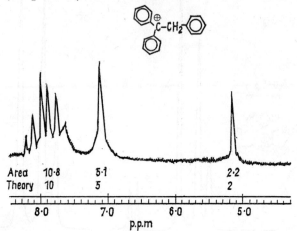

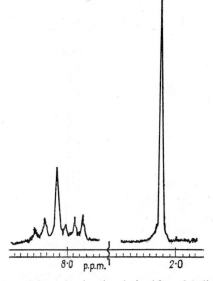

FIG. 3–9–3. Spectrum of diphenylbenzyl carbonium ion (XX) in SO_2–SbF_5–FSO_3H at —60°C. (External TMS.) (After Olah and Pittman[1902].)

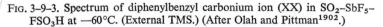

FIG. 3–9–4. Spectrum of the carbonium ion derived from 2,3-dimethyl-3-phenyl-butan-2-ol in SO_2–SbF_5–FSO_3H at —60°C. (External TMS.) (After Olah and Pittman[1903].)

We have seen that delocalization of a positive charge can readily be detected by the characteristic downfield shifts of protons attached to carbon atoms involved in the delocalization process. Figure 3–9–3 shows the spectrum of the diphenylbenzyl carbonium ion in SO_2–SbF_5–FSO_3H solution at $-60°$.[1902] It is clear from the chemical shifts that the positive charge is delocalized over only two aromatic rings and the ion is therefore formulated as (XX). Similarly, the phenylisopropylcarbonium ion is formulated as (XXI),

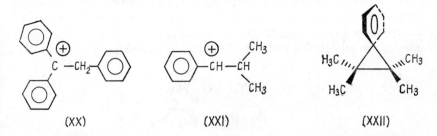

(XX)   (XXI)   (XXII)

since the aromatic protons are deshielded, whereas the methyl protons absorb at $1·2 \mp$ ppm indicating that they are remote from the positive charge. The ion derived from 2,3-dimethyl-3-phenylbutan-2-ol and 2-phenyl-3,3-dimethylbutan-2-ol gives rise to the spectrum shown in Fig. 3–9–4.[1903] We note that all four methyl groups are equivalent.

This observation is consistent with a bridged structure (XXII), a pair of rapidly equilibrating open ions (XIX) or a rapid equilibrium of the type (XXIII) ⇌ (XXIV). The chemical shifts of the protons of the aromatic ring

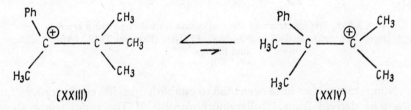

(XXIII)   (XXIV)

clearly indicate substantial delocalization of the positive charge over the ring, so that equilibrating ions of type (XIX) can only be present to a minor extent. Both the bridged ion and the equilibrium mixture (XXIII) ⇌ (XXIV) are consistent with these observations, provided the latter is well to the left as would be predicted on the basis of the expected stabilities of the two ions. A similar situation exists for the ions derived from *erythro-* and *threo-*3-phenylbutan-2-ol (Fig. 3–9–5). In this case, distinction between the bridged ion and the equilibrium system, (XXV) ⇌ (XXVI) ⇌ (XVII) is possible since, if the latter is sufficiently rapid to average the chemical shifts of the methyl groups, it will also result in averaging of spin–spin coupling between the methyl protons and the remaining two protons of the butane skeleton. In this case, the methyl groups are predicted to absorb as a triplet, whereas the observed absorption is two doublets as expected for the bridged ion.

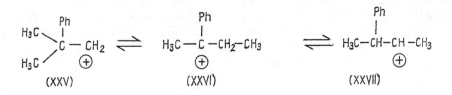

(XXV)   (XXVI)   (XXVII)

A comparison of the chemical shifts in the two series suggests that open ions (XIX) might be present to some extent in the phenyldimethyl system, since the aromatic protons are at higher fields and the methyl groups at lower fields than analogous absorptions in the ions derived from 3-phenyl-butan-2-ol. A bridged structure has also been demonstrated for the ion derived from 2-(9-anthryl)ethanol.[737] In this case the shifts of the aromatic protons are very similar to those of the anthracenonium ion.

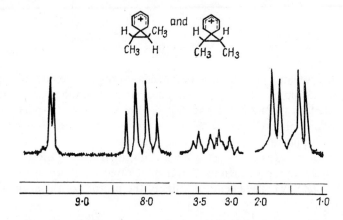

FIG. 3–9–5. The spectrum of the carbonium ions derived from *erythro-* and *threo*-3-phenylbutan-2-ol in SO_2–SbF_5–FSO_3H. (External TMS.) (After Olah and Pittman[1903].)

N.m.r. rate studies have been used to establish open ion configurations for the ions derived from 1,2-diarylnorborneols.[2202] The room temperature spectra are consistent with either bridged ions or rapidly equilibrating open ions, but at −70°C the absorptions of the aromatic protons broaden. This indicates that the aromatic rings are no longer equivalent and strongly favours equilibrating open ions in this system. The 2-norbornyl cation (XXVIII) has been studied by several groups of investigators.[875,2169,2203] Saunders, Schleyer and Olah[2203] have found that this ion gives rise to a singlet at room temperature, which broadens as the temperature is lowered and eventually changes to three singlets (δ = 5·2, 3·0, 2·0*; intensities 4:1:6). The low temperature spectrum is explained in terms of rapid 2,6-hydrogen shifts (XXVIII B) and Wagner–Meerwein rearrangements (XXVIII A). These processes would result in averaging of the shifts of protons at carbon atoms 1, 2, 6 and of 3, 5, 7, leaving the 4-proton unaffected. The coalescence to a singlet at higher temperatures is then attributed to a slower 2,3-hydrogen shift (XXVIII C). The rates of this process at various temperatures have been determined by line shape analysis (Chapter 2–1) and found to have $k(12\cdot9°C) = 10^4 \, sec^{-1}$ and $E_A = 10\cdot8 \pm 0\cdot6$ kcal/mole. It was estimated that, at −120°C, the 2,6-hydrogen shift and the Wagner–Meerwein rearrangement are 10^9 times faster than the 2,3-shift. Winstein[2599] has used this fact, in conjunction with other data, to show that the rate of the

Wagner–Meerwein rearrangement is of the order of kT/h and hence has $\Delta G_A = 0$.

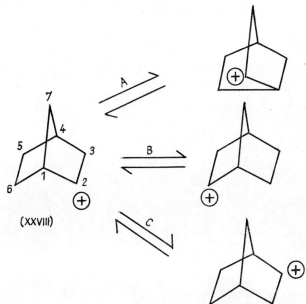

(XXVIII)

The 7-norbornadienyl cation in the open form has a plane of symmetry which includes carbon atoms 1, 4 and 7. The spectrum of this ion has been observed and very carefully analysed, and conclusively indicates that this symmetry does not apply.[2362,2363] The results are consistent with the structure (XXIX). The spectrum of the 7-norbornenyl ion shows similar features.[2111,350]

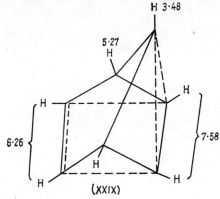

(XXIX)

(viii) AROMATIC CATIONS

The protons of aromatic cations, such as the tropylium ion usually absorb at very low fields, because the effect of the positive charge and the diamagnetic anisotropy of the π-electron system operate in the same direction.[872] Some representative examples are given in Table 3–9–6.

TABLE 3–9–6. CHEMICAL SHIFTS FOR SOME AROMATIC CATIONS

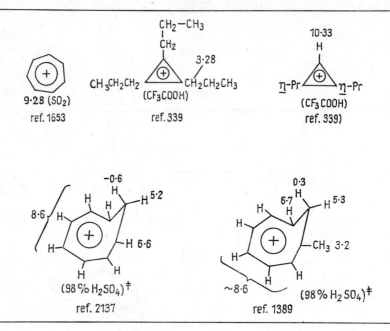

9·28 (SO₂)

ref. 1653

ref. 339

(CF₃COOH)
ref. 339)

(98 % H₂SO₄)‡
ref. 2137

(98 % H₂SO₄)‡
ref. 1389

(ix) ACYL AND RELATED CATIONS

The acylium ion, $[RC{=}O]^{\oplus}$ apparently has a charge density at the carbon atom which is very nearly the same as a carbonium ion since it has been shown[2442,630] that the shifts of α-protons are similar in the two species [cf. (XXX) and (XXXI)]. In contrast, in acidium ions[630] [e.g. (XXXII)] and acetoxonium ions[68] [e.g. (XXXIII)] the charges reside largely on the oxygen atoms. The conjugate acids of ketones also have shifts characteristic of carbonium ions.[631]

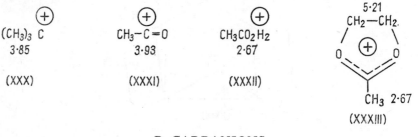

B. CARBANIONS

It is expected that the presence of a negative charge will, in general, result in appreciable upfield shifts in the same way, as we have already seen, as a positive charge causes deshielding. This is borne out by data for the triphenyl methyl cation and anion shown in Table 3–9–7. Accurate chemical shifts are readily obtained as the usual solvents for carbanion salts are ethers, dimethyl sulphoxide, etc., in which tetramethylsilane is readily soluble. On the other hand, carbanions are less well-defined structures than carbonium ions and, in some cases, it is not immediately obvious whether the species under investigation is essentially ionic or whether there is covalent bonding between the assumed ion and counterion. In addition, truely ionic species can exist as tight ion pairs or solvent-separated ion pairs depending on the solvent, temperature, the counterion and the structure of the anion, and this can influence chemical shifts (Table 3–9–7).

TABLE 3–9–7. RELATIVE CHEMICAL SHIFTS
FOR TRIPHENYLMETHYL CATION AND ANION

Proton	$Ph_3C^\oplus$ (ref. 793)	$Ph_3C^\ominus \| Na^\oplus$ D.M.E. (ref. 1056)	$Ph_3C^\ominus Cs^\oplus$ (THF) (ref. 1056)
o	7·69	7·27	7·24
m	7·87	6·51	6·65
p	8·24	5·94	6·09

Table 3–9–8 gives the chemical shifts of a series of neohexylorganometallic compounds.[2606] It is clear that there is no reason to suppose that any of these compounds are fully ionic. However, we will include lithium and magnesium compounds in this discussion because they are of considerable intrinsic interest and, at least in certain cases, there is strong evidence of complete ionic dissociation.

TABLE 3–9–8. CHEMICAL SHIFTS OF α- AND β-PROTONS IN NEOHEXYL-
ORGANOMETALLIC COMPOUNDS,
$(CH_3)_3 \cdot C \cdot CH_2 \cdot CH_2 \cdot M$, IN ETHER[2606]

M	Li	Mg	Zn	Al	Hg
α-H	−1·08	−0·68	0·15	−0·20	1·01
β-H	1·35	1·39	1·47	1·25	1·58
Electronegativity	1·0	1·2	1·5	1·5	1·9

(i) GRIGNARD REAGENTS AND ALKYLMAGNESIUM COMPOUNDS

The spectrum of the Grignard reagent (XXXIV) from ethyl bromide clearly indicates an appreciable electron density at the α-carbon atom. However, an examination of other primary derivatives has demonstrated that in

$$CH_3\text{—}CH_2\text{—}MgBr$$
1·20 −0·65

ref. 2179

(XXXIV)

$$CH_3\text{—}CH=CH\text{—}CH_2\text{—}MgBr$$
1·55 4·48 5·88 0·72

ref. 1878

(XXXV)

neither the Grignard reagent nor the corresponding dialkylmagnesium is the organic residue present to any significant extent as a free carbanion. Thus, the four methylene protons of neohexylmagnesium halides give rise to an AA'BB' spectrum (p. 134) which changes to an A_2B_2 spectrum with increasing temperature.[2553,2554] This observation indicates that, at the lower temperatures, the α-methylene group is configurationally stable which would not be true of the free carbanion. Similar results have been obtained for 2-methylbutylmagnesium bromide, in which the methylene protons are non-equivalent at $-30°$ and become equivalent at higher temperatures through a process which has an activation energy of 12 kcal/mole.[871] It is of interest that the corresponding dialkylmagnesium has a very similar spectrum but the activation energy is 18·2 kcal/mole.

Even in the case of allyl Grignard reagents, in which the free carbanion is expected to be much more stable, the evidence is strongly against a carbanionic structure. Nordlander, Young and Roberts[1878] have examined the room temperature spectrum of butenylmagnesium bromide and found it to be consistent with the structure (XXXV). The spectra of allyl[1877] (XXXVI) and γ,γ-dimethylallyl[2551] Grignard reagents have also been reported.

(ii) Lithium Alkyls and Related Compounds

The first reported data for organolithium compounds were for allyl- and vinyl-lithium. The spectrum of the former (XXXVII) is very similar to that of allylmagnesium bromide (XXXVI). Both compounds give rise to AX_4 spectra. This is interpreted[1877] as indicating that these compounds are equilibrating mixtures of covalent species, as the corresponding anion should give rise to an AA'BB'X spectrum unless there is an accidental degeneracy in the chemical shifts of the two types of terminal protons. It is assumed that

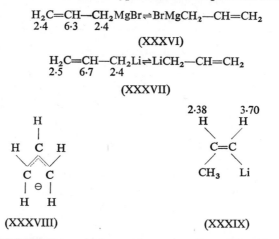

$$H_2C=CH-CH_2MgBr \rightleftharpoons BrMgCH_2-CH=CH_2$$
$$\quad\; 2\cdot4 \qquad 6\cdot3 \quad\; 2\cdot4$$

(XXXVI)

$$H_2C=CH-CH_2Li \rightleftharpoons LiCH_2-CH=CH_2$$
$$\quad\; 2\cdot5 \qquad 6\cdot7 \quad\; 2\cdot4$$

(XXXVII)

(XXXVIII)

(XXXIX)

the anion (XXXVIII) would be configurationally stable. Absolute chemical shifts of vinyllithium were not reported, but data are available for both *cis*-(XXXIX) and *trans*-propenyllithium.[2245]

Alkyllithiums have also been investigated.[364,365,475,2235] It is clear that these compounds have a complex structure in solution which depends on the nature of the solvent. Evidence for dimers,[475] tetramers[2235] and hexamers[365] has been given. These structures involve Lewis bases, if they are present, so that the observed chemical shifts are dependent on the nature of the solvent. (Table 3–9–9.)

TABLE 3–9–9. PROTON CHEMICAL SHIFTS FOR ETHYLLITHIUM IN THREE SOLVENTS[365]

Solvent	$\delta_{(CH_2)}$	$\delta_{(CH_3)}$
Toluene	−0·83	1·4
Triethylamine	−1·06	—
Diethyl ether	−1·10	—

Sandel and Freedman[2160] have reported the spectra of benzyl-, benzhydryl- and trityl-lithium in tetrahydrofuran (Table 3–9–10). There seems little doubt that tityllithium is completely ionized as its spectrum is virtually independent of solvent[2160,1056] and is identical with that of sodium triphenylmethide. In view of the covalent structure of allyllithium, there is some doubt as to the structures of the other two members of this series. In this connection, it is noteworthy that the ^7Li chemical shift for benzyllithium is 1 ppm to lower field than in the other two compounds and that the lithium resonance of trityllithium is considerably less broad than either benzyllithium or benz-

TABLE 3–9–10. PROTON CHEMICAL SHIFTS FOR PHENYLMETHYL CARBANIONS IN TETRAHYDROFURAN[2160]

Compound	δ_o	δ_m	δ_p
Ph_3CLi	7·31	6·52	5·96
Ph_2CHLi	6·51	6·54	5·65
$PhCH_2Li$	6·09	6·30	5·50

hydryllithium.[2490] Line broadening could be due either to quadrupole broadening or to a slow exchange. In either case, the observation is consistent with a change in bonding from a completely ionic bond in trityllithium to partial covalency in the other two compounds. There appears to be considerable charge delocalization in 1,1-diphenyl-*n*-hexyllithium which is substantially reduced by the addition of a Lewis acid, diethylzinc.[2489] There seems little doubt that 9-fluorenyllithium is ionic and exists as tight ion pairs or solvent separated ion pairs depending on the cation solvating power of the solvent.[676]

(iii) AROMATIC ANIONS

A number of aromatic mono- and di-anions, all of which obey Hückel's rule, have been studied and Table 3–9–11 gives the chemical shifts of ions of this type. In this series, the presence of the negative charge(s) opposes the deshielding effect associated with the diamagnetic anisotropy of the π-electron system. This is most striking in the case of the cyclo-octatetraene di-anion[1376,922] in which the opposing influences cause the protons to have almost the same chemical shift as cyclo-octatetraene itself.

TABLE 3–9–11. CHEMICAL SHIFTS FOR SOME AROMATIC ANIONS

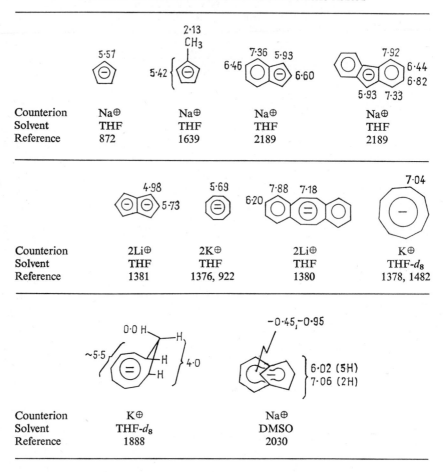

Counterion	$Na^{\oplus}$	$Na^{\oplus}$	$Na^{\oplus}$	$Na^{\oplus}$
Solvent	THF	THF	THF	THF
Reference	872	1639	2189	2189

Counterion	$2Li^{\oplus}$	$2K^{\oplus}$	$2Li^{\oplus}$	$K^{\oplus}$
Solvent	THF	THF	THF	THF-d_8
Reference	1381	1376, 922	1380	1378, 1482

Counterion	$K^{\oplus}$	$Na^{\oplus}$
Solvent	THF-d_8	DMSO
Reference	1888	2030

(iv) MISCELLANEOUS CARBANIONS

Meisenheimer complexes, formed by addition of alkoxide ions to *sym*-tri-nitrobenzene, are stable analogues of the postulated intermediates in nucleo-philic aromatic substitution of activated aromatic molecules. These structures have been studied by proton magnetic resonance spectroscopy.[2238] The anion derived from *sym*-trinitrobenzene and methoxide ion has the shifts indicated (XL). It has been shown that 2,4,6-trinitroanisole initially forms the anion (XLI) but that this is rapidly converted to (XLII).

Maercker and Roberts[1647] have examined the spectrum of potassium diphenylcyclopropylmethide (XLIII) and have concluded that the negative charge is delocalized over the aromatic rings and possibly to some extent over the cyclopropane ring, since the protons of the latter are centred near -0.27 ppm.

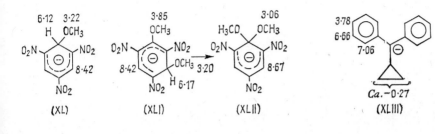

(XL)   (XLI)   (XLII)   (XLIII)

PART 4

SPIN–SPIN COUPLING

In CHAPTER 2–3 we have dealt principally with the basic mechanism of spin–spin coupling and with the extraction of the values of the coupling constants from the experimentally observed multiplets.

As spin–spin coupling involves the electrons in the bonds separating the atoms of the nuclei involved, it is expected that the number of intervening bonds, as well as any factors affecting the distribution of electrons in the "coupling path" (such as bond multiplicity, electronegativity of substituents, stereochemistry, polarization due to external or internal factors etc.), will affect the value of the coupling constant between them.

This expectation is in fact realized, and it is found that coupling constants are very characteristic of the environment of the nuclei concerned. In most cases variations of the coupling constants with structure and stereochemistry can be satisfactorily explained on the basis of what is known about the mechanism of spin–spin coupling, but *accurate* theoretical prediction of the numerical values of the coupling constants in molecules of interest to organic chemists is not feasible at present. In the treatment given below we shall rely heavily on empirical correlations but shall consider also the relevant theories, because they give some insight into the variations to be expected between the model system chosen and the situation actually encountered. Next to the type of nucleus, the most important influence on the magnitude of coupling constants is the number of intervening bonds and it will therefore be convenient to divide the discussion below into chapters dealing with coupling across a specified number of bonds.

GEMINAL INTERPROTON COUPLING

GEMINAL coupling is defined as the interaction between nuclei bound to the same atom, i.e. across two bonds. Clearly, of most interest are the geminal interactions of two protons bound to the same carbon atom and, although the coupling is formally across two σ-bonds, the hybridization of the central carbon atom strongly influences the magnitude of the coupling constant. Accordingly, we shall subdivide this chapter into sections dealing with geminal coupling across sp^3 and sp^2 carbon atoms, respectively.

A. GEMINAL COUPLING ACROSS AN sp^3 CARBON ATOM

Calculations for strongly coupled systems or double irradiation experiments have shown conclusively that the geminal coupling constants in acyclic systems,† in cyclopropanes,[1241,2567,1946,1666,1826,1242,2536,2267,904] in metacyclophanes,[1067,2536] dioxolanes,[890] β-lactams,[182] six-membered rings,[1838,95,2453] indene oxide,[757] ethylene sulphide[1239,1666] and cyclobutanes[1587] are opposite in sign[1353] to both vicinal H—C—C—H and direct ^{13}C—H coupling constants, and hence *negative* (p. 117). Geminal coupling constants in epoxides and ethyleneimines are positive.[2263,754,1666,2592,1239,757] We shall therefore assume that, with the exception of these two systems, the geminal coupling constants are negative when the sign has not been determined. Some typical geminal coupling constants are shown in Table 4–1–1.

The range of observed values varies from $+6\cdot3$ Hz for glycidic acid[2592] to $-21\cdot5$ Hz for cyclopentene-3,5-dione,[1069,177] includes some near-zero coupling constants, and the values are well correlated with certain structural features (see below).

A number of theoretical papers have been published concerning geminal coupling.[1350,1069,2052,2001]

Of those, only the valence bond treatment of Barfield and Grant[176,177,178] had led to predictions of immediate interest in the present context, until the publication of the work[2001] of Pople and Bothner-By who have developed a molecular orbital theory of geminal coupling applicable to both sp^3 and sp^2 hybridized carbon atoms.

This treatment, while still not leading to a satisfactory calculation of the absolute magnitude of geminal coupling constants, *predicts all the trends* noted below and is also of considerable theoretical interest, because it leads

† References 96, 820, 821, 889, 903, 905, 907, 908, 1330, 1515, 1631, 1927 and 2365.

TABLE 4–1–1.[a,b,c] GEMINAL COUPLING CONSTANTS ACROSS AN sp^3 HYBRIDIZED CARBON ATOM

Compound	Coupling constants (Hz)	References
CH_4	$-12 \cdot 4 \pm 0 \cdot 6$	231, 233
$(CH_3)_4Si$	$-14 \cdot 15 \pm 0 \cdot 08$	1621
CH_3F	$-9 \cdot 6$	233
CH_3OH	$-10 \cdot 8$	233
CH_3Cl	$-10 \cdot 8$	233
CH_3I	$-9 \cdot 2$	233
CH_2Br_2	$-5 \cdot 5$	1858
CH_2Cl_2	$-7 \cdot 5$	1858
$CH_2(CN)_2$	$-20 \cdot 3$	177
$PhCH_2CN$	$-18 \cdot 5$	177
CH_3CN	$-16 \cdot 9$	177
$CH_3—CO—CH_3$	$-14 \cdot 9$	177
$CH_3—COOH$	$-14 \cdot 5$	177
$ArCH_3$	$-13 \cdot 8$ to $-14 \cdot 8$	1620
	$J_{AB} = -12$	2389
	$J_{AB} = -17 \cdot 6$	2389
	$J_{AB} = -21 \cdot 5$	177

TABLE 4–1–1 *(cont.)*

Compound	Coupling constants (Hz)			References
	$J_{Hax, Heq} = -12\cdot6$			1811
$CH_3CH(OCH_AH_BCH_3)_2$	$J_{AB} = -9\cdot4$			2194
	$J_{AB} = -3\cdot1--9\cdot1$			See text
	$J_{AB}(gem)$	$J_{AX}(trans)$	$J_{BX}(cis)$	
	$+5\cdot66$	$+2\cdot52$	$+4\cdot06$	1666
	$+0\cdot97$	$+3\cdot19$	$+6\cdot03$	1666
	$-1\cdot38$	$+5\cdot60$	$+6\cdot54$	1666
	$-4\cdot9$	$+6\cdot8$	$+12\cdot6$	2592
	$-9\cdot1$	$+5\cdot4$	$+8\cdot0$	2592
	$+4\cdot5$	$+1\cdot4$	$+2\cdot2$	2592
	$+6\cdot3$	$+1\cdot9$	$+5\cdot0$	2592

TABLE 4–1–1 *(cont.)*

Compound	Coupling constants (Hz)	References
	$J_{AB} = -6{\cdot}0$ to $-6{\cdot}2$; $J_{CD} = -10{\cdot}9$ to $-11{\cdot}5$; $J_{EF} = -12{\cdot}6$ to $-13{\cdot}2$	621, 2036
	$0-\pm2$	See text

^a A number of vicinal coupling constants are included.

^b The values quoted for geminal coupling constants between equivalent protons were obtained from deuterated compounds (*cf.* Chapter 2–3).

^c Some of the signs of the coupling constants quoted in this table have not actually been determined, but can probably be assumed to be as stated.

to the conclusion that values of the geminal coupling constants provide a means of distinguishing between inductive and hyperconjugative effects.

It will be convenient to discuss separately the various factors which influence magnitude of the geminal coupling constants, but it should be remembered that, while the various effects will obviously tend to cancel or reinforce each other, *strict* additivity cannot be always expected.

(i) INFLUENCE OF THE H—C—H ANGLE

An early theoretical treatment[1069] predicted that the geminal coupling constants for protons bound to both sp^3 and sp^2 hybridized carbon atoms should be strongly dependent on the H—C—H angle. The values found for methane, epoxides, substituted β-propiolactones and ethylene derivatives fitted the theoretical curve, but the agreement is in fact fortuitous because the theoretically derived signs are incorrect. On present evidence it is difficult to decide whether the magnitude of H—C—H angle, *per se*, influences the geminal coupling constant.[233,1584,2001]

(ii) INFLUENCE OF NEIGHBOURING π-BONDS

For freely rotating systems, each adjacent π-bond decreases the algebraic magnitude of the geminal coupling constant by 1·9 Hz, in good agreement with the theoretical prediction of 1·5 Hz,[177] and the increments are additive. Thus, the geminal coupling constants in 13 compounds, including methane, fall on a straight line[177] when plotted (see Fig. 4–1–1) against the number of

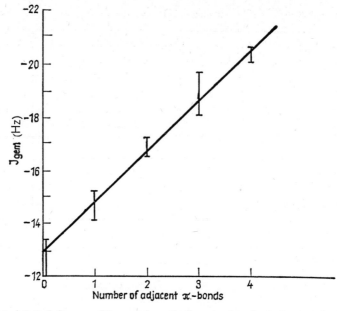

FIG. 4–1–1. Influence of the number of adjacent π-bonds on the magnitude of J_{gem} in non-cyclic systems. (After Barfield and Grant[177].)

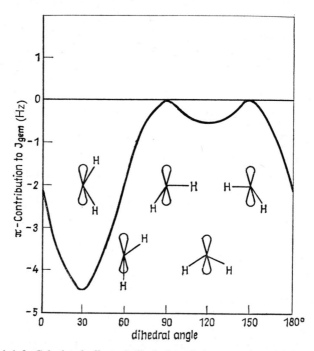

FIG. 4–1–2. Calculated effect of dihedral angle between the methylene group and the π-lobes of adjacent π-bonds on J_{gem}. The small drawings illustrate the projections of the methylene groups on the π-bond. (After Barfield and Grant[177].)

adjacent π-bonds. Considering that the average deviations for the values in Fig. 4–1–1 vary from ± 0.2 to ± 0.6 Hz and that electronegativity effects exist (see below), the correlation appears to be excellent. The effect of an adjacent carbonyl group on the magnitude of the geminal coupling constant has also been discussed by Takahashi[2389] who lists a large number of data for methylene sidechains in steroid and heterocyclic compounds, which are made non-equivalent by being attached to an asymmetric centre (see Chapter 5–2), as well as data for some cyclic systems. It is possible that a cyclopropyl substituent may act as a π-bond.[780]

The theory further predicts[177] that the effect of the adjacent π-bonds should be dependent upon their orientation with respect to the methylene group, in the manner shown in Fig. 4–1–2. Clearly, the value of 1.5 Hz, calculated[177] for the freely rotating group, is an average derived from the curve in the Figure. The experimentally obtained geminal coupling constants in [2.2]metacyclophane[1067] (see reference 2536 for paracyclophane), acenaphthene derivatives,[798,1917,643,641,642] miscellaneous cyclic compounds[2536,192,177,1214,1584] and a number of steroidal ketones[241] appear to confirm the relation shown in Fig. 4–1–2. This orientation dependence of geminal coupling should also be reflected in the variation of the magnitude of J_{gem} in acylic, or flexible systems with solvent or temperature and has been used for conformational analysis.[2576] Small differences between predictions based on the valence bond[177] and molecular orbital[2001] treatments have not yet been tested empirically.

(iii) INFLUENCE OF RING SIZE

Although it is difficult to obtain values for geminal coupling in unsubstituted alicyclic compounds, and substituents (and possibly also their orientation[2001]) are known to modify their magnitude (see below), sufficient data are available to show that the geminal coupling constants may vary with ring size. The geminal coupling constants in some cycloheptanes[1319] are in the range of -13.5 to -14 Hz. The geminal coupling constant in cyclohexane,[1811] obtained from the spectra of partially deuterated derivatives, is -12.6 Hz, i.e. the same as that in methane. Introduction of β-substituents (generally hydroxyl, acetoxyl or chloro groups) appears to alter the magnitude of J_{gem} only slightly, values for various cyclohexane derivatives,[1543,2441,1542,884,1818,1838,955,780] 1,3-dithians,[24] some pyranosides[2615] and 1,3-dioxan[621] all being in the range of -11 to -14 Hz. The direction of the changes appears to be only poorly correlated with those obtained in some systematic studies (see below), but the actual range of values is too small to permit detailed interpretation.

No values for unsubstituted cyclopentane are available, but the results obtained with prolines,[17,18] sultones,[1889] deoxyribose,[1266] lactones,[1584,300,1214] heterocyclic compounds,[1002,2265,1028,2459] hexachlorobicyclo-[2.2.1]heptenes[2588,1508] and miscellaneous systems[642,1372,1917,780] suggest that, after making allowances for substituent effects, we may assume that the geminal coupling constants in five-membered rings are not appreciably different[1584] from those in six-membered rings or freely rotating systems.

A number of values are available for cyclobutane systems† and they range from $-10 \cdot 9$ to -14 Hz for compounds without π-bonds on adjacent carbon atoms and from $-15 \cdot 3$ to -17 Hz for those with adjacent carbonyl groups. Again, we may conclude that the basic geminal coupling constant in cyclobutanes is not very different from the normal values. J_{gem} in cyclobutene[296] is -12 Hz. Geminal coupling constants in the four-membered rings of some bicyclo[2.1.1]hexanes[2563] are approximately -7 Hz and in β-lactams[182] the geminal coupling constants in the methylene groups attached to the nitrogen are only 5–6 Hz. Numerous data obtained for substituted cyclopropanes‡ show that the geminal coupling constants are in the range of $-3 \cdot 1$ to $-9 \cdot 1$ Hz, i.e. considerably larger algebraically than those in other systems. This is clearly connected with the sp^2-like hybridization in 3-membered rings.[2001] Further, because much systematic work has been done in this series,[2592,1946,1009,2567,2519] it is possible to correct for the effect of substituents (see below). Thus the magnitude of the geminal coupling constants, together with the characteristic chemical shifts of cyclopropyl protons (see Chapter 3–8 D) can be used diagnostically[1420] in structural work.

Geminal coupling constants in heterocyclopropanes are greatly influenced by the nature of the heteroatom. The results for styrene oxide, styrene sulphide and phenyl aziridine are given in Table 4–1–1. Data on substituted heterocyclopropanes are available,[347,1796,1571,1831,2592,291,757] but only those for epoxides (see below) are sufficiently numerous to enable us to draw firm conclusions with respect to substituent effects.

(iv) INFLUENCE OF SUBSTITUENT ELECTRONEGATIVITY

In a series of substituted methanes,[233] the geminal coupling constants appear to *increase* algebraically with the increasing electronegativity of substituents (see Table 4–1–1). This correlation is not simple and does not hold well for methyl halides[233,2001] but the substituent effects are additive[1858] in mono- and disubstituted ethanes. Electronegative substituents in the β-position appear (see Table 4–1–1) to *decrease* the algebraic magnitude of the geminal coupling constants in 1,1-dichloropropanes,[2592] epoxides,[2592] hexachlorobicyclo[2.2.1]heptenes,[2588,1508] and probably also in simple ethanes[20,233,1174,1229,909,2365] (in which typical values are between -9 and -14 Hz). Interesting examples of methylene groups substituted by both carbonyl and electronegative groups are listed by Takahashi.[2389]

From the very few examples available[15,1584,17,20,2001] it would appear that the configuration of the β-substituent does not lead to very marked effects; however the direction of the change is in accord with theory.[2001]

---

† References 177, 296, 332, 341, 1020, 1587, 1668, 1705, 1776, 2195, 2239, 2388 and 2585.
‡ References 125, 780, 1009, 1241, 1420, 1534, 1655, 1946, 2021, 2267, 2517, 2519, 2536, 2567 and 2592.

(v) EFFECT OF ORIENTATION OF α-SUBSTITUENTS

The geminal coupling constants for a series of 1,3-dioxans[2036,621] are shown in Table 4–1–1. The range of geminal coupling constants (obtained only in terms of their differences)[2318] in a number of 1,4-diheterocyclohexanes is small. Thus it appears that α-electronegative substituents in six-membered rings[24] have a similar effect on the magnitude of J_{gem} as in methanes.[233]

Methylene groups adjacent to one oxygen atom in five-membered rings have been reported[890,2023,4,2611,817,466,2001] to have geminal coupling constants in the range of -9.6 to -7.5 Hz, although at least one value of -12 Hz is known.[1790] Protons in the methylenedioxy group[983,1001,240,2626,41] couple with J's of only 0–2 Hz (sign unknown). The above suggests that the effect of α-oxygen substituents, although formally not containing π-bonds, is also orientation dependent.[2001]

(vi) MISCELLANEOUS EFFECTS

A nearly vanishing geminal coupling has been reported[1399] for an unusual metal organic compound where the methylene group concerned is directly bonded to manganese. A geminal coupling of only 1.7 Hz has been reported[1705] for a bicyclo[1.0.1]butane derivative, and of only 3.13 Hz in a tricyclopentane,[505] possibly reflecting unusual hybridization due to strain.

A significant solvent and temperature dependence of geminal coupling constants in some freely rotating systems has been reported.[820,821,20,1064,1063] In styrene oxide[2313] J_{gem} varies from 5.3 to 6.0 Hz and appears to be related to the dielectric constant of the solvent.

B. GEMINAL COUPLING ACROSS AN sp^2 CARBON ATOM

In the majority of organic compounds, the geminal coupling constants across sp^2 hybridized carbon atoms are within the range of $+2$ to -2 Hz (i.e. are much smaller in absolute magnitude than the common range of geminal coupling constants across sp^3 hybridized carbon atoms) but values of up to 42 Hz have been reported.[2250] The unusual values are well correlated with certain structural features, so that in practice the small geminal coupling constants are characteristic of a $C=CH_2$ group.

The most important influence on the magnitude of the geminal coupling across an sp^2 hybridized carbon atom is the nature of the element to which carbon is attached. Thus, in compounds containing $-N=CH_2$ groups, such as formaldoxime, its methyl ether, Schiff's bases of formaldehyde etc.,[2248,2249] the geminal coupling constants are 7.6 to 17 Hz (sign unknown) and in formaldehyde[2250,2001] $+42$ Hz, the largest interproton coupling known except that found in H—D,[2045] which corresponds to 285 Hz for $J_{H,H}$ in the hydrogen molecule.

Geminal coupling constants in cumulenes, allenes and ketene are predicted to have large absolute magnitudes (sign presumed -ve), which is theoretically analogous[61] to the π-bond effect on geminal coupling across an sp^3 carbon,[177] and the values found in 1,1-dimethylallene and ketene are 9·0 and 15·8 Hz, respectively.[61]

TABLE 4–1–2. AVERAGE VALUES OF COUPLING CONSTANTS IN VINYL COMPOUNDS $CH_2=CHX$ (IN Hz)[2182]

X	$E_X{}^a$	$J_{geminal}$	J_{cis}	J_{trans}
—F	3·95	−3·2	4·65	12·75
—Cl	3·2	−1·4	7·3	14·6
Br	3·0	−1·8	7·1	15·2
—OR	3·5	−1·9	6·7	14·2
—OAr	3·5	−1·5	6·5	13·7
—OCOR	3·5	−1·4	6·3	13·9
—Phosphate	3·5	−2·3	5·8	13·2
—NO$_2$	3·35	−2·0	7·6	15·0
—NR$_2$	3·0	0	9·4	16·1
—COOR	2·5	1·7	10·2	17·2
—CN	2·5	1·3	11·3	18·2
—COR	2·5	1·8	11·0	18·0
—R	2·5	1·6	10·3	17·3
—Ar	2·5	1·3	11·0	18·0
—Py	2·5	1·1	10·8	17·5
—Sulphone	3·0	−0·6	9·9	16·6
—Sn	1·9	2·8	14·1	20·3
—As	2·1	1·7	11·6	19·1
—Sb	2·0	2·0	12·6	19·5
—Pb	1·9	2·0	12·1	19·6
—Hg	1·9	3·5	13·1	21·0
—Al	1·5	6·3	15·3	21·4
—Li	1·0	7·1	19·3	23·9

^{a} Electronegativity of substituent X.

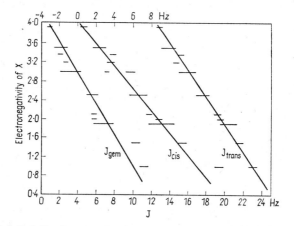

FIG. 4–1–3. Relation between the electronegativity of X and coupling constants in vinyl derivatives XCH=CH$_2$. (After Schaefer[2182].) The upper scale of J-values applies to J_{gem}, the lower scale to $J_{vicinal}$ (*cis* and *trans*).

In vinyl compounds, X—CH=CH$_2$, the geminal coupling constants as well as other coupling constants have an approximately linear dependence[2182,799,1802] on the electronegativity of X (Table 4–1–2 and Fig. 4–1–3).

Table 4–1–2 shows that conjugation of the vinyl group leaves J_{gem} virtually unaffected. This is further borne out by the results obtained with 1,3-butadiene derivatives[1178,255,1436] where values of J_{gem} are very close to those in propenes[311,2541,2542] and to that in ethylene itself[1601] where J_{gem} is +2·5Hz.

Where the C=CH$_2$ group is exocyclic, the magnitude of J_{gem} may depart considerably from that expected from electronegativity considerations alone. Thus, in avenaciolide[348] (I) J_{gem} is only *ca.* 0·5 Hz, in itaconic anhydride[1244] (II) *ca.* 0 Hz, not detectible in the cyclobutene (III)[2569] −3·87 Hz in diketene (IV)[1777,1226,348] and 5 Hz in (V).[1677]

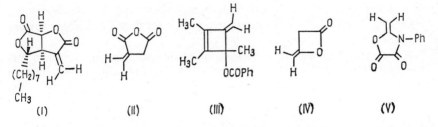

(I)   (II)   (III)   (IV)   (V)

Geminal coupling constants across an sp^2 carbon atom are very significantly solvent dependent.[2250,2248,2249,2503,2501] For instance, J_{gem} for α-chloroacrylonitrile[2503] is ±1·96 in cyclohexane and ±3·24 Hz in DMSO.

C. GEMINAL COUPLING ACROSS A HETEROATOM

This type of spin–spin coupling is not commonly encountered; J_{H-P-H} in some phosphine derivatives[1667] is 12 Hz and J_{H-S-H} (obtained from J_{H-S-D})[2382,2206] is 13·7 Hz. In some amides[1326] $J_{H-N(COR)-H}$ is in the range of 2·2–2·5 Hz, possibly reflecting the sp^2-like hybridization of the nitrogen atom.

VICINAL INTERPROTON COUPLING

VICINAL coupling is defined as the interaction between nuclei bound to contiguous atoms, i.e. a coupling across three bonds. The vast majority of vicinal interproton coupling constants involve the systems H—C—C—H and H—C=C—H and, although it has been shown that the interactions are of very similar type[1348] and involve mainly the σ-electrons, the geometries and other properties of the two systems are so different that it will be convenient to discuss them separately.

A number of theoretical investigations of vicinal coupling have been carried out[1348,1354,1628a,1350,1171,2051,45,1349] some of which[1348,1354] have had a considerable impact on the applications of vicinal coupling data to the solution of structural and stereochemical problems.

A. VICINAL COUPLING ACROSS THREE SINGLE BONDS

The absolute magnitude of vicinal coupling constants in saturated systems varies from 0 to 16 Hz (generally 5–8 Hz† for the averaged coupling constants in freely rotating systems, cf. Chapter 2–1), and numerous sign determinations (Chapter 4–1 A) have shown them to be positive, with the possible exception of two examples,[821,1064] which will be discussed in detail below.

It is convenient to discuss separately the factors which are known to affect the magnitude of vicinal coupling constants, before considering applications resulting from the established correlations between the magnitude of the vicinal coupling constants and structure. It may, however, be stated at the outset that essentially all applications of interest in organic chemistry stem from the relation between J_{vicinal} and dihedral angle‡ (the Karplus rule, see below) which is one of the most powerful stereochemical generalizations in organic chemistry. The remaining effects are important mainly because they set certain limitations on the application of the Karplus rule to structural and conformational problems.

---

† Propane[2257]: 7·4 Hz, ethane:[1601,1008] 8·0 Hz, from ^{13}C satellite analyses.
‡ The dihedral angle, Φ, is the angle subtended by the two CH bonds when viewed in the Newman projection (cf. Fig. 4–2–1).

(i) The relation between vicinal coupling and dihedral angle—the Karplus rule

On the basis of valence bond calculations,[1348] which agree reasonably well with calculations based on a molecular orbital approach,[531] Karplus[1348] predicted an approximate relation between the dihedral angle Φ and the vicinal coupling constant $J_{H-C-C-H'}$ which is usually expressed in the form

$$J = J^0 \cos^2 \Phi - C \qquad [0° \leqslant \Phi \leqslant 90°] \qquad (4\text{-}2\text{-}1)$$

$$J = J^{180} \cos^2 \Phi - C \qquad [90° \leqslant \Phi \leqslant 180°] \qquad (4\text{-}2\text{-}2)$$

and which is often referred to as the *Karplus equation* or the Karplus rule. J^0, J^{180} and C are constants, for which the values $J^0 = 8.5$ Hz, $J^{180} = 9.5$ Hz and $C = -0.3$ Hz were originally chosen[1348] as most appropriate for an "unperturbed" H—C—C—H' fragment.

Dependence of the vicinal coupling constants on factors *other than the dihedral angle* is implicit in the original treatment[1348] and has been since stated quite clearly and explicitly by Karplus.[1354] Nevertheless, the *magnitude* of the other influences is such that they can be rightfully regarded as "perturbations" of the Karplus rule. These perturbations concern

(a) the electronegativities of substituents,
(b) the orientation of substituents,
(c) hybridization of the carbon atoms,
(d) bond angles,
(e) bond lengths,

and are discussed below.

Empirical confirmation of the Karplus rule was obtained from the close fit of a number of experimentally determined vicinal coupling constants in camphane-2,3-diols,[90] some sugar derivatives,[20] adamantane derivatives,[861,862] codeine derivatives[192] and other compounds, to the original Karplus curve or the closely similar Conroy curve.[531] An even larger number of data can be accommodated if the parameters J^0 and J^{180} are varied over the range of 10–16 Hz[20,15,1542,2591,551,1550,1501] while C remains at -0.3 Hz, i.e. is negligible for all practical purposes.

The use of these "modified" versions of the Karplus rule implies that while factors unrelated to the dihedral angle (see below) may influence the value of the constants in the Karplus equation, the *form* of the equation remains the same[1550] and thus the Karplus rule may be represented by a family of curves shown in Fig. 4–2–1. It would thus appear that, provided suitable values of the parameters J^0 and J^{180} are chosen (either by referring to a closely related compound or by solving simultaneous equations to satisfy the criterion of internal consistency[15,551,17,18,13,23,1550,1549,108]), the *exact* magnitude of the dihedral angle can be derived from the magnitude of the vicinal coupling constant. Unfortunately, although such a procedure is

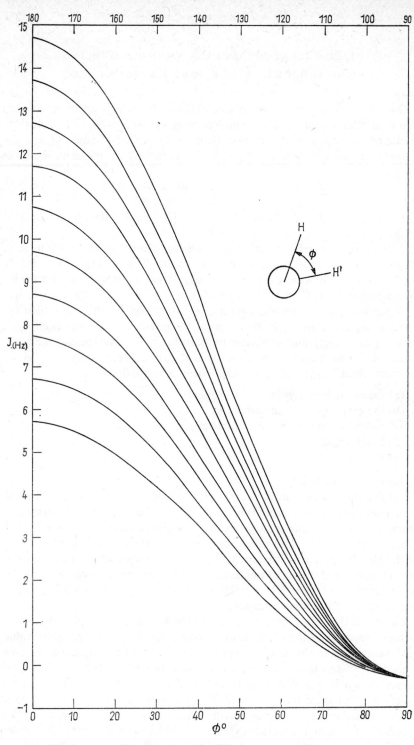

FIG. 4–2–1. Plot of function $J = J^0 \cos^2\varphi - 0.3$ $(J = J^{180} \cos^2\varphi - 0.3)$ for values of $J^0 = (J^{180})$ of 5.7–14.7 Hz.

highly attractive, it is unlikely to be accurate in many cases because (see below) some of the factors independent of the Karplus relation appear to be themselves orientation dependent.

(ii) INFLUENCE OF SUBSTITUENT ELECTRONEGATIVITY

The existence of an approximately linear relation between substituent electronegativities and the vicinal coupling constants in ethyl derivatives was realized[989] in 1956, but the analysis of a large number of data[162,20,1723,451] showed clearly that the effect of electronegativity, which may be expressed in the form

$$J = A - B.E. \tag{4-2-3}$$

where A and B are constants (A of the order of 9·5 and B of the order of 0·8 Hz) and E is the Huggins electronegativity, is different in various systems, i.e. A and B are not general constants. Thus, for ethyl derivatives[20,2353,366,374,732] the electronegativity dependence is rather low, the vicinal coupling constants in ethyl fluoride and ethyl lithium being 6·9 and 8·4 Hz respectively, but it is considerably higher in some rigid systems (see below). Nevertheless, equation (4–2–3) shows clearly that J_{vicinal} decreases with increasing electronegativity of the substituents.[2504,162]

It was also found[1064,20] that the effect of the electronegativity of substituents is not strictly additive in, for example, 1,2-disubstituted ethanes. Investigation of vicinal coupling constants in the conformationally stable hexachlorobicyclo[2.2.1]heptenes,[2588,1508] cyclopropanes[2592] and epoxides[2592,1534] showed that, while an *approximately* linear correlation between substituent electronegativity and J_{vicinal} exists, the slopes of the straight lines differ from system to system and also for different geometrical relations within the same system.[2588,1508,2592,1500]

These results, together with anomalies of a similar type which operate in other cyclic systems,[108,4,2318,1233,25] suggest strongly that the magnitude of vicinal coupling constants in saturated systems is influenced, not only by the electronegativity of substituents, but also by their *orientation*. A convincing proof of this hypothesis was produced by Williams and Bhacca[2580] who reported that $J_{\text{axial,equatorial}}$ in ring A of steroids of structure (I) is in the range of $5 \cdot 5 \pm 1$ Hz for R = OH or OAc, while $J_{\text{equatorial,axial}}$ in ring A of steroids of structure (II) is in the range of 2·5–3·2 Hz for R = OH or OAc. These ranges are substantially different, although in the absence of distortion, which is not considered likely, the dihedral angles are identical in both cases. The same phenomenon may be observed in the spectra of 4-t-butyl-cyclohexanols[95] and *m*-dioxans.[2036]

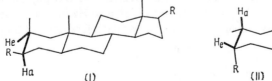

(I) (II)

Booth[291] extended the results of Williams and Bhacca[2580] to other systems and showed that, as a general rule, the electronegative substituent will exert its maximum effect (resulting in minimum $J_{vicinal}$) where an antiperiplanar relation exists between a part of the "coupling path" (indicated in heavy line) and the bond by which the electronegative substituent is attached [as in (II) but not in (I)]. It is also probable that a second maximum in the effect of electronegativity occurs for fully a eclipsed configuration. It is gratifying that the effect of electronegativity of substituents and their orientation is in good accord with theoretical predictions,[1354] but some anomalies may still exist.[25]

Although the magnitude of the "orientation effect" does not appear to be very large at first sight, its existence has a profound bearing on the applications of the Karplus rule, because it introduces a factor which itself depends upon the dihedral angle. Thus a straightforward use of the Karplus rule to obtain the dihedral angle from the magnitude of $J_{vicinal}$ could lead to very large errors under some circumstances. For instance, in one of the examples quoted by Williams and Bhacca,[2580] the axial–equatorial coupling constant of type (I) was found to be approximately 6·4 Hz, while the corresponding equatorial–axial coupling constant (identical substituent) of type (II) was 2·5 Hz. Taking the reasonable value of 10 Hz as J^0, Fig. 4–2–1 could be used to derive dihedral angles of 35° and 58°, respectively, for compounds where the dihedral angles, in the absence of conformational distortion, would be expected to be 60° in both cases. Even allowing for the fact that this pair of values represents an extreme in the range found for substances of this type,[2580] and without entirely rejecting the possibility of conformational distortion, the potential error remains very large, because a difference of 1 Hz in the value of $J_{vicinal}$ in the 60° region (with J^0 taken as 10 Hz) is seen from the Figure to correspond to an error of approximately 7°.

(iii) INFLUENCE OF HYBRIDIZATION

Although in this section we have limited the discussion of the vicinal coupling constants to coupling across three single bonds, systems of this sort may still vary, in so far as one or both of the intervening carbon atoms may be sp^2 hybridized. Bothner-By and Harris[306] have discussed this problem in some detail.

In freely rotating systems, including alkyl allenes† $J_{H-C-\overset{\parallel}{C}-H}$ is in the range of 5–8 Hz and varies in the expected manner when the population of certain conformers‡ alters.[313,302,305] The coupling between aldehydic protons and the protons on the carbon α to the carbonyl group is generally in the range of 1–3 Hz[1337] but depends strongly on the population of conformers[1337,2011,21,1535] and may be as high as 6 Hz in some cases.[1337] The coupling is positive in sign[1188] in all cases (see references in Chapter 4–4C).

† References 158, 203, 311, 346, 809, 1189, 1313, 1467, 1665, 1795, 2003, 2020, 2115, 2184 and 2324.

‡ Bothner-By and Günther[305] quote $J_{gauche} = 3·7$ Hz and $J_{trans} = 11·5$ Hz.

In cyclo-olefins,† the corresponding coupling constants show strong depend-ence on ring size in the general direction predicted by the Karplus rule, the maximum reported value being 13 Hz,[150] although the coupling in cyclo-butene[296,467] (0·80 Hz, probably negative sign)[296] indicates that, at least in small rings, other factors may be important. It thus appears that $J_{vicinal}$ in the fragment H—$\overset{\|}{C}$—C—H is analogous to that in H—C—C—H. A detailed investigation by Garbisch[950] showed however that an appreciable π-contri-bution of positive sign may be important for cases with dihedral angles near 90° (maximum contribution was estimated at +2·6 Hz)[950] and thus the mini-mum value of $J_{vicinal}$ in the fragment H—$\overset{\|}{C}$—C—H may be higher than ex-pected. Garbisch also set up empirical equations, relating $J_{H-\overset{\|}{C}-C-H}$ with dihedral angle, which contain π-terms as well as the usual σ-terms, but in view of the presence of other influences (see above) on vicinal coupling, values so obtained can be regarded as approximate only. In practical terms, the possibility of π-contributions would be important only for dihedral angles close to 90°, because such terms vanish both at 0° and at 180° and because they could only form a relatively small contribution to fairly substantial values of $J_{vicinal}$ in the intermediate range.

Fewer data are available for the fragment H—$\overset{\|}{C}$—$\overset{\|}{C}$—H. In butadiene derivatives‡ and some large rings,[1762] where the dihedral angle is likely to be close to 180°, $J_{H-\overset{\|}{C}-\overset{\|}{C}-H}$ is in the range of 10–12 Hz. In cyclopentadiene it is only 1·94 Hz,[1486] in cyclohex-1,3-diene 5·14 Hz,[1486] in a series of 1,3-cyclohexadienones[2531] 6–7 Hz, in various cycloheptatriene derivatives[1567, 924,1486,2480] 5–7 Hz, and in cycloheptadienone[466] 8 Hz. In cyclo-octatetra-ene,[93] where the dihedral angle is close to 90°, it is quite small. In α,β-un-saturated aldehydes†† the aldehydic proton is coupled to the α-proton with $J = 8$ Hz for the *transoid* conformation[308] and probably with $J = 2·6$ Hz for the *cisoid* conformation.[308,1690]

(iv) NEGATIVE VICINAL COUPLING CONSTANTS

In 1,2-dichloropropane and 1,2-dibromopropane, the indirectly obtained values[821] (see below) for J_{gauche} (dihedral angle = 60°) and J_{trans} (dihedral angle = 180°) are −1·5 Hz and +15·5 Hz respectively. It was considered[821] that the negative values for J_{gauche} are due to the parameter C in equa-tion (4–2–1) taking a value of −3 Hz instead of the commonly accepted value of −0·3 Hz. These results suggest that the parameter C, as well the con-stants J^0 and J^{180}, also depend on structure. The range of values of C could be quite large, because an *upward* displacement of the Karplus curve has also been considered.[1549,1543,108]

So far, the results obtained for J_{gauche} by the analysis of data for the con-formationally mobile 1,2-dihalogeno-propanes[821] have not been confirmed

---

† References 150, 296, 334, 756, 899, 940, 1486, 1507, 1760, 2302, 2473 and 2565.
‡ References 255, 306, 764, 1178, 1436, 1690 and 1862.
†† References 308, 689, 690, 911, 982, 1323, 2003 and 2184.

for any rigid systems. In 6,7-dichlorosteroids with ring A aromatic,[1914] the vicinal coupling constants corresponding to dihedral angles of approximately 70° and 50° were found to be 2·2 and 4·0 Hz, respectively, which can only be reconciled with the value of $-1·5$ Hz for J_{gauche} if the smaller coupling constant is negative. The absolute magnitude of the vicinal coupling constants in other 6,7-disubstituted steroids with ring A aromatic[2602] is similar to those in 6,7-dichloro derivatives[1914] and thus the results of sign determinations in these or analogous systems may have wider implications.

In a number of tetrahalogenoethanes,[1063,1064] the differences between the indirectly obtained *gauche* and *trans* vicinal coupling constants are also larger than expected from the Karplus equation† and include one negative J_{gauche}. In this case, the cumulative effect of the electronegative substituents (which are so oriented in the rotamers in which a *gauche* interaction occurs as to exert the maximum effect on the magnitude of vicinal coupling) may be responsible. Thus, while no final evidence exists that the parameter C in the Karplus equation is always close to $-0·3$ Hz, it cannot yet be considered as proven that it assumes large negative values.‡

(v) INFLUENCE OF OTHER FACTORS

It is expected on theoretical grounds[1354] that an increase in the magnitude of the angles H—C—C′ and C—C′—H′ in the fragment H—C—C′—H′ or H—C=C′—H′ will lead to a diminution of the vicinal coupling constant $J_{\text{HH}'}$. This effect is important[670] in the case of cyclic olefins (see below) and is probably reflected in the results obtained with cyclopentadiene,[1486] cyclohexadiene,[1486] cyclohexadienones[2531] and cyclobutene.[296] So far it has not been tested in the saturated fragment, but may be important in strained saturated systems. Similarly, the length of the central C—C′ bond (single or double) is expected to be inversely proportional to $J_{\text{HH}'}$. Other effects,[1073] has yet not extensively investigated, have been considered.[1354]

(vi) VICINAL COUPLING IN CYCLIC SYSTEMS

A large number of spectra of substituted cyclopropane derivatives, including fused ring systems, have been recorded†† and the ranges for vicinal coupling constants are $J_{\text{cis}} = 7·0–12·6$ Hz and $J_{\text{trans}} = 4·0–9·6$ Hz (see Table 4–1–1). Although the ranges overlap considerably, in accord with the Karplus rule J_{trans} is never larger than J_{cis} for any given pair and the effect of substituents can be estimated fairly accurately from the known dependence upon their electronegativity.[2592] Most of the references recorded above contain large

---

† A ratio of J^{180}/J^0 of approximately 2 would be required.

‡ The authors in collaboration with Dr. D.P. Kelly have found from the analysis of the spectra of 1,2-dihalogeno-3,3- dimethylbutanes, which appear to be conformationally pure, that the *gauche* vicinal coupling constants are positive, thus casting doubt on the assumptions involved in the indirect method[821].

†† References 1009, 1241, 1242, 1500, 1654, 1655, 1946, 2021, 2213, 2241, 2267, 2519, 2567 and 2592.

collections of data, and it is likely that if the substitution pattern of a cyclopropane is known, the configuration of substituents may be fairly reliably estimated, even if only one isomer is available, by referring to vicinal coupling constants in related cyclopropanes. The chemical shifts of the cyclopropyl protons also vary characteristically with the orientation of substituents (cf. Chapter 3–8 D) and thus afford an independent check on the assignments.

In heterocyclopropanes, J_{cis} is again always greater than J_{trans} as expected from the Karplus relation, but the whole range is displaced toward lower values, presumably due to the presence of the electronegative substituent on both carbon atoms. Some values for these systems are quoted in Table 4–1–1, and a large number of data are available for epoxides[291,1534,2091,2092,1796, 1571,1534,757] and other heterocyclopropanes.[1571,1796,1831,650,347] The variation of J_{vic} with electronegativity has been explored[2592,757] for epoxides. For some disubstituted epoxides[1534,2115] *trans* coupling constants of less than 1 Hz have been recorded. In episulphides and epoxides, fused to five- and six-membered rings, $J_{vicinal}$ is often small[2428] but is diagnostic in the steroid series.[2428]

Vicinal coupling constants in cyclobutanes,[719,1587,2585,962,1589] their silicon analogs,[985] and cyclobutanones[2239,1230,332,524,2195] are in the range of 4 to 13 Hz and no clear pattern emerges regarding the relative values of J_{cis} and J_{trans}. This is most likely connected with conformational preferences, but as with other vicinal coupling constants, substituent effects are superimposed upon them. In cyclobutenes,[341,467,296,1705,880] the vicinal coupling constants in the saturated fragment are in the range of approximately 1·6–4·9 Hz, i.e., considerably smaller than in the saturated analogues. It is possible that this may be connected with the effect[1354] of angular distortion mentioned above. Data for a series of substituted β-lactams[1320,182] are also available; J_{cis} is in the range of 4·9–5·9 Hz and J_{trans} 2·2–2·8 Hz.

A large amount of data on vicinal coupling in five-membered rings is available, but, as with cyclobutanes, no pattern can be discerned regarding the relative magnitudes of *cis* and *trans* coupling constants. This is obviously connected with the flexibility of cyclopentane rings, and in fact, the magnitude of vicinal coupling constants in five-membered rings has been used to obtain some information about their conformation.†

The actual range of values of vicinal coupling constants encountered in five-membered rings (generally 5–10 Hz) suggests that no special effects operate. A number of systems, e.g. dioxolans and carbonates,[94,4,890,1549] indolines,[107,1638,9] pyrazolines,[2459,1122,2265,1123] indenes and acenaphthenes,[1917,643,641,642,798] sultones,[1889] indanes,[2135] succinic anhydrides,[777] tetrahydrofurans,[941,207,13] oxazolines,[246] dihydrobenzofurans,[246,1002] γ-lactones,[1584] tetrahydropyrroles,[23,18,17] and furanosides,[1541,1525,14,2115, 1084,1552,13] among others, have been investigated fairly extensively and these data may prove of use for empirical assignments of configuration. A discussion of substituent effects on vicinal coupling constants in five-membered rings is given by Erickson.[777]

† References 13, 17, 18, 19, 23, 890, 941, 1084, 1549 and 1584.

Some points of interest which emerge are that in five-membered rings which cannot deviate appreciably from planarity,[107] such as acenaphthene,[641, 642,798] J_{cis} is always appreciably larger than J_{trans} as expected from the Karplus rule, and that the introduction of a double bond into the ring does not appear to influence appreciably the magnitude of the vicinal coupling constants in the saturated fragment as appears to be the case with cyclobutenes (see above).

Innumerable results obtained from the spectra of compounds containing six-membered rings[1233] or six-membered rings fused to other structures (e.g. steroids) establish firmly that, for conformations which do not depart appreciably from the chair form, $J_{\text{axial, axial}}$ is in the range of 8–13 Hz and $J_{\text{axial, equatorial}}$ is in the range of 2–6 Hz. In comparable systems† $J_{\text{equatorial, equatorial}}$ is invariably significantly smaller than $J_{\text{axial, equatorial}}$, the difference being of the order of 1 Hz. It has been proposed[2610] that this difference reflects the flattened form of the cyclohexane ring, but in view of the established effect of substituent orientation[291] (see above), it is difficult to evaluate this hypothesis in quantitative terms. The existence of well defined limits for diaxial vicinal interactions on one hand and axial–equatorial and equatorial–equatorial interactions on the other, has led to innumerable configurational and conformational assignments, inparticular in the sugar[811,551,1087,2615,935,1784] and steroid[241] fields, none of which has yet been challenged. Even where, due to virtual coupling (Chapter 2–3 C (xii)), the signal cannot be resolved (as is often the case with steroids and terpenes), the width of the multiplet at half height (W_H) is highly characteristic, an axial proton generally having W_H larger than 15 Hz while W_H of an equatorial proton is generally below 12 Hz.[1121] This permits the configurational assignment of any proton situated on a rigid cyclohexane ring in the chair form, provided its absorption can be separately observed. Even where conformational purity is uncertain, the assignment of a signal to an "essentially equatorial" or "essentially axial" proton is generally permissible.[1121,2276]

Some examples of configurational and conformational analysis by n.m.r. deal with rings larger than six-membered.[1067,1372,1454,1674,2215] In all such cases it has been assumed that no specific "ring-size" effects on vicinal coupling exist and there is no reason to expect that these assumptions are likely to be challenged.

In structures containing rigidly fused rings, the values of vicinal coupling constants, while varying with the substitution pattern, fall into characteristic ranges. Together with certain characteristic shielding effects, the values of J_{vicinal} thus afford a useful means of establishing stereochemistry, especially in the well investigated norbornane and norbornene series. The data quoted in Table 4–2–1 were taken from references 1501, other collections of data in this and related (e.g. bicyclo[2.2.2]octane and bicyclo[2.1.1]hexane) series may be found in the literature‡ and in references listed in connection with long-range coupling across four single bonds (Chapter 3–4 G). For vicinal

† References 9, 106, 167, 244, 495, 621, 806, 1542, 1543, 1550 and 2610.

‡ References 824, 831, 894, 937, 1465, 1508, 1760, 1827, 1828, 1857, 1943, 2133, 2323, 2363, 2366, 2434, 2438, 2563 and 2588.

coupling constants in other fused ring-systems involving 3-, 4- or 5-membered rings see references 505, 719, 2428.

TABLE 4–2–1. VICINAL COUPLING CONSTANTS (Hz) IN BICYCLO[2.2.1]HEXANES
AND BICYCLO[2.2.1]HEXENES

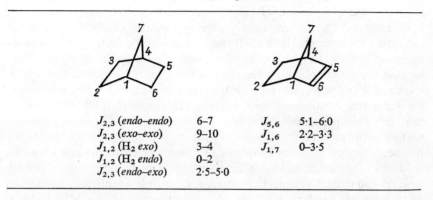

$J_{2,3}$ (endo–endo)	6–7		$J_{5,6}$	5·1–6·0
$J_{2,3}$ (exo–exo)	9–10		$J_{1,6}$	2·2–3·3
$J_{1,2}$ (H$_2$ exo)	3–4		$J_{1,7}$	0–3·5
$J_{1,2}$ (H$_2$ endo)	0–2			
$J_{2,3}$ (endo–exo)	2·5–5·0			

(vii) VICINAL COUPLING IN FLEXIBLE SYSTEMS[744,804,882]

As shown in Chapter 2–1, the chemical shifts and coupling constants obtained from the n.m.r. spectra of a species which consists of a mixture of rapidly interconverting forms are *average* values. In this section, we shall discuss the implications of this general phenomenon affecting the vicinal coupling constants in species which consist of mixtures of conformers. This particular case is important, because while any temperature (or solvent) variations in other coupling constants are essentially second-order effects, the proportions of conformers present in a rapidly equilibrating mixture are quite sensitive to temperature (and solvent) changes and hence the *average* values of $J_{vicinal}$ in flexible systems may vary markedly. Conversely, the average values of $J_{vicinal}$ may be used to obtain information about the proportion of conformers present under any specific set of conditions. As implied by equations (2–1–7) and (2–1–8) (p. 60), the average parameter (J or δ) is the arithmetic mean of the parameters for the components of the rapidly equilibrating mixture, and hence if the parameters for each component in a two-component mixture are accurately known, the proportion of each can be quite simply extracted. In cases of more than two components, considerable complications[1063] arise but we shall limit ourselves here essentially to the two component problem, which is the more commonly encountered situation in conformational investigations of interest in organic chemistry.

It is important to realise that the average vicinal coupling constant is an average of two (or more) vicinal coupling constants and *not in any way related to the coupling constant in a single intermediate conformation* with an intermediate value of the dihedral angle.

Some of the problems involved in distinguishing between a single conformer and a mixture of rapidly interconverting conformers have been discussed,[2441,24,290,1610,534,804] and a detailed discussion of the application of n.m.r. to conformational problems in cyclohexanes is given by Franklin and Feltkamp.[882]

The principal difficulty in such investigations is the determination of the parameters (here $J_{vicinal}$) in the pure conformers. The most satisfactory procedure is the observation of the "frozen out" spectra which, as shown in Chapter 5–2 (p. 55), are simply superimpositions of the spectra of the individual components (here conformers) and which directly yield the parameters of interest. In practice, however, such spectra can be rarely obtained, because the magnitude of barriers to conformational inversion are generally too small. In principle, it is possible to extract the values of n.m.r. parameters for individual conformers by observing the temperature[1064,1063] or solvent[1063,821] dependence of the averaged spectrum. This approach appears to be fruitful when applied to simple substituted ethanes, but it has so far not been applied to many problems of direct interest to organic chemists.

Most of the numerous studies involving conformational problems in both acyclic† and cyclic‡ systems have utilized, more or less explicitly, the method of "reasonable fixed models" for obtaining the magnitudes of the vicinal coupling constants (and chemical shifts; Chapter 3–8 G) in the individual conformers. Clearly, the validity of the results obtained in this manner depends entirely upon the appropriateness of the model, which may be a very closely related substance or merely a general value of coupling constants. The examples given below illustrate some common approaches to conformational problems:

(a) *Conformation of cyclohexanol*[95]

The n.m.r. spectrum of cyclohexanol (or its acetate) is impossible to analyse because of the presence of a large number of overlapping lines. The spectra of the 3,3,4,4,5,5-hexadeutero derivatives can however be analysed to give the conformationally averaged values of "large" and "small" vicinal coupling constants. At the same time *cis*- and *trans*-3,3,4,5,5-pentadeutero-4-t-butylcyclohexanols were prepared and their spectra were analysed in the same way to give the values for $J_{axial,axial}$ (11·07 Hz), $J_{axial,equatorial}$ and $J_{equatorial,axial}$ (3·00 and 4·31 Hz) and $J_{equatorial,equatorial}$ (2·72 Hz). The nonequality of the three gauche interactions is of the type expected from the configuration of the electronegative substituent (the hydroxyl group) or from ring flattening (see above).

It was reasonably assumed that the 4-t-butyl derivatives exist in only one conformation (*viz.* that with the t-butyl group equatorial), and that the

† For some references dealing with acyclic systems see 302, 312, 313, 369, 815, 820, 821, 929, 933, 1315, 1337, 1342, 1524, 1535, 1641, 1642, 1869, 1872, 1927, 2011, 2174, 2224, 2256 and 2401.

‡ For some references dealing with flexible cyclic systems see 24, 95, 103, 104, 229, 288, 289, 290, 293, 294, 295, 298, 370, 373, 423, 434, 487, 534, 712, 713, 743, 746, 747, 800, 804, 805, 806, 807, 822, 884, 917, 918, 919, 947, 955, 1232, 1234, 1289, 1319, 1531, 1543, 1549, 1559, 1591, 1610, 1704, 1771, 1825, 1859, 1934, 2014, 2015, 2016, 2082, 2084, 2220, 2349, 2350, 2441 and 2465.

magnitudes of the four vicinal coupling constants, in the *individual conformers* of the flexible deuterated cyclohexanol and its acetate, are the same as those in the conformationally fixed 4-t-butyl derivatives. This allows the calculation of the proportions of the two conformers from the spectra of the flexible compounds, for any particular solution.

In an analogous example, the conformation of 1,2-cyclohexanediols and their derivatives in solution was established[1542] using isopropylidene derivatives as "fixed models".

(b) *Conformation of 2-bromocyclohexanone*[952]

This example illustrates the use of vicinal coupling constants in conformational analysis without the necessity of obtaining the actual values of the coupling constants. The 2-proton in 2-bromocyclohexanone gives rise to the X part of an ABX pattern, which may be further complicated by virtual and long-range coupling, but in which the separation of the strong outer lines will always be equal to the sum of the two vicinal coupling constants ($J_{AX} + J_{BX}$) (see Chapter 2–3). As both vicinal coupling constants would be averages of those in the two postulated conformers (chair forms with bromine axial and equatorial) the sum would also be an average value. Similarly, $J_{AX} + J_{BX}$ was also determined for a number of conformationally fixed derivatives (e.g. *cis*- and *trans*-4-t-butyl-2-bromocyclohexanones) and were found to be 18·2 Hz for an axial proton and 5·7 Hz for an equatorial proton. The average values obtained from the spectra of 2-bromocyclohexanone itself at various concentrations in carbon tetrachloride were then interpreted in terms of percentages of conformers with the 2-proton axial and equatorial, respectively. An independent check on these figures was then obtained by performing similar calculations using the average chemical shifts in conjunction with the model chemical shifts obtained from the same fixed derivatives; the two methods gave similar results.

Numerous papers deal with the configuration and conformation of α-halogenoketones in rings A of steroids and triterpenes.[299,1549,15,2276,890]

(c) *Assignment of* threo *and* erythro *configurations*

This example illustrates the possibility of assigning the configuration in a conformationally mobile molecule without the use of fixed models.

Consider the Newman projections of the staggered conformations of a generalized *erythro* (IIIa, b, c) and *threo* (IVa, b, c) diastereoisomers where L denotes a large and S a small substituent (each pair of L's and S's are two different groups).

It is quite obvious from a consideration of non-bonded interactions that the conformer (IIIa) will contribute more heavily to the mixture III than conformer (IVa) to the mixture IV. Knowing that J_{trans} ($\Phi = 180°$) will invariably be larger than J_{gauche} ($\Phi = 60°$), in accordance with the Karplus equation, we predict that, at ordinary temperatures, the average vicinal coupling constant (J_{AB}) in the *erythro* isomer will tend to be larger than that in the *threo* isomer,[1402] thus enabling us to assign their relative configurations.

As an example,[1870] the vicinal coupling constants (at room temperature; 10% solutions in $CDCl_3$) of the *erythro* and *threo* isomers of *p*-methoxycinnamic acid dibromides (where bromine is considered as the large group) were found to be 11·6 and 9·9 Hz, respectively. Other examples have been reported.[1402]

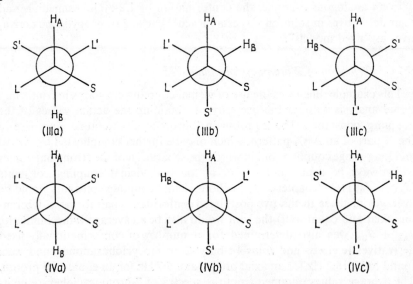

In case where the spatial requirements of the "large" and "small" groups are very different, it could be perhaps assumed that J_{AB} in the *erythro* isomer would be essentially J_{trans} (in the range above 10 Hz) and that in the *threo* isomer essentially J_{gauche} (in the range below 5 Hz). It would then be possible to assign the configuration of a *single* isomer with some confidence.

If possibilities for intramolecular hydrogen bonding exist,[2049,961,1711, 2634,37] this factor may outweigh simple steric considerations and lead to a larger $J_{vicinal}$ for the *threo* isomer, but even under those circumstances n.m.r. can be used for assignment of configuration, although in a more indirect manner.[2049] A completely independent argument can also be constructed on the basis of shielding differences[169] if one or more of the substituents is significantly magnetically anisotropic.

(viii) USES AND ABUSES OF THE KARPLUS EQUATION

Having considered the factors which influence the magnitude of vicinal coupling constants in saturated systems, it may be useful to recapitulate at this stage the status of the Karplus equation before going on to some examples of its application. It must be remembered that some of the applications of the Karplus equation to stereochemical problems were made before its limitations were fully realized[1354] and attempts were made to "read off" the values of dihedral angles directly from the magnitudes of the vicinal coupling constants, using equations (4–2–1) and (4–2–2) with their original parameters. Later, "modified parameters" and "self consistent analyses" were introduced, but, in view of the marked dependence of J_{vicinal} upon the *orientation* of electronegative substituents, even these procedures are likely to involve significant approximations.

It is obviously desirable to decide the scope and limitations of the Karplus rule and consider ways in which quantitatively reliable data can be obtained from it.

One approach to the problem would be to construct extended forms of the equations (4–2–1) and (4–2–2) or their equivalents,[1354] which would contain functions allowing for the influence of substituent electronegativity, substituent orientation, decay of additivity of substituent influence, π-terms, angle distortion terms etc.[1354] These could then be tested against experimental results and the various constants could be adjusted to give a best fit relation. As far as the authors are aware, no work on these lines has yet appeared in the literature, but certain limitations can be pointed out. Firstly, the amount of data for systems where conformational distortion can be ignored completely is rather small and such data refer mainly to only a few values of the dihedral angle. Secondly, even in the case of very closely related compounds, the dependence of J_{vicinal} on electronegativity is not always prefectly linear. Finally, the contribution of π-terms, and probably other factors as well, appears to lead only to approximate corrections. Because errors are as likely to be cumulative as cancelling, there is no reason to suppose that an accuracy of better than about 10° is likely to be achieved by this approach.

While the above may seem unduly pessimistic, it can be shown that, in practice, the problems which cannot be solved accurately are mainly limited to the determination of the *exact* degree of conformational distortion in systems which exist in the form of a single, strained conformer rather than in the form of an equilibrium mixture of two or more forms (see above). This is so, because of the undoubtedly reliable properties of the Karplus rule; in particular:

(i) The values of J_{vicinal} will always diminish as Φ approaches 90° from either the direction of 0° or 180°.

(ii) The values of J_{vicinal} will generally be either "large" or "small", and reference to molecular models and to the family of curves in Fig. 4–2–1 will usually allow a choice to be made on inspection, even if the J^0 and J^{180} parameters have to be estimated quite crudely by reference to similar compounds, some values being obviously incompatible with some configurations.

(iii) It is probably safe to assume that, to a first approximation, *small* deviations in J_{vicinal} in fragments H—C—C—H correspond to small changes in dihedral angles in a manner shown by equations (4–2–1) and (4–2–2). Thus, the latter can be *approximately* estimated and gives the direction of the suspected conformational distortions. It is doubtful whether this procedure is sound when applied over a larger range of Φ-values.

These properties, when used with discretion, permit the solution of an enormous number of problems by procedures which will be exemplified below.

It is essential that when a procedure involving the formal calculation of

Karplus coefficients is used, and such procedures are quite convenient and add to the clarity of the interpretation, the *quantitative* results must be heavily qualified.

We shall now illustrate the scope of configurational and conformational analysis utilizing the Karplus rule by means of a few examples. The principal criterion for choosing these particular cases from the very large number available was the simplicity of presentation and caution in making extensive conclusions. On the other hand, completely obvious cases such as differentiation between axial and equatorial protons in cyclohexanes were avoided as too trivial.

(ix) EXAMPLES OF APPLICATION OF THE KARPLUS RULE

(a) *Assignment of configuration and conformation in the addition product of 16-dehydropregnenolone acetate and methyl vinyl ether*[1976]

The gross structure of the product was found to be V. The configuration at the newly created asymmetric centre (C_{16b}) and the conformation of the dihydropyran ring were determined as follows: the proton at C_{16b} gives rise to a doublet of doublets 291 Hz downfield from TMS (at 60 MHz) with splittings of 8·6 and 4·4 Hz. Treating this as the X part of an AMX system, i.e. equating the splittings with coupling constants, and converting them to dihedral angles using modified J^0 and J^{180} values derived from carbohydrate series,[1550] gives the following possible dihedral angles: For $J = 8\cdot6$ Hz, $\Phi_{acute} = 11\cdot15°$ and $\Phi_{obtuse} = 158°$ while for $J = 4\cdot4$ Hz, $\Phi_{acute} = 45°$ and $\Phi_{obtuse} = 132°$.

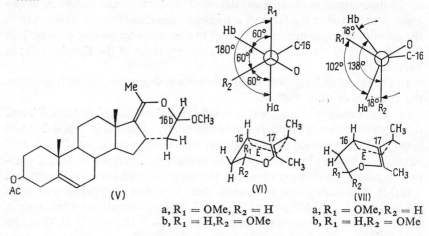

a, R_1 = OMe, R_2 = H   a, R_1 = OMe, R_2 = H
b, R_1 = H, R_2 = OMe   b, R_1 = H, R_2 = OMe

The dihydropyran ring can exist in two forms, the half-chair form (VI) and the half-boat form (VII), each with two possible configurations of the methoxyl group (a and b). A study of Dreiding models leads to a number of values for the dihedral angles which are shown in Table 4–2–2 and in the projections of (VI) and (VII), together with the four possible values of the dihedral angle calculated from the coupling constants (see above). It can be seen that only one of the configurations of the methoxyl group in the half-boat form is in approximate agreement (starred values in Table 4–2–2) with a self-consistent set of dihedral angles calculated from the vicinal coupling constants.

TABLE 4–2–2. DIHEDRAL ANGLES OBSERVED IN DREIDING MODELS
COMPARED WITH ANGLES CALCULATED FROM SPECTRAL DATA

| | | Degrees | | |
Structure	Φ	Observed[a]	Calculated[b]	Δ
VIa	R_2Ha	60	11·5	+48·5
	R_2Hb	60	45	+15
VIb	R_1Ha	180	158	+22
	R_1Hb	60	45	+15
			132	+48
			11·5	+48·5
VIIa	R_2Ha	18	45	−27
	R_2Hb	138	158	−20
			11·5	+6·5
			132	+6·0
VIIb	R_1Ha	102	158	−56
	R_1Hb	18	45	−27
			132	−30
			11·5	+6·5

[a] From Dreiding models (cf. projection structures VI and VII).
[b] From modified Karplus equation (cf. text).

It is important to remember that this type of analysis can only be considered reliable if the variations in the deviations between the expected and determined results are large. In the particular example quoted above, a further uncertainty is introduced in the first-order analysis of the spectrum, but it is unlikely to be large because of the self-consistency of the results.

(b) *Stereochemistry of some santonin derivatives*

The stereochemistry of a number of sesquiterpenes related to santonin has been determined (or confirmed) by n.m.r. spectroscopy[1980] without directly invoking any numerical parameters for J^0 and J^{180} in equations (4–2–1) and (4–2–2). For instance, the following vicinal coupling constants can be obtained by inspection of the spectrum of artemisin acetate (VIII). $J_{6,7} = 11\cdot6$ Hz;

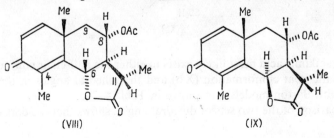

(VIII)   (IX)

$J_{8,9axial} = J_{8,7} = 10.9$ Hz; $J_{7,11} = 11.5$ Hz. All these values are too large to be accommodated in any other manner than in the all *trans* relationship shown in structure (VIII). On the other hand, in 6-epiartemisin acetate (IX), $J_{8,9axial} = J_{8,7} = 10.5$ Hz; $J_{7,11} = 0$ Hz; $J_{6,7} = 5.7$ Hz. Molecular models show that epimerization at C-6 decreases the dihedral angles H_6—C_6—C_7—H_7 and H_7—C_7—C_{11}—H_{11} from nearly 180° to approximately 25° and 100° respectively, i.e. in agreement with the changes in the magnitudes of vicinal coupling constants actually observed. Epimerization at C_8 or C_{11}, on the other hand, would alter the magnitude of only one vicinal coupling constant, while epimerization at C_7 would lead to a much smaller dihedral angle H_7—C_7—C_{11}—H_{11} and hence a non-vanishing $J_{7,11}$.

The stereochemical assignments were then further confirmed by referring to vicinal coupling constants in closely related compounds of known stereo-chemistry[1980] and by reference to the steric dependence of the long-range coupling between the methyl at C_4 and H_6 (see Chapter 4–4).

Often, fairly complex stereochemical problems may be solved in this manner without any necessity for comparison between the actual coupling constants and those calculated on the basis of the Karplus equation, the observation of molecular models allowing one to assign "large" or "small" values to a number of vicinal coupling constants.

(c) *Conformation of shikimic acid*[1085a]

The spectrum of shikimic acid in D_2O yielded the parameters shown in Table 4–2–3. By using the Karplus equation with either the original (un-modified) or modified parameters ($J^0 = 9.3$ Hz, $J^{180} = 10.4$ Hz), the vicinal coupling constants could be converted into the dihedral angles shown in Table 4–2–4.

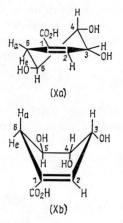

(Xa)

(Xb)

Now, shikimic acid most likely exists in either the half-chair conformation (Xa) or the boat conformation (Xb) and the dihedral angles in these (obtained with Barton models) are shown in Table 4–2–5.

Comparison of the two sets of dihedral angles shows that the data derived from vicinal coupling constants favour the half-chair form. $J_{2,3}$ was not considered because of sp^2 hybridization of C_2, but the dihedral angles H_2—C_2—C_3—H_3 calculated on the basis of $J_{2,3} = 4.0$ Hz and either of the sets of Karplus parameters are approximately 44° (or 136°) or 47° (or 133°) i.e. also in accord with the half-chair form.

TABLE 4–2–3. SPECTRAL DATA FOR SHIKIMIC ACID

	Chemical shifts (δ-values)†						
	H-2	H-3	H-4	H-5	H-6e	H-6a	
First order	6·88	4·41	3·69	4·00	2·87	2·14	
From analysis	6·88	4·41	3·76	4·18	2·81	2·11	

	Coupling constants (Hz)							
	$J_{2,3}$	$J_{2,6}$	$J_{3,4}$	$J_{3,6}$	$J_{4,5}$	$J_{5,6a}$	$J_{5,6e}$	$J_{6a,6e}$
First order	4·0	1·8	3·9	1·5	8·4	6·2	5·0	18·5
From analysis	4·0	1·8°	3·8	1·5°	8·4	5·9	4·8	18·5

† Since spectra were measured in deuterium oxide solution, no signals were observed from—OH or —CO_2H.
° Uncorrected values.

TABLE 4–2–4. EXPERIMENTAL DIHEDRAL ANGLES
FOR SHIKIMIC ACID (DEG.)

Parameters	3, 4	4, 5	5, 6a	5, 6e
Unmodified	46	163	144	39
Modified	48	156	141	42

TABLE 4–2–5. DIHEDRAL ANGLES† (DEG.) FOR SHIKIMIC ACID, AND THE VALUES (Hz) OF J^0 AND J^{180} NECESSARY TO REPRODUCE THE OBSERVED COUPLING CONSTANTS

	H-2, H-3	H-3, H-4	H-4, H-5	H-5, H-6a	H-5, H-6e
Half-chair	43	50 $(J^0 = 10)$	180 $(J^{180} = 8·7)$	170 $(J^{180} = 6·4)$	50 $(J^0 = 12·4)$
Boat	115	60 $(J^0 = 16·4)$	120 $(J^{180} = 34·8)$	180 $(J^{180} = 6·2)$	60 $(J^0 = 20·4)$

† Using Barton models.

This example illustrates that a fairly definite choice can sometimes be made between two conformations even if neither of the expected sets of dihedral angles is in very good agreement with those obtained from the Karplus relation. The alternative interpretation, involving the presence of distorted forms, was also considered.[1085a]

It can also be seen that the *exact* magnitudes of the coupling constants and the exact values chosen for the parameters J^0 and J^{180} in the Karplus equation need not be critical to the solution of a problem in which a choice can be made between two conformations (or configurations) with sufficiently widely differing dihedral angles. For instance, to reproduce the angles for the boat conformation would require some unreasonable values for J^0 and J^{180} (Table 4-2-5).

(x) COUPLING IN THE SYSTEM CH_3—CH

In acyclic fragments, spin–spin coupling in this system can always be easily interpreted and needs no special discussion. However, numerous reports suggest that secondary methyl groups attached to cyclic systems are split by vicinal methine protons in a manner which is characteristic of their orientation. Thus in six-membered rings, the splitting of an axial methyl group by an equatorial methine group always appears to be larger than the splitting of an equatorial methyl by an axial methine group for any given pair of compounds.†

The origin of this effect is not entirely clear. In some cases it was recognized[2303,955] that the different appearance of the signals of the axial and equatorial methyl groups was due to second-order characteristics in the spectrum, i.e. that the equatorial methyl group *appears* to be less split because the axial methine proton is more strongly coupled to other protons in the ring, thus giving rise to virtual coupling (Chapter 2–3C(xii)). It is however possible that a genuine difference in the *coupling constants* also exists. Methyl groups attached to heterocyclopropanes[1831,347] and to the α-carbons in five-membered saturated heterocycles[107,94] appear to be less split than those attached to β-carbons in five-membered saturated heterocycles,[107] presumably due to the effect of the electronegative heteroatom. Similar effects have been observed in β-lactones[182] and may be connected with rehybridization of the ring carbon atom.

(xi) COUPLING IN THE SYSTEM H—C—X—H
WHERE X IS A HETEROATOM

The heteroatoms commonly found in structures containing the fragment H—C—X—H are nitrogen, oxygen and sulphur and hence, with the exception of nitrogen, the vicinal coupling constants $J_{H,H}$ would mainly be values averaged by rotation about the C—X bond. Further, under most conditions, rapid prototropic exchange takes place, removing any coupling to protons in groups such as —OH, —NH, and —SH (Chapter 2–1). Nevertheless, by appropriate choice of solvents the relevant coupling constants can often be obtained and may be of considerable diagnostic importance.

Where coupling constants $J_{H-C-N-H}$, $J_{H-C-S-H}$[1672] and $J_{H-C-O-H}$ can be

† References 535, 955, 1289, 1290, 1317, 1800, 1829, 2297 and 2303.

obtained, they generally range from 5 to 9 Hz for freely rotating systems. In some alcohols, variations of $J_{H-C-O-H}$ with solvent have been attributed to changes in conformation about the C—O bond.[1772] In some cases where an antiperiplanar conformation predominates, because of hydrogen bonding, values of 10–12 Hz for $J_{H-C-O-H}$ have been reported.[949] Coupling of this sort can always be positively identified either by exchanging the labile proton with D_2O or by promoting faster exchange through the addition of a trace of acid (trifluoroacetic acid is suitable), and hence no attempts are generally made to correlate particular ranges of values assumed by $J_{H-C-X-H}$ with structure.[1772]

Analysis of the spectrum of ethanol as an A_3B_2C system[1848] showed that the signs of the two vicinal coupling constants (J_{CH_2,CH_3} and $J_{CH_2,OH}$) are the same and double irradiation experiments[1229] confirm that the signs of the geminal and vicinal ($J_{H-C-O-H}$) coupling constants in 2-nitropropylene glycol are opposite. Thus, there is little doubt that $J_{H-C-O-H}$ (and by implication $J_{H-C-X-H}$) is positive in sign.

The principal utility of this type of coupling has been in determining the position of tautomeric equilibria[701,956,707,709] or sites of protonation. Thus the Schiff's base from benzylamine and 2-acetyl-α-naphthol was shown to exist in the form (XIa) rather than (XIb) because the benzylic methylene group was split[701] by the NH proton with the normal coupling constant $J_{H-N-C-H}$ = 5 Hz. On exchange with D_2O the doublet assigned to the benzylic protons collapsed, thus confirming the assignment.

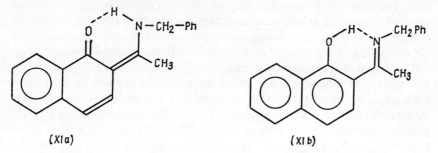

(XIa)   (XIb)

The spectrum of 4-methyl-2-pyridone[1366] in sulphuric acid shows that H_6 is coupled to *one* NH proton, thus confirming that protonation occurs on the oxygen.

The multiplicity of the signal assigned to the hydroxyl proton under conditions where exchange is very slow, can be used[464] to distinguish primary, secondary and tertiary alcohols (triplet, doublet and singlet, respectively). The best solvent for this purpose appears to be dry dimethyl sulphoxide,[464,112,445] but it is interesting to note that the same effect is found in carbon tetrachloride or deuterochloroform[375,840,377] which have been rigorously freed from traces of hydrochloric acid. An analogous approach is useful with amine salts[84,1602] which can often be prepared *in situ* by using CF_3COOH as solvent.

Other examples of vicinal coupling in saturated systems with heteroatoms forming part of the coupling path are shown in Table 4–2–6.

TABLE 4–2–6. MISCELLANEOUS EXAMPLES OF
VICINAL INTERPROTON COUPLING IN SATURATED
SYSTEMS ACROSS HETEROATOMS

System	$J_{H-X-Y-H}$ (Hz)	References
H—C—P—H	7·7–8·2	2546, 1667
H—C—P—H $^\oplus$	5·5	2285
H—P—P—H	6·8	1667
H—C—Si—H	3·30–4·68[a]	734, 594
H—C—Sn—H	2·4–2·7	832
H—C—Pb—H	1·5	832
H—Si—Si—H	2·5–4·0[b]	2466

[a] Shows linear dependence on the electronegativity of the group attached to silicon.
[b] Poor correlation with the electronegativity of the group attached to silicon.

(xii) VICINAL COUPLING AS AN AID IN STRUCTURE DETERMINATION

From the discussion in sections (i)–(xi) above, it is clear that the consideration of vicinal coupling constants is a powerful aid in the determination of structure and stereochemistry of unknown substances. In this section we shall enumerate briefly some common errors which must be avoided in connection with the most widely used aspect of vicinal coupling, namely the determination of the number of protons on adjacent carbon atoms.

The degree of splitting of any resolved resonance is clearly connected with the number of magnetic nuclei which are spin–spin coupled to the proton (or a group of equivalent protons) responsible for it. Because it is commonly assumed that vicinal coupling constants are "in the range of 3–10 Hz", geminal coupling constants "in the range of 12–15 Hz" and long-range coupling constants "are negligible" there has been a tendency to arrive at the number of protons vicinally (or geminally) coupled after only a cursory inspection of spectra. As shown in Chapter 4–1, and in sections (i)–(xi) above, these ranges, while representing coupling constants in commonly found fragments, *are not in fact* representative and careless interpretation is also open to other pitfalls. Some of the main sources of error are enumerated below:

(a) The range of geminal coupling constants completely overlaps the range of vicinal coupling constants, if only absolute magnitudes (which are usually the only available quantity) are considered. Fortunately, although vicinal coupling constants in olefinic systems (see below) also overlap the range of $J_{H-C-C-H}$, confusion rarely arises because of the characteristic chemical shifts of olefinic protons.

(b) Very small, or zero, vicinal coupling constants are not common (dihedral angles close to 90°).

(c) Quite appreciable long-range coupling constants are common (Chapter 4–4) and may be (cf. examples 2 and 3 in Table 4–4–1, p. 317) larger (absolute magnitude) than vicinal coupling constants in the same molecule.

(d) Coupling to other magnetic nuclei, in particular ^{19}F and ^{31}P must also be considered (Chapter 4–5).

(e) As with all applications of coupling constants, pitfalls associated with equating experimentally obtained *splittings* with coupling constants must be avoided (Chapter 2–3).

Other difficulties arise in stereochemical and conformational problems (see above) and from mistaken assumptions about symmetry and equivalence (Chapter 5–2). To sum up, while the use of vicinal coupling in structural work can lead to valuable results, it should be approached with more caution than has often been the practice in the past.

B. VICINAL INTERPROTON COUPLING ACROSS ONE DOUBLE AND TWO SINGLE BONDS

Although the mechanism of vicinal coupling in the system H—C=C—H is generally considered to be closely related[1354,670,1348] to that in the system H—C—C—H, the fact that in the former the dihedral angle is limited to two values (0° and 180°) has focussed attention on the variations of $J_{H-C=C-H}$ with factors such as electronegativity of substituents and H—C=C angles, which are of secondary importance in the H—C—C—H case (see Section A above).

Ethylenic coupling constants are of the same sign[1515] as $J_{13C,H}$ and hence positive.

The literature contains a very large amount of scattered data for coupling in olefinic systems, but as a number of valuable correlations have been established, the vast majority of them need not be consulted for the purpose of interpreting the spectra of unknown compounds.

Valuable collections of data for acyclic systems can be found in the work of Banwell and Sheppard,[159] Brügel, Ankel and Krückberg,[380] Schaefer[2182] and Laszlo and Schleyer,[1510] and for cyclic system in the work of Chapman,[461] Garbisch,[953] Smith and Kriloff,[2302] Tori and Nakagawa[2435] and Laszlo.[1500]

(i) *Cis* AND *trans* COUPLING

It was noticed quite early that *trans* ethylenic coupling constants are always larger than *cis* ethylenic coupling constants in any vinyl compound or for any pair of 1,2-disubstituted ethylenes. This relationship can be readily verified by inspection of Tables 4–1–2 and 4–2–7, respectively; it can also be seen that the two ranges overlap so that, in the absence of further information, assignments can not always be made unless a pair of 1,2-disubstitited ethylenes is available. Fortunately (see below), the correlations of $J_{ethylenic}$ with substituent electronegativity are so reliable that, in fact, assignments can be made with confidence even when only one isomer of a 1,2-disubstituted ethylene is available, provided that the nature of the substituents is known.

TABLE 4–2–7. VICINAL COUPLING CONSTANTS IN
DISUBSTITUTED OLEFINS X—CH=CH—Y†

X	Y	J_{cis} (Hz)	J_{trans} (Hz)
F	F	−2·0	9·5
F	Br	3·5	11·0
Cl	OEt	6·0	—
Cl	Cl	5·3	12·1
F	Me	4·5	11·1
OCOMe	C_5H_{11}	7·0	12·5
Br	Br	4·7	11·8
OR	R	6·2–6·7	12·0–12·6
OMe	C≡CH	6·8	—
Ph	SPh	11·0	14·4
Cl	CH_2Cl	7·2	13·1
Ph	SCH_2Ph	11·6	15·5
Ph	COOH	12·3	15·8
Ph	CN	—	17·1
COOEt	COOEt	11·9	15·5
COOMe	Me	11·4	15·5
Me	CN	11·0	16·0
H	H	11·7	19·0

† These data are selected from ref. 1510.

The relationship $J_{trans} > J_{cis}$ is predictable from theoretical considerations[1348] and corresponds to the larger J^{180} than J^0 in the system H—C—C—H.

(ii) INFLUENCE OF ELECTRONEGATIVITY OF SUBSTITUENTS

Electronegative substituents tend to diminish the magnitude of J_{trans} and J_{cis} in vinyl compounds[2182] (cf. Table 4–1–2 and Fig. 4–1–3) and 1,2-disubstituted ethylenes[1500,1510] (cf. Table 4–2–7). The relationship is approximately linear with the electronegativity, or the sum of the electronegativities, respectively, and hence can be expressed by an equation of the type $J = A - B \cdot E$ (p. 283). However, as with the system H—C—C—H the lines of best fit (cf. Fig. 4–1–3) for various coupling constants are of a different slope,[1500,2182,1510,162,48] and this is presumably connected with the orientation of the electronegative substituent with respect to the coupling path, and is in accord with theory.[1354]

In spite of some scatter, the above relationship exemplified by the data in Tables 4–1–2 and 4–2–7 should make it possible to assign the configuration of any 1,2-disubstituted ethylene in an unequivocal manner.

(iii) INFLUENCE OF H—C=C ANGLE

Theory predicts[1354] that $J_{H-C=C'-H'}$ (and $J_{H-C-C'-H'}$) should decrease with an increase of the angles H—C=C' and C=C'—H' (or H—C—C' and C—C'—H'). This effect is difficult to confirm experimentally for vicinal coupling across three single bonds (cf. Chapter 4–2A(v)) but it is extremely important in the case of *cis* olefinic coupling in cyclic structures, where a strong dependence of $J_{H-C=C-H}$ on ring size (and hence on the angle H—C=C) has been established.[461,2302,1507,670,2435]

The ranges quoted in Table 4–2–8 are based on results obtained with simple cyclo-olefins,[2302,1507,953] some fused ring systems and α,β-unsaturated carbonyl compounds.[461]

TABLE 4–2–8. *Cis* OLEFINIC COUPLING
CONSTANTS IN CYCLIC SYSTEMS

Ring size	$J_{H-C=C-H}$ (Hz)
3	0·5–1·5
4	2·5–3·7
5	5·1–7·0
6	8·8–11
7	9–12·5
8	10–13
cis-Cyclononene	10·7
cis-Cyclodecene	10·8

While the correlation embodied in Table 4–2–8 is of obvious utility, particularly in distinguishing between 4-, 5- and 6-membered rings, caution should be exercised for cases where the double bond is substituted by electron withdrawing groups and/or delocalized (cf. below) as in compounds XII–XXI. Generally,[1500] ether functions diminish $J_{vicinal}$ in cyclic olefins but ketones **do not**.

(iv) INFLUENCE OF BOND-ORDER OF THE DOUBLE BOND

Although the vicinal coupling H—C=C—H is considered to be essentially transmitted through the σ-electron framework, the bond-order or bond length of the double bond may also constitute one of the factors determining the magnitude of the olefinic coupling constant.[2435,1500] The relation between hybridization of the olefinic carbon atoms (as determined from $J_{13C,H}$) and the olefinic coupling constant has also been investigated.[670,2435] However, as valence angles, bond orders, bond lengths and hybridization are closely inter-related in cyclic olefins, it is unlikely that, at the present state of knowledge, these considerations could be directly applied to the solutions of organic chemical problems.

The theoretically related dependence of *ortho* coupling on the bond order in aromatic systems is probably of more immediate interest (see below). On the other hand, observed decreases in vicinal olefinic coupling constants on formation of π-complexes with transition metals to indicate that participation of a double bond in the formation of a π-complex might be detected by observing changes in olefinic coupling constants. If the method is to be used, it is crucial to determine whether, in fact, a π-complex or a σ-complex is involved.[2358,1066]

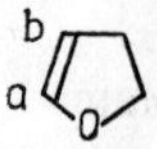

(XII)

$J_{a,b} = 2$

ref. 1432,1500,2419

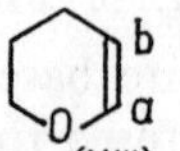

(XIII)

$J_{a,b} = 6 \cdot 0$

ref. 270,1500

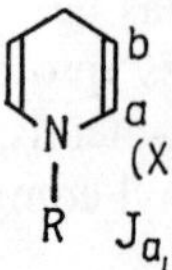

(XIV)

$J_{a,b} = 7 \cdot 9 - 9 \cdot 0$

·ref. 1748,664,334

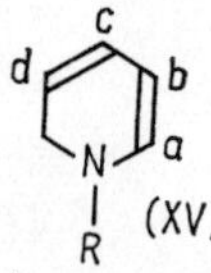

(XV)

$J_{a,b} = 6 \cdot 9$

$J_{c,d} = 9 \cdot 7 - 10 \cdot 0$

ref. 664

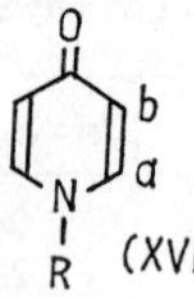

(XVI)

$J_{a,b} = 8$

ref. 201

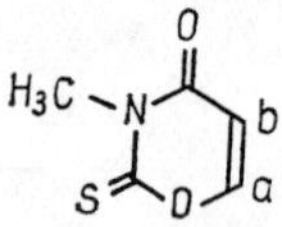

(XVII)

$J_{a,b} = 10$

ref. 2497

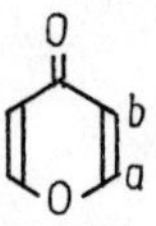

(XVIII)

$J_{a,b} = 6 \cdot 3$

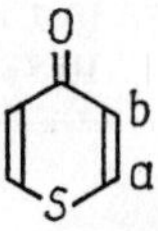

(XIX)

$J_{a,b} = 10 \cdot 7$

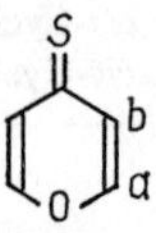

(XX)

$J_{a,b} = 5 \cdot 8$

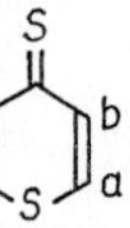

(XXI)

$J_{a,b} = 10 \cdot 4$

ref. 1297, 359

In at least one case (*cis*-1-chloro-2-ethoxyethylene) the olefinic coupling constant was found to vary significantly with solvent[1505] and concentration, a total range of 4·2–6·3 Hz being observed.

† References 414, 868, 1237, 1612, 1778, 1814, 2029, 2253, 2428 and 2516.

INTERPROTON COUPLING IN AROMATIC AND HETEROCYCLIC SYSTEMS

IN THIS Chapter we shall be concerned with spin–spin interactions between protons directly attached to the same aromatic or heterocyclic ring. We shall defer the consideration of coupling between side-chain and ring protons and between aromatic protons situated on different rings in polycyclic systems to Chapter 4–4.

Even within this restricted range we must deal with coupling across three, four, or five bonds, some of which exhibit a range of bond orders. Further, as ring sizes vary so will certain angular relationships which we have just shown to be important in the case of olefinic coupling (Chapter 4–2B) and lastly the electronegativities of the heteroatoms and interactions with substituents also add to the complexity of these systems. In Chapter 3–7 we have seen that *chemical shifts* of aromatic and heterocyclic protons are similarly determined by a number of factors.

Fortunately, the actual range of magnitudes of interproton coupling constants in aromatic and heterocyclic systems are fairly characteristic and thus, while the theoretical picture may appear to be complicated, in practice useful correlations may be safely made.

It appears that the relative signs of the *ortho* (three-bond), *meta* (four-bond) and *para* (five-bond) interactions in benzenoid compounds are the same (presumed positive)† and this has also been observed in quinolines,[1947] furans and thiophenes,[1015,971] benzofuroxan,[1113] furan,[911,2127] and pyridines.[979] J_{meta} and $J_{13C,H}$ (direct) in 2,4,6-trichlorobenzene are of the same sign.[902]

It is generally considered[162,1803,431,254,1187,2583] that the *ortho* coupling constants in aromatic and heterocyclic molecules are closely related to *cis*-vicinal coupling constants in olefins. The *meta* coupling constants have been recognized[162] as abnormally large for an interaction taking place across 4 bonds and not involving π-electrons (see Chapter 4–4). It is now considered likely[173] that the relatively large *meta* coupling is in fact due to the favourable stereochemistry for an interaction across four σ-bonds. *Para* coupling constants are often observable only as line broadening in routine spectra.

A number of coupling constants in aromatic and heterocyclic systems are listed in Table 4–3–1 and although useful correlations are most likely to be made on a purely empirical basis, certain general trends become apparent, and are commented on below.

† References 10, 156, 158, 162, 845, 1015, 1323, 1684 and 2172.

TABLE 4–3–1. INTERPROTON COUPLING CONSTANTS IN AROMATIC AND HETEROCYCLIC SYSTEMS[a,b,c]

Whenever possible the coupling constants refer to the parent compound.
A range refers to a series of substituted compounds; where no range is given
it can usually be assumed that it spans approximately 1–1.2 Hz

System	J_{ortho} (Hz)	J_{meta} (Hz)	J_{para} (Hz)	References
	6.0–9.4	1.2–3.1	0.2–1.5	156, 158, 326, 549, 1015, 1427, 1684, 2160, 2306
	1, 2: 8.3–9.1 2, 3: 6.1–6.9	1, 3: 1.2–1.6	1, 4: approx 1	644, 700, 760, 1299, 1367, 2522, 2525
	1, 2: 8.3 2, 3: 6.5			1299
	1, 2: 8.0–9.0 2, 3: 6.9–7.3 3, 4: 8.0–8.5	1, 3: 0.9–1.6 2, 4: 1.2–1.8	1, 4: 0.3–0.7	198, 1299, 1700
	1, 2: 8–8.4 4, 5: 9.2–9.5			1701
	2, 3: 1.8 3, 4: 3.5	2, 4: 0.8 2, 5: 1.6		9, 1015, 1043, 1940, 2069, 2316, 2441
	2, 3: 2.6 3, 4: 3.4	2, 4: 1.1 2, 5: 2.2		9, 1038, 1039, 1146, 1910

TABLE 4–3–1 *(cont.)*

System	J_{ortho} (Hz)	J_{meta} (Hz)	J_{para} (Hz)	References
(thiophene ring; positions 4, 3, 5, 2; S)	2, 3: 4·7 3, 4: 3·4	2, 4: 1·0 2, 5: 2·9		1015, 1031, 1037, 1040, 1043, 1146, 1196, 1197, 1446, 2334, 2386
(selenophene ring; positions 4, 3, 5, 2; Se)	2, 3: 5·4 3, 4: 3·6	2, 4: 1·1 2, 5: 2·5		1146
(pyrazole ring; positions 3, 4, 5; N, N, H)	3, 4: 1·6 4, 5: 2·9	3, 5: 0·6–0·7		518, 819, 1077, 2584
(imidazole ring; positions 4, 5, 2; N, NH)	4, 5: 1·6	2, 4 ~ 2, 5: 0·8–1·5		1669, 2074, 2075
(thiazole ring; positions 4, 5, 2; N, S)	4, 5: 3·1–3·6	0–2·2		137, 492, 1076, 2395
(pyridine ring; positions 4, 5, 3, 6, 2; N)	2, 3: 4·0–5·7 3, 4: 6·8–9·1	2, 4: 0–2·5 3, 5: 0·5–1·8 2, 6: 0–0·6	2, 5: 0–2·3	379, 979, 1197, 1442, 1444, 1450, 1837, 2617
(pyridazine ring; positions 4, 5, 3, 6, 2; N, N)	3, 4: 5·1 4, 5: 8·0–9·6	3, 5: 1·8	3, 6: 3·5	533, 616, 1815, 2435
(pyrimidine ring; positions 4, 5, 3, 6, 2; N, N)	4, 5: 5	4, 6: 2·5 2, 4: 0·6	2, 5: 1·5	616, 766, 1036, 1041, 1887, 2074, 2075, 2435, 2477
(pyrazine ring; positions 5, 3, 6, 2; N, N)	2, 3: 1·8–2	2, 6: 0·5	2, 5: 1·8	766, 2435

TABLE 4–3–1 *(cont.)*

System	J_{ortho} (Hz)	J_{meta} (Hz)	J_{para} (Hz)	References
	2, 3: 3·1 4, 5: 7·8 5, 6: 7·1 6, 7: 8·1	4, 6: 1·2 5, 7: 1·3	4, 7: 0·9	28, 259, 1678
	2, 3: 2–3·3 4, 5: 8·0 5, 6: 7·3 6, 7: 8·0	4, 6: 1·1 5, 7: 1·2	4, 7: 0·7	259, 563
	2, 3: 5·5–6·0 4, 5: 8·9 6, 7: 8·8–9·2	4, 6: 2·0–2·6 5, 7: 2·5	4, 7: 0·7	1703, 1752, 2383, 2385, 2387
	4, 5: 8·2 5, 6: 6·9 6, 7: 8·4	4, 6: 1·0 5, 7: 0·8	4, 7: 1·0	257
	4, 5: 8·2 5, 6: 7·1	4, 6: 1·4	4, 7: 0·7	256, 2075
	4, 5: 8·6 5, 6: 7·2	4, 6: 0·8	4, 7: 1·0	256
	2, 3: 4·3 3, 4: 8·3 5, 6: 8·2 6, 7: 6·8 7, 8: 8·2	2, 4: 1·8 6, 8: 1·1 5, 7: 1·6	5, 8: 0·3	258, 669, 1947, 2083
	3, 4: 5·6			2435

TABLE 4–3–1 *(cont.)*

System	J_{ortho} (Hz)	J_{meta} (Hz)	J_{para} (Hz)	References
	3, 4: 5·8 5, 6: 7·9 6, 7: 6·9 7, 8: 8·6	5, 7: 1·6 6, 8: 1·3	5, 8: 0·9	260
	5, 6: 8·0 6, 7: 7·0 7, 8: 8·6	2, 4: 0 5, 7: 1·4 6, 8: 0·9	5, 8: 0·6	260, 1717
	5, 6: 8·2 6, 7: 6·8 7, 8: 8·2	5, 7: 1·2 6, 8: 1·2	5, 8: 0·6	260
	2, 3: 1·8 5, 6: 8·4 6, 7: 6·9 7, 8: 8·4	5, 7: 1·6 6, 8: 1·6	5, 8: 0·6	260, 2435
	2·3–3·1	1·0–1·6		1556, 1833, 2041, 2117, 2118
	1, 2: 8 4, 5: 4·2			277, 1255

[a] A number of references contain general collections of data: 2435, 2545, 1197.

[b] In indole and pyrrole $J_{NH,H_\alpha} \sim J_{NH,H_\beta} \sim 2$ Hz.

[c] Data on other systems: Annulenes and macro-rings;[2333,2332,133,2613] dithiofulvenes;[1590] benzofuroxan;[1113] pteridine;[1717] pentalenyl dianion;[1381] indolizines;[885,1657,262] cyclopentadienyl anion;[557] thienopyrrole;[1911,2149,1205,1202,2448] phenanthroline;[1387,431,1758] furanoquinolines;[2122] thienothiophene;[965] borazanaphthalene;[646] purines;[1718,1268,507,2075] pyrylium salts;[143] pyrone,4-thiapyrone,4-thiopyrone and 4-thiathiopyrone;[1297,202,359] biphenylene;[1367,1702] tropolone;[2368] imidazo[1,2-a]pyridine;[1951] 1,2-dithiole-3-ones and 1,2-dithiole-3-thiones;[362] furazan and thiadiazole;[1907] isothiazole;[2347a] oxides of heterocyclic *N*-aromatics;[1886,2436,1887] s-triazolo[1,5-a]pyrimidine;[1657] polycyclic aza compounds;[669] miscellaneous polynuclear aromatic compounds;[1742,1699,1695,1699a,1694,2346] oxazole;[1076] isoxazole;[2435] azaindolizines;[1657,262] pyridones;[765] fulvenes;[2317,1862,1405,1746] tropylium ion;[924] benz[a]anthracenes;[193] oxadiazoles;[1799] azulenes;[595,1745]

A. INFLUENCE OF RING SIZE

It can be clearly seen from the data in Table 4–3–1 that there is a marked decrease in the *ortho* coupling constants on passing from 6-membered to 5-membered rings, but the *meta* coupling constants, while exhibiting considerable scatter in values, do not show this trend. This is not unexpected as the *ortho* coupling is related to vicinal coupling in olefinic systems (see above) and has been discussed in relation to the angle H—C=C,[1146,670,2435] while the planar W-path which is probably essential for effective *meta* coupling[173] (see Chapter 4–4) is similar in 5- and 6-membered rings.

B. INFLUENCE OF BOND ORDER

Although *ortho* coupling is considered to be dominated by σ-electrons, it appears to show a marked dependence on bond order,[1299] which is especially noticable in the case of naphthalene.[1299,760,1369] Whatever the mechanism of this dependence, which could actually reflect bond-length,[1354] it has been observed in heterocyclic compounds,[2477] biphenylene,[1367] tropolones,[2368] acenaphthalyne,[1507] chelated hydrogen-bonded compounds[2006] and miscellaneous polycyclic compounds.[1299,2319,2398] It is quite likely that bond fixation is involved in the variation in *ortho* coupling constants in a number of heteroaromatic molecules (cf. Table 4–3–1) and in strained species, such as benzocyclopropane.[111]

C. EFFECTS DUE TO HETEROATOMS

In general, substitution of a heteroatom for the —CH= group in heteroaromatic systems appears to lead to a decrease in *ortho* coupling constants across the adjacent bond (Table 4–3–1), in line with the effects of electronegative substituents noted above. However, as the introduction of heteroatoms also undoubtedly alters both bond orders and bond angles it is difficult to separate these effects. For the purpose of identifying and correlating spectra, it is important to note the remarkably small magnitudes ($J \sim 1 \cdot 5\,\mathrm{Hz}$) of *ortho* coupling constants in pyrazines, imidazoles, furazan[1907] etc. (Table 4–3–1).

D. EFFECTS DUE TO SUBSTITUENTS

The effects of substituents on the coupling constants discussed here have not been sufficiently thoroughly investigated to permit routine structure–spectra correlations. However, it has been noted that in a series of *p*-disubstituted benzenes,[549] both J_{ortho} and J_{meta} vary significantly with the electronegativities of the substituents, although the relationship need not always be simple.[1684] Variations of this type have also been noted in naphthalenes,[2522] pyridines,[1444,379] five-membered heterocyclics,[2545,2069] pyridazines,[616] and in other series (cf. references in Table 4–3–1).

Minor solvent effects have been occassionally noted,[260] while protonation in the case of heterocyclic compounds usually has well defined[2347a] effects on coupling constants.

LONG-RANGE INTERPROTON COUPLING

THE so-called "long-range" interproton coupling is simply defined[2358,241] as coupling between protons separated by more than three bonds. Long-range coupling constants are generally considerably smaller in absolute magnitude (usually 0–3 Hz) than the geminal or vicinal coupling constants, but with care can be easily observed with modern instruments. Although long-range interactions have been reported in some of the early collections of n.m.r. data, for several years the phenomenon was considered exceptional and little attempt was made either to systematize the data or to relate them with theory. It was only when the scattered results had been collected[1187,162,2358] that it became apparent that the majority of the reported long-range interactions could be easily correlated with structure and stereochemistry, and also fitted into the body of theory.

Although unexplained phenomena still exist and the present classification (cf. Sections C–I below) is unlikely to be final, many correlations based on long-range coupling constants are no less reliable than those based on other J values.

A. OBSERVATION OF LONG-RANGE COUPLING

From the point of view of spectral *analysis* (cf. Chapter 2–3) there are no unusual difficulties associated with the determination of long-range coupling constants from the appearance of spin multiplets and, in fact, such splittings can often be easily interpreted using only first-order theory, because the small magnitude of the coupling constants ensures that at least *some* of the $\Delta v/J$ ratios are favourable. However, because the long-range coupling constants may be vanishingly small while the experimentally available resolution † is at present between 0·3 and 0·6 Hz, long-range coupling of significant magnitude may not result in actual splitting of resonances and may thus escape detection. In the remainder of this section we shall therefore concern ourselves with the experimental determination of small splittings. It should be quite obvious that the procedures outlined below are not restricted to long-range coupling, but may be used to unravel any multiplicity (fine structure) which is difficult to measure experimentally. Further, it must always be remembered that any line separations, or indirect measures thereof, obtained by the procedures outlined below are not *directly* related to the actual coupling constants unless first-order rules apply (cf. Chapter 2–3).

† For definition of resolution in terms of experimental variables see Chapter 1–2B.

The first and most obvious necessity is to obtain the relevant portions of the spectra at optimum homogeneity and greatest sweep width available with the instrument. It is very commonly found that signals appearing as singlets at sweep widths corresponding to 1 Hz per 1 mm of chart length will show appreciable structure at a sweep width corresponding to 1 Hz per 10 mm of chart length.

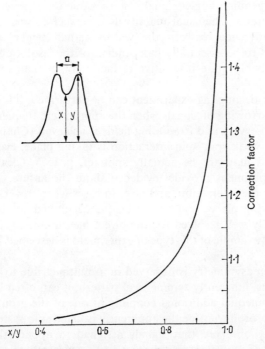

Fig. 4–4–1. The effect of imperfect resolution on the apparent magnitude of splitting of resonances.

If the coupling is sufficiently large, in relation to the resolution of the instrument, to cause a degree of splitting in the resonance concerned without affording *complete* resolution of lines (as in the case of the schematic doublet in Fig. 4–4–1), the separation between the peaks (distance a) will be smaller than the actual splitting by a factor which has been calculated for perfect Lorenzian line shapes[2166] and which is shown on the graph in Fig. 4–4–1. It can be seen that for a resolution [which we define for this purpose as the ratio $(y - x)100/y$] better than 60 per cent, the correction amounts to less than 2 per cent of the magnitude of a and is thus negligible considering the general accuracy of measurement, but that for a resolution of less than 10 per cent a serious error could result from the neglect of imperfect resolution. Kowalewski and Kowalewski[1446] describe related methods of measurement, and the general principles for unravelling overlapping signals are discussed by Allen.[51]

Turner[2452] has described a method for estimating small splittings from the observation of the wiggle-beat patterns (Fig. 1–2–18, p. 44), which can be applied for any reasonably strong signal, provided that the spectrometer is capable of very fast scanning speeds. When applicable, the wiggle-beat method yields results which may be significant to as little as 0·01 Hz.

Even in cases where no splitting at all is apparent it is possible to obtain evidence for the presence of small coupling by means of several methods. The most obvious is the comparison of the width of the apparent "singlet" with that of a genuine singlet signal, e.g. that due to tetramethylsilane used as internal reference. The usual quantity used for such comparisons[2277,2278, 2125] is the difference between the widths at half-height† which can be shown[2125,2278] to be essentially independent of the resolution of the instrument and probably quite insensitive to factors other than spin–spin coupling.‡

When a spin-decoupling experiment can be performed, it is often possible to observe a narrowing of signals when the appropriate frequency difference between the observing and irradiating fields is achieved (Chapter 3–4), thus proving that a weak spin–spin interaction is taking place, even though the signal observed may not be actually split (cf. p. 157). Clearly, the spin-decoupling experiment provides evidence about the nature of the protons responsible for the interaction, and also, in cases where several small interactions are taking place simultaneously, the decoupled spectrum may become sufficiently well resolved for the direct measurement of all splittings in turn. A good example of this type of experiment is described by Osawa and Neeman.[1914]

Finally, when asymmetry is observed in a multiplet due to an AB, A_2B_2 or an analogous, inherently symmetrical pattern, it can often be traced[2142, 1870,430] to an unequal additional coupling of one of the groups of protons involved with protons outside the apparently isolated spin system. Similarly, it is commonly observed that signals assigned to apparently completely isolated groups of equivalent protons (e.g. methoxyl, acetoxyl, tertiary methyl etc.) are of unequal height, presumably due to very small long-range interactions of unequal magnitude. In the case of angular methyl groups (cf. Section G) such interactions are sufficiently significant to be used in structural and stereochemical assignments.

If the last type of measurement (and comparisons of W_H with that of tetramethylsilane) are to be meaningful, it is essential to ascertain whether the relative differences are not due to different saturation factors, Z_0 (p. 10). A crude, but probably generally valid, method of achieving this is to obtain the measurements of, for example, asymmetry or differences in W_H at two widely differing values of radio frequency field strengths. If the result remains essentially unaltered (as is usually the case within the normal range of radio-frequency power used[1870,2278]), it is probably safe to assume that the phenomena observed are due to small long-range interactions.

It should be obvious from the above that unless special precautions are

---

† Width at half-height is often denoted by the symbols W_H or $W_{1/2}$.

‡ This assumes that line-widths are essentially controlled by the inhomogeneity of the applied magnetic field, which seems to be true for most molecules of moderate molecular weight, except where quadrupole broadening (p. 10) occurs.

taken, splittings due to long-range interproton coupling may not be observed. It follows that signals reported in the literature as devoid of fine structure may in fact show further splitting under more careful examination, and hence evidence for the *absence* of long-range coupling should always be treated with caution.

B. THEORY OF LONG-RANGE COUPLING

The few existing treatments in this area (see below) do not differ in kind from those concerned with short-range coupling, but must, perforce, take account of a larger number of bonding electrons. The principal difficulty associated with the calculation of (semi-empirical) values of long-range coupling constants is the fact that a number of separate interactions involving various electrons must be considered. These have a characteristic dependence on steric and electronic factors and result in coupling constants of either like or unlike signs. Thus, the agreement with experimental results may depend upon the correctness of treatments and assumptions made in *more than one* set of calculations, each of which may be of considerable sophistication and involve quite drastic simplifications. Further, because of the relatively small range of values of long-range coupling constants and the difficulty in obtaining accurate data (see above), it is not always possible to test experimentally any secondary effects, for example, those associated with the type of substituents.

In spite of the above considerations, there is little doubt that the theoretical approaches to, at least, the two most common types of long-range interactions are fundamentally sound because they have been amply confirmed experimentally. The first of these two treatments involves coupling via the π-electron framework in unsaturated systems, often referred to as σ–π configuration interaction[1351,1352]† and is sometimes discussed in terms of hyperconjugation.[2324,1198,1187,1194] The second interaction involves coupling across four single bonds.[173,1422] Extensive discussions of these and other modes of long-range coupling can be found in the literature.‡

The above treatments predict certain stereospecific pathways necessary for effective coupling between protons involving particular electrons in bonds separating them. However, it is now widely thought†† that any pair of protons or groups of protons in a real molecule may interact simultaneously via more than one path, with the result that the actually observed coupling constant may be a sum (if the signs are like) or a difference (if the signs are unlike) of the interactions expected from each path in isolation.

We shall discuss the relevant aspects of the theoretical predictions in connection with the various groups of structures in which well-documented examples of long-range coupling may be found.

† For the sake of brevity we shall refer to this as the "σ–π mechanism".
‡ References 162, 254, 306, 950, 951, 966, 1013, 1189, 1198, 1500, 1803, 2127 and 2358.
†† References 173, 577, 757, 950, 966, 1208, 1870, 2144, 2324 and 2358.

C. ALLYLIC AND HOMOALLYLIC COUPLING

Coupling across one double and three single bonds in the system (I) is defined as allylic coupling.

Structure (I) implies a number of variables: (a) *Cisoid* (i.e. $J_{1,3}$) or *transoid* (i.e. $J_{2,3}$) coupling paths. (b) The "allylic" angle, Θ. (c) Substituents at C_1, C_2 and C_3. (d) The bond order of the double bond. (e) The incorporation of the fragment (I) into various cyclic structures.

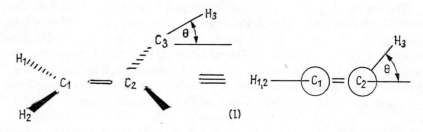

(I)

Theory predicts[1351,1352,1614,1842] that the contribution to allylic coupling involving the σ–π mechanism should result in a coupling constant which is:

(a) Negative in sign with values of 0 to -3 Hz.

(b) Similar for *cisoid* and *transoid* coupling.

(c) Zero at $\Theta = 0$ and 180°, and at a maximum (largest negative value) at $\Theta = 90°$. In between these extreme values the function should show a $\cos^2\Theta$ dependence, i.e. it is related to the degree of σ–π orbital overlap.[2358]

(d) Dependent on the π-bond order of the double bond.

As will be shown below, a large body of experimental data agrees with the above predictions and hence it is reasonable to say that allylic coupling is dominated by this mechanism, with the exception of certain narrowly defined configurations (see below) and conjugated systems (cf. Section H) where other modes of interaction are important.

Analysis of strongly coupled systems or double irradiation experiments, have shown conclusively that the signs of allylic coupling constants in many substituted propenes† and in a number of cyclic systems (cf. Table 4–4–1) are opposite to those of vicinal coupling constants and hence negative. Very commonly,‡ for pairs of propenes of the general form (II), (III) and (IV), J_{cisoid} is larger (in absolute magnitude) by approximately 0·3–0·6 Hz than $J_{transoid}$. This correlation is however not very reliable, because a number of actual reversals[2144,1691,1983,1325,649] have been reported and in a larger number of instances (e.g. 1,3-dichloropropene,[158] allylamine,[44] fluoropropene,[203,302] β,β-dimethylstyrenes, [2144]butenes,[1259,1682] α,β-unsaturated esters[891]) the values are within approximately 0·1 Hz. Examples of allylic

† References 44, 158, 161, 203, 302, 305, 311, 516, 649, 689, 814, 851, 887, 891, 911, 1323, 1691, 2003, 2541 and 2542.

‡ References 168, 187, 313, 346, 995, 1192, 1194, 1259, 1260, 1795, 1983, 2087 and 2105.

TABLE 4–4–1. ALLYLIC AND HOMOALLYLIC COUPLING CONSTANTS IN SOME CYCLIC SYSTEMS[a,b,c]

	Structure	Coupling constants (Hz)	References
1		$J_{AB} = +5 \cdot 58$ $J_{AX} = -1 \cdot 98$ $J_{BX} = +2 \cdot 02$	225a, 756 2511
2		$J_{AB} = +5 \cdot 8$ $J_{AX} = +1 \cdot 7$ $J_{BX} = -2 \cdot 1$	899
3		$J_{AB} = +2 \cdot 5$ $J_{AX} = -1 \cdot 6$ $J_{BX} = +2 \cdot 7$	940
4		$J_{AX} = 1 \cdot 1$ $J_{BX} = 2 \cdot 0$	938, 2229
5		$J_{AB} = -3 \cdot 87$ $J_{AX} = -1 \cdot 36$ $J_{BX} = -1 \cdot 94$	348, 1226, 1777
6		$R = CN, CH_3 \quad R' = OR, Ph, CN$ $J_{2,4} = -0 \cdot 6 - +0 \cdot 5$	334, 758
7		$R = CN, CH_3 \ R' = OR, Ph, CN$ $J_{2,4} = +0 \cdot 9$ to $-0 \cdot 1$	334
8		$J_{A,X} = 0 \cdot 8$	1181, 2139

See references 2358, 950 and 2143. [b] Where signs of coupling constants are given, the sign of the vicinal coupling constants is *assumed* to be positive. [c] Arrows show possible ground-state electron delocalization (see text).

TABLE 4–4–1 *(cont.)*

	Structure	Coupling constants (Hz)	References
9		$J_{AX} = 1 \cdot 7$	497, 502
10		$J_{AX} = 0 \cdot 8$	1364
11		$J_{AX} = 0 \cdot 9$	200
12		$J_{AX} = 0 \cdot 8$	200
13		$J_{AX} = 1 \cdot 8$ $J_{BX} = 1 \cdot 1$	938
14		$J_{AX} = 1 \cdot 5$	1461
15		$J_{AX} = 1 \cdot 6$	1881
16		$J_{AX} = 1 \cdot 5$	853, 1461

TABLE 4–4–1 *(cont.)*

	Structure	Coupling constants (Hz)	References
17		$J_{AX} = 1.6$	1461
18		$J_{AX} = 1.5$	1461
19		$J_{AX} = 0.8$	1461
20		$J_{BX} = 1.5$ $J_{AX} = 2.5$	
21		$J_{BX} = 2.8$ $J_{AX} = 1.9$	1035, 1211 1212, 1213
22		$J_{AX} = 1.5$ $J_{AB} = 1.5$	
23		$J_{AX} \sim 3$ $J_{BX} \sim 3$	477
24		$J_{AB} = 1.0$ $J_{MX} = 1.5$ $J_{MY} = 1.5$	1931

TABLE 4–4–1 *(cont.)*

	Structure	Coupling constants (Hz)	References
25		$J_{AX} \sim 0$ $J_{BX} \sim 1\cdot5$	1748
26		$J_{AX} = 1\cdot4$ $J_{BX} = 1\cdot4$	938
27		$J_{AB} = +3\cdot1$ $J_{AC} = +4\cdot5$ $J_{AD} = -1\cdot3$ $J_{CD} = +0\cdot8$ $J_{BD} = +2\cdot9$	938
28		$J_{AB} = 2\cdot8$ $J_{BD} \sim J_{BC} \lesssim 0\cdot5$ $J_{BD} = 2\cdot8$ $J_{CD} = 2\cdot8$ $J_{AC} = 1\cdot5$	1932
29		$J_{12} = +2\cdot70$ $J_{34} = -12\cdot00$ $J_{13} = -0\cdot80$ $J_{35} = +1\cdot65$ $J_{15} = +1\cdot55$ $J_{36} = +4\cdot35$	296, 1668
30		$J_{AB} = 2\cdot5$	2544
31		$J_{AB} = 2\cdot5$	794

TABLE 4–4–1 *(cont.)*

	Structure	Coupling constants (Hz)	References
32		$J_{AX} = 2 \cdot 1$	1118
33		$J_{AB} = 7 \cdot 5 - 11$	708
34		$J_{H_{9a}, H_{12a}} = 3 \cdot 5$ $J_{H_{9a}, H_{12e}} = 2 \cdot 9$	422
35		$J_{H_{9e}, H_{12a}} = 2 \cdot 0$ $J_{H_{9e}, H_{12e}} < 1$	422
36		$J_{7,11} = 5$	115
37		$J_{1,2} = 1 \cdot 7$ $J_{1,3} = 2 \cdot 5$ $J_{2,3} = 18$ $J_{3,4} = 2 \cdot 5$ $J_{2,4} \sim 0$	2021

coupling in propenes are shown on Table 4–4–2. In di- and trichloroprope-
nes[1795] allylic coupling constants are in the range of 0·4–0·8 Hz.

In six- and five-membered rings of type (V), it appears[1870,348,300] that the
transoid coupling constants are consistently larger (by approximately 0·4 Hz)
and it is possible that this relation may hold for four-membered and three-
membered rings as well.[348,1244,149,1009,1777]

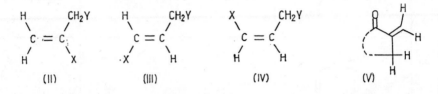

(II)   (III)   (IV)   (V)

The lack of clearly defined differences between *cisoid* and *transoid* allylic
coupling constants is not unexpected from theoretical considerations (see
above).

Because steric factors are critical in determining the magnitude of allylic
coupling constants (see below) and in conformationally mobile systems, such
as propenes, the average conformation (and hence the average value of the
allylic angle) may be substitution dependent. The *direct* effect of substituents
(say X and Y in structures II, III and IV) is not easy to evaluate. However,
from the examples shown in Table 4–4–2, it can be readily seen that even
where conformational effects are unlikely, allylic coupling constants in propenes exhibit a significant range of values. Some of the substituent effects
could involve lowering the bond-order of the double bond (see below), or
other "disturbances" of the electronic path, or they may effect[649] the small
contribution to allylic coupling which involves only σ-electrons. Unfortunately, little systematic work has appeared in this area.

The steric requirements predicted for effective allylic coupling via the
σ–π path have been amply demonstrated for several series of 6-substituted-
Δ^4-3-oxosteroids[521,2432,1608,520] (VI), where it was found that the olefinic

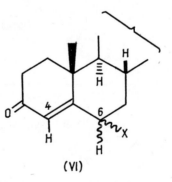

(VI)

protons at C_4 couple with the β-proton at C_6 in the 6α-substituted series
(*cisoid* coupling, allylic angle Θ approximately 85°) but not with the 6α-
proton in the 6β-substituted series (*cisoid* coupling, allylic angle Θ approx-

TABLE 4–4–2. ALLYLIC COUPLING CONSTANTS IN SOME NON-CYCLIC SYSTEMS[a]

Structure	Allylic coupling constants (Hz)		References
	J_{cisoid}	J_{transoid}	
$\begin{array}{c} H \\ \end{array} C=C \begin{array}{c} CH_3 \\ F \end{array}$ (H below left)	−1·0	−0·4	649
$\begin{array}{c} H \\ H \end{array} C=C \begin{array}{c} CH_3 \\ Br \end{array}$	−1·4	−0·8	649
$\begin{array}{c} Br \\ H \end{array} C=C \begin{array}{c} CH_3 \\ H \end{array}$	—	−1·8	649
$\begin{array}{c} H \\ Br \end{array} C=C \begin{array}{c} CH_3 \\ H \end{array}$	−1·7	—	649
$X-C_6H_4-\overset{H}{\underset{}{C}}=C\begin{array}{c} CH_3 \\ CH_3 \end{array}$	1·29–1·41	1·20–1·35	2144
$\begin{array}{c} H \\ H \end{array} C=C \begin{array}{c} CH_3 \\ COCH_3 \end{array}$	1·3	0·7	1259
$\begin{array}{c} H \\ H \end{array} C=C \begin{array}{c} CH_3 \\ CHO \end{array}$	1·45	1·05	689
$\begin{array}{c} CH_3OOC \\ H \end{array} C=C \begin{array}{c} CH_3 \\ CH_3 \end{array}$	1·3	1·3	1259
$\begin{array}{c} CH_3 \\ Cl \end{array} C=C \begin{array}{c} H \\ COOCH_2CH_3 \end{array}$	1·16	—	1983
$\begin{array}{c} Cl \\ H_3C \end{array} C=C \begin{array}{c} H \\ COOCH_2CH_3 \end{array}$	—	1·20	1983
$\begin{array}{c} (CH_3)_2CH \\ H \end{array} C=C \begin{array}{c} H \\ H \end{array}$	−1·7	−1·2	44, 311, 313, 2135
$\begin{array}{c} CH_3CH_2 \\ H \end{array} C=C \begin{array}{c} H \\ H \end{array}$	−1·9	−1·3	

[a] Negative sign is relative to J_{vicinal}, which is assumed to be positive. Where no sign is given, it has not been determined, but may be assumed to be negative.

imately 10°).† These findings have been confirmed by the examination of a large number of compounds of known stereochemistry[520,950,173,2358] and indicate strongly that allylic coupling is dominated by the σ–π mechanism.

An exception is compounds where the allylic coupling is *transoid* with Θ equal to zero or approximately zero. In those cases[173,950,2358,1509,2323,1500] it has been generally found that a coupling of *positive* sign and a magnitude of approximately 1 Hz is observed. This may be rationalized[173,950] in terms of a contribution of positive sign involving σ-electrons only, which is at a maximum for this particular configuration (cf. Section G). Unfortunately, it is difficult to obtain a fragment with this configuration without resorting to a bicyclic structure[1509,2323] which may be strained and hence may exhibit special phenomena. However, as the available examples[173,950,2358] include 1,3-cyclohexadiene itself[1486] (see Section H for data) it seems likely that the above rationalization is in fact correct. A special case may be observable in bicyclo[2.1.3]octenes[152,1273,1274,1275] and bicyclo[3.1.3]nonenes,[523] where in some examples there appears to be no observable *transoid* allylic coupling, presumably as a result of approximate cancellation of the two effects.

For practical purposes, the special case of *transoid* allylic coupling with Θ equal to approximately 0° is not very common and the correlation between the magnitude of allylic (and homoallylic, cf. below) coupling constants and configuration at C_3 in structures incorporating the fragment (I) has been successfully used in structural work.‡

The contribution of an interaction via the σ-path may give an explanation[649,2144] for the *tendency* of the *cisoid* allylic coupling constants in propenes to exceed, in absolute magnitude, the *transoid* coupling constants, because only some conformations in the former configuration would give rise to an appreciable coupling (of opposite sign) via the σ-path. If this is so, it is not surprising that cyclic systems of type (V), where this path is sterically supressed, do not show the same trend.

Besides determination of stereochemistry in compounds of defined structure, the magnitude of allylic coupling constants may be utilized for the determination of the average conformation[2358,305] in flexible systems (e.g. propenes) where the substituent at C_3 in the fragment (I) may influence the average value of Θ through steric or polar effects.[2541,1795,305,313,346,302]

In α,β-unsaturated aldehydes, allylic coupling to the aldehydic proton is generally unobservable,[2184,2003,308,1323,982,689,911] which is possibly because of the tendency of such systems to assume the S-*trans* conformation.

The dependence of the absolute magnitude of the allylic coupling constants on the angle Θ clearly leads us to expect that, in certain cyclic structures, allylic coupling constants should be large. This is found in several examples shown in Table 4–4–1, but it can also be seen that in a number of cases (entries 8, 10, 11, 12, 19; see reference 2143 for other examples) the allylic

---

† In unsubstituted compounds and some other cases, complications may arise due to virtual coupling.[2280,2608,520,521]

‡ References 131, 192, 344, 418, 577, 679, 759, 1004, 1131, 1132, 1149, 1159, 1855, 1931, 2392, 2393, 2478 and 2533.

coupling constants are markedly lower than expected on purely steric grounds.

A rationalization of this phenomenon can be sought[2143] in a dependence of J_{allylic} on the bond-order of the double bond,[1614] which is theoretically predictable (see above) and which will also be considered in connection with the bond-order of aromatic bonds and benzylic coupling (see Section E). It is thus possible that allylic coupling may be useful as a probe for the determination of bond-orders. Clearly, a standardized group e.g. $H_3C—C=C—H$ would have to be used[2143,1881,362] to avoid variation with steric factors.

The low allylic coupling constants[334] in entries 6 and 7 in Table 4–4–1 may be caused by conformational effects; allylic coupling in simple dihydro-pyridines[664] and related systems[2391] has also been reported.

The coupling across four single and one double bond in the system (VII) has been defined as *homoallylic coupling.*[1979]

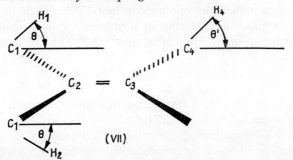

(VII)

The possible variable factors are those enumerated above for allylic coupling with the difference that *two* allylic angles (Θ and Θ') are involved. Theory predicts[1351,1352] that the contribution to homallylic coupling involving the σ–π path should result in a coupling constant which is:

(a) Positive in sign with values of 0–4 Hz.

(b) Similar for *cisoid* ($J_{1,4}$ in VII) and *transoid* ($J_{2,4}$ in VII) coupling.

(c) Dependent on the magnitude of Θ and Θ' in a manner analogous to that described for allylic coupling. It must be noted that if any one of the angles Θ or Θ' takes up the values of either 0° or 180°, the total predicted interaction becomes zero, i.e. a term involving the product of $\cos^2\Theta$ and $\cos^2\Theta'$ is present.

(d) Dependent on the π-bond order of the double bond.

Further, in comparable systems (e.g. I and VII) there should be little difference between the absolute magnitudes of allylic and homallylic coupling constants when a methyl group is substituted for the vinylic hydrogen in the allylic fragment,[1351,1197,1194,1187,162] e.g. J_{AB} in (VIII) should be of opposite sign but of approximately the same absolute magnitude as J_{AB} in (IX).

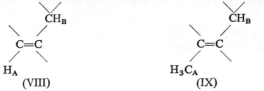

(VIII)   (IX)

This "methyl replacement technique", as well as the generally observable sign alteration with the number of bonds, has been used as a diagnostic characteristic for the σ–π mechanism in several investigations[471,1189,966, 967,968,2127,851] into the mechanism of long-range coupling.

Relative sign determinations in tiglaldehyde[851] and angelica lactone[940] (see Table 4–4–1) show that allylic and homoallylic coupling constants are of opposite sign, as predicted.

From the results obtained with a limited number of isomeric 2-butenes (tiglic and angelic acid and their derivatives,[887] nidulin and isonidulin,[199] 2-bromobut-2-enes,[2105,1682] α,β,β-trimethylstyrenes,[1870] α,β-dialkylstyrenes[168] and other compounds[1908]) it appears that homallylic coupling constants between methyl groups are within the range of 1·0–1·5 Hz, i.e. close to the allylic coupling constants in propenes (cf. Table 4–4–2) and that the *transoid* coupling constants are *consistently* higher (by approximately 0·3 Hz) than the *cisoid* homallylic coupling constants. This correlation could be useful, but it obviously needs further confirmation. Coupling constants between vicinal methyl groups in cyclic structures[2230,33,2358,1870] appear to be in the range of 1–1·5 Hz.

As with allylic coupling, the origin of the differences between *cisoid* and *transoid* homoallylic coupling constants may also lie[2144] in a small, positive contribution to the overall interaction transmitted through the σ-electrons along the "extended zig-zag path" (cf. Section F) which can only be assumed in certain conformations of the *trans* dialkyl groups. Unlike the case of allylic coupling, in which the two contributions are of opposite sign, here the signs are the same and a reinforcement results for the *transoid* situation.

The dependence of homoallylic coupling on the dihedral angles Θ and Θ' has been confirmed in derivatives of α-santonin,[1979,1980] codeine[192] and in a number of 6-substituted-4-methylcholest-4-enes.[2279] In all cases homallylic coupling was negligible when either Θ or Θ' approached 0° or 180°.

In a number of cyclic systems (cf. Table 4–4–1, entries 3, 4, 20), homoallylic coupling constants are actually larger than the allylic coupling constants although a second $\cos^2\Theta$ term (which must be less than unity for even the most favourable allylic angle) is involved. It thus appears that homoallylic coupling constants may be intrinsically larger (absolute values) than allylic coupling constants. The very large values in entries 34, 35 and 36 (Table 4–4–1) appear to support this view. It is interesting that the largest of these (entry 36) appears in a *transoid* system.[115] The remarkably large homoallylic coupling in dihydrobenzenes[708] (entry 33 in Table 4–4–1) can be rationalized by the presence of two equivalent coupling paths.

Scattered examples[2358] (e.g. entries 3, 4 and 25 in Table 4–4–1) appear to indicate that homoallylic coupling constants are sensitive to the bond-order of the double bond in a manner analogous to allylic coupling constants.

Insufficient data are available to test rigorously any secondary effects of substituents on homoallylic coupling, but they may exist.[938,1932]

In a number of cases, long-range coupling apparently analogous to allylic and homoallylic coupling has been observed[2358] in systems where one or

TABLE 4-4-3. ALLYLIC AND HOMOALLYLIC COUPLING CONSTANTS ACROSS NITROGEN
(See also reference 2358)

Structure	Coupling constants (Hz)	References
	$J_{AB} = 2$	1553
	$J_{AB} \sim J_{AB'} \sim 1$ (The $A_3A_3'B_3B_3'$ pattern was not analysed)	1553
	$J_{AB} = 0.85-2.3$	2534
	$J_{AB} = 1.38-1.65$	2512
	$J_{AX} = 1.0$ $J_{BX} \sim 0$	2348
	$J_{AB} = 1.4$	701
	$J_{AB} = 1.1$	1431
	$J_{AB} = 0.67$	463

more of the intervening carbon atoms have been replaced by nitrogen. Some examples of such structures are given in Table 4–4–3. Other compounds exhibiting this phenomenon are the amides,† which have been extensively studied by n.m.r. because of the presence of a partial double bond and hence restricted rotation (cf. Chapters 2–1 and 5–1). It appears that here both allylic and homoallylic coupling constants are larger for the *transoid* configuration. An example of long-range coupling in a lactam[463] is shown in Table 4–4–3.

Included in the examples listed in Table 4–4–1 are a number of structures incorporating four-membered rings and hence presumably appreciably strained. Although at least two sets of data (entries 5 and 27) indicate that allylic coupling in these structures does not differ appreciably from their unstrained analogs, because the sign identity is preserved, other examples shown in Table 4–4–1, and listed in the literature,[2358,1500] suggest that special effects related to ring strain may be present.

In general, ring strain might be expected to affect coupling constants because of alteration of bond angles and lengths, and partial rehybridization.

Certain five-membered ring systems, such as 2,5-dihydrofuran[1293,941] and 2,5-dihydropyrrole (e.g. in dehydroproline derivatives[1293] and pyrrolizidine alkaloids[578]) show remarkably large coupling constants between the formally homoallylic H_2 and H_5, which appear to vary with the *cis–trans* stereochemistry. It is very likely[578,1293] that the interaction observed is in fact a sum of interactions across the homoallylic system and across the four-sigma-bond pathway via the heteroatom.

D. LONG-RANGE COUPLING IN ACETYLENES, ALLENES AND CUMULENES

Long-range coupling constants in these systems are in good agreement with those predicted on the basis of the σ–π mechanism[1351] and have also been discussed in detail in terms of a mechanism involving second-order hyperconjugation.[2324] The theoretically predicted alternation of sign (negative coupling constants for coupling across an even number of bonds, positive coupling constants for coupling across an odd number of bonds) has been confirmed.[1351,2324,1665,1849,2322,1173,809] The examples shown on Table 4–4–4 were drawn from the considerable amount of data available.‡

It can be seen immediately that conjugated acetylenic systems are capable of transmitting appreciable spin–spin interactions across up to as many as nine bonds! This phenomenon is related to the fact that the triple bond provides a "multiple path" for spin–spin coupling.[1351] Entries 14 and 15 in Table 4–4–4 show that, in allenes, when the angle corresponding to the allylic angle is closer to 90°, a more effective coupling results, in accord with theory.

† References 224, 250, 319, 336, 1007, 1125, 1208, 1326, 1438, 1497, 1498, 1785, 1786, 1787, 1865, 1962, 2047, 2373, 2557, 2558 and 2559.

‡ References 331, 560, 699, 733, 996, 1067, 1106, 1173, 1192, 1194, 1311, 1351, 1433, 1455, 1665, 1849, 2020, 2076, 2273, 2287, 2322, 2324, 2358 and 2540.

TABLE 4-4-4. LONG-RANGE COUPLING CONSTANTS IN ACETYLENES AND CUMULENES

	Structure	Coupling constants (Hz)	References
1	H_3C—C≡C—H	2·93	2273
2	X—CH_2—C≡C—H, X=Cl, Br, I	2·6–2·8	2539
3	H_3C—C≡C—CH_3	2·7	2324
4	H—C≡C—C≡C—H	2·2	2324
5	H_3C—C≡C—C≡C—H	1·27	2324
6	EtO—CH_2—C≡C—C≡C—H	1·1	1311
7	$(EtO)_2CH$—C≡C—C≡C—H	0·7	1311
8	H_3C—C≡C—C≡C—CH_3	1·3	2324
9	H_3C—C≡C—C≡C—C≡C—H	0·65	1311
10	H_3C—C≡C—C≡C—C≡C—CH_2—OH	0·4	2324
11	R—CH=C=CH—R′	~6	1433, 1467, 2324
12	H_2C=C=CHX, X=Cl, Br, I	6·1–6·3	996
13	X A B H_3C—CH=C=CHCl	J_{BX} = 2·4, J_{AB} = −5·8	1665, 2324
14	$(CH_3)_2C$=C=CH_2	3·03	2324
15	⬠=C=CH_2 (A, B)	J_{AB} = 4·58	560, 1106
16	(structure with $COOCH_3$, CH_3, H, X, C, A, B, H)	J_{AB} = 6·5 J_{BX} = 1·1 J_{BC} = 1·3	2020
17	H—C≡C—C (structure with H, C, X, A, B)	J_{AB} = −2·17 J_{AC} = 0·70 J_{AX} = 0·92	1173, 2322
18	H—C≡C—C (with H, =O)	0·58	2322
19	H_3C (A), H_3C (B), CHO, H, X — C=C=C=C	J_{AX} ∓ J_{BX} ~ 1·2	1433

11a   AN

It can also be seen (e.g. entries 1 and 3) that the substitution of a methyl group for a proton may not materially affect the absolute magnitude of the long-range coupling constant in analogy with the allylic and homoallylic case.

In a number of acetylenes[2287] of the general formula

$$Ph_{n-1}—X—CH_2—C\equiv C—H$$

where X is an element (O, N, S, Se, Te, As, Sn) of valence n, the coupling between the acetylenic proton and the methylene groups appears to depend on the electronegativity of X with a total range of 2·4–2·9 Hz.

In silyl acetylene,[733] $H_3Si—C\equiv C—H$, the long-range interproton coupling constant is 1·47 Hz, i.e. appreciably different from methylacetylene (entry 1, Table 4–4–4).

A large number of examples of long-range coupling in conjugated poly-acetylenes can be found in the work of Böhlmann and his collaborators[280] which also contains data on other unusual types of long-range interactions.

In the only available example of a cumulene higher than an allene (entry 19, Table 4–4–4), it is interesting to note that the coupling of H_X to the two methyl groups appears to be unequal, in possible analogy to the *cisoid* and *transoid* coupling in the allylic and homoallylic classes.

E. BENZYLIC COUPLING

We shall define coupling between protons attached to sp^3 hybridized benzylic carbon atoms and ring protons in aromatic and heterocyclic rings as benzylic coupling, although it will be shown (cf. Section F) that protons attached to sp^2 hybridized benzylic carbon atoms, and indeed to hetero-atoms, may also be appreciably coupled to ring protons.

Clearly, in the case of a substituted benzene, one could distinguish benzylic coupling (sometimes also referred to as sidechain coupling[1197,2143,1032,2127]) to the *ortho, meta* and *para* protons, i.e. across 4, 5 or 6 bonds.[2143] Available evidence shows† that, for a methyl group attached to a benzene ring, the coupling to the *ortho* proton is generally in the range of 0·6–0·9 Hz and that to the *para* proton approximately 0·5–0·6 Hz, i.e. the *ortho* coupling tends to be larger, but may be identical with the *para* coupling in some cases.[2143] The *meta* coupling is smaller,[2142] values of 0·36[2143] and 0·37[513] Hz having been reported. Thus, clearly a potentially useful correlation exists.

The magnitudes of the *ortho* and *para* benzylic coupling constants are quite close to those calculated[644,513] on the basis of a formula due to McCon-nell[1614] for a mechanism involving π-electrons, but it is uncertain whether the *meta* benzylic coupling should be merely smaller than the *ortho* and *para* analogs or actually zero.[513,644] It is possible[1870] that the *meta* benzylic coupling does not in fact involve the aromatic π-electrons, because it appears to be most effective for compounds which can assume[1870] the extended zig-zag conformation (cf. Section F). This would explain the relatively large magnitude of *ortho* and *para* benzylic coupling constants in acenaphthenes[644] where the "allylic" angle is higher than the average allylic angle for a freely

† References 31, 225a, 513, 644, 1185, 1194, 1427, 1842, 1843, 2142 and 2143.

rotating methyl group and where the benzylic *meta* coupling is unobservable. It has in fact been suggested[1870] that steric factors favouring effective inter-action of benzylic protons with the *ortho* and *para* ring protons (i.e. an allylic angle close to 90°) suppress coupling to the *meta* position and *vice versa*.

In analogy to allylic coupling, it is theoretically predictable[1185,1614,513,644,1842,1843,2143] and experimentally verified[1842,1843,2143] that, at least the *ortho* benzylic coupling depends upon the bond-order of the aromatic bond involved. A striking example[2143] is 9-methylphenanthrene where the methyl group gives rise to a doublet with a splitting of 1.05 ± 0.05 Hz. The relation between *ortho* benzylic coupling constants and bond-order has now been investigated[1842] quantitatively and thus opens up the possibility of probing the bond-orders in aromatic systems. Clearly, as with allylic coupling, a standardized fragment (presumably $H—C=C—CH_3$) should be employed to avoid complications due to steric factors.

Benzylic coupling in thiophenes† and furans[2127,1043,2315] has been in-vestigated in greater detail than in benzenoid compounds. The coupling con-stants depend on double bond order in the expected manner. For instance, in 3-methylfuran[2127] the methyl group couples to H_2 with $J = 1.20$ Hz and to H_4 with $J = 0.45$ Hz, and an almost identical set of results has been obtained for 3-methylthiophenes.[1713] These coupling constants also exhibit more subtle effects attributable to ground state electron shifts.[2143]

Some substituent effects in furans[2315] and in a series of substituted cresols[1427] have been noted.

The relative sign determinations of benzylic coupling in methylthio-phenes[966,971,967,968] and furans[2127] show that (assuming the vicinal inter-actions to be positive) the *ortho* benzylic coupling is negative in sign while the coupling to a proton further removed (analogous to the *meta* position) is positive. This constitutes some evidence for the dual nature of the mecha-nism involved. The reader is directed to a series of papers by Scandinavian workers[966,967,968,971,1043,2127] for a thorough discussion of the mechanism of side-chain coupling. It appears that the —SH group behaves in a manner analogous to the methyl groups.[967,968]

Side-chain coupling has also been noted in pyrimidines,[1041,2074] pyri-midine-*N*-oxides,[1887] phenanthrolines,[431] azaindolizines,[1657] azulenes,[1745,595] cinnoline-*N*-oxides,[1886] pyridazine-*N*-oxides,[2436] diazines,[2435] isothia-zoles,[2347a] thiazoles,[2395,69,1076] oxazoles,[1076] pyrroles and indoles[2143,1038,1039,9] and other heterocyclic systems.[1951,2397,488,415]

In most heterocyclic compounds the *ortho* benzylic coupling constant to a methyl group is in the range of $0.5–1.3$ Hz and often shows obvious depend-ence on the bond-order.

Small but observable coupling (ca. 0.5 Hz) *between* adjacent benzylic methyl groups has been noted.[513,2384,1032] Such "homobenzylic" coupling is not unexpected in view of the many analogies between allylic coupling and benzylic coupling.[2358,2143] An example of this type of interaction occurs[1291] between H_6 and H_9 in steroids with ring A aromatic.

† References 544, 966, 967, 968, 971, 1030, 1187, 1194, 1197, 1713 and 2334.

TABLE 4–4–5. EXAMPLES OF LONG-RANGE COUPLING VIA THE "EXTENDED ZIG-ZAG PATH"[a]

Structure	Coupling constants (Hz)	References
	$J_{AB} \sim J_{AC} \sim 2$	620
	$J_{AB} = 2$	620
	$J_{AB} = 1$	387
	$J_{AB} = 1{\cdot}5$	1846
	$J_{AB} = 1{\cdot}4$	280 281
	$J_{AB} = 0{\cdot}5$	2523

[a] See also ref. 2358.

F. COUPLING BETWEEN PROTONS SEPARATED
BY FIVE BONDS ALONG AN EXTENDED ZIG-ZAG PATH

It has been noted[162,2358,763] that several groups of compounds exhibit appreciable long-range coupling between protons at the ends of a (generally conjugated) chain of configuration (X), and qualitative discussions of the probable mechanism have appeared.[162,306,513,763,845,847,1189,2448]

To this category appear to belong the coupling between H_3 and H_7 in compounds of type XI† (e.g. benzofuran,[259,762,763,513] indole,[259,2143,763] indazole,[257] indolizine and azaindolizines,[262] benzothiophene,[763,1703,2383,2385] indene[762] and miscellaneous heterocyclic systems[1662,2437]) and between H_4 and H_8 in compounds of type (XII)† (e.g. chromene,[763,1978] quinoline,[258,89,550] phenanthrene[1698] and coumarin[1978]), the coupling

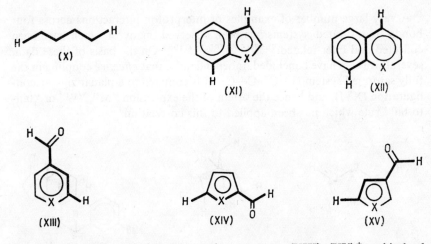

between aldehydic and ring protons in structures (XIII)–(XV)‡ and indeed between ring protons and coplanar protons on sp^2 hybridized benzylic carbon atoms or heteroatoms.[1345,2184,906,1870,843,846] The last type includes phenolic hydroxyl groups[846,843,906] constrained in the configuration (X) by hydrogen bonding. Thienopyrroles and related systems[2448,1911,1202,965,1662] also show a similar type of long-range coupling, and analogous, miscellaneous examples are listed in Table 4–4–5.

Coupling constants in all these categories range from 0·4–2·0 Hz for essentially planar configurations of type (X), and fall off rapidly with departures from planarity,[1345,1870] a fact which has been utilized for the assignment of conformation in some aromatic aldehydes and their derivatives.[1345,843]

The signs of several long-range coupling constants of this type have been determined in benzofuran,[513] 2-furanacrolein,[2184] furanaldehydes,[1189] thiophenealdehydes[513,848,509] and benzaldehydes,[845,1446] and were found to be the same as vicinal coupling constants between ring protons, i.e. positive.

† Other examples are listed in references 2358, 1703 and 1453.
‡ References 509, 843, 846, 847, 848, 970, 1033, 1034, 1038, 1189, 1443, 1446 and 1695.

In many of the above references considerably smaller (0·1–0·4 Hz) inter-actions between the aldehydic (or equivalent) protons and *ortho* ring protons were also observed. Where signs were determined, these secondary inter-actions appear to be of opposite sign to the larger coupling constants across the path (X), and it is likely[1870] that they involve σ–π overlap. In fact, it has been suggested[1870] that the hybridization of the benzylic carbon atom is not of decisive importance in determining the preferred long-range coupling path. In other words, coupling along the path (X) may involve protons bound to sp^3 hybridized carbon atoms and benzylic-type coupling (cf. Section E) could involve protons bound to sp^2 hybridized carbon atoms, provided that steric requirements were satisfied.

G. COUPLING ACROSS FOUR SINGLE BONDS

A very large number of examples of interproton interactions across four bonds in saturated systems has been observed, many of which have been collected and inter-related.[2358,173,1500,162,2062] On the basis of these data, several workers have concluded, independently, that *effective* coupling in the fully saturated system H—C—C—C—H is confined to a planar zig-zag con-figuration (XVI), and hence the origin of the expression "M", "W" or "tail-to-tail" rule which has been applied to this correlation.

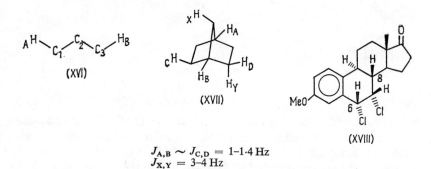

$$J_{A,B} \sim J_{C,D} = 1\text{–}1\cdot4\,\text{Hz}$$
$$J_{X,Y} = 3\text{–}4\,\text{Hz}$$

Few theoretical investigations have been carried out on long-range coup-ling in saturated systems[1422,173,162,254] but Barfield[173] has been able to obtain semi-empirical estimates of an "indirect" contribution to such coup-ling, which predicts a positive maximum for the W-configuration (XVI). Most of the relative sign determinations for structures with significant spin-spin coupling across four single bonds have been obtained either with sys-tems incorporating three-membered rings[757,2091,2092,2358] (which may be anomalous due to strain) or with flexible or semi-flexible systems[821,905, 909,94,312] (see below). However, results with allylic systems with *transoid* coupling and $\Theta = 0°$ and inferences which can be drawn from sign deter-minations in flexible systems[2358] (see below), suggest strongly that coupling across four single bonds in the W-configuration is in fact positive in sign, in accord with Barfield's mechanism.

For essentially rigid systems which do not incorporate *highly* strained structures (e.g. four-membered rings), the long-range coupling constants across the W-path are generally within the range of 1–2 Hz. In bicyclo[2.2.1]-heptanes (XVII) larger values have been reported.

A large amount of data is available for bicyclo[2.2.1]heptanes, bicyclo-[2.2.2]octanes, bicyclo[2.1.1]hexanes, bicyclo[2.1.3]octanes, their unsaturated and heterocyclic analogs and other systems of well defined configuration.† No exceptions from the W-rule have been noted, although the variations of coupling constants with substitution and introduction of unsaturation show clearly[1500] that factors other than purely geometrical ones must be involved. It is important to note that some of these interactions are of the allylic type (see below and Section C) but are most unlikely to involve the σ–π mechanism.[173,950,2358,2323]

Of the very few apparent exceptions to the W-rule,[2358] the coupling between H_6 and H_8 ($J = 1.8$ Hz) in the steroid (XVIII)[1914] is most disturbing,[2358,1013,241,173] because there is no obvious reason (such as strain, unusual substitution patterns etc.) to account for this anomaly. However, in some very closely related steroids (where acetyl groups and or bromine atoms were substituted for the chlorine at C_6 and C_7), no analogous long-range coupling has been observed.[1291] On the other hand, *small* interactions (J approximately 0.5 Hz) between protons not separated by four bonds along the W-path, have been observed in some *m*-dioxans,[2036] carbohydrates[1547] and in several acyclic systems (see below).

Thus at present, correlations based on the W-rule should be limited to coupling constants in excess of 1 Hz.

In strained rigid systems (Table 4–4–6) quite substantial long-range interactions (up to 18 Hz) have been noted. These also appear to conform to the W-rule, and the larger interactions appear to be associated with a multiplicity of equivalent paths. However, as some of the coupling constants are far in excess of any calculated maximum values,[173] strained systems should be, for the time being, considered purely on an empirical basis.

In examining the steric effects on long-range coupling in the fragment (XVI), we have to consider *two* dihedral angles, H_A—C_1—C_2—C_3 and C_1—C_2—C_3—H_B, i.e. rotation about two bonds C_1—C_2 and C_2—C_3. Barfield[173] has produced calculated values for the "indirect effect" for a family of curves, keeping one of the dihedral angles constant at selected values and varying the other one continuously between 0° and 360°. Of this family of curves, the one where the fixed dihedral angle corresponds to the average value taken up by the protons in the freely rotating methyl group, i.e. for the special case of the fragment H—C_1—C_2—CH_3, is particularly interesting (Fig. 4–4–2). It can be seen that the indirect effect predicts a positive maximum for the dihedral angle, defined by rotation about the bond C_1—C_2, equal to 180°. A number of results[821,905,608,312,94,609,909] appear to support this, values of between +0.4 and +0.8 Hz having been reported for a *trans*

† Besides the collections of data listed above, see also references 90, 384, 481, 554, 613, 635, 831, 937, 1086, 1273, 1274, 1275, 1465, 1509, 1546, 1547, 1737, 1738, 1827, 1828, 1857, 1981, 2323, 2363, 2366, 2417, 2420, 2427, 2434, 2438, 2460, 2491, 2492 and 2614.

TABLE 4–4–6. LONG-RANGE COUPLING CONSTANTS IN STRAINED SYSTEMS

Structure	Coupling constants (Hz)	References
	$J_{AB} = 0.5–2.0$	182, 962, 1020
	J_{AB} trans $= 1.0 \pm 0.1$ J_{AB} cis $= 1.9 \pm 0.1$	332
	$J_{AB} = 6–8$ $J_{A'B} \sim J_{A'B'} \sim 0$	2561, 2563
	$J_{AB} = 4.4–10$	1705, 2561
	$J_{AB} = 6.65$ $J_{AX} = 2.45$ $J_{BX} = 1.25$	2595
	$J_{AB} = 5.80$ $J_{AB'} = 0.9$	505
	$J_{AB} = 10$ $J_{XY} = 110$	2561

methyl–proton interaction, e.g. for J_{AB} in dibromopropane (XIX).[905] Further, a number of smaller values have been found for the gauche methyl–proton interactions, e.g. J_{AC} in (XIX), ranging from 0 to -0.2 Hz, in reasonable agreement with the trend shown in Fig. 4–4–2, especially if it is remembered that the calculated values refer only to one coupling path, albeit apparently the most important one. Thus the smaller gauche coupling constants may be dominated by other mechanisms.[173,2358]

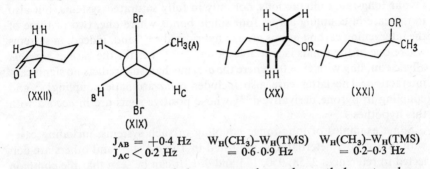

(XXII) (XIX) (XX) (XXI)

$J_{AB} = +0.4$ Hz $W_H(CH_3)-W_H(TMS)$ $W_H(CH_3)-W_H(TMS)$
$J_{AC} < 0.2$ Hz $= 0.6$–0.9 Hz $= 0.2$–0.3 Hz

The relative magnitudes of the *trans* and *gauche* methyl–proton long-range interactions have been well established from the study of the width of the signals due to tertiary methyl groups in steroids,[2579,238,2277,2125,1300] decalones,[2125] diterpenes,[2125] 1-methyl-4-t-butylcyclohexanols,[2278] and miscellaneous compounds.[1475] It was found that tertiary methyl groups in structures where a conformation with a favourable coupling path can be assumed (e.g. XX), give rise to significantly broader signals than those where such a path is not possible (e.g. XXI). This correlation has obvious applications in stereochemical problems and, in fact, arguments of this type have already been used in deciding the conformation of a camphor derivative,[1060] some diterpenes[430] and bromocyclohexanones.[954] The average values of long-range coupling constants have also been used[712,713] in a conformational study of 2,2-dimethylcyclohexanones and related compounds.

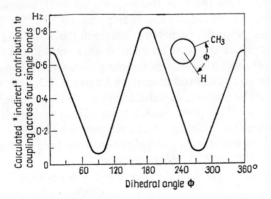

Fig. 4–4–2. Calculated values of the "indirect contribution" to long-range spin–spin coupling involving a methyl group. (After Barfield[173].)

The points corresponding to other geometries (cf. Fig. 4–4–2) cannot yet be considered established experimentally with any degree of certainty. Data on a number of methyl substituted dioxolans and carbonates,[94,41,757] indene oxide,[757] propylene oxide,[757] epichlorohydrin[2092] and glycidaldehyde[2091] are available, but uncertainties associated with conformational distortion and ring strain makes the interpretation of these results difficult.

It has been pointed out[951] that the planar W-configuration appears to favour long-range interactions, not only in fully saturated systems, but also in the case of coupling across four single bonds where one, two or three of the intervening carbon atoms are sp^2 hybridized,[2358] and, indeed, in systems where one of the bonds is a double bond,[173,950] i.e. for the case of *transoid* allylic coupling with $\Theta = 0°$, where the σ–π mechanism predicts no significant interaction. The latter case also includes *meta* aromatic coupling[173] and coupling in pyrone derivatives[1297] whose positive signs are in accord with this hypothesis.

Some examples of long-range coupling in such systems, including cases with intervening heteroatoms, are listed in Table 4–4–7 and others are collected in references 2358, 950, 173 and 951. It can be seen that the common range of values is 1–3 Hz for structures where there is essentially no departure from planarity in the W-path. The generality of these examples appears to be sufficient for use in structural and stereochemical assignments, particularly for molecules where all three of the intervening carbon atoms are sp^2 hybridized.

A special case is encountered in the system where only the middle carbon is sp^2 hybridized (i.e. H—C—$\overset{\|}{C}$—C—H). While numerous examples exist showing that effective coupling occurs in the W-configuration,[2062,2276,1506,951,2358] e.g. between the equatorial protons in (XXII), at least one series of carefully studied examples[1506] shows that significant coupling also occurs between the axial protons in the fragment (XXII). This is not unreasonable considering that, besides the transmission across four single bonds through the indirect mechanism (best for the W-configuration), another interaction, involving the central π-bond has also been proposed,[1204,2552,2358] for which the diaxial configuration, permitting maximum σ–π orbital overlap,[2358] would be favoured. If the total interaction were transmitted through these paths, one might predict significant coupling between the two axial protons and between the two equatorial protons in fragment (XXII) but not between axial and equatorial protons, as indeed appears to be the case in some steroidal haloketones[2276] (see however reference 951). The considerably smaller (generally 0·3–0·6 Hz) coupling constants between freely rotating groups separated by a carbonyl or a similar group have been discussed in some detail.[1204,2552,2380,2381,958,2358]

Occasionally, quite appreciable ($J = 0·4$–$1·2$ Hz) interactions[2358,361,477] have been noted in the fragment —CH—$\overset{\|}{\underset{\|}{C}}$—CH—, but they appear to depend on substitution in an, as yet, undetermined manner and are often

TABLE 4-4-7. LONG-RANGE COUPLING IN UNSATURATED SYSTEMS INCORPORATING A PLANAR W-PATH

Structure	J (Hz)	References
	$J_{A(equatorial),B} = 1\cdot4\text{–}2\cdot2$	951, 1880, 1881
	$J_{AB} = 1\cdot9$	550
	$J_{AB} = 1\cdot7$	30
	$J_{AB} = 1\cdot5$ $J_{BC} = 10$	30
	$J_{AX} = 1\cdot05$ $J_{AB} = 1\cdot05$	360
	$J_{AB} = 2\cdot1$	1928
	$J_{AB} = 1\cdot2\text{–}1\cdot6$	664

TABLE 4–4–7 *(cont.)*

Structure	J (Hz)	References
	$J_{AB} = 1·7$ $J_{CD} = 2·9–3·0$	180
  X=O,S ; Y=O,S	$J_{AB} = +0·9–+2·0$ $J_{CD} = +2·2–+4·1$	202, 359, 1297
  X=O, NOH, NOR,	$J_{AB} = 2·0–2·8$	510, 1161, 1880, 1881
	$J_{AC} = 1·5–1·9$ $J_{AD} = 1·9–2·4$	1405, 1862, 2317
	$J_{AB} = 2·4–3·2$	1366
	$J_{AB} = 2·3–2·5$	2164
	$J_{AB} = 1·3$ $J_{AC} = 8·0$ $J_{BC} = 6·0$	1915

TABLE 4–4–7 *(cont.)*

Structure	J (Hz)	References
	$J_{AB} = 2\text{–}3$	1160, 2531
	$J_{AB} = 2 \cdot 2\text{–}2 \cdot 4$	200, 415
	$J_{AB} = 1 \cdot 5$	1780

not observable.[2144] In the system $H_3C-\overset{\text{C}}{\underset{\|}{}}-CH_3$, this coupling is often in the range $0 \cdot 3\text{–}0 \cdot 6$ Hz.[361,2144]

An appreciable coupling across four σ-bonds involving an aliphatic —OH group has been reported.[1909]

H. LONG-RANGE COUPLING IN 1,3-BUTADIENE DERIVATIVES

From the above discussions of the probable mechanisms of long-range coupling in various structures, it is clear that long-range coupling in 1,3-butadienes may present a very complex picture, because a number of paths, known to be favourable for effective long-range coupling, could be present in this group of compounds and could depend critically on conformation.

Further, there are considerable difficulties associated with the analysis of n.m.r. spectra of lightly substituted 1,3-butadienes and the number of studies which have so far appeared† is limited.[764,255,1178,1436,306,2187,1404,1486] Most of these results have been summarized by Bothner-By and Harris[306] who also give a detailed qualitative discussion of the probable mechanisms of spin–spin coupling in these systems and consider conformational effects.

† Some of the structures listed in Table 4–4–5 and described in reference 280, may also be regarded as butadiene derivatives incorporated into cyclic systems.

In structures (XXIII) and (XXIV) we summarize the results obtained with 1,3-butadiene itself[1178] and 1,3-cyclohexadiene[306,1486] where the conformations are known with reasonable certainty.[306]

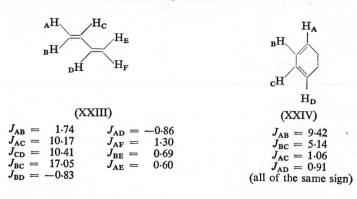

(XXIII)

$J_{AB} =$	1·74	$J_{AD} =$	−0·86
$J_{AC} =$	10·17	$J_{AF} =$	1·30
$J_{CD} =$	10·41	$J_{BE} =$	0·69
$J_{BC} =$	17·05	$J_{AE} =$	0·60
$J_{BD} =$	−0·83		

(XXIV)

$J_{AB} =$	9·42
$J_{BC} =$	5·14
$J_{AC} =$	1·06
$J_{AD} =$	0·91

(all of the same sign)

I. MISCELLANEOUS EXAMPLES OF LONG-RANGE COUPLING

Although many of the examples of long-range interproton coupling encountered in this section may be rationalized in terms of the reasonably well-established mechanisms discussed above, we shall confine ourselves to listing in Table 4–4–8 some of the apparently more common examples with only the minimum of discussion.

The coupling between methoxy and ring protons (entry *a*) appears to be structurally dependent and although the coupling constants are very small, they can be conveniently studied by indirect methods (cf. Section A) and afford useful correlations. Similar coupling has been noted in an enol ether.[954]

Long-range coupling in formates (entry *b*) appears to be general and has been used in structural investigations.[142] The special case of vinyl formate (entry *c*) could involve ground-state electron delocalization (arrows) and thus be analogous to long-range coupling in 1,3-butadienes (Section H). This explanation has also been invoked for the case of methyl vinyl sulphide (entry *d*) and some furenidones (entries *e* and *f*).

Coupling along the path indicated in entry (*g*) may also occur in thieno-pyrroles and related compounds.[1662,965,1911,1202,2448] Other modes of long-range coupling in polycyclic aromatic compounds include appreciable ($J = 0·8$ Hz) coupling between H_5 and H_8 in phenanthrenes[1698] and several examples[2358,2184,1698] of weaker coupling (J approximately 0·4 Hz) between *peri*-protons. It is attractive to consider this type of coupling in terms of conjugated paths in important mesomeric structures,[763] but with the exception of the cross-ring coupling discussed in Section F, little systematic work has been done in this area.

Spin–spin coupling across five single bonds appears to be quite rare; entries (*h*), (*i*) and (*j*), where multiple extended zig-zag paths appear to be

TABLE 4-4-8. MISCELLANEOUS EXAMPLES OF LONG-RANGE COUPLING

	Structure	J (Hz)	References
a		$J_{AB} = 0\text{--}0\cdot8$	842, 846, 1015, 1622
b		$J_{AX} = 0\cdot8\text{--}1\cdot0$ $J_{BX} = 0\cdot5\text{--}0\cdot6$	142, 1445
c		$J_{AD} = -0\cdot7$ $J_{BD} = +1\cdot6$ $J_{CD} = 0\cdot8$	1226, 1831
d		$J_{AD} = 0$ $J_{BD} = 0\cdot2$ $J_{CD} = 0\cdot4$	1180
e		$J_{AB} = 0\cdot9$	1181, 2139
f		$J_{AB} = 0\cdot8$	1181, 2139
g		$J_{AB} \sim 0\cdot4$	262, 1703, 2385
h		$J_{AB} = 1\cdot2$	2358

TABLE 4–4–8 *(cont.)*

	Structure	J (Hz)	References
i		$J_{AB} \sim 0.7$	1547
j		$J_{2e,6e} = 1.5$ $J_{2e,5e} = 0.9$ $J_{2e,6a}; J_{6e,4a};$ $J_{4a,6a}; J_{6e,2a};$ and $J_{6a,2a}$ $= 0.3–0.5$	2036
k		$J_{AB} \sim 0.5$	378
l		$J_{AB} = 3.5$	1628

involved, and the highly stereospecific long-range coupling in *N*-benzyl-β-lactams[182] are some of the few reported cases.

Small coupling (*J* approximately −0·5 Hz) across five bonds in *p*-benzo-quinones and their derivatives[1881] may be related to *para* coupling in aromatic systems. Stereospecific long-range coupling between protons separated by up to six bonds has been reported[1344,2251] in hydrazones. Some unusual examples also occur in 3,3-dimethylcyclopropene[502] and some furanosenes.[87,1540]

The hydrido protons of metal hydrides are often appreciably coupled to ligand protons,[2358] coupling constants of 1·2–1·8 Hz having been reported[580] in some protonated ferrocenes.

As has been noted in several instances above, substitution of heteroatoms for carbon atoms[2358] often does not affect long-range interproton coupling.

COUPLING BETWEEN PROTONS
AND OTHER NUCLEI[1803]

With the exception of ^{13}C which is present to the extent of only 1·108 per cent at natural abundance and ^{14}N where $J_{N,H}$ is generally unobservable, the vast majority of organic substances do not contain magnetic nuclei other than protons. Thus for the purpose of structure determinations, coupling between protons and other nuclei is only of either marginal importance or of interest in connection with specialized topics, such as organic fluorine or phosphorus chemistry.

Nevertheless, a large number of investigations have been carried out in this area, and many of the results are of considerable theoretical importance.

A thorough discussion of this work lies outside the scope of this book and would indeed be of comparatively little interest to the majority of organic chemists. In the brief treatment presented below, we shall confine ourselves to results which are of obvious applicability.

A. COUPLING BETWEEN ^1H AND ^{13}C

We have already referred to the use of ^{13}C satellites as an aid in the analysis of proton spectra (Chapter 2–3) and in fact, most of the data on spin–spin coupling between ^1H and ^{13}C have been obtained from the *proton* spectra of compounds containing ^{13}C either at natural abundance or isotopically enriched. For various reasons, principally of an experimental nature,[2364] the ^{13}C spectra are inferior to the proton spectra as regards measurement of splittings due to ^1H, ^{13}C interactions, but indirect methods[903,78,904] based on double resonance have certain advantages over both and may indeed be used to obtain ^{13}C chemical shifts as well as ^{13}C, H coupling constants.

In general, only the direct ^{13}C,H coupling constants can be conveniently obtained from the proton spectra of compounds containing ^{13}C at natural abundance, because only these coupling constants are large enough to place the satellites outside the very much stronger central signals due to protons attached to ^{12}C (Chapter 2–3). Fortunately, it is also these direct coupling constants which are of most theoretical and practical interest.

Not unexpectedly, coupling between magnetic nuclei separated by only one bond (e.g. direct coupling between ^{13}C and ^1H) is much more amenable to theoretical treatment than coupling across more than one bond and several such calculations have in fact been performed,† all of which predict large positive values for direct $J_{13C,H}$, sometimes in close agreement with experimental values.

Table 4–5–1 shows some values of direct (and longer range) $^{13}C,H$ coupling constants which have been abstracted from the much larger collections[+] found in the literature. The most striking result is the apparent linear dependence[1807,2272] of $J_{13C,H}$ on the per cent of s-character of the carbon atom (Fig. 4–5–1) with no coupling predicted for vanishing s-character. This relation constitutes a strong vindication of the importance of the Fermi contact term in direct spin–spin interactions, because this term appears only in the s-orbitals.

TABLE 4–5–1. ^{13}C, ^1H COUPLING CONSTANTS

Unless otherwise indicated, the coupling constants refer to direct (one bond) coupling. For references see text

Structure	$J_{13C,H}$ (Hz)
CH_4	125
CH_3Cl	151
CH_2Cl_2	178
$CHCl_3$	209
CH_3CH_3	126
$^{13}CH_3CH_2Cl$	128
$CH_2{=}CH_2$	157
$CHCl{=}^{13}CH_2$	160
$CH_2{=}^{13}CHCl$	195
$CH_3{-}C{\equiv}^{13}CH$	$J_{13C-H} = 248, J_{13C\equiv C-C-H} = 4{\cdot}8$
$H{-}C{\equiv}C{-}C{\equiv}C{-}H$	259
$H{-}^{13}CO{-}R$	168–172
$H{-}^{13}CO{-}CCl_3$	207
$H{-}^{13}CO{-}F$	267
Cyclohexane	124
Cyclopentane	128
Cyclobutane	134
Cyclopropane	161
Norbornane (Bridgehead)	142
Norbornadiene (Bridgehead)	146
Benzene	159
Pyridine	$\alpha = 179; \beta = 163; \gamma = 152$
$(CH_3)_2CH{-}^{13}\overset{\overset{\displaystyle O}{\|}}{C}{-}CH(CH_3)_2$	$J_{13C-C-H} = 4{\cdot}1, J_{13C-C-C-H} = 5{\cdot}1$
$(CH_3CH_2)_2{}^{13}CDOH$	$J_{13C-C-H} = 4{\cdot}0, J_{13C-C-C-H} = 5{\cdot}3$
Ethyl maleate	$J_{13C=C-H} = 1{\cdot}5$
Ethyl fumarate	$J_{13C=C-H} = 4{\cdot}4$

However, while the direct $J_{13C,H}$ is not very sensitive to the introduction of substituents at remote positions (cf. Table 4–5–1) it is extremely sensitive to substituents on the carbon atom actually involved in the spin–spin inter-

† References 185, 695, 1068, 1312, 1411, 1994, 2002, 2050, 2051 and 2633.
[+] References 670, 696, 1016, 1094, 1095, 1100, 1312, 1660, 1719, 1804, 1807, 2085 and 2435.

action. The relation between the electronegativity of these substituents and the deviation from the magnitude of the direct ^{13}C,H coupling in the parent hydrocarbon, is itself regular and approximately additive.[1658,1661] The finer points and theoretical significance of such variations have been the subject of considerable research,† which is outside the scope of this work, but sufficient data are now available to enable the organic chemist to use empirically the values of direct ^{13}C,H coupling constants as an adjunct to the chemical shifts in identifying the environment of a proton bound to a carbon atom.

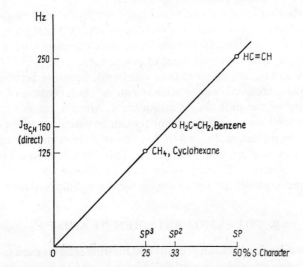

FIG. 4–5–1. The relation between direct $J_{13_{C,H}}$ and the hybridization of the carbon atom.

Middaugh and Drago[1754] give an example of such an approach. In CH$_3$—CO—S—CH$_3$ the two methyl groups give rise to two singlets at $\delta = 2\cdot30$ and $2\cdot27$ ppm and it is thus impossible to assign them. However, the ^{13}C satellites of the resonance at $2\cdot30$ ppm are 131 Hz apart, while those of the resonance at $2\cdot27$ ppm are 143 Hz apart. Examination of a range of model compounds revealed that direct $J_{13C,H}$ in CH$_3$—S is in the range of 138–140 Hz, while direct $J_{13C,H}$ in CH$_3$—CO fragments is in the range of 127–130 Hz thus allowing the assignment of the resonance at $2\cdot30$ ppm to CH$_3$CO— and of the resonance at $2\cdot27$ ppm to CH$_3$—S—. While this particular problem might be considered of merely academic interest, the identification of CH$_3$—CO— and CH$_3$—S— groups in general could be important. This technique could also be used to distinguish between accidental and symmetry equivalence.

The relation between the direct ^{13}C,H coupling constant and hybridization of the carbon atom has led to attempts[836] to correlate ring-strain and the resulting partial rehybridization with the magnitude of $J_{13C,H}$. While the

† References 221, 881, 1013, 1016, 1094, 1095, 1228, 1312, 1338, 1514, 1624, 1660, 1661, 1681, 1713, 1719, 1726, 1804, 1808, 1810, 2054, 2085, 2153, 2204, 2205, 2304, 2435 and 2436.

quantitative "reading off" of the percentage of *s*-character in strained systems directly from the magnitude of $J_{13C,H}$ is not now considered justified,[1763, 2561,1338] the considerable amount of data collected for various cyclic and strained systems,† some of which are quoted in Table 4–5–1, could be useful in identifying protons attached to ring-systems of various sizes. For example, in cyclobutanes the values of direct $J_{13C,H}$ are fairly characteristic while the proton chemical shifts are not.

In addition, the values of direct coupling constants for protons in various positions in aromatic and heterocyclic systems[2435,670,998,2072,2074,1238,1713] are also fairly characteristic and could undoubtedly be used empirically.

So far, we have been dealing only with the direct $^{13}C,H$ coupling which can be seen to have considerable interest for organic chemists both as an aid in spectral analysis and for assignment of resonances.

The *indirect*, i.e., across more than one bond, coupling between ^{13}C and protons has also received considerable attention‡, but, because of the relative inaccessibility of the data and a considerably smaller range of values, no obvious applications to organic chemistry can be easily perceived at present. The few values quoted in Table 4–5–1 are merely illustrative. It can be seen that $J_{13C-C-H}$ is smaller than $J_{13C-C-C-H}$. This phenomenon is not confined to coupling between ^{13}C and protons but can also be observed with other heavy, magnetic nuclei, in, for example, some ethyl derivatives.[454]

B. COUPLING BETWEEN ^{1}H AND ^{19}F

Spin–spin coupling between protons and fluorine exhibits even more variety than interproton coupling, in particular with regard to long-range coupling and theoretical implications. Further, although the amount of data available is considerably smaller than the results on interproton coupling, it is quite obvious that proton-fluorine coupling constants could provide at least as much structural information as interproton coupling constants.

A complete discussion of this topic would, however introduce a marked imbalance in this book because fluorine-containing compounds are, of course, comparatively rare in organic chemistry. We shall therefore present a number of results in tabular form (Table 4–5–2) and confine the discussion mainly to an indication of major literature sources. The data and references given in Table 4–5–2 should enable the reader to obtain most of the information pertinent to the solution of structural problems.

Some collections of data on coupling constants involving ^{19}F[212,1803,241] are available and most of the investigations concerning relative signs[2353, 1948,786,175,649,755] have been summarized by Evans, Manatt and Elleman.[786]

N.m.r. spectra of fluorophosphoranes,[2208] sulphur pentafluorides[272,1209] and $J_{H,F}$ in flexible systems[1491,1490,274,1869] have been covered separately.

---

† References 197, 496, 501, 505, 836, 953, 1500, 1507, 1571, 1666, 1705, 1763, 2232, 2319, 2434, 2435 and 2561.

‡ References 247, 454, 696, 699, 832, 902, 904, 914, 1238, 1332, 1334, 1335, 1338, 1339, 1601, 1603, 1805, 1897 and 2054.

TABLE 4–5–2. PROTON–FLUORINE COUPLING CONSTANTS

Structure	Coupling constants (Hz)	References
$$\underset{a}{H-\overset{\displaystyle \mid}{\underset{\displaystyle \mid}{C}}-F}$$	45–81 (generally 45–50)	58, 121, 302, 715, 752, 755, 1088, 1611, 1803, 2254, 2353, 2463, 2609
CH_3F	81	121
CH_3CH_2F	$J_{CH_2,F} = 46 \cdot 7$; $J_{CH_3,F} = 25 \cdot 2$	2353
$$\underset{\text{and } H-\overset{\displaystyle \|}{C}-C-F}{\overset{b}{H-\overset{\displaystyle \mid}{\underset{\displaystyle \mid}{C}}-\overset{\displaystyle \mid}{\underset{\displaystyle \mid}{C}}-F}}$$	0–30; generally $J_{gauche} = 0\text{–}5^c$ $J_{trans} = 10\text{–}25^{\,c}$	7, 302, 715, 752, 753, 795, 815, 825, 1063, 1075, 1088, 1869, 1901, 2365, 2463, 2609
	$J_{F_{ax},H_{ax}} = 34 \cdot 3$ $J_{F_{ax},H_{eq}} = 11 \cdot 5$ $J_{F_{eq},H_{eq}}$ and $J_{F_{eq},H_{ax}} < 5\text{–}8$	1296
$H-C\equiv C-F$	21	1755
	$J_{gem} = 85\ (72\text{–}90)$ $J_{trans} = 52\ (12\text{–}52)$ $J_{cis} = 20\ (-3\text{–}+20)$	161, 203, 274, 454, 552, 624, 649, 835, 1626, 2240, 2501
	$J_{ortho} = 6 \cdot 2\text{–}10 \cdot 3$ $J_{meta} = 3 \cdot 7\text{–}8 \cdot 3$ $J_{para} = 0\text{–}2 \cdot 5$	120, 785, 823, 1305, 1391, 1398, 1523, 1707, 1948, 1950, 2056, 2114, 2183, 2190, 2252
	$J_{ortho} = 2 \cdot 5$ $J_{para} = 1 \cdot 5$ $J_{meta} = 0$	1707

TABLE 4–5–2 *(cont.)*

Structure	Coupling constants (Hz)	References
H_3C—CH=CH—F	$J_{CH_3, F} = 2 \cdot 4\text{–}3 \cdot 3$	203, 302, 649, 1419, 2609
H—C≡C—CF_3	0–1	825
	$J_{H_4, F_{6\beta}} = 5\text{–}5 \cdot 5$ $J_{H_4, F_{6\alpha}} \sim 0$	759, 1419, 2609
Ph—CH_2—$\overset{\overset{\textstyle O}{\|}}{C}$—$CH_2$—$F$	$J_{H-C-C-C-F} = 3 \cdot 2$ (with $\overset{\|}{O}$)	2254
H—$\overset{\|}{\underset{\|}{C}}$—$\overset{\|}{\underset{\|}{C}}$—$\overset{\|}{\underset{\|}{C}}$—$F$	0–3·6	753, 795
	$J_{H_A, F_{ortho}} \sim 0$ $J_{H_A, F_{para}} = 0 \cdot 8$	532

    [a] The values of geminal fluorine–proton coupling constants increase with increasing electronegativity of substituents.[121,1803]

    [b] The values of vicinal fluorine–proton coupling constants appear to decrease with increasing substituent electronegativity[1803] and vary with dihedral angle in a manner analogous to vicinal interproton coupling constants.[1869]

    [c] These ranges apply to polyhalogenated compounds and therefore it is to be expected that the values are lower than the average coupling constant in ethyl fluoride.

    [d] Values in brackets are ranges encountered in olefinic compounds. Invariably, $J_{gem} > J_{trans} > J_{cis}$ in any particular series. Solvent dependence has been observed.[2501]

    [e] Coupling constants between the trifluoromethyl group attached to a benzene ring and ring protons are sensitive to conformation.[1298] A small (0–1 Hz) coupling between fluorine and t-butyl groups attached to benzene rings has been observed.[2114] Fluorine attached to benzene rings may also be appreciably coupled to —X—CH_3[405] or even —CO—N—CH_3[1560] groups in a highly stereospecific manner.

TABLE 4-5-3. PROTON–PHOSPHORUS COUPLING CONSTANTS

Structure	Coupling constants (Hz)	References
CH_3PH_2	$J_{P-H} = 186.4 \ (180-200)^a$	693, 776, 1663, 1667, 2546
EtO, EtO—P(=O)(H)	$J_{P-H} = 630 \ (630-707)^b$	1098, 1721, 1723
$(CH_3)_3P$	2·7	
$(CH_3)_3P{=}O$	13·4	
$(CH_3)_4P^{\oplus}I^{\ominus}$	14·4	472, 789, 1023, 1151, 1667, 2353, 2546
$(CH_3CH_2)_3P$	$J_{CH_3, P} = 13.7; \ J_{CH_2, P} = 0.5$	
$(CH_3CH_2)_3P{=}O^c$	$J_{CH_3, P} = 16.3; \ J_{CH_2, P} = 11.9$	
$(CH_3CH_2)_4P^{\oplus}I^{\ominus}$	$J_{CH_3, P} = 18.0; \ J_{CH_2, P} = 13.0$	
$(CH_3)_2CH-P(=O)(Y)(X)$	$J_{CH_3, P} = 15-30; \ J_{CH, P} = 12-18$	1683, 2282
$(CH_3)_3P^{\oplus}HX^{\ominus}$	$J_{CH_3, P} = 16.6; \ J_{P-H} = 515$	2285
$H_3C-C(=O)-P(OEt)(OEt)$; $H_3C-C-P(OR)(OR)$; $H_3C-C(=O)-P(OR)(R)$	$J_{CH_3, P} = 10.5-18$	222, 602, 828
$H_3C-C(=O)-CH_2-P(=O)(OR)(OR)$	$J_{CH_2, P} = 23$	546

TABLE 4–5–3 *(cont.)*

Structure	Coupling constants (Hz)	References
R, R' CH—S—P with S (=S on P), OR, OR	$J_{CH-S-P} = 15$	1916
(ring structure with O, P, Ph, Ph, Ph, $_AH$)	$J_{H_A, P} = 4.0$	1150
H, H C=C $1/3P$, H (marked d)	$J_{P-C-H} = 11.7\text{–}22.3$ $J_{P-C=C-H \ (cis)} = 13.6\text{–}19.5$ $J_{P-C=C-H \ (trans)} = 30.2\text{–}40.3$	38, 81, 454, 1667
Ph, Ph—P=C Ph, with CO—NH$_2$ and CH$_2$—Br	$J_{CH_2, P} = 8.0$	2449
H_3C—O—P (marked e)	11.4–13	222, 130, 1099, 2037, 2038, 2039, 2040, 2042, 2043, 2044
R—CH$_2$—O—P (marked e)	6.5–10	130, 546, 602, 1683, 1721, 1725, 1916, 2283
R, R' CH—O—P (marked e)	5–7	130, 546, 2283
MeO, MeO—P, MeO, with O—, O—, H_A (ring)	$J_{H_A, P} = 18$	2042
H_3C—N—P	8.5–25	407, 548, 789, 1206, 1331, 1676, 2290
$(Me_2N)_3P$	8.82	
$(Me_2N)_3P{=}O$	9.47	
$(Me_2N)PCl_2$	13.07	548
$(Me_2N)PCl_2$ with ‖ S	17.69	

TABLE 4–5–3 *(cont.)*

Structure	Coupling constants (Hz)	References
¹⁄₃PO *f* X	$J_{P,H_{ortho}} = 10\cdot5-11\cdot5$ $J_{P,H_{meta}} = 2\cdot1-3\cdot4$	1023
H—Pt—P	13–16	474
P—N—C—C—H	0·6–1·2	1331
P—N—N—C—H	<0·4	1331
	$J_{HA,P} \neq 0$	473

a Range in phosphine derivatives.
b In related compounds.
c Similar values are found in related compounds[1151] (Et₃PXY and Et₃PS).
d And derivatives.
e In phosphates, phosphites, phosphinates, etc.
f The values in related phosphonium salts[1023] are similar.

Proton–fluorine coupling constants in four-membered rings† exhibit certain peculiarities and the long-range coupling constants are often larger (7–16 Hz) than the vicinal coupling constants (1–20 Hz).

Long-range proton–fluorine coupling in aliphatic systems (across up to 5 or 6 single bonds) is highly stereospecific and has been thoroughly investigated[1417,2124,569,1419,570] and discussed[570,241] in connection with fluorinated steroids. It also occurs in alkadienes,[302,2240,307] bicycloheptenes[2589] and other systems.[808,274,2570]

C. COUPLING BETWEEN ¹H AND ³¹P

The introductory remarks in the preceding section also apply to proton–phosphorus coupling constants, and hence we shall restrict ourselves to a tabular presentation (cf. Table 4–5–3).

Due to, *inter alia*, variable valence states, phosphorus compounds exhibit considerable ranges in coupling constants to protons removed by the same number of bonds; the ranges quoted in some of the categories listed in

† References 436, 1088, 1277, 1403, 1491, 1668, 2388, 2514 and 2532.

Table 4–5–4. Coupling Between Protons and Miscellaneous Nuclei[1803]

Structure *or* coupling path	Coupling constants (Hz)	References
R $\quad \diagdown$ $\qquad {}^{14}N—H$ $\quad \diagup$ R $\qquad R$ $\qquad \vert$ $R—{}^{14}N—H$ $\qquad \vert$ $\qquad R$	40–68	[a]
$(CH_3CH_2)_4{}^{14}N^\oplus \; I^\ominus$	$J_{CH_3, N} = 1.9$; $J_{CH_2, N} < 0.4$	[b] 73, 401, 877, 1481, 1759
$—CH_\beta—CH_\alpha—{}^{14}N \equiv C$	$J_{H_\alpha, N} = 1.8$–2.7; $J_{H_\beta, N} = 2.6$–3.5	56, 1470
${}^{15}N—H$ ${}^{15}N—C—H$ and ${}^{15}N=C—H$ ${}^{15}N—C—H$ $\qquad \parallel$ $\qquad O$ ${}^{15}N—C—C—H$ and ${}^{15}N—C=C—H$	51–94 0–3.9 15.6–19 0.7–2.7	[c] 247, 319, 703, 704, 705, 2120, 2820
${}^{29}Si—H$	200–300	[d] 454
${}^{29}Si—C—H$	6.6–7.2	2204
${}^{29}Si(CH=CH_2)_4$	$J_{Si, H}(cis) = 8$ $J_{Si, H}(trans) = 22$	594
$(CH_3CH_2)_4{}^{117}Sn$ $(CH_3CH_2){}^{119}Sn$	$J_{CH_3, Sn} = 68.1$; $J_{CH_2, Sn} = 30.8$ $J_{CH_3, Sn} = 71.2$; $J_{CH_2, Sn} = 32.2$	[e] 2353
${}^{117}Sn(CH=CH_2)_4$	$J_{Sn, H}(gem) = 96.0$; $J_{Sn, H}(cis) = 86.1$ $J_{Sn, H}(trans) = 174.1$	454
${}^{119}Sn(CH=CH_2)_4$	$J_{Sn, H}(gem) = 97$; $J_{Sn, H}(cis) = 90.4$ $J_{Sn, H}(trans) = 183$	
$(CH_3CH_2)_4{}^{207}Pb$	$J_{Pb, CH_3} = 125$; $J_{Pb, CH_2} = 41.0$	[f] 2353
${}^{207}Pb(CH=CH_2)_4$	$J_{Pb, H}(gem) = 212.4$; $J_{Pb, H}(trans)$ $= 330.1$; $J_{Pb, B}(cis) = 161.7$	454
$Pb—C—C—H$	0–22.5	926
$(CH_3CH_2)_2{}^{199}Hg$	$J_{Hg, CH3} = 115.2$; $J_{Hg, CH2} = 87.6$	[g] 2353
${}^{199}Hg(CH=CH_2)_2$	$J_{Hg, H}(gem) = 128.4$; $J_{Hg, H}(trans)$ $= 296$; $J_{Hg, H}(cis) = 159.5$	454
${}^{11}B—H$	$J_{B–H}(direct) = 80$–190	[h]

TABLE 4–5–5. LIST OF REFERENCES TO PROTON SPECTRA OF
COMPOUNDS CONTAINING LESS COMMON MAGNETIC NUCLEI

Isotope	Spin	Natural abundance (%)	References
^{3}H	$\frac{1}{2}$	0	2409
^{17}O	$\frac{5}{2}$	$3\cdot7\times10^{-2}$	2100, 1572
^{103}Rh	$\frac{1}{2}$	100	337, 558
^{111}Cd	$\frac{1}{2}$	12·86	1412, 1411
^{113}Cd	$\frac{1}{2}$	12·34	
^{183}W	$\frac{1}{2}$	14·28	614
^{195}Pt	$\frac{1}{2}$	33·7	2126, 923, 2308
^{205}Tl	$\frac{1}{2}$	70·48	1650, 1651, 100,
^{203}Tl	$\frac{1}{2}$	29·52	1649, 1124, 2353, 454

Footnotes to Table 4–5–4

a This is rarely observed in practice because of rapid exchange of N–H protons;[1437] signals due to protons attached to ^{14}N are also broadened due to quadrupole relaxation.[1989] It must also be remembered that ^{14}N, unlike all nuclei discussed so far, has spin = 1 rather than $\frac{1}{2}$ and thus will cause a splitting into triplets of equal intensity (cf. analogous splitting by ^{2}H discussed in Chapter 2–3).

b This relation is general for tetra-alkylammonium salts and the coupling appears to be stereospecific.[2103,957] No ^{14}N, H coupling has been noted in compounds other than those listed here except in certain protonated β-amino-α,β-unsaturated carbonyl compounds.[1452]

c ^{15}N has spin = $\frac{1}{2}$ and natural abundance of only 0·365 per cent. However, a number of data are available (references cited here contain a larger number of examples and back references) which could be useful in structural applications in enriched compounds.[704,703,2293]

d For proton spectra of compounds containing ^{29}Si, see general references to Group IV and references: 382, 383, 593, 594, 691, 733, 736, 1068, 1312, 1661, 2204, 2205 and 2466.

For specific references to spectra of compounds containing germanium, see the general references to Group IVB elements and references 2148 and 572.

e For proton spectra of compounds containing tin, see general references to Group IVB elements and references 65, 614, 410, 490, 491, 601, 832, 963, 1203, 1384, 1411, 1412, 1707, 1850 and 2204. Direct J_{Sn-H} has values of the order of 2000 Hz.

f See also general references to Group IVB and references 601, 926, 1412, 1850 and 2204.

g ^{199}Hg has spin = $\frac{1}{2}$ and natural abundance of 16·86 per cent. For proton spectra see also references 512, 636, 1130, 1388, 1412, 1850, 1852, 2128, 2526, 2527 and 2528.

h ^{11}B has spin = $\frac{3}{2}$ and natural abundance of 81·17 per cent. The remaining isotope ^{10}B is also magnetic and thus considerable complications result. A large amount of data on boron-containing compounds are available, most of which refer to inorganic compounds. Some useful results can be found in references 942, 943, 1000, 1972 and 2587.

Table 4–5–3 should therefore be taken as indicative only and original references must be consulted in cases of unusual substitution. The signs of many proton–phosphorus coupling constants have been determined[1667,2546,81] and some discussion of the influence of electronic factors appears in references 1723, 1725, 1683, 1151 and 1023. From data collected in Table 4–5–3 it also appears that steric effects are operative.

D. COUPLING BETWEEN PROTONS AND OTHER ELEMENTS

Table 4–5–4 lists mainly some of the typical values of spin–spin coupling constants for the elements in Group IVB, which are:

$$^{29}Si, \quad spin = \tfrac{1}{2} \qquad natural\ abundance = 4 \cdot 70\%$$
$$^{73}Ge, \quad spin = \tfrac{9}{2} \qquad natural\ abundance = 7 \cdot 61\%$$
$$^{117}Sn, \quad spin = \tfrac{1}{2} \qquad natural\ abundance = 7 \cdot 67\%$$
$$^{119}Sn, \quad spin = \tfrac{1}{2} \qquad natural\ abundance = 8 \cdot 68\%$$
$$^{207}Pb, \quad spin = \tfrac{1}{2} \qquad natural\ abundance = 21 \cdot 11\%$$

Some general discussions of proton spectra of compounds containing Group IVB elements are available.[2085,1624,2244,832,2304,694]

The remaining magnetic isotopes which are occasionally found incorporated into organic structures are listed in Table 4–5–5.

The very brief treatment afforded here does not reflect the considerable theoretical interest[1013,1083] attached to some of this work but merely the rarity of this group of compounds in the ordinary practice of organic chemistry. The potential usefulness of enriched ^{15}N compounds (Table 4–5–4) should however be recognized.

PART 5

APPLICATIONS OF TIME-DEPENDENT PHENOMENA

THE theory of time-dependent effects[1575,623] in n.m.r. spectroscopy (Chapter 2–1) has been applied to a number of problems of interest to organic chemists, although generally work of this type is classified as "physical" or "physical-organic". In this Part we shall briefly examine some examples of applications of time-dependent phenomena and list others according to the type of kinetic effect involved.

We shall then examine in some detail the effects of internal rotation and molecular symmetry on the appearance of some types of spectra. Tautomeric equilibria are also discussed here, because their study by n.m.r. spectroscopy requires an appreciation of time-dependent phenomena.

GENERAL APPLICATIONS

A. EXPERIMENTAL METHODS INVOLVED IN THE STUDY OF TIME-DEPENDENT PHENOMENA IN N.M.R. SPECTROSCOPY

Although very often a single measurement may give valuable information about a time-dependent phenomenon (e.g. the presence of averaged or other such spectra sets limits to the rates at which an exchange process takes place), normally it is necessary to alter experimental conditions in such a way as to *alter* the rate of the process involved.

The most obvious way of altering reaction rates is to alter the temperature. With variable temperature probes as standard accessories to most modern n.m.r. spectrometers, such experiments are easily performed and constitute the bulk of studies in this area. Other methods involve changes in pH, solvent or concentration, but it must be realized that any variation of conditions is likely to cause changes in chemical shifts, and to a lesser extent in coupling constants, as well as altering rates of exchange processes.

In practically all cases, the measurement of the activation energy associated with an exchange process involves observation of a number of spectra near the point of coalescence (cf. Chapter 2–1). This is relatively straightforward to interpret in the case of a doublet collapsing to a broad singlet, but can involve considerable difficulty when signals are more complex in shape. In a number of examples cited below, computerized matching procedures have been devised to deal with results of this type.

The above remarks apply to reactions, or other exchange processes, which are relatively fast on the n.m.r. time scale. Perhaps far more common in organic chemistry are reactions which are slow on the n.m.r. time scale and which may be followed with the aid of n.m.r. spectroscopy by simply record- ing spectra at intervals and obtaining the concentration of reactants and products by integration. While experiments of this type involve neither theoretical nor instrumental difficulties, and hence will not be discussed any further, it is the authors' opinion that this technique represents one of the more important applications of n.m.r. spectroscopy. Unlike i.r. spectroscopy and mass spectrometry which are difficult to apply on a quantitative basis, and u.v. spectroscopy, optical rotation and gas–liquid chromatography which are restricted to fairly limited groups of compounds, n.m.r. represents a convenient and quantitative analytical method which is, at least in principle, applicable to all compounds containing protons or other magnetic nuclei, the only drawbacks being limited sensitivity and the expense of instrumentation.

A specific application, particularly suited to analysis by n.m.r., is protium–deuterium exchange, for example, in enolization reactions and many examples of this type are recorded in the literature.[226,504,1382,1594,2064,2099,2605]

At the other extreme of sophistication, special methods, using modified experimental conditions, such as multiple irradiation[849,850,852] and the spin-echo experiment,[54,1296,55] have been used to obtain kinetic and thermodynamic data, but to date these have not found routine applications in organic chemistry.

B. HYDROGEN EXCHANGE AND HYDROGEN BONDING

One of the most commonly encountered reactions which may be "fast" on the n.m.r. time scale is intermolecular exchange of labile protons between groups such as —OH, —NH and —SH, and kinetic n.m.r. studies of these and the related phenomenon of intermolecular hydrogen bonding have been the subject of a number of reviews.[1734,1686,1724,1570]

The averaged spectrum corresponding to rapid exchange is generally encountered with high concentrations, with some polar solvents, and in the presence of traces of acidic or basic catalysts, such as mineral acids which are usually present in commercial solvents as an impurity. Under suitable conditions (which may involve high dilution in non-polar media, low temperatures or carefully controlled pH) it is often possible to obtain separate signals for each type of labile proton present (cf. Fig. 1–1–11a, p. 18) and even to observe spin–spin splitting in systems H—C—X—H where X = O, N or S (cf. Fig. 1–1–10, p. 16, and Chapter 4–2A(xi)).

The appearance, under routine conditions, of n.m.r. signals for the protons of commonly encountered R—X—H groups has been discussed by Rae.[2034] As expected, the rates of exchange are usually in the order —SH < —NH— < —OH. The —SH group generally does not exchange rapidly enough to give an averaged spectrum whereas amines exhibit averaged, partially averaged or non-averaged spectra depending on the pK_a of the particular base. Thus, the methyl group in *N*-methylaniline ($pK_a = 4·85$) gives rise to a singlet, the residence time of the NH proton being too short to permit the observation of spin–spin coupling, while the methyl group in 2,5-dichloro-*N*-methylaniline ($pK_a = 1·55$) gives rise to a doublet, $J_{H–N–C–H} = 5·38$ Hz.

Generally, in a protonated amine (e.g. in solution in trifluoroacetic acid[1602]) proton exchange is slow on the n.m.r. time scale and this nostic value, in combination with the characteristic changes in chemical shifts[1602] which take place when the group —NH is transformed to —NH$_2^\oplus$, is of diagnostic utility.

A number of detailed quantitative studies involving proton transfer and hydrogen bonding in amines have been reported † and in some cases activation energies of these processes have been evaluated from spectra obtained

† References 527, 772, 773, 802, 1046, 1047, 1052, 1574, 1593, 2101 and 2104.

for a range of temperatures. It is of interest that the exchange reactions involving —NH groups can be conveniently studied[703] by observing the effect of experimental conditions on the direct coupling $J^{15}{}_{N-H}$ in suitably labelled compounds.†

The exchange reactions involving hydroxylic protons in water, alcohols and carboxylic acids have also been the subject of numerous communications,‡ but in general, these results have little direct application in organic chemistry.

Uncatalysed proton exchange in amides[2154] is almost always slow on the n.m.r. time-scale, thus leading to characteristic broad signals due to partial averaging of the spin–spin coupling to nitrogen by quadrupole relaxation.[2408] As a result, the detection of these signals is sometimes difficult. In such cases, it may be found that recording of the spectrum in the integral mode will indicate the position of the amide resonance. Further, this signal can be removed by *in situ* exchange with D_2O in a manner similar to that used for labile protons in alcohols etc. (p. 53), but *only* if a trace of a strong base (triethylamine is convenient) is added to the sample.

Quantitative studies have also been carried out on the exchange rates of protons in phosphonium ions,[2285] acetylenes,[471,470] amino acids,[2197, 2198,459] amidinium ions,[1863] azoles[1306] and thiols.[2199] Some of these investigations have been concerned with the estimation of dimeric and polymeric hydrogen-bonded species and the equilibria between them.

C. LIGAND EXCHANGE

The metal–carbon bonds in organometallic compounds, metal–ligand bonds in coordination compounds, and bonds in π-complexes, are often relatively weak and may break and reform at a rate which leads to observable results in n.m.r. spectra.

In the cases of aluminium,†† cadmium,[637,63,1618] mercury,[2159] magnesium[2159,637] and zinc[1618,637] compounds, rapid ligand exchange, leading to disproportionation in suitable cases, can be observed from averaging effects in n.m.r. spectra and the rates of some of these reactions have been determined. Exchange rates of ligands of other types, for example, ethylenediamine tetraacetate complexes[1466] and zirconium chelates,[34] have also been studied and in some cases[1960] exchange rates have been obtained from line broadening. In at least one example (π-complex between cyclopentadienyl rhodium and ethylene[558]) two separate processes can be inferred from the temperature dependence of the n.m.r. spectrum.

The important problem of the structure of the Grignard reagents (p. 264) as reflected in their n.m.r. spectra, has received a great deal of attention,[128, 873,867,2094] especially by Roberts and his co-workers.[1877,1878,2550,2551,

---

† The use of ^{15}N avoids complications (generally in the form of extreme broadening) associated with the quadrupole moment of ^{14}N.[2408]

‡ References 611, 893, 1048, 1050, 1051, 1054, 1592, 1635, 1722, 1735, 1939 and 2078.

†† References 63, 1182, 1184, 1769, 1770 and 1809.

[2553,2554] In particular, some light has been thrown upon the inversion rates at the carbon–magnesium bond and the structures of vinylic,[128] allylic[1877, 1878,2551] and indenyl[2094] Grignard reagents by kinetic n.m.r. studies, and by comparison of their spectra with those of related compounds such as dialkyl magnesium deriviatives and lithium alkyls (cf. Chapter 3–9).

D. PARTIAL DOUBLE BOND CHARACTER

The rotational barriers about formal single bonds in certain classes of compounds are so much larger than those normally encountered (cf. conformational effects discussed below) that they become particularly amenable to n.m.r. investigations. Thus, although the distinction is purely quantitative, such phenomena are usually considered separately as involving "partial double bond character".

The most common and thoroughly investigated examples of this type are amides, where the process (Ia) ⇌ (Ib) leads to a doubling up of signals due to R, R′ and R″, when the rates are slow, and to averaged signals when the rates are fast on the n.m.r. time scale. The many investigations in this area† include not only determinations of equilibria between Ia and Ib (from spectra where exchange is slow) and the magnitude of rotational barriers (from spectra near the coalescence temperatures), but also extensive studies of effects due to solvation, protonation and complex formation with Lewis acids.

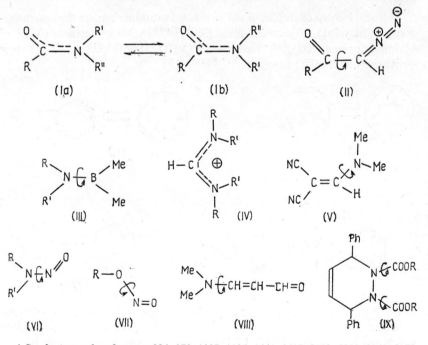

† See for example references 224, 878, 1007, 1125, 1441, 1497, 2132, 2293, 2343, 2371, 2557, 2558 and 2559 and references listed in connection with allylic coupling in amides (Chapter 4–4).

Closely related are studies on cyclic amides,[1786,2373] diazoketones (II),[1329] sulphinamides,[1787] sulphonamides,[2338] mesitoyl compounds,[1670] amino-boranes (III),[2151,2499] formamidinium salts (IV)[2055] and other amidinium ions,[1864,1096] dicyanoketamine (V),[2443] *N*-nitrosoamines (VI),[403,85] alkyl nitrites (VII),[1975,1018] *β*-amino-*α*,*β*-unsaturated carbonyl compounds (VIII), and their higher vinylogues,[1690,1451] *p*-nitrosodimethylaniline[1643] and ben-zaldehydes.[101]

In some cases, for example Schiff's bases[587] and quinone monoximes,[1881] isomerization by rotation about the formal *double bond*, C=N, appears to be sufficiently rapid under some conditions attainable in an n.m.r. probe, to lead to averaged spectra. In at least one case (IX), the existence of two restricted rotation processes accompanied by conformational inversion, can be deduced from the temperature variation of the n.m.r. spectrum.[336]

E. VALENCE-BOND TAUTOMERISM[2219]

The term "valence-bond tautomerism" is used to describe intramolecular reactions involving rearrangements of bonds and is usually reserved for reactions which are relatively fast and reversible. In fact, the range of reaction rates is such that n.m.r. becomes practically the only means of detecting such processes,[2219] which are also often termed "degenerate Cope rearrangements".

The best known examples of valence-bond tautomerism are the rearrangements of bicyclo[5.1.0]octa-2,5-diene (X),[687,683] cyclo-octatetraene (XI),[2482] cycloheptatriene (XII),[485] "bullvalene" (XIII),[2217,2218,2167,1485,2216] their derivatives, and related systems.[2219,2140,2217]

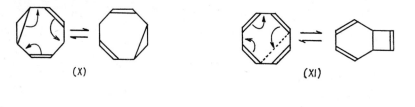

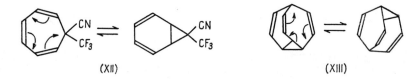

The equilibria indicated by arrows in structures (X)–(XIII) are not the only possible ones in most instances and, in fact, the number of possible structures in bullvalene[2217] is 1,209,600. At 120°, however, all bond shifts in bullvalene are sufficiently fast to give a single narrow line at 4·22 ppm (Fig. 5–1–1),[2217] while the spectrum at −85° consists of separate signals for the six vinylic protons at 5·65 ppm and four cyclopropane and methine protons at 2·58 ppm. It is interesting that the two latter classes appear to resonate at the same frequency and it has been suggested[2217] that the relatively high shielding of the triply allylic methine proton is due to

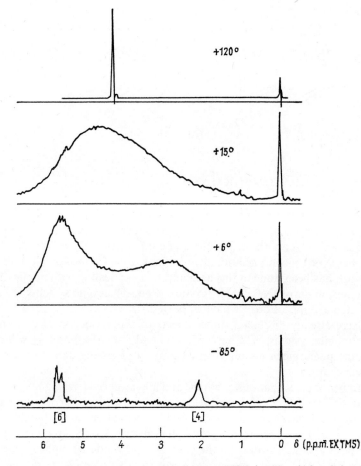

FIG. 5–1–1. 60 MHz spectrum of bullvalene (XIII) in CS_2. (After Schröder *et al.*[2217].)

its position directly above the plane of the cyclopropane ring (cf. Chapter 2–2C(xii)). It is also important to note that the chemical shift of the averaged spectrum (4·22 ppm) is almost exactly equal to the arithmetical average $[(6 \times 5·65 + 4 \times 2·58)/10 = 4·42 \text{ ppm}]$, the small discrepancy probably arising from the inexact juxtaposition of the cyclopropane and methine protons. In between the two extremes, the n.m.r. spectra of bullvalene take on the typical exchange-broadened appearance (Fig. 5–1–1) from which, at least in principle, it should be possible to obtain the energy of activation for the exchange process.

Closely resembling the degenerate Cope rearrangements is the equilibrium between the two forms of furoxan[1113,775] (XIV) where the high temperature spectra take the form of a symmetrical AA'BB' pattern, while the low temperature spectra consist of a complex ABCD pattern. The energy of activation for the process (XIV) has been calculated from the n.m.r. data to be approximately 15 kcal/mole.[1113]

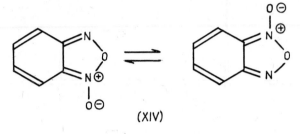

(XIV)

F. CONFORMATIONAL CHANGES

Conformational inversions are clearly processes by which nuclei can interchange and the rates are such that the appearance of n.m.r. spectra may be affected by them. As both the positions of conformational equilibria and the energy barriers to conformational inversions are of considerable interest, much work has been done in this area and reviews dealing with applications to cyclohexane derivatives,[744,882] cyclic systems,[70] acyclic systems[1063] and miscellaneous applications[663] have appeared.

We have already discussed some aspects of n.m.r. spectra of conformationally mobile systems (cf. Chapters 3–8 G and 4–2 A (vii)) and we will consider some phenomena arising from symmetry of ethane derivatives in the following chapter.

A number of investigations of ethane and alkene derivatives † and carbonyl compounds[2011,21,343,1948] have yielded data for energy barriers and conformer distributions. Normally, rotational barriers about single bonds are so low that exchange-broadened spectra are observed only at temperatures below about −60°. However, in the case of some hindered biphenyl derivatives (e.g. (XV)[1750]) and dineopentyltetramethylbenzene (XVI),[675] the methylene protons H_A and H_B show non-equivalence (cf. the following chapter) at room temperature or slightly below, slow rotation about the bonds indicated being responsible for the generation of asymmetry. In both cases, the barriers to the rotation processes were calculated from the spectra near the regions of coalescence.

The n.m.r. spectrum of cyclohexane consists of a single sharp line ($\delta = 1\cdot36$ ppm) at room temperature, in keeping with the expected fast chair–chair interconversion and the resulting interchange of axial and equatorial protons. Between −70 and −100°, the rate of conformational exchange approximates to the frequency separation between axial and equatorial pro-

† References 7, 274, 815, 1233, 1298, 1524, 1868, 1869 and 2224.

tons and a complex spectrum results, from which the free energy, enthalpy and entropy associated with the inversion have been obtained.[1278,1279,2255] However, because of the complexity of the low temperature spectra, caused by the presence of a twelve-spin (albeit fairly symmetrical) system, more accurate values have been obtained from variable temperature spectra of cyclohexane-d_{11}.[323,102,324] This species also gives rise to a single resonance at 1·36 ppm at room temperature, but below $-100°$ two lines, separated by 0·48 ppm and symmetrically disposed about 1·36 ppm, are observed. The upfield line is broader, thus identifying it as that arising from the axial proton,[102] because vicinal diaxial $J_{H,H}$ (and hence vicinal diaxial $J_{H,D}$) is always larger than axial–equatorial or equatorial–equatorial coupling (cf. Chapter 4–2A(vi) and Chapter 2–3C(viii)). Heteronuclear deuterium spin–spin decoupling leads to narrow lines in the low temperature spectra and measurements near the coalescence point give accurate values for the free energy etc., by a straightforward process.

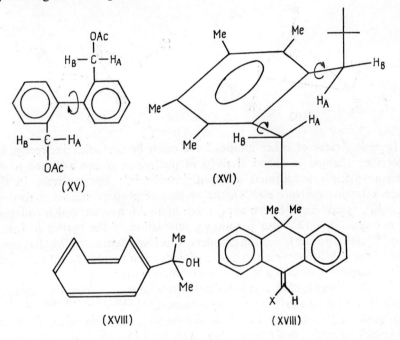

(XV)

(XVI)

(XVIII)

(XVIII)

Analogous studies[2465,801,370,322,2406] have been performed with substituted cyclohexanes, particularly by Reeves and his co-workers[2015,2016,46,2014,2084,2082] (cf. also references listed in connection with vicinal coupling in flexible systems, Chapter 4–2A(vii)).

Other conformational processes which have been studied in detail include those in cyclobutanes,[1489,1490] cycloheptatriene and its derivatives,[530,98,1280,1486] *cis-cis-cis*-1,4,7-cyclononatriene,[2031,2457] [14]- and [18]-annulenes,[946] cyclic oxides and sulphides,[1591,918,487,489,917,919] piperidines,[1487] benzocycloheptane,[1055] bridged biphenyls,[1893,1764] cyclo-octane,[93,105] and cyclo-octatetraene.[93,103] The inversion of cyclo-octatetraene was studied with the aid of ^{13}C satellite analysis[93] (cf. Chapter 2–3C(vii)) and by considering the symmetry properties (cf. the following chapter) of the derivative (XVII).[103] Inversion rates in fluorocyclo-octatetraene have also been investigated.[1075] The methyl groups in (XVIII) exhibit non-equivalence in low temperature spectra[581] due to slow boat–boat inversion of the central ring.

G. MISCELLANEOUS PROCESSES

Of the remaining processes which have been studied by means of time-dependent n.m.r. phenomena, the inversion of amine nitrogen is of most interest. With aziridines[315,172,2396,1576] inversion rates are comparatively slow, for example (XIX) gives rise to two methyl signals at room temperature in dry carbon tetrachloride.[172]

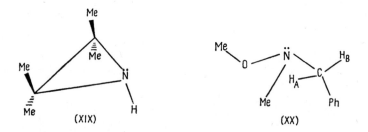

Inversion rates of other amines have often been studied indirectly, by observing changes in chemical shifts of methylene groups attached to an unsymmetrically substituted nitrogen.[2170,1025,2103] For example, in the hydroxylamine derivative (XX), the methylene protons are nonequivalent at $-40°$, as evidenced by the appearance of an AB quartet, which collapses to a singlet at $-12°$, thus permitting calculation of the barrier to inversion[1025] (cf. also the following chapter). It is important to realize that when one of the substituents on the amine nitrogen is a proton, *two* kinetic processes, namely inversion and proton exchange must be considered.[2170,172, 2101,2104,2103] Hydration and ketalization reactions of carbonyl compounds have been studied fairly extensively by n.m.r.,[113,210,1167,1168,1045,932,1579] but most of the work has been concerned with determining the positions of equilibria under conditions (slow exchange) for which discrete species could be observed.

In general, complexing and dissociative phenomena are difficult to study by n.m.r. because the equilibria tend to be one-sided. This makes the estimation of the degree of dissociation difficult, from either the spectra corresponding to the slow exchange case (one set of signals may be too weak to detect) or from averaged spectra (displacement from the spectrum expected from the predominant species may be too small to be estimated), while the calculation of energies of activation would be even more difficult. Nevertheless, some studies of charge-transfer complexes[692,1499,1105] and of the ionization of tropyl azide[2618] have appeared. A special case of equilibrium arises when one of the components of the system is paramagnetic (e.g. a transition metal ion or aryloxy radical). In such instances, a broadening may be induced in the n.m.r. signals due to the complex, from which information about the degree of association with the paramagnetic species, lifetimes of the complex, electron transfer processes etc., may be obtained.[1959,1456,525,1783,667,345,1751]

It should also be noted that most of the intermolecular interactions in the liquid phase which influence the appearance of n.m.r. spectra,[1724] including the ubiquitous solvent effects on chemical shifts, may be treated in terms of "collision complexes", and as such involve time-dependent effects.

INTERNAL ROTATION AND EQUIVALENCE OF NUCLEI

THE n.m.r. spectral consequences of internal rotation in ethanes (other cases may be considered as extensions of the treatment of ethanes) follow in a reasonably straightforward manner from the theory of the time-dependent phenomena in n.m.r.[1575] (Chapter 2–1) and symmetry considerations, but because the subject is often imperfectly understood, we shall treat it here in some detail.

In 1958, Pople[1988] classified the n.m.r. spectra expected for substituted ethanes and little fundamental work[1524,1062,1063,2224,1064] has since appeared. Pople's classification (Table 5–2–1) enbraces two types of spectra: "the slow rotation" case which corresponds to a rate of internal rotation which is slow on the n.m.r. time-scale and thus results in superimposition of the spectra of individual rotamers, and the "rapid rotation" case where the rate of internal rotation is fast on the n.m.r. time-scale, and which gives an averaged spectrum.

It is important to define carefully the physical implications of these descriptions; in particular it must be understood that it is *assumed* that an ethane derivative is essentially always present as one, two or three of the three fully staggered conformers (rotamers) and that the nature of these conformers is essentially independent of temperature,† solvents etc. Thus, both the "slow rotation" and "rapid rotation" cases in Table 5–2–1 refer to mixtures of the same types of conformers which are interconverting at different rates. The only assumption made about the *relative populations* of the conformers is that all the conformers are present, which would certainly be true at any temperature above 0°K.

To illustrate the reasoning involved in the predictions made in Table 5–2–1, we shall consider entry number 15 in detail. The molecule CHX_2—CHY_2, like all ethanes, can exist in three staggered conformers, two of which (the *gauche* forms corresponding to the dihedral angles of 60 and 300° between the protons) must be energetically equivalent. The internal energy diagram of a molecule CHX_2—CHY_2 must therefore have the symmetrical form shown in Fig. 5–2–1, although the heights of barriers to interconversion (E_A), the energy differences between the two forms (ΔE), the relative energy of the

---

† This assumption is at best only approximately correct. As the temperature increases the exact nature of the individual conformers, and not only their distribution, will almost certainly vary and appreciable populations of conformers of intermediate configuration may also appear. However, the deviation from the assumption made would result principally in quantitative rather than qualitative errors.

TABLE 5–2–1. FORM OF SPECTRA OF SUBSTITUTED ETHANES[a,b]

No.	Type	Form of spectrum	
		Slow rotation	Rapid rotation
1	$CH_3—CH_2X$	ABB′CC′	A_2B_3
2	$CH_3—CHX_2$	ABC_2	AB_3
3	$CH_3—CHXY$	ABCD	AB_3
4	$CH_2X—CH_2X$	A_4(*trans*), AA′BB′(*gauche*)	A_4
5	$CH_2X—CH_2Y$	AA′BB′(*trans*), ABCD(*gauche*)	AA′BB′
6	$CH_3—CX_3$	A_3	A_3
7	$CH_3—CX_2Y$	AB_2	A_3
8	$CH_3—CXYZ$	ABC	A_3
9	$CH_2X—CHY_2$	AB_2 and ABC	AB_2
10	$CH_2X—CHYZ$	Three ABC	ABC
11	$CH_2U—CX_3$	A_2	A_2
12	$CH_2U—CX_2Y$	A_2 and AB	A_2
13	$CH_2U—CXYZ$	Three AB	AB
14	$CHX_2—CHX_2$	Two A_2	A_2
15	$CHX_2—CHY_2$	Two AB	AB
16	$CHX_2—CHYZ$	Three AB	AB
17	CHXY—CHXY(*meso*)	A_2 and AB	A_2
18	CHXY—CHXY(*dl*)	Three A_2	A_2
19	CHUV—CHXY (two isomers)	Six AB	Two AB

[a] After Pople.[1988]
[b] If substituents U, V, X, Y, Z contain magnetic nuclei, further complications will result.

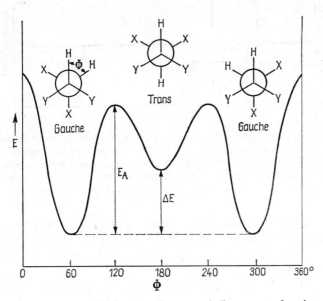

FIG. 5–2–1. Internal potential energy (free energy) diagram as a function of the dihedral angle Φ for a species $CHX_2—CHY_2$.

two forms (*gauche* form energetically favoured) and the actual shape of the curves are, of course, arbitrary.

As the temperature is varied, the rates of interconversions of the rotamers and their relative populations are varied, but their nature is assumed unchanged.

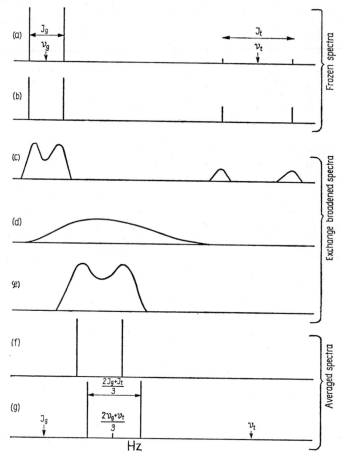

FIG. 5–2–2. A schematic representation of one half of the temperature dependent spectra of the species CHX_2—CHY_2. The temperature increases from (a) to (g). Although entry 15 in Table 5–2–1 and the discussion in the text refer to AB systems, in this figure (for ease of representation) they are assumed to be AX systems. This does not affect the validity of the arguments developed in the text.

The n.m.r. spectra expected from the species CHX_2—CHY_2 are shown in Fig. 5–2–2, only one half of the spectra are shown (say those due to the fragment CHX_2—) but, by symmetry, the remaining portion is absolutely identical (cf. Chapter 2–3C(v)).

At low temperatures (Fig. 5–2–2a and b), the rate of interconversion is low (Pople's "slow rotation" case) and we obtain "frozen spectra", i.e. simple superposition of the two spectra, one for the two identical *gauche* rotamers and one for the *trans* rotamer. Each rotamer has two nonequivalent protons which couple vicinally, and the half spectra (a) and (b) accordingly show two halves (two A portions of two AB quartets, cf. p. 129) of the two AB spectra predicted by Pople.

On purely statistical grounds, the lines due to the *gauche* rotamer should be twice as intense as those due to the *trans* rotamer, but, as we have arbitrarily chosen the *gauche* form as energetically favourable, the intensities in spectra (a) and (b) were arbitrarily set at 14:1 and 2·8:1, respectively, reflecting the increasing proportion of the less favourable *trans* rotamer with the increase of temperature. By simple integration and the use of the basic equilibrium equation (5–2–1) (ignoring partition functions[1063] and applying

$$\frac{\text{population of } gauche}{\text{population of } trans} = K \cdot \exp\left(\frac{E_{gauche} - E_{trans}}{RT}\right) \qquad (5\text{–}2\text{–}1)$$

appropriate statistical weighting factors), we could obtain the free energy difference between the two rotamers (ΔE), and by analysing two spectra taken at different temperatures, we could also obtain the entropy and enthalpy differences. Obviously, spectra of this type are very informative, although difficult to obtain experimentally, because the low barriers to conformational inversion necessitate the attainment of very low temperatures.

From the point of view of the matter under discussion, i.e. the effect of symmetry on the appearance of spectra, the important result is that the *form* of the spectra (two AB) will remain unaltered as long as the interconversion of the rotamers is slow on the n.m.r. time scale, i.e. it is independent of the rotamer distribution which will affect only the relative intensities of the two AB quartets.

As the temperature is increased, the rate of rotamer interconversion will become comparable with the frequency difference between the *gauche* and *trans* forms and we shall expect a series of complex, broadened (by exchange) spectra [spectra (c), (d) and (e)], from which information can be obtained about the activation energy of the exchange process. This aspect of temperature dependence of n.m.r. spectra has already been considered (Chapter 2–1) and does not enter into this discussion.

With a further increase of temperature, the rate of interconversion of the rotamers becomes greater than the frequency difference between the *gauche* and *trans* forms and an averaged spectrum (f) results. However, the distribution of the isomers is still governed by equation (5–2–1) and can be simply determined from spectra of the type (f) by equations (2–1–7) and (2–1–8) (p. 60), *provided* that we have been able to obtain the necessary data for the individual rotamers from the frozen spectra (see however references 1063, 1064, 821).

As the temperature is raised still further, the proportion of the energetically less favourable *trans* rotamer tends towards the statistical ratio of 1:2. At the hypothetical upper temperature limit (Fig. 5–2–2g), this ratio will in fact be achieved, but the form of the spectrum remains the same, i.e. one half of an AB quartet as predicted by Pople (Table 5–2–1). Thus, throughout the temperature range for which conformer interconversion is fast on the n.m.r.

time-scale (rapid rotation), the *form* of the spectrum remains the same (here an AB quartet) and only the *position* of the lines alters with changing proportions of the rotamers, reflecting the changing average chemical shifts and coupling constants.

The remaining types in Table 5–2–1 can also be deduced in an analogous manner, but it has often been questioned why *even at the upper temperature limit*, the spectra corresponding to entries (10) and (13) and their analogs remain unsymmetrical. In particular, the case of the methylene group attached to an asymmetric centre (i.e. entry 13; entry 10 is an obvious extension of entry 13) has led to some difficulty, it being felt that at the upper temperature limit the two protons should become equivalent because "they describe the same circle in space at a fast rate". However, as explicitly stated above, even at the upper temperature limit we treat a substituted ethane as a mixture of separate rapidly interconverting rotamers, and a uniform circular motion of a "windmilling" kind, without any effects due to non-bonded interactions between substituents, is not considered here.

The three staggered conformers corresponding to entry (13), Table 5–2–1 are (Ia), (Ib) and (Ic). In each of them, the protons H_A and H_B reside in different environments and hence the frozen spectrum of $CH_AH_BU—CXYZ$ will consist of three superimposed AB quartets. It further follows, that in an averaged spectrum ("fast rotation") with unequal populations of (Ia), (Ib) and (Ic), the average chemical shifts of H_A and H_B will not be the same, unless by accidental equivalence, and hence the form of the spectrum will be an AB quartet.

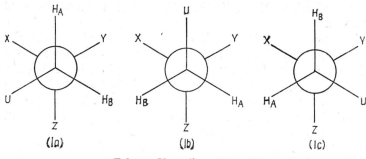

(Z, large: X, medium; Y, small)

It might be argued that, at the upper temperature limit [populations of (Ia), (Ib) and (Ic) equal], each of the protons H_A and H_B will spend an equal time in the same spatial environment, *viz.* between either X and Y, Y and Z or Z and X, and that hence the average environments of H_A and H_B should be the same, resulting in an A_2 spectrum.

However, a careful inspection of the rotamers (Ia), (Ib) and (Ic) shows that only the *primary* environments of H_A and H_B are thus averaged, because while, e.g., H_A is between X and Y, U is between X and Z (rotamer Ia), whereas when H_B is between X and Y, U is between Y and Z (rotamer Ic). Therefore the environments of H_A in (Ia) and H_B in (Ic) are not *exactly* the same and hence the chemical shifts of H_A and H_B will not be expected to be exactly averaged even when the populations of (Ia) and (Ic) are equal. This will result in an AB spectrum *for all population distributions* of rapidly interconverting rotamers ("fast rotation") of CH_AH_BU—$CXYZ$. It must, however, be noted that Table 5–2–1 does not state that the AB spectra for the rapid rotation case of CH_AH_BU—$CXYZ$ will be *identical* at all temperatures. In fact, as we have recognized that unequal population of conformers would be expected to cause nonequivalence between H_A and H_B, it follows that the total nonequivalence between H_A and H_B would be due to components associated with *unequal conformer population*† and to an irreducible component due to the low symmetry of CH_AH_BU—$CXYZ$. We shall refer to the latter as *intrinsic nonequivalence*[1062] and an analogous result can be obtained by considering algebraically the departure of the symmetry of a system from the unique, three-fold axial symmetry of the methyl group.[1062]

It is interesting to speculate on the physical origin of intrinsic asymmetry. One simple cause could be steric distortion. If we set, arbitrarily, the steric requirements of $Z > X > Y$ and of U as much larger than H_A (which is of course equal to H_B), it is highly unlikely that the three staggered rotamers of CH_AH_BU—$CXYZ$, as depicted in (Ia), (Ib) and (Ic), would represent the actual geometries, which would be far more likely to be distorted towards the (exaggerated) representations (IIa), (IIb) and (IIc). It is immediately apparent that when H_A is between X and Y, as in conformer (IIa), it will be closer to Y than to X, while when H_B is between X and Y, as in conformer (IIc), it is closer to X than to Y. Therefore, if either or both of the groups

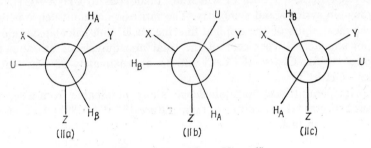

(Z, large; X, medium; Y, small)

X and Y have any long-range effect on chemical shifts, the chemical shifts of H_A and H_B, in surroundings which appear to be primarily the same (i.e. between X and Y in this case), would not be identical, hence causing a contribution towards intrinsic nonequivalence. Similar arguments could be constructed on non-steric grounds. Let us assume that the bonds C—U and C—Y are significantly polarized and polarizable. It would be reasonable to expect that in rotamer (Ia), where U and Y are *trans*, the electron distribu-

---

† The statement is sometimes made that nonequivalence is due to "hindered rotation" or "steric hindrance to rotation". This is not correct, because this phenomenon is associated with interconversion *rates*, which, once recognized as "fast" on the n.m.r. time-scale, do not affect the appearance of spectra. Population distribution for the fast rotation case is purely a function of temperature and energy differences between conformers (equation 5–1–1), if the usual approximations are made.[1063]

tion in bonds C—U and C—Y (and in fact in all other bonds) would be different to that in rotamer (Ic) where U and Y are *gauche*. The electron distribution in the bonds C—H_A and C—H_B could then be affected unequally in the rotamers (Ia) and (Ic) with consequences on chemical shifts.

The existence of nonequivalence due to *U*nequal *C*onformer *P*opulations (which we shall shorten here to *UCP nonequivalence*) and *intrinsic nonequivalence* being thus recognized, it becomes of some interest to consider the *relative contributions* of the two to the total nonequivalence, for the following reasons. First, if the contribution of intrinsic nonequivalence were known, the observed total nonequivalence could be used as a measure of inequality in conformer populations. On the other hand, intrinsic nonequivalence itself could possibly give information concerning the actual shape and electronic environment in individual conformers, about which little is known.

The most direct method of attempting to obtain a measure of intrinsic nonequivalence is to obtain the data for the individual conformers from frozen spectra, and, by simple arithmetical averaging, predict the chemical shifts (and coupling constants where applicable) for the hypothetical averaged spectrum at the upper temperature limit, where conformer populations are equal.

The experimental difficulties in obtaining frozen spectra are however so formidable that this procedure has, so far, been applied to only two compounds with the required degree of symmetry, *viz.* CF_2Br—$CFBrCl$[1868] and CF_2Br—$CHBrCl$,[1869] both of which are analogous to CH_2X—$CHYZ$ as regards spin systems and symmetry. The intrinsic nonequivalence between the chemical shifts of the geminal fluorine nuclei thus calculated, amounts to approximately 10 per cent of the total nonequivalence at room temperature[1064] for CF_2Br—$CFBrCl$ and to approximately 33 per cent for CF_2Br—$CHBrCl$.

In principle, intrinsic nonequivalence is also obtainable from a series of averaged spectra taken at different temperatures[1062,1063,1064] (cf. also Chapter 4–2) and, in fact, for CF_2Br—$CFBrCl$ the values obtained by the two methods are remarkably close. All of the above results are unlikely to be very accurate, because of the assumptions made regarding the unalterable nature of the individual conformers with temperature, and other simplifications.[1063,1967,389]

It must not be assumed that at any temperature the observed nonequivalence is a maximum value, i.e. that the total nonequivalence decreases steadily with the increase in temperature until at the hypothetical upper temperature limit it becomes equal to the intrinsic nonequivalence, because the *sign* of the UCP nonequivalence need not *necessarily* be the same as that of the intrinsic nonequivalence. In terms of the example CH_AH_BU—$CXYZ$, this means that UCP nonequivalence may cause H_A to appear at a higher field, while intrinsic nonequivalence may cause H_B to appear at a higher field. Thus theoretically, over any particular temperature range the observed nonequivalence could appear to be either increasing or decreasing, although, not unexpectedly, the latter is usually the case (see references below).

In the molecules, $Me_2C(CN)—CMe(CN)—CH_AH_B—C(CN)Me_2$,[2311] and $PhCHBr—CH_AH_BBr$ in certain solvents,[1257] nonequivalence between the geminal protons was found to increase in certain temperature ranges possibly due to the above cause.† In other solvents, the spectra of $PhCHBr—CH_AH_BBr$ appear to show equivalence in the chemical shifts of the methylene protons at an intermediate point in the temperature range, but H_A and H_B appear to be nonequivalent both below and above this temperature.[1257]

It is also probably unsafe to assume that any apparent lack of change in nonequivalence with temperature implies that the observed effect must be entirely due to intrinsic nonequivalence. This is because the build-up in the population of a conformer (or conformers) could be slow if appreciable energy differences existed,‡ and furthermore, changes in conformer populations over appreciable temperature ranges could take place in such a manner as not to affect UCP nonequivalence appreciably. Moreover, when nonequivalent coupling constants, as well as chemical shifts, are observed (e.g. entry 10 in Table 5–2–1), changes in nonequivalence of coupling constants and of chemical shifts need not be parallel,[1257] and deceptively simple spectra (e.g. the AA′X type) have been observed.[1257]

To sum up, the problem of obtaining a reliable quantitative distribution of the observed nonequivalence between UCP and intrinsic contributions is quite difficult, but the mere observation of nonequivalence in acyclic systems is an indication of the symmetry of the molecule.†† However, absence of nonequivalence need not necessarily be associated with the absence of an asymmetric centre, because the UCP and intrinsic effects may cancel, or simply be unobservably small. Thus a number of derivatives of malic acid[1257] give simple AX_2 spectra.

On the other hand, the acyclic methylene group (or its equivalent) need not be directly attached to the asymmetric centre, as can be shown by a straightforward extension of the above arguments and as is very often observed in practice (see below).

Literally hundreds of examples corresponding to entries 10 and 13 in Table 5–2–1 may be found in the literature. Another common situation involves an isopropyl group attached to an asymmetric centre,[2321] i.e. the system $CMe_2H—CXYZ$, where the methyl groups appear as *two* pairs of doublets separated by a frequency difference corresponding to their total nonequivalence, with each doublet showing a vicinal splitting due to the methine proton. The form of the spectrum for the rapid rotation case is

† However variations in solvent effects and association effects with temperature would also have to be considered.

‡ This of course can be always calculated by making reasonable assumptions about energy differences between conformers.

†† The overall symmetry, rather than the presence of an asymmetric centre, has to be considered; e.g., when two identical asymmetric centres are present the possibility of internal compensation has to be taken into consideration. On the other hand, in molecules of the type $U—CH_AH_B—CXY—CH_AH_B—U$, which do not posses an asymmetric centre, the methylene protons will, in general, be nonequivalent, because "from the point of view" of either of the pairs of methylene protons, the central carbon atom bears three different substituents, thus making this case equivalent to $U—CH_AH_B—CXYZ$.

generally A_3M_3X rather than the general A_3B_3C because J_{AM}, the coupling between the nonequivalent methyl groups, is usually negligible. Further, because J_{AX} and J_{MX} would for all practical purposes be identical, the methine proton will give rise to an AX_3 quartet, each of whose components will be in turn split into an identical AX_3 quartet thus leading to a septet with a first-order distribution of intensities. An example of a spectrum of this type is shown in Fig. 5–2–3, and a study of this phenomenon has been made by van der Vlies.[2464]

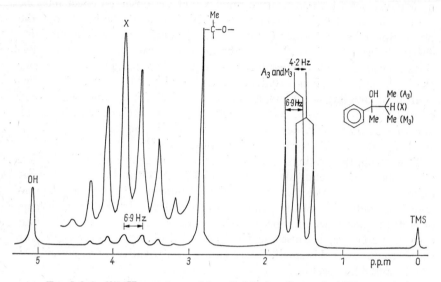

FIG. 5–2–3. 60 MHz spectrum of 3-methyl-2-phenylbutan-2-ol (50 per cent in CCl_4). The resonances due to the aromatic protons are not shown.

An ethyl group, either directly attached to an asymmetric centre or some bonds removed,[1330,2549,763,2194,1747] merits special comment. The appearance of such spectra can be quite complicated (ABC_3 pattern[1330]) but, particularly with ethoxy groups where the methylene protons are usually observed without overlap with other resonances. the departures from the simple four-line pattern of the A part of an A_2X_3 system can be easily observed.

Besides these straightforward cases, many interesting investigations have been carried out involving considerations of symmetry and equivalence in n.m.r., some of which will be briefly mentioned below. Most of the authors have discussed the nonequivalence effects in terms of unequal conformer population (UCP), but it must be remembered that a structural feature which can be postulated to cause UCP, and hence UCP nonequivalence, could also cause conformer distortion or a similar phenomenon, and hence intrinsic nonequivalence.

Little work has appeared on the extension of the above treatment from substituted ethanes to acyclic systems with two identical asymmetric centres†

---

† Clearly, two non-identical asymmetric centres (e.g. in steroids[2389]) can be treated independently.

where internal compensation is possible, but a series of results on 2,4-disub-stituted pentanes[2024,1642,1641,688] and 2,5-disubstituted hexanes[1315] could serve as a model for problems of this type, especially regarding the equivalence or otherwise of the central methylene group.

Roberts and his co-workers[2547,2548] have carried out systematic investiga-tions of the factors associated with the nonequivalence of acyclic methylene groups. In compounds of the type (III) and (IV) it was found that the chemical shift differences between H_A and H_B depended on the size, proximity and type (related to long-range shielding effects) of the groups R.

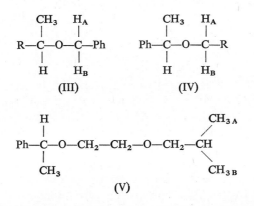

Interestingly, the nonequivalence does not decrease monotonically with the distance from the asymmetric centre, but appears to exhibit a secondary maximum for certain configurations where steric effects related to Newman's "rule of six" would be expected to operate.[2548] Further, observable non-equivalence was detected between the methyl groups in (V), although the asymmetric centre is removed by seven bonds. Similarly, long range effects of this sort were also observed in N,N-dimethylbenzylamines carrying asym-metric ring substituents.[2048]

Solvent effects on the magnitude of nonequivalence in acyclic compounds are general[2547,2321,1257] and have been correlated in some cases[2547] with the dielectric constant of the solvent, which in turn influences the population distribution of conformers.

In a number of cases[2505,2500,2603,1788,2064,574] the asymmetric centre is a sulphur atom, e.g., in sulphoxides,[2064,2500] sulphinates,[2603] sulphin-amides,[1788] sulphites,[1330] etc. Particularly interesting is the case of benzyl methyl sulphoxide[2064] where the protons of the methylene group were shown to be not only nonequivalent by n.m.r., but also to exchange with deuterium at different rates, presumably due to differences in accessibility.

In principle, the asymmetric centre can also be provided by an sp^3 nitrogen atom, and in fact such nonequivalence has been observed between the methylene protons in structures of the type (VI) when the inversion rate of the nitrogen was sufficiently low.[2170,2103,1025,336] Conversely, nonequival-ence was used to obtain inversion or exchange rates in amines and the con-formational inversion rate in a substituted cyclo-octatetraene.[103]

It has been pointed out[1165] that nonequivalence provides a simple method of assigning configurations to heterocyclic bases of type (VII); thus the benzylic methylene protons in (VIIa) give rise to an AB quartet, while those in (VIIb) give rise to an A_2 singlet.

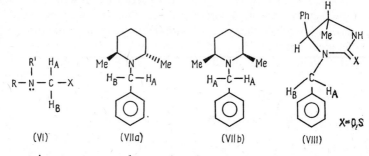

(VI) (VIIa) (VIIb) (VIII)

In certain structures, where other factors can be kept constant, non-equivalence can be used as a probe into comparative shielding contributions by various groups. Thus, in (VIII) it was found that sulphur produced greater nonequivalence between the methylene protons H_A and H_B than oxygen.[2337]

In compounds of type (IX)[381] and (X),[1561] very appreciable nonequivalence has been observed, presumably due to the proximity of strongly aniso-tropic groups. Thus in (XI), the methyl groups resonate 0·73 ppm apart, and this was attributed to intrinsic nonequivalence because of lack of temperature effects.[381]

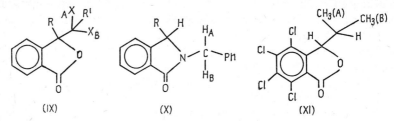

(IX) (X) (XI)

Methylene protons in certain substituted diphenyls, e.g. (XII), also give rise to AB quartets in their n.m.r. spectra,[1750] because of asymmetry induced through restricted rotation about the aryl–aryl bond. As expected, this dis-appears with increasing temperature and the exchange broadened spectra can be used to determine the rotational barrier in the aryl–aryl bond. A related phenomenon occurs with certain bridged diphenyls[1764,1893,1892], for example (XIII), where the bridge methylene protons are nonequivalent.

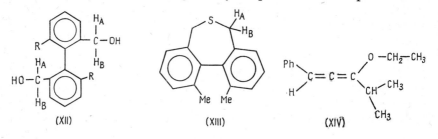

(XII) (XIII) (XIV)

An asymmetric centre may also be generated by an allenic system,[1688] the ethyl protons of (XIV) for instance giving rise to an ABC_3 rather than an A_2C_3 spin system and the methyl groups of the isopropyl function also being nonequivalent.

The papers describing the examples cited above also contain extensive discussions of the problems of symmetry and equivalence, and other references[†] include further examples and/or discussion, while innumerable cases[‡] testify to the common occurrence of the phenomenon.

† References 254, 327, 574 763, 777, 925, 1283, 1415, 1675, 1747, 1933 and 2321.
‡ Some of the more interesting and/or better documented examples can be found in references 40, 123, 146, 354, 414, 424, 430, 576, 596, 829, 929, 1002, 1017, 1022, 1210, 1215, 1229, 1231, 1288, 1496, 1529, 1680, 1753, 1866, 1889, 2196, 2282, 2284, 2374, 2464 and 2638.

TAUTOMERISM

THE term tautomerism is usually defined as "structural isomerism with a low barrier to interconversion between the isomers". As the actual range of values can only be arbitrarily defined, the "phenomenon" of tautomerism simply "fades" into ordinary isomerism for high barriers and into resonance for a zero barrier.[308] Further, although this is rarely explicitly stated, the position of equilibrium is tacitly taken into consideration, the textbook examples of tautomerism being usually those where appreciable concentrations of at least two tautomers can be detected or inferred. Systems where one of the forms is present in very low concentrations (e.g. acetone or phenol) are often considered as only "potentially tautomeric".

In spite of the artificiality of the concept, tautomeric equilibria cover an important area of organic chemistry and n.m.r. spectroscopy has been widely applied to their study. We shall present here a brief résumé of the applicability of the method, with emphasis on its limitations, and list some of the results. The reasons for the inclusion of this Chapter under the heading of time-dependent phenomena is that the principal source of errors in the study of tautomeric equilibria by n.m.r. derives from overlooking the possibility of tautomerizations which are "fast" on the n.m.r. time-scale.

In any consideration of a system which might exhibit tautomerism, it is generally possible to postulate all reasonable forms (e.g. keto and enol), and the equilibrium between three such hypothetical species (A, B and C) can be described by the scheme (i).

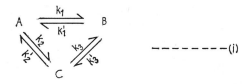

$$-------(i)$$

Clearly, a two-species system is only a special case of scheme (i) and a system of more than three species represents only a straightforward extension. By definition, under equilibrium conditions $k_1 = k_1'$ etc.

Now, it follows from the theory of time-dependent phenomena that if, say, k_1 is a "fast" reaction on the n.m.r. time-scale, we cannot observe the n.m.r. spectra of A and B separately, but instead we shall observe the averaged spectral parameters (chemical shifts and coupling constants) of A and B weighted by their respective mole fractions. If, in the same system, k_2 and k_3 are "slow" reactions, then the averaged spectrum of A and B will be accompanied by a spectrum of C. It follows that, unless signals attributable to *all* the likely tautomeric structures are present in the n.m.r. spectrum, the possibility of "fast" interconversions between two or more tautomers must be taken into consideration.

A common consequence of rapid exchange is the appearance of a more symmetrical spectrum[1373] (cf. examples below) and hence structural deductions based on apparent symmetry must be treated with caution when applied to potentially tautomeric systems.

A further limitation is due to the relative insensitivity of n.m.r. spectroscopy. Under the usual experimental conditions, the detection of a minor tautomer is difficult when it is present as less than 5 per cent of the mixture.[1042,818]

It may be possible by careful choice of experimental conditions, such as solvent and temperature, to slow down the tautomeric interconversions and thus obtain the n.m.r. spectrum of a mixture of tautomers. This approach was successful in a study of the tautomeric equilibrium between *p*-nitrosophenol and quinone monoxime[1881] where, under certain conditions, the n.m.r. spectrum shows a symmetrical AA'BB' pattern (16 per cent of total intensity) due to the nitroso tautomer superimposed upon the unsymmetrical ABMX pattern associated with the quinone monoxime tautomer (Fig. 5–3–1a), although under most conditions[1881,827] the exchange rate between the two forms is "fast" on the n.m.r. time-scale and an averaged spectrum results, which, because of its symmetrical appearance (Fig. 5–3–1b), would tend to be interpreted as evidence for the predominance of the nitroso tautomer.

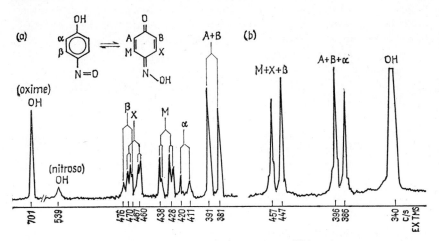

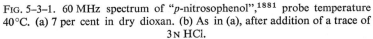

FIG. 5–3–1. 60 MHz spectrum of "*p*-nitrosophenol",[1881] probe temperature 40°C. (a) 7 per cent in dry dioxan. (b) As in (a), after addition of a trace of 3N HCl.

In most cases it is not practicable to slow down the tautomerization rate until two superimposed spectra appear, and an approach involving "reasonable fixed models" must be adopted.[1373] In the case of many tautomeric equilibria, the models are *C*-, *O*- or *N*-methylated derivatives, which are not only reasonably easily obtained, but also eminently suitable because in most molecular environments the substitution of a methyl group for a proton leads only to minor changes in chemical shifts, which may be compensated for in favourable cases.[1375] It will be recognized that the problem is similar in type to the determination of the proportion of conformers in flexible derivatives of cyclohexane.

The "fixed derivative" approach has, however, serious quantitative limitations, because the chemical shifts and/or coupling constants in the (possibly hypothetical) "pure tautomers" cannot be expected to be *exactly* the same as those in the fixed derivatives. Hence it would not be generally possible to decide the *exact* proportion of the tautomers present in the equilibrium from the averaged spectrum. Clearly, the method will work best when the spectral parameters of the fixed models for the tautomers are widely different, because the average parameter depends on the mole fraction *and* the separation between the individual parameters. Examples of some successful applications will be given below, but the limitation in the precision will always make it difficult to distinguish between a single tautomer, a rapidly equilibrating mixture with one tautomer greatly predominating, and a slowly equilibrating mixture with one tautomer present at a level much below 5 per cent. In many cases even the assignment of the structure to a single tautomer on the basis of n.m.r. spectra alone may be difficult.

We shall now discuss the major fields of application of n.m.r. spectroscopy to the study of tautomeric equilibria. The classical examples of tautomerism in β-dicarbonyl compounds are particularly well-suited to investigation by n.m.r. spectroscopy, because the interconversion rates between the tautomers are generally slow, thus permitting direct determination of the proportions of the keto and enol tautomers by integration of n.m.r. spectra. Many systematic investigations, often dealing with the effects of structure and solvent on the position of the tautomeric equilibrium, may be found in the literature[2129,404,977,420,123,50,210,818,1468,1909] (cf. also Table 3–2–8). In some favourable cases several tautomeric forms of a β-triketone may also be observable by n.m.r. Thus, 2,4,6-heptanetrione has been shown[1671] to exist in deuterochloroform solution as a mixture of 7 per cent of the triketone (I), 68 per cent of the mono-enol (II) and 25 per cent of the di-enol (III), because the exchange between the tautomers was slow on the n.m.r. time-scale thus allowing separate estimation of all forms in spite of some overlap of signals.

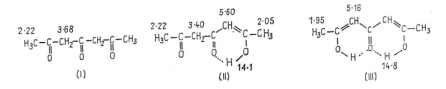

(I)   (II)   (III)

However, with most triketones which are capable of giving several enols† the situation is more complex and indirect deductions and model compounds

† References 245, 841, 853, 854, 855, 856, 1147, 1148 and 1757.

have to be used. Similarly the hydroxymethylene, keto and aldo-enol equilibrium (IV)[949,953,639,853] presents a complicated picture and the exact tautomeric forms of several cyclopentapolyones[1172] can only be obtained by refering to fixed (*O*-methyl) derivatives. On the other hand, in several investigations of ring–chain tautomerism, fairly unequivocal answers have been obtained from n.m.r. data alone.[1495,2636,1942]

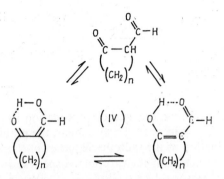

In the extensive series of investigations of equilibria of type (V)[956,2294,2226,2604] carried out principally by Dudek, Holm and their co-workers† much useful information was derived from coupling between the protons in the R group and NH (which should be appreciable only in structure Vb) and from $J_{15_{N,H}}$ in labelled compounds[703,704,705] where the average values of this large coupling constant give reliable information about the proportion of species of the type (Va) and (Vb).

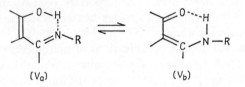

Many studies of tautomeric equilibria in heteroaromatic compounds have been carried out. These include pyrrolones,[1984] pyrazoles,[1775,1077] isoxazolones,[1365,316] hydroxypyrimidines,[2630] pyrazolones,[1301,741,1363] amino- and hydroxypyridines,[211,1303] maleic hydrazide,[1890,1375] aminopyrimidones[1374,1756,2454] and various nucleosides,[959] but in the majority of instances a single spectrum was observed and reference to model compounds and other spectroscopic data was necessary. Only in exceptional cases, for instance isoindoles[2475] and the azidomethine-tetrazole (VI),[2397,2398] could a direct estimate of the tautomeric equilibrium be made from n.m.r. spectra alone.

The equilibrium (VIa) ⇌ (VIb) is particularly instructive, because solvent changes are capable of shifting it from one species to the other to the extent of making one *or* the other form undetectable by n.m.r. It is interesting to note that the magnitudes of the benzylic coupling constants (cf.

† References 701, 702, 703, 704, 705, 706, 707, 709 and 710.

Chapter 4–4E) give evidence of bond-fixation in (VIa) and this is confirmed by the magnitudes of the vicinal coupling constants (cf. Chapter 4–3B) in the parent system.[2398]

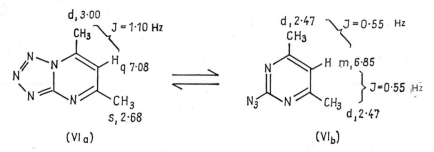

$$\text{(VI}_a) \rightleftharpoons \text{(VI}_b)$$

Investigations concerning the site of protonation in compounds where more than one site is possible are clearly similar, the problem being equivalent to the determination of the tautomeric nature of the protonated species. In many cases the structure of the protonated species can be deduced from n.m.r. spectra in a reasonably straightforward manner, for example with many aromatic compounds,[1964] pyrroles,[11,2537,480] indoles,[1170] azulenes[1745,1744,1736] and indolizines.[885] In other cases, for instance pyridones,[1368,2461,1361,1366] the information derived from $J_{H-N-C-H}$ (cf. Chapter 4–2A(xi)) proved valuable. Labelling with ^{15}N is a very useful technique[2120] in problems of this type, as the large direct coupling between ^{15}N and protons, as well as other information, becomes available. Other investigations include protonation studies on aminoazobenzenes,[1011] pyrimidines,[1269,2120] amino acids[459] and imidazoles[2347] and the important conclusion about the predominance of *O*-protonation of amides has been reached mainly on the grounds of n.m.r. evidence.[250,874,1154,984,1498,1362]

Other notable examples of investigations of tautomeric systems by n.m.r. spectroscopy are polyhydroxy- and polyamino-naphthalenes[1104] and anthracenes,[26] cyclobutane-1,3-dione,[2498] the nitrile–thiol–iminothiolactone system[2352] and tropolone derivatives.[1882,2368] Tautomerism in azonaphthols has been investigated by means of [17]O spectroscopy[1572] which was also used in a study of the benzofurazanoxide system.[659] The mobile equilibria between carbonyl compounds and their ketals or hydrates[932,113] are sufficiently slow for the various species to be estimated directly from the n.m.r. spectra of the reaction mixtures, and these investigations are similar in type to those involving tautomeric equilibria.

REFERENCES

1. ABRAGAM, A., *The Principles of Nuclear Magnetism*, Clarendon Press, 1961, p. 480.
2. ABRAHAM, R.J., *J. Chem. Phys.*, **34,** 1062 (1961).
3. ABRAHAM, R.J., *Mol. Phys.*, **4,** 145 (1961).
4. ABRAHAM, R.J., *J. Chem. Soc.*, 256 (1965).
5. ABRAHAM, R.J. and H.J.BERNSTEIN, *Canad. J. Chem.*, **37,** 1056 (1959).
6. ABRAHAM, R.J. and H.J.BERNSTEIN, *Canad. J. Chem.*, **37,** 2095 (1959).
7. ABRAHAM, R.J. and H.J.BERNSTEIN, *Canad. J. Chem.*, **39,** 39 (1961).
8. ABRAHAM, R.J. and H.J.BERNSTEIN, *Canad. J. Chem.*, **39,** 216 (1961).
9. ABRAHAM, R.J. and H.J.BERNSTEIN, *Canad. J. Chem.*, **39,** 905 (1961).
10. ABRAHAM, R.J., E.O.BISHOP and R.E.RICHARDS, *Mol. Phys.*, **3,** 485 (1960).
11. ABRAHAM, R.J., E.BULLOCK and S.S.MITRA, *Canad. J. Chem.*, **37,** 1859 (1959).
12. ABRAHAM, R.J., R.FREEMAN, L.D.HALL and K.A.McLAUCHLAN, *J. Chem. Soc.*, 2080 (1962).
13. ABRAHAM, R.J., L.D.HALL, L.HOUGH and K.A.McLAUCHLAN, *J. Chem. Soc.*, 3699 (1962).
14. ABRAHAM, R.J., L.D.HALL, L.HOUGH, K.A.McLAUCHLAN and H.J.MILLER, *J. Chem. Soc.*, 748 (1963).
15. ABRAHAM, R.J. and J.S.E.HOLKER, *J. Chem Soc.*, 806 (1963).
15a. ABRAHAM, R.J., A.H.JACKSON, G.W.KENNER and D.WARBURTON, *J. Chem. Soc.*, 853 (1963).
16. ABRAHAM, R.J., A.H.JACKSON and G.W.KENNER, *J. Chem. Soc.*, 3468 (1961).
17. ABRAHAM, R.J. and K.A.McLAUCHLAN, *Mol. Phys.*, **5,** 195 (1962).
18. ABRAHAM, R.J. and K.A.McLAUCHLAN, *Mol. Phys.*, **5,** 513 (1962).
19. ABRAHAM, R.J., K.A.McLAUCHLAN, L.D.HALL and L.HOUGH, *Chem. and Ind.*, 213 (1962).
20. ABRAHAM, R.J. and K.G.R.PACHLER, *Mol. Phys.*, **7,** 165 (1963–64) and references herein.
21. ABRAHAM, R.J. and J.A.POPLE, *Mol. Phys.*, **3,** 609 (1960).
22. ABRAHAM, R.J., R.C.SHEPPARD, W.A.THOMAS and S.TURNER, *Chem. Comm.*, 43 (1965).
23. ABRAHAM, R.J. and W.A.THOMAS, *J. Chem. Soc.*, 3739 (1964).
24. ABRAHAM, R.J. and W.A.THOMAS, *J. Chem. Soc.*, 335 (1965).
25. ABRAHAM, R.J. and W.A.THOMAS, *Chem. Comm.*, 431 (1965).
26. ABRAHAM, R.J. and W.A.THOMAS, *J. Chem. Soc.*, (B) 127 (1966).
27. ABRAMOVITCH, R.A., D.J.KROEGER and B.STASKUN, *Canad. J. Chem.*, **40,** 2030 (1962).
28. ACHESON, R.M., *J. Chem. Soc.*, 2630 (1965).
29. ACHESON, R.M., R.S.FEINBERG and J.M.F.GAGAN, *J. Chem. Soc.*, 948 (1965).
30. ACHESON, R.M., M.W.FOXTON and G.R.MILLER, *J. Chem. Soc.*, 3200 (1965).
31. ACRIVOS, J.V., *Mol. Phys.*, **5,** 1 (1962).
32. ACRIVOS, J.V., *J. Chem. Phys.*, **36,** 1097 (1962).
33. ADAM, W., *Chem. Ber.*, **97,** 1811 (1964) and W. ADAM and S.STERNHELL, unpublished data.
34. ADAMS, A.C. and E.M.LARSEN, *J. Amer. Chem. Soc.*, **85,** 3508 (1963).
35. AFTALION, F., D.LUMBROSO, M.HELLIN and F.COUSSEMANT, *Bull. Soc. Chim. France*, 1958 (1965).
36. AGAHIGIAN, H., H.GAUTHIER, H.HOBERECHT and L.RAPHAELIAN, *Canad. J. Chem.*, **41,** 2913 (1963).

37. AGAHIGIAN, H., J.F.MORAVECK and H.GAUTHIER, *Canad. J. Chem.*, **41**, 194 (1963).
38. AGUIAR, A.M. and D.DAIGLE, *J. Org. Chem.*, **30**, 3527 (1965).
39. AKAGAWA, M., M.SASAMOTO and M.ONDA, *Chem. Pharm. Bull.*, **12**, 1497 (1964).
40. ALBERTY, R.A. and P.BENDER, *J. Amer. Chem. Soc.*, **81**, 542 (1959).
41. ALDERWEIRLDT, F. and M.ANTEUNIS, *Bull. Soc. Chim. Belg.*, **73**, 889, 903 (1964).
42. AL-JALLO, H.N.A. and E.S.WAIGHT, *J. Chem. Soc.*, (B) 73, 75 (1966).
43. ALEXAKOS, L.G. and C.D.CORNWELL, *Rev. Sci. Instr.*, **34**, 790 (1963).
44. ALEXANDER, S., *J. Chem. Phys.*, **32**, 1700 (1960).
45. ALEXANDER, S., *J. Chem. Phys.*, **34**, 106 (1961).
46. ALLAN, E.A., E.PERMUZIC and L.W.REEVES, *Canad. J. Chem.*, **41**, 204 (1963).
47. ALLAN, E.A. and L.W.REEVES, *J. Phys. Chem.*, **67**, 591 (1963).
48. ALLEN, G., D.J.BLEARS and K.H.WEBB, *J. Chem. Soc.*, 810 (1965).
49. ALLEN, G. and J.M.BRUCE, *J. Chem. Soc.*, 1757 (1963).
50. ALLEN, G. and R.A.DWEK, *J. Chem. Soc.*, (B)161 (1966).
51. ALLEN, L.C., *Nature*, **196**, 663 (1962).
52. ALLEN, P., P.J.BERNER and E.R.MALINOWSKI, *Chem. and Ind.*, 208 (1963).
53. ALLEN, L.C. an.d L.F.JOHNSON, *J. Amer. Chem. Soc.*, **85**, 2668 (1963).
54. ALLERHAND, A, F.M.CHEN and H.S.GUTOWSKY, *J. Chem. Phys.*, **42**, 3040 (1965).
55. ALLERHAND, A. and H.S.GUTOWSKY, *J. Amer. Chem. Soc.*, **87**, 4092 (1965).
56. ALLERHAND, A. and P. VON R.SCHLEYER, *J. Amer. Chem. Soc.*, **85**, 886 (1963).
57. ALLINGER, N.L., M.A.DAROOGE and R.B.HERMANN, *J. Amer. Chem. Soc.*, **83**, 1974
    28, (1961).
58. ALLINGER, N.L., M.A.DA ROOGE, M.A.MILLER and B.WAEGELL, *J. Org. Chem.*,
    780 (1963).
59. ALLINGER, N.L. and G.A.YOUNGDALE, *J. Amer. Chem. Soc.*, **84**, 1020 (1962).
60. ALLISON, J.A.C. and G.H.CADY, *J. Amer. Chem. Soc.*, **81**, 1089 (1959).
61. ALLRED, E.L., D.M.GRANT and W.GOODLETT, *J. Amer. Chem. Soc.*, **87**, 673 (1965).
62. ALLRED, A.L. and A.L.HENSLEY JR., *J. Inorg. and Nuclear Chem.*, **17**, 43 (1961).
63. ALLRED, A.L. and C.R.MCCOY, *Tetrahedron Letters*, 25 (1960).
64. ALLRED, A.L. and E.G.ROCHOW, *J. Amer. Chem. Soc.*, **79**, 5361 (1957).
65. AMBEREGER, E., H.P.FRITZ, C.G.KREITER and M.R.KULA, *Chem. Ber.*, **96**, 3270
    (1963).
66. ANBAR, M., A.LOEWENSTEIN and S.MEIBOOM, *J. Amer. Chem. Soc.*, **80**, 2630 (1958).
67. ANDERSON, JR., A.G., W.F.HARRISON and R.G.ANDERSON, *J. Amer. Chem. Soc.*, **85**,
    3448 (1963).
68. ANDERSON, C.B., E.C.FRIEDRICH and S.WINSTEIN, *Tetrahedron Letters*, 2037 (1963).
69. ANDERSON, H.J., D.J.BARNES and Z.M.KHAN, *Canad. J. Chem.*, **42**, 2375 (1964).
70. ANDERSON, J.E., *Quart. Rev.*, **19**, 426 (1965).
71. ANDERSON, J.E., *Tetrahedron Letters*, 4713 (1965).
72. ANDERSON, J.M., *Mol. Phys.*, **8**, 505 (1964).
73. ANDERSON, J.M., J.D.BALDESCHWIELER, D.C.DITTMER and W.D.PHILLIPS, *J. Chem.
    Phys.*, **38**, 1260 (1963).
74. ANDERSON, R.C., *J. Heterocyclic Chem.*, **1**, 279 (1964).
75. ANDERSON, W.A., *Phys. Rev.*, **102**, 151 (1956).
76. ANDERSON, W.A., *Phys. Rev.*, **102**, 153 (1956).
77. ANDERSON, W.A., *Rev. Sci. Instr.*, **32**, 241 (1961).
78. ANDERSON, W.A., *J. Chem. Phys.*, **37**, 1373 (1962).
79. ANDERSON, W.A., *Rev. Sci. Instr.*, **33**, 1160 (1962).
80. ANDERSON, W.A. and R.FREEMAN, *J. Chem. Phys.*, **37**, 85 (1962).
81. ANDERSON, W.A., R.FREEMAN and C.A.REILLY, *J. Chem. Phys.*, **39**, 1518 (1963).
82. ANDERSON, W.A., R.FREEMAN and C.A.REILLY, *J. Chem. Phys.*, **39**, 1520 (1963).
83. ANDERSON, W.A. and H.M.MCCONNELL, *J. Chem. Phys.*, **26**, 1496 (1957).
84. ANDERSON, W.R. and R.M.SILVERSTEIN, *Anal. Chem.*, **37**, 1417 (1965).
85. ANDREADES, S., *J. Org. Chem.*, **27**, 4163 (1962).
86. ANDREWS, E.R., *Nuclear Magnetic Resonance*, Cambridge University Press, 1955,
    p. 66.
87. ANET, E.F.L.J., *Aust. J. Chem.*, **18**, 837 (1965).

88. ANET, F.A.L., *Proc. Chem. Soc.*, 327 (1959).
89. ANET, F.A.L., *J. Chem. Phys.*, **32**, 1274 (1960).
90. ANET, F.A.L., *Canad. J. Chem.*, **39**, 789 (1961).
91. ANET, F.A.L., *Canad. J. Chem.*, **39**, 2262 (1961).
92. ANET, F.A.L., *Canad. J. Chem.*, **39**, 2316 (1961).
93. ANET, F.A.L., *J. Amer. Chem. Soc.*, **84**, 671 (1962).
94. ANET, F.A.L., *J. Amer. Chem. Soc.*, **84**, 747 (1962).
95. ANET, F.A.L., *J. Amer. Chem. Soc.*, **84**, 1053 (1962).
96. ANET, F.A.L., *J. Amer. Chem. Soc.*, **84**, 3767 (1962).
97. ANET, F.A.L., *Canad. J. Chem.*, **41**, 883 (1963).
98. ANET, F.A.L., *J. Amer. Chem. Soc.*, **86**, 458 (1964).
99. ANET, F.A.L., *Tetrahedron Letters*, 3399 (1964).
100. ANET, F.A.L., *Tetrahedron Letters*, 4001 (1964).
101. ANET, F.A.L. and M.AHMAD, *J. Amer. Chem. Soc.*, **86**, 119 (1964).
102. ANET, F.A.L., M.AHMAD and L.D.HALL, *Proc. Chem. Soc.*, 145 (1964).
102a. ANET, F.A.L. and A.J.R.BOURN, *J. Amer. Chem. Soc.*, **87**, 5250 (1965).
103. ANET, F.A.L., A.J.R.BOURN and Y.S.LIN, *J. Amer. Chem. Soc.*, **86**, 3576 (1964).
104. ANET, F.A.L. and M.Z.HAQ, *J. Amer. Chem. Soc.*, **87**, 3147 (1965).
105. ANET, F.A.L. and J.S.HARTMAN, *J. Amer. Chem. Soc.*, **85**, 1204 (1963).
106. ANET, F.A.L. and J.M.MUCHOWSKI, *Proc. Chem. Soc.*, 219 (1962).
107. ANET, F.A.L. and J.M.MUCHOWSKI, *Chem. and Ind.*, 81 (1963).
108. ANET, F.A.L., R.A.B.BANNARD and L.D.HALL, *Canad. J. Chem.*, **41**, 2331 (1963).
109. ANET, R., *Chem. and Ind.*, 897 (1960).
110. ANET, R., *Canad. J. Chem.*, **40**, 1249 (1962).
111. ANET, R. and F.A.L.ANET, *J. Amer. Chem. Soc.*, **86**, 525 (1964).
112. ANGYAL, S.J., V.A.PICKLES and O.RAJENDRA, *Angew. Chem.*, **76**, 794 (1964).
113. ANTEUNIS, M., F.ALDERWEIRELDT and M.ACKE, *Bull. Soc. Chim. Belg.*, **72**, 797 (1963).
114. ANTEUNIS, M. and D.TAVERNIER, *Tetrahedron Letters*, 3949 (1964).
115. APPEL, H.H., R.P.M.BOND and K.H.OVERTON, *Tetrahedron*, **19**, 635 (1963).
116. APPLEQUIST, D.E., G.F.FANTA and B.W.HENRIKSON, *J. Amer. Chem. Soc.*, **82**, 2368 (1960).
116a. APSIMON, J.W., W.G.CRAIG, P.V.DEMARCO, D.W.MATHIESON, L.SAUNDERS and W.B.WHALLEY, *Chem. Comm.*, 359 (1960).
117. ARATA, Y., H.SHIMIZU and S.FUJIWARA, *J. Chem. Phys.*, **36**, 1951 (1962).
118. ARATA, Y., H.SHIMIZU and S.FUJIWARA, *J. Chem. Phys.*, **36**, 1953 (1962).
119. ARCHER, E.D., J.H.SHIVELY and S.A.FRANCIS, *Anal. Chem.*, **35**, 1369 (1963).
120. ARULDHAS, G. and P.VENKATESWARLU, *Mol. Phys.*, **7**, 65 (1963–64).
121. ARISON, B.H., T.Y.SHEN and N.R.TRENNER, *J. Chem. Soc.*, 3828 (1962).
122. ARMSTRONG, R.S., private communications (1964).
123. ARNAL, E., G.LAMATY, C.TAPIERO, F.WINTERNITZ, J.WYLDE and R.WYLDE, *Bull. Soc. Chim. France*, 952 (1963).
124. ARNAL, E., J.ELGUERO, R.JAQUIER, C.MARZIN and J.WYLDE, *Bull. Soc. Chim. France*, 877 (1965).
125. ARNAL, E., A.A.PAVIA and J.WYLDE, *Bull. Soc. Chim. France*, 460 (1964).
126. ARNETT, E.M., M.E.STREM and R.A.FRIEDEL, *Tetrahedron Letters*, 658 (1961).
127. ARNOLD, J.T., *Phys. Rev.*, **102**, 136 (1956).
128. AUFDERMARSH, C.A., *J. Org. Chem.*, **29**, 1994 (1964).
129. AVRAM, M., I.G.DINULESCU, E.MARICA, G.MATEESCU, E.SLIAM and C.D.NENITZESCU, *Chem. Ber.*, **97**, 382 (1964).
130. AXTMANN, R.C., W.E.SCHULER, J.H.EBERLY, *J. Chem. Phys.*, **31**, 850 (1959).
131. AYER, W.A. and C.E.MCDONALD, *Canad. J. Chem.*, **43**, 1429 (1965).
132. AYER, W.A., C.E.MC DONALD and J.B.STOTHERS, *Canad. J. Chem.*, **41**, 1113 (1963).
133. BADGER, G.M., G.E.LEWIS, U.P.SINGH and T.M.SPOTSWOOD, *Chem. Comm.*, 492 (1965).
134. BAILEY, W.J. and S.S.MILLER, *J. Org. Chem.*, **28**, 802 (1963).
135. BAK, B., J.B.JENSEN, A.L.LARSEN and J.RASTRUP-ANDERSON, *Acta Chem. Scand.*, **16**, 1031 (1962).

136. BAK, B., J.JOHS, L.L.CRISTIANSEN, L.LIPSCHITZ and J.T.NIELSON, *Acta Chem. Scand.* **16,** 2318 (1962).
137. BAK, B., J.T.NIELSEN, J.RASTRUP-ANDERSON and M.SCHOTTLÄNDER, *Spectrochim. Acta,* **18,** 741 (1962).
138. BAKER, E.B., *J. Chem. Phys.,* **37,** 911 (1962).
139. BAKER, E.B. and L.W.BURD, *Rev. Sci. Instr.,* **34,** 238 (1963).
140. BAKER, E.B., L.W.BURD and G.N.ROOT, *Rev. Sci. Instr.,* **34,** 243 (1963).
141. BAKER, S.A., J.HOMER, M.C.KEITH and L.F.THOMAS, *J. Chem. Soc.,* 1538 (1963).
142. BALASUBRAMANIAN, S.K., D.H.R.BARTON and L.M.JACKMAN, *J. Chem. Soc.,* 4816 (1962).
143. BALABAN, A.T., G.R.BEDFORD and A.R.KATRITZKY, *J. Chem. Soc.,* 1646 (1964).
144. BALABAN, A.T., T.H.CRAWFORD and R.H.WILEY, *J. Org. Chem.,* **30,** 879 (1965).
145. BALDESCHWIELER, J.D., *J. Chem. Phys.,* **36,** 152 (1962).
146. BALDESCHWIELER, J.D., F.A.COTTON, B.D.N. RAO and R.A.SCHUNN, *J. Amer. Chem. Soc.,* **84,** 4454 (1962).
147. BALDESCHWIELER, J.D. and E.W.RANDALL, *Chem. Revs.,* **63,** 81 (1963).
148. BALDESCHWIELER, J.D. and E.W.RANDALL, *Proc. Chem. Soc.,* 303 (1961).
149. BALDWIN, J.E., *J. Org. Chem.,* **29,** 1882 (1964).
150. BALDWIN, J.E., D.H.R.BARTON, J.L.BLOOMER, L.M.JACKMAN, L.RODRIGUEZ-HAHN and J.K.SUTHERLAND, *Experientia,* **18,** 345 (1962).
151. BALDWIN, J.E. and J.D.ROBERTS, *J. Amer. Chem. Soc.,* **85,** 2444 (1963).
152. BALDWIN, J.E. and R.A.SMITH, *J. Amer. Chem. Soc.,* **87,** 4819 (1965).
153. BALSUBRAMANIAN, S.K., D.H.R.BARTON and L.M.JACKMAN, *J. Chem. Soc.,* 4816 (1962).
154. BANNARD, R.A.B., *Canad. J. Chem.,* **42,** 744 (1964).
155. BANNARD, R.A.B., *Canad. J. Chem.,* **44,** 775 (1966).
156. BANWELL, C.N., *Mol. Phys.,* **4,** 265 (1961).
157. BANWELL, C.N., A.D.COHEN and N.SHEPPARD, *Proc. Chem. Soc.,* 266 (1959).
158. BANWELL, C.N., A.D.COHEN and N.SHEPPARD, *Proc. Chem. Soc.,* 268 (1959).
159. BANWELL, C.N. and N.SHEPPARD, *Mol. Phys.,* **3,** 251 (1960).
160. BANWELL, C.N. and H.PRIMAS, *Mol. Phys.,* **6,** 225 (1963).
161. BANWELL, C.N. and N.SHEPPARD, *Proc. Royal Soc.* London, **263,** 136 (1961).
162. BANWELL, C.N. and N.SHEPPARD, *Disc. Farad. Soc.,* No. 34, 115 (1962).
163. BARBER, M.S., J.B.DAVIS, L.M.JACKMAN and B.C.L.WEEDON, *J. Chem. Soc.,* 2870 (1960).
164. BARBER, M.S., A.HARDISSON, L.M.JACKMAN and B.C.L.WEEDON, *J. Chem. Soc.,* 1625 (1961).
165. BARBER, M.S., L.M.JACKMAN, C.K.WARREN and B.C.L.WEEDON, *J. Chem. Soc.,* 4019 (1961).
166. BARBIER, C. and J.DELAMU, *Rev. L'Institut Français du Petrole,* **17,** 1 (1962).
167. BARBIER, C., J.DELMAU and J.RANFT, *Tetrahedron Letters,* 3339 (1964).
168. BARBIEUX, M., N.DAFAY, J.PECHER and R.H.MARTIN, *Bull. Soc. Chim. Belg.,* **73,** 716 (1964).
169. BARBIEUX, M. and R.H.MARTIN, *Tetrahedron Letters,* 2919 (1965).
170. BARCLAY, L.R.C., C.E.MILLIGAN and L.D.HALL, *Canad. J. Chem.,* **40.** 1664 (1962).
171. BARCZA, S., *J. Org. Chem.,* **28,** 1914 (1963).
172. BARDOS, T.J., C.SZANTAY and C.K.NEVADA, *J. Amer. Chem. Soc.,* **87,** 5796 (1965).
173. BARFIELD, M., *J. Chem. Phys.,* **41,** 3825 (1964).
174. BARFIELD, M. and J.D.BALDESCHWIELER, *J. Chem. Phys.,* **41,** 2633 (1964).
175. BARFIELD, M. and J.D.BALDESCHWIELER, *J. Mol. Spec.,* **12,** 23 (1964).
176. BARFIELD, M. and D.M.GRANT, *J. Amer. Chem. Soc.,* **83,** 4726 (1961).
177. BARFIELD, M. and D.M.GRANT, *J. Amer. Chem. Soc.,* **85,** 1899 (1963).
178. BARFIELD, M. and D.M.GRANT, *J. Chem. Phys.,* **36,** 2054 (1962).
179. BARLET, R., J.L.PIERRE and P.ARNAUD, *Compt. Rend.,* **262,** (C), 855 (1966).
180. BARNER, R., A.BOLLER, J.BORGULYA, E.G.HERZOG, W. VON PHILIPSBORN, C. VON PLANTA, A.FÜRST and H.SCHMID, *Helv. Chim. Acta,* **48,** 94 (1965).
181. BARON, M. and D.P.HOLLIS, *Rec. Trav. Chim.,* **83,** 391 (1964).

182. BARROW, K.D. and T.M.SPOTSWOOD, *Tetrahedron Letters*, 3325 (1965).

183. BARTLETT, P.D. and T.G.TAYLOR, *J. Amer. Chem. Soc.*, **83**, 856 (1961).

184. BARTON, D.H.R., H.T.CHEUNG, A.D.CROSS, L.M.JACKMAN and M.MARTIN-SMITH, *J. Chem. Soc.*, 5061 (1961).

185. BARTOW, D.S. and J.W.RICHARDSON, *J. Chem. Phys.*, **42**, 4018 (1965).

186. BARTZ, K.W. and N.F.CHAMBERLAIN, *Anal. Chem.*, **36**, 2151 (1964).

187. BASKOV, YU.V., T.URBAŃSKI, M.WITANOWSKI and L.STEFANIAK, *Tetrahedron*, **20**, 1519 (1964).

188. BATES, R.B., R.H.CARNIGHAN, R.O.RAKUTIS and J.H.SCHAUBLE, *Chem. and Ind.*, 1020 (1962).

189. BATES, R.B. and D.M.GALE, *J. Amer. Chem. Soc.*, **82**, 5749 (1960).

190. BATES, R.B., D.M.GALE and B.J.GRUNER, *J. Org. Chem.*, **28**, 1086 (1963).

191. BATES, R.B., V.PROCHAZKA and Z.CEKAN, *Tetrahedron Letters*, 877 (1963).

192. BATTERHAM, T.J., K.H.BELL and U.WEISS, *Aust. J. Chem.*, **18**, 1799 (1965).

193. BATTERHAM, T.J., L.TSAI and H.ZIFFER, *Aust. J. Chem.*, **18**, 1959 (1965).

194. BAUER, L. and A.L.HIRSCH, *J. Org. Chem.*, **31**, 1210 (1966).

195. BAUER, L., G.E.WRIGHT, B.A.MIKRUT and C.L.BELL, *J. Heterocyclic Chem.*, **2**, 447 (1965).

196. BAUMGARTEN, H.E., P.L.CREGER and R.L.ZEY, *J. Amer. Chem. Soc.*, **82**, 3977 (1960).

197. BAUMGARTEN, H.E., J.F.FUERHOLZER, R.D.CLARK and R.D.THOMSON, *J. Amer. Chem. Soc.*, **85**, 3303 (1963).

198. BAVIN, P.M.G., K.D.BARTLE and J.A.S.SMITH, *Tetrahedron*, **21**, 1087 (1965).

199. BEACH, W.F. and J.H.RICHARDS, *J. Org. Chem.*, **26**, 3011 (1961).

200. BEAK, P. and H.ABELSON, *J. Org. Chem.*, **27**, 3715 (1962).

201. BEAK, P. and J.BONHAM, *J. Amer. Chem. Soc.*, **87**, 3365 (1965).

202. BEAK, P. and G.A.CARLS, *J. Org. Chem.*, **29**, 2678 (1964).

203. BEAUDET, R.A. and J.D.BALDESCHWIELER, *J. Mol. Spec.*, **9**, 30 (1962).

204. BECCONSALL, J.K. and R.A.Y.JONES, *Tetrahedron Letters*, 1103 (1962).

205. BECKER, E.D., *J. Chem. Phys.*, **31**, 269 (1959).

206. BECKER, E.D., *J. Phys. Chem.*, **63**, 1397 (1959).

207. BECKER, E.D. and M.BEROZA, *Tetrahedron Letters*, 157 (1962).

208. BECKER, E.D., R.B.BREDLEY and C.J.WATSON, *J. Amer. Chem. Soc.*, **83**, 3743 (1961).

209. BECKER, H.D. and G.A.RUSSELL, *J. Org. Chem.*, **28**, 1896 (1963).

210. BECKER, M., *Ber. Bunseges. Phys. Chem.* **68**, 669 (1964).

211. BEDFORD, G.R., H.DORN, G.HILGETAG and A.R.KATRITZKY, *Rec. Trav. Chim.*, **83**, 189 (1964).

212. BEGUIN, C., *Bull. Soc. Chim. France*, 2711 (1964).

213. BELL, F. and K.R.BUCK, *J. Chem. Soc.*, (C)904 (1966).

214. BELLEAU, B. and N.L.WEINBERG, *J. Amer. Chem. Soc.*, **85**, 2525 (1963).

215. BENKESER, R.A., J.HOOZ, T.V.LISTON and A.E.TREVILLYAN, *J. Amer. Chem. Soc.*, **85**, 3984 (1963).

216. BENN, R.W. and R.M.DODSON, *J. Org. Chem.*, **29**, 1142 (1964).

217. BENOIT, H. and J.POMMIER, *Compt. Rend.*, **256**, 3647 (1963).

218. BENSON, R.E. and R.V.LINDSAY, *J. Amer. Chem. Soc.*, **81**, 4247 (1959).

219. BENSON, R.E. and R.V.LINDSAY, *J. Amer. Chem. Soc.*, **81**, 4250 (1959).

220. BENSON, W.R. and A.E.POHLAND, *J. Org. Chem.*, **30**, 1129 (1965).

221. BENT, H.A., *Chem. Revs.*, **61**, 275 (1961).

222. BENTRUDE, W.G. and E.R.WITT, *J. Amer. Chem. Soc.*, **85**, 2522 (1963).

223. BAGANZ, H., K.PRAEFCKE and J.ROST, *Chem. Ber.*, **96**, 2657 (1963).

224. BERGER, A., A.LOEWENSTEIN and S.MEIBOOM, *J. Amer. Chem. Soc.*, **81**, 62 (1959).

225. BERGQUIST, M.S. and T.NORIN, *Arkiv Kemi*, **22**, 137 (1964).

225a. BERGSON, G., *Acta Chem. Scand.*, **17**, 2691 (1963).

226. BERGSON, G. and A.M.WEIDLER, *Acta Chem. Scand.*, **17**, 862 (1963).

227. BERKOZ, B., A.D.CROSS, M.E.ADAMS, H.CARPIO and A.BOWERS, *J. Org. Chem.*, **28**, 1976 (1963).

228. BERLIN, A.J., L.P.FISCHER and A.D.KETLEY, *Chem. and Ind.*, 509 (1965).
229. BERLIN, A.J. and F.R.JENSEN, *Chem. and Ind.*, 998 (1960).
230. BERNHEIM, R.A. and T.P.DAS, *J. Chem. Phys.*, **33**, 1813 (1960).
231. BERNHEIM, R.A. and B.J.LAVERY, *J. Chem. Phys.*, **42**, 1464 (1965).
232. BERNSTEIN, H.J., W.G.SCHNEIDER and J.A.POPLE, *Proc. Roy. Soc.*, A**236**, 515 (1956).
233. BERNSTEIN, H.J. and N.SHEPPARD, *J. Chem. Phys.*, **37**, 3012 (1962).
234. BERRY, H.S., R.DEHL and W.R.VAUGHAN, *J. Chem. Phys.*, **34**, 1460 (1961).
235. BERSON, J.A. and E.S.HAND, *J. Amer. Chem. Soc.*, **86**, 1978 (1964).
236. BERTELLI, D.J. and C.GOLINO, *J. Org. Chem.*, **30**, 368 (1965).
237. BERTHIER, G., M.MAYOT, B.PULLMAN and A.PULLMAN, *J. Phys. Rad.*, **13**, 15 (1952).
238. BHACCA, N.S., J.E.GURST and D.H.WILLIAMS, *J. Amer. Chem. Soc.*, **87**, 302 (1965).
239. BHACCA, N.S., L.F.JOHNSON and J.N.SHOOLERY (Vol. 1), BHACCA, N.S., D.P.HOL-LIS, L.F.JOHNSON and E.A.PIER (Vol. II), *NMR Spectra Catalog*, Vols. I and II, Varian Associates 1962 and 1963.
240. BHACCA, N.S. and R.STEVENSON, *J. Org. Chem.*, **28**, 1638 (1963).
241. BHACCA, N.S. and D.H.WILLIAMS, *Application of NMR Spectroscopy in Organic Chemistry, Illustration from the Steroid Field*, 1964, Prentice-Hall, San Francisco.
242. BHACCA, N.S. and D.H.WILLIAMS, *Tetrahedron Letters*, 3127 (1964).
243. BHACCA, N.S. and D.H.WILLIAMS, *Tetrahedron Letters*, 3129 (1964).
244. BHACCA, N.S., M.E.WOLFF and R.KOWK, *J. Amer. Chem. Soc.*, **84**, 4976 (1962).
245. BICK, I.R.C. and D.H.S.HORN, *Aust. J. Chem.*, **18**, 1405 (1965).
246. BINSCH, G., R.HUISGEN and H.KÖNIG, *Chem. Ber.*, **97**, 2893 (1964).
247. BINSCH, G., J.B.LAMBERT, B.W.ROBERTS and J.D.ROBERTS, *J. Amer. Chem. Soc.*, **86**, 5564 (1964).
248. BIRCH, A.J., M.KOCOR, N.SHEPPARD and J.WINTER, *J. Chem. Soc.*, 1502 (1962).
249. BIRCHALL, T., A.N.BOURNS, R.J.GILLESPIE and P.J.SMITH, *Canad. J. Chem.*, **42**, 1433 (1964).
250. BIRCHALL, T. and R.J.GILLESPIE, *Canad. J. Chem.*, **41**, 2642 (1963).
251. BIRCHALL, T. and R.J.GILLESPIE, *Canad. J. Chem.*, **42**, 502 (1964).
252. BIRCHALL, T. and R.J.GILLESPIE, *Canad. J. Chem.*, **43**, 1045 (1965).
253. BICHO, J.G., E.ZAVARIN and N.S.BHACCA, *J. Org. Chem.*, **28**, 2927 (1963).
254. BISHOP, E.O., *Ann. Repts.*, **58**, 55 (1961).
255. BISHOP, E.O. and J.I.MUSHER, *Mol. Phys.*, **6**, 621 (1963).
256. BLACK, P.J. and M.L.HEFFERNAN, *Aust. J. Chem.*, **15**, 862 (1962).
257. BLACK, P.J. and M.L.HEFFERNAN, *Aust. J. Chem.*, **16**, 1051 (1963).
258. BLACK, P.J. and M.L.HEFFERNAN, *Aust. J. Chem.*, **17**, 558 (1964).
259. BLACK, P.J. and M.L.HEFFERNAN, *Aust. J. Chem.*, **18**, 353 (1965).
260. BLACK, P.J. and M.L.HEFFERNAN, *Aust. J. Chem.*, **18**, 707 (1965).
261. BLACK, P.J. and M.L.HEFFERNAN, *Aust. J. Chem.*, **19**, 1287 (1966).
262. BLACK, P.J., M.L.HEFFERNAN, L.M.JACKMAN, Q.N.PORTER and G.R.UNDERWOOD, *Aust. J. Chem.*, **17**, 1128 (1964).
263. BLOCH, F., *Phys. Rev.*, **70**, 460 (1946).
264. BLOCH, F., W.W.HANSEN and M.PACKARD, *Phys. Rev.*, **70**, 474 (1946).
265. BLOEMBERGEN, N., E.M.PURCELL and R.V.POUND, *Phys. Rev.*, **73**, 679 (1948).
266. BLOMQUIST, A.T. and R.A.VIERLING, *Tetrahedron Letters*, 655 (1961).
267. BLOODWORTH, A.J. and A.G.DAVIES, *J. Chem. Soc.*, (B)125 (1966).
268. BLOOM, S.M. and R.F.HUTTON, *Tetrahedron Letters*, 1993 (1963).
269. BLOOM, A.L. and J.N.SHOOLERY, *Phys. Rev.*, **97**, 1261 (1955).
270. BOBBITT, J.M., D.W.SPIGGLE, S.MAHBOOB, W. VON PHILIPSBORN and H.SCHMID, *Tetrahedron Letters*, 321 (1962).
271. BODEN, N., J.W.EMSLEY, J.FEENEY and L.H.SUTCLIFFE, *Chem. and Ind.*, 1909 (1962).
272. BODEN, N., J.W.EMSLEY, J.FEENEY and L.H.SUTCLIFFE, *Trans. Farad. Soc.*, **59**, 620 (1963).
273. BODEN, N., J.W.EMSLEY, J.FEENEY and L.H.SUTCLIFFE, *Mol. Phys.*, **8**, 133 (1964).
274. BODEN, N., J.W.EMSLEY, J.FEENEY and L.H.SUTCLIFFE, *Proc. Roy. Soc.*, A**282**, 559 (1964).

275. BODEN, N., J.FEENEY and L.H.SUTCLIFFE, *J. Chem. Soc.*, 3482 (1965).

276. BOEKELHEIDE, V. and J.P.PHILLIPS, *J. Amer. Chem. Soc.*, **85**, 1545 (1963).

277. BOEKELHEIDE, V., F.GERSON, E.HEILBRONNER and D.MEUCHE, *Helv. Chim. Acta*, **46**, 1951 (1963).

278. VOGEL, E., W.PRETZER and W.A.BÖLL, *Tetrahedron Letters*, 3613 (1965).

279. BÖHLMANN, F., C.ARNDT and J.STARNICK, *Tetrahedron Letters*, 1605 (1963).

280. BÖHLMANN, F. and collaborators, *Chem. Ber.*, **94**, 3193 (1961); **96**, 588, 1485 (1963); **97**, 801, 1179, 1193 (1964); **98**, 1736, 2596 (1965). *Annalen*, **668**, 51 (1963).

281. BÖHLMANN, F., W. VON KAP-HERR, L.FANGHÄNEL, C.ARNDT, K.M.KLEINE and S.KÖHN, *Chem. Ber.*, **98**, 1411, 1416 and 1616 (1965).

282. BÖHLMANN, F. and D.SCHUMANN, *Tetrahedron Letters*, 2435 (1965).

283. BÖHLMANN, F., D.SCHUMANN and C.ARNDT, *Tetrahedron Letters*, 2705 (1965).

284. BÖHLMANN, F., D.SCHUMANN and H.SCHULTZ, *Tetrahedron Letters*, 173 (1965).

285. BOHM-GOSSL, T., W.HUNSMANN, L.ROHRSCHNEIDER, W.M.SCHNEIDER and W.ZIEGAN-BEIN, *Chem. Ber.*, **96**, 2504 (1963).

286. BONNETT, R. and D.E.McGREER, *Canad. J. Chem.*, **40**, 177 (1962).

287. BOOTH, G.E. and R.J.OUELLETTE, *J. Org. Chem.*, **31**, 544 (1966).

288. BOOTH, H., *J. Chem. Soc.*, 1841 (1964).

289. BOOTH, H., *Tetrahedron*, **19**, 91 (1963).

290. BOOTH, H., *Tetrahedron*, **20**, 2211 (1964).

291. BOOTH, H., *Tetrahedron Letters*, 411 (1965).

292. BOOTH, H., *Tetrahedron*, **22**, 615 (1966).

293. BOOTH, H. and N.C.FRANKLIN, *Chem. and Ind.*, 954 (1963).

294. BOOTH, H., N.C.FRANKLIN and G.C.GIDLEY, *Tetrahedron*, **21**, 1077 (1965).

295. BOOTH, H. and G.C.GIDLEY, *Tetrahedron Letters*, 1449 (1964).

296. BORCIC, S. and J.D.ROBERTS, *J. Amer. Chem. Soc.*, **87**, 1056 (1965).

297. BORDWELL, F.G. and K.M.WELLMAN, *J. Org. Chem.*, **28**, 1347 (1963).

298. BORY, S., M.FETIZON, P.LASZLO and D.H.WILLIAMS, *Bull. Soc. Chim. France*, 2541 (1965).

299. BOSE, A.K., M.S.MANHAS, E.R.MALINOWSKI, *J. Amer. Chem. Soc.*, **85**, 2795 (1963).

300. BOSSHARD, P., S.FUMAGALLI, R.GOOD, W.TRUEB, W. VON PHILIPSBORN and C.H. EUGSTER, *Helv. Chim. Acta*, **47**, 769 (1964).

301. BOTHNER-BY, A.A., *J. Mol. Spec.*, **5**, 52 (1960).

302. BOTHNER-BY, A.A., S.CASTELLANO and H.GÜNTHER, *J. Amer. Chem. Soc.*, **87**, 2439 (1965).

303. BOTHNER-BY, A.A. and R.E.GLICK, *J. Chem. Phys.*, **26**, 1647 (1957).

304. BOTHNER-BY, A.A. and R.E.GLICK, *J. Chem. Phys.*, **26**, 1651 (1957).

305. BOTHNER-BY, A.A. and H.GÜNTHER, *Disc. Farad. Soc.*, **34**, 127 (1962).

306. BOTHNER-BY, A.A. and R.K.HARRIS, *J. Amer. Chem. Soc.*, **87**, 3451 (1965).

307. BOTHNER-BY, A.A. and R.K.HARRIS, *J. Amer. Chem. Soc.*, **87**, 3445 (1965).

308. BOTHNER-BY, A.A. and R.K.HARRIS, *J. Org. Chem.*, **30**, 254 (1965).

309. BOTHNER-BY, A.A. and C.NAAR-COLIN, *Ann. N.Y. Acad. Sci.*, **70**, 833 (1958).

310. BOTHNER-BY, A.A. and C.NAAR-COLIN, *J. Amer. Chem. Soc.*, **80**, 1728 (1958).

311. BOTHNER-BY, A.A. and C.NAAR-COLIN, *J. Amer. Chem. Soc.*, **83**, 231 (1961).

312. BOTHNER-BY, A.A. and C.NAAR-COLIN, *J. Amer. Chem. Soc.*, **84**, 743 (1962).

313. BOTHNER-BY, A.A., C.NAAR-COLIN and G.GÜNTHER, *J. Amer. Chem. Soc.*, **84**, 2748 (1962).

314. BOTHNER-BY, A.A. and J.A.POPLE, *Ann. Rev. Phys. Chem.*, **16**, 43 (1965).

315. BOTTINI, A.T., R.L.VANETTEN and A.J.DAVIDSON, *J. Amer. Chem. Soc.*, **87**, 755 (1965).

316. BOULTON, A.J., A.R.KATRITZKY, A.M.HAMID and S.OSKNE, *Tetrahedron*, **20**, 2835 (1964).

317. BOURN, A.J.R., D.G.GILLIES and E.W.RANDALL, *Proc. Chem. Soc.*, 200 (1963).

318. BOURN, A.J.R., D.G.GILLIES and E.W.RANDALL, *Tetrahedron*, **20**, 1811 (1964).

319. BOURN, A.J.R. and E.W.RANDALL, *Mol. Phys.*, **8**, 567 (1964).

320. BOWIE, J.H., D.W.CAMERON, P.E.SCHÜTZ, D.H.WILLIAMS and N.S.BHACCA, *Tetrahedron*, **22**, 1771 (1966).

321. BOVEY, F.A., *Chem. Eng. News*, 98 (August 1965).
322. BOVEY, F.A., E.W.ANDERSON, F.P.HOOD and R.L.KORNEGAY, *J. Chem. Phys.*, **40**, 3099 (1964).
323. BOVEY, F.A., F.P.HOOD, E.W.ANDERSON and R.L.KORNEGAY, *J. Chem. Phys.*, **41**, 2041 (1964).
324. BOVEY, F.A., F.P.HOOD, E.W.ANDERSON and R.L.KORNEGAY, *Proc. Chem. Soc.*, 146 (1964).
325. BOVEY, F.A., F.P.HOOD, E.PIER and H.E.WEAVER, *J. Amer. Chem. Soc.*, **87**, 2060 (1965).
326. BOVEY, F.A., F.P.HOOD, E.PIER and H.E.WEAVER, *J. Amer. Chem. Soc.*, **87**, 2062 (1965).
327. BOWMAN, N.S., D.E.RICE and B.R.SWITZER, *J. Amer. Chem. Soc.*, **87**, 4477 (1965).
328. BOYD, G.V. and L.M.JACKMAN, *J. Chem. Soc.*, 548 (1963).
329. BOZAK, R.E. and K.L.RINEHART JR., *J. Amer. Chem. Soc.*, **84**, 1589 (1962).
330. BRACE, N.O., *J. Org. Chem.*, **28**, 3093 (1963).
331. BRAILLON, B., *J. Chim. Phys.*, **58**, 495 (1961).
332. BRAILLON, B., J.SALAÜN, J.GORE and J.M.CONIA, *Bull. Soc. Chim. France*, 1981 (1964).
333. BRAME, E.G., *Anal. Chem.*, **37**, 1183 (1963).
334. BRAMLEY, R. and M.D.JOHNSON, *J. Chem. Soc.*, 1372 (1965).
335. BRAUN, R.A., *J. Org. Chem.*, **28**, 1383 (1963).
336. BRELIERE, J.C. and J.M.LEHN, *Chem. Comm.*, 426 (1965).
337. BRENNER, K.S., E.O.FISCHER, H.P.FRITZ and C.G.KREITER, *Chem. Ber.*, **96**, 2632 (1963).
338. BRESLOW, R. and L.J.ALTMAN, *J. Amer. Chem. Soc.*, **88**, 504 (1966).
339. BRESLOW, R., H.HOVER and H.W.CHANGE, *J. Amer. Chem. Soc.*, **84**, 3168 (1962).
340. BRESLOW, R., H.HOVER and H.W.CHANGE, *J. Amer. Chem. Soc.*, **84**, 3170 (1962).
341. BRESLOW, R., J.LOCKHART and A.SMALL, *J. Amer. Chem. Soc.*, **84**, 2793 (1962).
342. BRESLOW, R., J.POSNER and A.KREBS, *J. Amer. Chem. Soc.*, **85**, 234 (1963).
343. BREY, JR., W.S. and K.C.RAMEY, *J. Chem. Phys.*, **39**, 844 (1963).
344. BRIGGS, L.H., R.C.CAMBIE, B.J.CANDY, G.M.O'DONOVAN, R.H.RUSSELL and R.N.SEELYE, *J. Chem. Soc.*, 2492 (1965).
345. BRITT, A.D. and W.M.YEN, *J. Amer. Chem. Soc.*, **83**, 4516 (1961).
346. BROADDUS, C.D., T.J.LOGAN and T.J.FLAUTT, *J. Org. Chem.*, **28**, 1174 (1963).
347. BROIS, S.J., *J. Org. Chem.*, **27**, 3532 (1962).
348. BROOKES, D., S.STERNHELL, B.K.TIDD and W.B.TURNER, *Aust. J. Chem.*, **18**, 373 (1965) and unpublished data.
349. BROOKES, D., B.K.TIDD and W.B.TURNER, *J. Chem. Soc.*, 5385 (1963).
350. BROOKHART, M., A.DIAZ and S.WINSTEIN, *J. Amer. Chem. Soc.*, **88**, 3135 (1966).
351. BROUWER, D.M. and E.L.MACKOR, *Proc. Chem. Soc.*, 147 (1964).
352. BROUWER, D.M., E.L.MACKOR and C.MACLEAN, *Rec. Trav. Chim.*, **85**, 109 (1966).
353. BROUWER, D.M., E.L.MACKOR and C.MACLEAN, *Rev. Trav. Chim.*, **85**, 114 (1966).
354. BROWN, D., B.T.DAVIS, T.G.HALSALL, A.R.HANDS, J.V.HATTON and R.E.RICHARDS, *J. Chem. Soc.*, 4492 (1962).
355. BROWN, H.W. and D.P.HOLLIS, *J. Mol. Spec.*, **13**, 305 (1964).
356. BROWN, J.M., *Tetrahedron Letters*, 2215 (1964).
357. BROWN, K.S. and C.DJERASSI, *J. Amer. Chem. Soc.*, **86**, 2451 (1964).
358. BROWN, K.S. and S.M.KUPCHAN, *J. Amer. Chem. Soc.*, **84**, 4592 (1962).
359. BROWN, N.M.D. and P.BLADON, *Spectrochim. Acta*, **21**, 1277 (1965).
360. BROWN, R.F.C. and I.D.RAE, *Aust. J. Chem.*, **18**, 1071 (1965).
361. BROWN, R.F.C., I.D.RAE and S.STERNHELL, *Aust. J. Chem.*, **18**, 61 (1965).
362. BROWN, R.F.C., I.D.RAE and S.STERNHELL, *Aust. J. Chem.*, **18**, 1211 (1965).
363. BROWN, S.E., D.A.MAGUIRE and D.O.PERRIN, *Chem. and Ind.*, 1974 (1965).
364. BROWN, T.L., D.W.DICKERHOCF and D.A.BAFUS, *J. Amer. Chem. Soc.*, **84**, 1371 (1962).
365. BROWN, T.L., R.L.GERTEIS, D.A.BAFUS and J.A.LADD, *J. Amer. Chem. Soc.*, **86**, 2135 (1964).

366. BROWN, T.L., D.W.DICKERHOOF and D.A.BAFUS, *J. Amer. Chem. Soc.*, **84**, 1371 (1962).
367. BROWN, T.L. and K.STARK, *J. Phys. Chem.*, **69**, 2679 (1965).
368. BROWNSTEIN, S., *J. Amer. Chem. Soc.*, **81**, 1606 (1959).
369. BROWNSTEIN, S., *Canad. J. Chem.*, **39**, 1677 (1961).
370. BROWNSTEIN, S., *Canad. J. Chem.*, **40**, 870 (1962).
371. BROWNSTEIN, S., *J. Org. Chem.*, **28**, 2919 (1963).
372. BROWNSTEIN, S. and K.U.INGOLD, *J. Amer. Chem. Soc.*, **84**, 2258 (1962).
373. BROWNSTEIN, S. and R.MILLER, *J. Org. Chem.*, **24**, 1886 (1959).
374. BROWNSTEIN, S., B.C.SMITH, G.EHRLICH and A.W.LAUBENGAYER, *J. Amer. Chem. Soc.*, **81**, 3826 (1959).
375. BRUCE, J.M. and P.KNOWLES, *Proc. Chem. Soc.*, 294 (1964).
376. BRUCE, J.M. and P.KNOWLES, *J. Chem. Soc.*, 4046 (1964).
377. BRUCE, J.M. and P.KNOWLES, *J. Chem. Soc.*, 5900 (1964).
378. BRÜGEL, W., *Ber. Bunsenges. Phys. Chem.*, **64**, 1121 (1960).
379. BRÜGEL, W., *Ber. Bunsenges. Phys. Chem.*, **66**, 159 (1962).
380. BRÜGEL, W., TH.ANKEL and F.KRÜCKBERG, *Ber. Bunsenges. Phys. Chem.*, **64**, 1121 (1960).
381. BRUMLIK, G.C., R.L.BAUMGARTEN and A.I.KOSAK, *Nature*, **201**, 388 (1964).
382. BRUNE, H.A., *Chem. Ber.*, **97**, 2829 (1964).
383. BRUNE, H.A., *Chem. Ber.*, **98**, 1998 (1965).
384. BRUNEL, Y., H.LEMAIRE and A.RASSAT, *Bull. Soc. Chim. France*, 1895 (1964).
385. BRYCE-SMITH, D. and A.GILBERT, *J. Chem. Soc.*, 2428 (1964).
386. BÜCHI, G., R.E.MANNING and F.A.HOCHSTEIN, *J. Amer. Chem. Soc.*, **84**, 3393 (1962).
387. BÜCHI, G., J.D.WHITE and G.N.WOGAN, *J. Amer. Chem. Soc.*, **87**, 3484 (1965).
388. BUCKINGHAM, A.D., *Canad. J. Chem.*, **38**, 300 (1960).
389. BUCKINGHAM, A.D., *J. Chem. Phys.*, **36**, 3096 (1962).
390. BUCKINGHAM, A.D and K.P.LAWLEY, *Mol. Phys.*, **3**, 219 (1960).
391. BUCKINGHAM, A.D. and E.G.LOVERING, *Trans. Farad. Soc.*, **58**, 2077 (1962).
392. BUCKINGHAM, A.D. and K.A.MCLAUCHLAN, *Proc. Chem. Soc.*, 144 (1963).
393. BUCKINGHAM, A.D. and J.A.POPLE, *Proc. Phys. Soc.*, **69** B, 1133 (1956).
394. BUCKINGHAM, A.D. and J.A.POPLE, *Trans. Farad. Soc.*, **59**, 2421 (1963).
395. BUCKINGHAM, A.D., W.H.PRICHARD and D.W.WHIFFEN, *Chem. Comm.*, 51 (1965).
396. BUCKINGHAM, A.D., T.SCHAEFER and W.G.SCHNEIDER, *J. Chem. Phys.*, **32**, 1227 (1960).
397. BUCKINGHAM, A.D., T.SCHAEFER and W.SCHNEIDER, *J. Chem. Phys.*, **34**, 1064 (1961).
398. BUCKINGHAM, A.D. and P.J.STEPHENS, *J. Chem. Soc.*, 2747 (1964).
399. BULLOCK, E., *Canad. J. Chem.*, **41**, 711 (1963).
400. BULLOCK, E., B.GREGORY and A.W.JOHNSON, *J. Amer. Chem. Soc.*, **84**, 2260 (1962).
401. BULLOCK, E., D.G.TUCK and E.J.WOODHOUSE, *J. Chem. Phys.*, **38**, 2318 (1963).
402. BUMGARDNER, C.L., *J. Org. Chem.*, **28**, 3225 (1963).
403. BUMGARDNER, C.L., K.S.MCCALLUM and J.P.FREEMAN, *J. Amer. Chem. Soc.*, **83**, 4417 (1961).
404. BURDETT, J.L. and M.T.ROGERS, *J. Amer. Chem. Soc.*, **86**, 2105 (1964).
405. BURDON, J., *Tetrahedron*, **21**, 1101 (1965).
406. BURG, A.B. and W.MAHLER, *J. Amer. Chem. Soc.*, **83**, 2388 (1961).
407. BURGADA, R., G.MARTIN and G.MAVAL, *Bull. Soc. Chim. France*, 2154 (1963).
408. BURGE, E.J. and O.SNELLMAN, *Phil. Mag.*, **40**, 994 (1949).
409. BURGSTAHLER, A.W., P.L.CHIEN and M.O.ABDEL-RAHMAN, *J. Amer. Chem. Soc.*, **86**, 5281 (1964).
410. BURKE, J.J. and P.C.LAUTERBUR, *J. Amer. Chem. Soc.*, **83**, 326 (1961).
411. BURKE, J.J. and P.C.LAUTERBUR, *J. Amer. Chem. Soc.*, **86**, 1870 (1964).
412. BURRELL, J.W.K., L.M.JACKMAN and B.C.L.WEEDON, *Proc. Chem. Soc.*, 263(1959).
413. BURSTEIN, S.H. and H.J.RINGOLD, *J. Org. Chem.*, **28**, 3103 (1963).
414. BURTON, R., L.PRATT and G.WILKINSON, *J. Chem. Soc.*, 594 (1961).
415. BUTT, M.A. and J.A.ELVIDGE, *J. Chem. Soc.*, 4483 (1963).

416. BUTTE, W.A. and C.C.PRICE, *J. Amer. Chem. Soc.*, **84**, 1367 (1962).

417. CABARET, F., J.R.DIDRY and J.GUY, *Comptes Rend.*, **255**, 1090 (1962).

418. CAINE, D. and J.B.DAWSON, *J. Org. Chem.*, **29**, 3108 (1964).

419. CALDER, I.C., T.M.SPOTSWOOD and W.H.F.SASSE, *Tetrahedron Letters*, 95 (1963).

420. CALLEJA, F.J.B., *Compt. Rend.*, **249**, 1102 (1959).

421. CAMBIE, R.C. and W.R.J.SIMPSON, *Tetrahedron*, **19**, 209 (1963).

422. CAMERON, D.W., D.G.I.KINGSTON, N.SHEPPARD and LORD TODD, *J. Chem. Soc.*, 98 (1964).

423. CAMPAIGNE, E., N.F.CHAMBERLAIN and B.E.EDWARDS, *J. Org. Chem.*, **27**, 135 (1962).

424. CAMPAIGNE, E. and B.E.EDWARDS, *J. Org. Chem.*, **27**, 4488 (1962).

425. CAMPAIGNE, E. and M.GEORGIADIS, *J. Org. Chem.*, **28**, 1044 (1963).

426. CAMPBELL, I.G.M., R.G.COOKSON, M.B.HOCKING and A.N.HUGHES, *J. Chem. Soc.*, 2184 (1965).

427. CARALP, L. and J.HOARAU, *J. Chim. Phys.*, **60**, 884 (1963).

428. CARALP, L. and J.HOARAU, *J. Chim. Phys.*, **60**, 886 (1963).

429. CARMAN, R.M., *Aust. J. Chem.*, **19**, 1535 (1966).

430. CARMAN, R.M. and N.DENNIS, *Aust. J. Chem.*, **17**, 395 (1964).

431. CARMAN, R.M. and J.R.HALL, *Aust. J. Chem.*, **17**, 1354 (1964).

432. CARPINO, L.A., *J. Amer. Chem. Soc.*, **84**, 2196 (1962).

433. CARPINO, L.A., *J. Amer. Chem. Soc.*, **85**, 2144 (1963).

434. CARR, J.B. and A.C.HUITRIC, *J. Org. Chem.*, **29**, 2506 (1964).

435. CARROL, F.I., *J. Org. Chem.*, **31**, 366 (1966).

436. CASERIO, M.C., H.E.SIMMONS JR., A.E.JOHNSON and J.D.ROBERTS, *J. Amer. Chem. Soc.*, **82**, 3102 (1960).

437. CASPI, E., T.A.WITTSTRUCK and P.K.GROVER, *Chem. and Ind.*, 1716 (1962).

438. CASPI, E., T.A.WITTSTRUCK and N.GROVER, *J. Org. Chem.*, **28**, 763 (1963).

439. CASPI, E., T.A.WITTSTRUCK and D.M.PIATAK, *J. Org. Chem.*, **27**, 3183 (1962).

440. CASTELLANO, S. and A.A.BOTHNER-BY, *J. Chem. Phys.*, **41**, 3863 (1964).

441. CASTELLANO, S. and G.CAPORICCIO, *J. Chem. Phys.*, **36**, 566 (1962).

442. CASTELLANO, S. and J.S.WAUGH, *J. Chem. Phys.*, **34**, 295 (1961).

443. CASTELLANO, S. and J.S.WAUGH, *J. Chem. Phys.*, **35**, 1900 (1961).

444. CASTELLANO, S. and J.S.WAUGH, *J. Chem. Phys.*, **37**, 1951 (1962).

445. CASU, B., M.REGGIANI, G.C.GALLO and A.VIGEVANI, *Tetrahedron Letters*, 2839 (1964).

446. CAUGHEY, W.S. and P.K.IBER, *J. Org. Chem.*, **28**, 269 (1963).

447. CAVA, M.P., R.POHILKE and M.J.MITCHELL, *J. Org. Chem.*, **28**, 1861 (1963).

448. CAVA, M.P., S.K.TALAPATRA, J.A.WEISBACH, B.DOUGLAS and G.O.DUDEK, *Tetrahedron Letters*, 53 (1963).

449. CAVANAUGH, J.R., *J. Chem. Phys.*, **39**, 2378 (1963).

450. CAVANAUGH, J.R., *J. Chem. Phys.*, **40**, 248 (1964).

451. CAVANAUGH, J.R. and B.P.DAILEY, *J. Chem. Phys.*, **34**, 1094 (1961).

452. CAVANAUGH, J.R. and B.P.DAILEY, *J. Chem. Phys.*, **34**, 1099 (1961).

453. CAWLEY, S. and S.S.DANYLUK, *J. Chem. Phys.*, **38**, 285 (1963).

454. CAWLEY, S. and S.S.DANYLUK, *J. Phys. Chem.*, **68**, 1240 (1964).

455. CHAN, S.I., R.T.IWAMASA and T.P.DAS, *Bull. Amer. Phys. Soc.*, **8**, 329 (1962).

456. CHAN, S.I. and T.P.DAS, *J. Chem. Phys.*, **37**, 1527 (1962).

457. CHAN, W.R., C.WILLIS, M.P.CAVA and R.P.STEIN, *Chem. and Ind.*, 495 (1963).

458. CHAPMAN, D., *J. Chem. Soc.*, 131 (1963).

459. CHAPMAN, D., D.R.LLOYD and R.H.PRINCE, *Proc. Chem. Soc.*, 336 (1962).

460. CHAPMAN, D. and R.K.HARRIS, *J. Chem. Soc.*, 237 (1963).

461. CHAPMAN, O.L., *J. Amer. Chem. Soc.*, **85**, 2014 (1963).

462. CHAPMAN, O.L. and P.FITTON, *J. Amer. Chem. Soc.*, **83**, 1005 (1961).

463. CHAPMAN, O.L. and E.D.HOGANSON, *J. Amer. Chem. Soc.*, **86**, 498 (1964).

464. CHAPMAN, O.L. and R.W.KING, *J. Amer. Chem. Soc.*, **86**, 1256 (1964).

465. CHAPMAN, O.L. and D.J.PASTO, *J. Amer. Chem. Soc.*, **82**, 3642 (1960).

466. CHAPMAN, O.L., D.S. PASTO and A.A.GRISWOLD, *J. Amer. Chem. Soc.*, **84**, 1213 (1962).

467. CHAPMAN, O.L., D.S.PASTO, G.W.BORDEN and A.A.GRISWOLD, *J. Amer. Chem. Soc.*, **84**, 1220 (1962).
468. CHAPMAN, O.L., H.G.SMITH and R.W.KING, *J. Amer. Chem. Soc.*, **85**, 803 and 806 (1963).
469. CHAPMAN, O.L., H.G.SMITH, R.W.KING, D.S.PASTO and M.R.STONER, *J. Amer. Chem. Soc.*, **85**, 2031 (1963).
470. CHARMAN, H.B., G.V.D.TIERS, M.M.KREEVOY and G.FILIPOVICH, *J. Amer. Chem. Soc.*, **81**, 3149 (1959).
471. CHARMAN, H.B., D.R.VINARD and M.M.KREEVOY, *J. Amer. Chem. Soc.*, **84**, 347 (1962); **83**, 1978 (1961).
472. CHATT, J., *Proc. Chem. Soc.*, 318 (1962).
473. CHATT, J. and R.G.HAYTER, *J. Chem. Soc.*, 2605 (1961).
474. CHATT, J. and B.L.SHAW, *J. Chem. Soc.*, 5075 (1962).
475. CHEEMA, Z.K., G.W.GIBSON and J.F.EASTHAM, *J. Amer. Chem. Soc.*, **85**, 3517 (1963).
476. CHEN, H.Y., *Anal. Chem.*, **34**, 1793 (1962).
477. CHERRY, W., Q.N.PORTER and S.STERNHELL, unpublished data.
478. CHESICK, J.P., *J. Amer. Chem. Soc.*, **84**, 3250 (1962).
479. CHESICK, J.P., *J. Amer. Chem. Soc.*, **85**, 2720 (1963).
480. CHIANG, Y. and E.B.WHIPPLE, *J. Amer. Chem. Soc.*, **85**, 2763 (1963).
481. CHIAVARELLI, S. and G.SETTIMJ, *J. Org. Chem.*, **30**, 1969 (1965).
482. CHOPRA, C.S., M.W.FULLER, K.J.L.THIEBERG, D.C.SHAW, D.E.WHITE, S.R.HALL and E.N.MASLEN, *Tetrahedron Letters*, 1847 (1963).
483. CHRISTENSEN, G.B., R.G.STRACHAM, N.R.TRENNER, B.H.ARISON, R.HIRSCHMANN and J.M.CHEMERDA, *J. Amer. Chem. Soc.*, **82**, 3995 (1960).
484. CHUPP, J.P. and A.J.SPEZIALE, *J. Org. Chem.*, **28**, 2592 (1963).
485. CIGANEK, E., *J. Amer. Chem. Soc.*, **87**, 1149 (1965).
486. GIOMOISIS, G. and J.D.SWALEN, *J. Chem. Phys.*, **36**, 2077 (1962).
487. CLAESON, G., G.M.ANDROES and M.CALVIN, *J. Amer. Chem. Soc.*, **82**, 4428 (1960).
488. CALDER, I.C. and W.H.F.SASSE, *Tetrahedron Letters*, 3871 (1964).
489. CLAESON, G., G.ANDROES and M.CALVIN, *J. Amer. Chem. Soc.*, **83**, 4357 (1961).
490. CLARK, H.C., J.T.KWON, L.W.REEVES and E.J.WELLS, *Canad. J. Chem.*, **41**, 3005 (1963).
491. CLARK, H.C., J.T.KWON, L.W.REEVES and E.J.WELLS, *Inorg. Chem.*, **3**, 907 (1964).
492. CLARKE, G.M. and D.H.WILLIAMS, *J. Chem. Soc.*, 4597 (1965).
493. CLARKE, R.T., *J. Amer. Chem. Soc.*, **83**, 965 (1961).
494. CLARK-LEWIS, J.W., L.M.JACKMAN and T.M.SPOTSWOOD, *Aust. J. Chem.*, **17**, 632 (1964).
495. CLARK-LEWIS, J.W., L.M.JACKMAN and L.R.WILLIAMS, *J. Chem. Soc.*, 3858 (1962).
496. CLOSS, G.L., *Proc. Chem. Soc.*, 152 (1962).
497. CLOSS, G.L. and W.BÖLL, *Angew. Chem.*, **75**, 640 (1963).
498. CLOSS, G.L. and L.E.CLOSS, *J. Amer. Chem. Soc.*, **81**, 4996 (1959).
499. CLOSS, G.L. and L.E.CLOSS, *J. Amer. Chem. Soc.*, **83**, 599 (1961).
500. CLOSS, G.L. and L.E.CLOSS, *J. Amer. Chem. Soc.*, **85**, 99 (1963).
501. CLOSS, G.L. and L.E.CLOSS, *J. Amer. Chem. Soc.*, **85**, 2022 (1963).
502. CLOSS, G.L., L.E.CLOSS and W.A.BÖLL, *J. Amer. Chem. Soc.*, **85**, 3796 (1963).
503. CLOSS, G.L. and H.B.KLINGER, *J. Amer. Chem. Soc.*, **87**, 3265 (1965).
504. CLOSS, G.L., J.J.KATZ, F.C.PENNINGTON, M.R.THOMAS and H.H.STRAIN, *J. Amer. Chem. Soc.*, **85**, 3809 (1963).
505. CLOSS, G.L. and R.B.LARRABEE, *Tetrahedron Letters*, 287 (1965).
506. CLOSS, G.L. and R.A.MOSS, *J. Amer. Chem. Soc.*, **86**, 4042 (1964).
507. COBURN, W.C., M.C.THORPE, J.A.MONTGOMERY and K.HEWSON, *J. Org. Chem.*, **30**, 1110 (1965).
508. COHEN, L.A., J.W.DALY, H.KNY and B.WITKOP, *J. Amer. Chem. Soc.*, **82**, 2184 (1960).
509. COHEN, A.D., R.FREEMAN, K.A.McLAUCHLAN and D.H.WHIFFEN, *Mol. Physics*, **7**, 45 (1963–64).
510. COHEN, L.A. and W.M.JONES, *J. Amer. Chem. Soc.*, **85**, 3397 (1963).

511. COHEN, L.A. and W.M.JONES, *J. Amer. Chem. Soc.*, **85,** 3402 (1963).
512. COHEN, A.D. and K.A.McLAUCHLAN, *Mol. Physics*, **7,** 11 (1963–64).
513. COHEN, A.D. and K.A.McLAUGHLAN, *Mol. Physics*, **9,** 49 (1965).
514. COHEN, A.I. and S.ROCK, *Steroids*, **3,** 243 (1964).
515. COHEN, A.I., D.ROSENTHAL, G.W.KRAKOWER and J.FRIED, *Tetrahedron*, **21,** 3171 (1965).
516. COHEN, A.D. and N.SHEPPARD, *Proc. Roy. Soc.*, **252A,** 488 (1959).
517. COHEN, A.D., N.SHEPPARD and J.J.TURNER, *Proc. Chem. Soc.*, 118 (1958).
518. COLA, M. and A.PEROTTI, *Gazz. Chim. Ital.*, **94,** 1268 (1964).
519. COLEBROOK, L.D. and D.S.TARBELL, *Proc. Natl. Acad. Sci., USA*, **47,** 993 (1961).
520. COLLINS, D.J., J.J.HOBBS and S.STERNHELL, *Aust. J. Chem.*, **16,** 1030 (1963).
521. COLLINS, D.J., J.J.HOBBS and S.STERNHELL, *Tetrahedron Letters*, 197 (1963).
522. COLLINS, D.J., J.J.HOBBS and S.STERNHELL, *Tetrahedron Letters*, 623 (1963).
523. COLVIN, E.W. and W.PARKER, *J. Chem. Soc.*, 5764 (1965) and private communication, 1966.
524. CONIA, J.M. and J.GORE, *Tetrahedron Letters*, 1379 (1963).
525. CONNICK, R.E. and C.P.COPPEL, *J. Amer. Chem. Soc.*, **81,** 6389 (1959).
526. CONNOLLY, J.D. and R.McCRINDLE, *Chem. and Ind.*, 379 (1965).
526a. CONNOLLY, J.D. and R.McCRINDLE, *Chem. and Ind.*, 2066 (1965).
527. CONNOR, T.M. and L.A.LOEWENSTEIN, *J. Amer. Chem. Soc.*, **83,** 560 (1961).
528. CONNOR, T.M. and C.REID, *J. Mol. Spec.*, **7,** 32 (1961).
529. CONROW, K., *J. Amer. Chem. Soc.*, **83,** 2958 (1961).
530. CONROW, K., M.E.H.HOWDEN and D.DAVIS, *J. Amer. Chem. Soc.*, **85,** 1929 (1963).
531. CONROY, H.H., in *Advances in Organic Chemistry*, Volume II (ed. R.A.Raphael), Interscience, N.Y., 1960, p.265.
532. COOK, C.D. and S.S. DANYLUK, *Tetrahedron*, **19,** 177 (1963).
533. COOKSON, R.C. and N.S.ISAACS, *Tetrahedron*, **19,** 1237 (1963).
534. COPE, A.C. and D.M.GALE, *J. Amer. Chem. Soc.*, **85,** 3743 (1963).
535. COPE, A.C., P.SCHEINER and M.J.YOUNGQUIST, *J. Org. Chem.*, **28,** 518 (1963).
536. COPPINGER, G.M. and J.L.JUNGNICKEL, *J. Chem. Phys.*, **38,** 2589 (1963).
537. COREY, E.J. and R.L.DAWSON, *J. Amer. Chem. Soc.*, **85,** 1782 (1963).
538. COREY, E.J., R.HARTMANN and P.A.VÁTAKENCHERRY, *J. Amer. Chem. Soc.*, **84,** 2611 (1962).
539. COREY, E.J., E.M.PHILBIN and T.S.WHEELER, *Tetrahedron Letters*, 429 (1961).
540. CORIO, P.L., *Chem. Revs.*, **60,** 363 (1960).
541. CORIO, P.L., *J. Mol. Spec.*, **8,** 193 (1962).
542. CORIO, P.L. and B.P.DAILEY, *J. Amer. Chem. Soc.*, **78,** 3043 (1956).
543. CORIO, P.L., R.L.RUTLEDGE and J.R.ZIMMERMAN, *J. Mol. Spec.*, **3,** 592 (1959).
544. CORIO, P.L. and I.WEINBERG, *J. Chem. Phys.*, **31,** 569 (1959).
545. COTTON, F.A., J.H.FASSNACHT, W.D.HORROCKS and N.A.NELSON, *J. Chem. Soc.*, 4138 (1959).
546. COTTON, F.A. and R.A.SCHUNN, *J. Amer. Chem. Soc.*, **85,** 2394 (1963).
547. COVTIZ and F.H.WESTHEIMER, *J. Amer. Chem. Soc.*, **85,** 1775 (1963).
548. COWLEY, A.H. and R.P.PINNELL, *J. Amer. Chem. Soc.*, **87,** 4533 (1965).
549. COX, P.F., *J. Amer. Chem. Soc.*, **85,** 380 (1963).
550. COX, J.M., J.A.ELVIDGE and D.E.H.JONES, *J. Chem. Soc.*, 1423 (1964).
551. COXON, B. and L.D.HALL, *Tetrahedron*, **20,** 1685 (1964).
552. COYLE, T.D., S.L.STAFFORD and F.G.A.STONE, *J. Chem. Soc.*, 743 (1961).
553. COYLE, T.D. and F.G.A.STONE, *J. Amer. Chem. Soc.*, **83,** 4138 (1961).
554. CRAIG, J.C., A.R.NAIK, R.PRATT, E.JOHNSON and N.S.BHACCA, *J. Org. Chem.*, **30,** 1573 (1965).
555. CRAM, D.J., C.K.DALTON and G.R.KNOX, *J. Amer. Chem. Soc.*, **85,** 1088 (1963).
556. CRAM, D.J. and M.GOLDSTEIN, *J. Amer. Chem. Soc.*, **85,** 1063 (1963).
557. CRAM, D.J. and R.D.PARTOS, *J. Amer. Chem. Soc.*, **85,** 1273 (1963).
558. CRAMER, R., *J. Amer. Chem. Soc.*, **86,** 217 (1964).
559. CRAWFORD, R.J. and C.WOO, *Canad. J. Chem.*, **43,** 3178 (1965).
560. CRISTOL, S.J. and J.K.HARRINGTON, *J. Org. Chem.*, **28,** 1413 (1963).

561. CRISTOL, S.J., T.W.RUSSELL, J.R.MOHRIG and D.E.PLORDE, *J. Org. Chem.*, **31**, 581 (1966).
562. CROMBIE, L. and J.W.LOWN, *J. Chem. Soc.*, 775 (1962).
563. CROMBIE, L. and D.A.WHITING, *J. Chem. Soc.*, 1569 (1963).
564. CROSS, A.D., *J. Amer. Chem. Soc.*, **84**, 3206 (1962).
565. CROSS, A.D. and C.BEARD, *J. Amer. Chem. Soc.*, **86**, 5317 (1964).
566. CROSS, A.D. and P.CRABBE, *J. Amer. Chem. Soc.*, **86**, 1221 (1964).
567. CROSS, A.D. and L.J.DURHAM, *J. Org. Chem.*, **30**, 3200 (1965).
568. CROSS, A.D. and I.T.HARRISON, *J. Amer. Chem. Soc.*, **85**, 3223 (1963).
569. CROSS, A.D. and P.W.LANDIS, *J. Amer. Chem. Soc.*, **84**, 3784 (1962).
570. CROSS, A.D. and P.W.LANDIS, *J. Amer. Chem. Soc.*, **86**, 4005, 4011 (1964).
571. CROSS, A.D., P.W.LANDIS and J.W.MAPHY, *Steroids*, **5**, 655 (1965).
572. CROSS, R.J. and F.GLOCKLING, *J. Chem. Soc.*, 4125 (1964).
573. CROSSLEY, N.S. and C.DJERASSI, *J. Chem. Soc.*, 1459 (1962).
574. CSAKVARY, F. and D.GAGNAIRE, *J. Chim. Phys.*, **60**, 546 (1963).
575. CULVENOR, C.C.J., *Tetrahedron Letters*, 1091 (1966).
576. CULVENOR, C.C.J. and W.G.WOODS, *Aust. J. Chem.*, **18**, 1625 (1965).
577. CULVENOR, C.C.J., M.L.HEFFERNAN and W.G.WOODS, *Aust. J. Chem.*, **18**, 1605 (1965).
578. CULVENOR, C.C.J., M.L.HEFFERNAN and W.G.WOODS, *Aust. J. Chem.*, **18**, 1607 (1966).
579. CUPAS, C.A., M.B.COMISAROW and G.A.OLAH, *J. Amer. Chem. Soc.*, **88**, 361 (1966).
580. CURPHEY, T.J., J.O.SANTER, M.ROSENBLUM and J.H.RICHARDS, *J. Amer. Chem. Soc.*, **82**, 5249 (1960).
581. CURTIN, D.Y., C.G.CARLSON and C.G.McCARTY, *Canad. J. Chem.*, **42**, 565 (1964).
582. CURTIN, D.Y. and S.DAYAGI, *Canad. J. Chem.*, **42**, 867 (1964).
583. CURTIN, D.Y. and D.H.DYBVIG, *J. Amer. Chem. Soc.*, **84**, 225 (1962).
584. CURTIN, D.Y., J.A.GROURS, W.H.RICHARDSON and K.L.RINHART, *J. Org. Chem.*, **24**, 93 (1959).
585. CURTIN, D.Y., H.GRUEN, Y.G.HENDRICKSON and H.E.KNIPMEYER, *J. Amer. Chem. Soc.*, **83**, 4838 (1961), *ibid* **84**, 863 (1962).
586. CURTIN, D.Y., H.GRUEN and B.A.SHOULDERS, *Chem. and Ind.*, 1205 (1958).
587. CURTIN, D.Y. and C.G.McCARTY, *Tetrahedron Letters*, 1269 (1962).
588. CURTIS, M.D. and A.L.ALLRED, *J. Amer. Chem. Soc.*, **87**, 2554 (1965).
589. CYMERMANN-CRAIG, J. and M.MOYLE, *J. Chem. Soc.*, 4402 (1963).
590. DAILEY, B.P., *J. Chem. Phys.*, **41**, 2304 (1964).
591. DAILEY, B.P. and J.N.SHOOLERY, *J. Amer. Chem. Soc.*, **77**, 3977 (1955).
592. DANYLUK, S.S., *Canad. J. Chem.*, **41**, 387 (1963).
593. DANYLUK, S.S., *J. Amer. Chem. Soc.*, **86**, 4504 (1964).
594. DANYLUK, S.S., *J. Amer. Chem. Soc.*, **87**, 2300 (1965).
595. DANYLUK, S.S. and W.G.SCHNEIDER, *Canad. J. Chem.*, **40**, 1777 (1962).
596. DARMS, R., T.THRELFALL, M.PESARO and A.ESCHENMOSER, *Helv. Chim. Acta*, **46**, 2893 (1963).
597. DAUBEN, H.J. and D.J.BERTELLI, *J. Amer. Chem. Soc.*, **83**, 4659 (1961).
598. DAUBEN, W.G., K.KOCH, O.L.CHAPMAN and S.L.SMITH, *J. Amer. Chem. Soc.*, **83**, 1768 (1961).
599. DAUBEN, W.G., R.M.COATES, N.D.VIETMEYER, L.J.DURHAN and C.DJERASSI, *Experientia*, **21**, 565 (1965).
600. DAUBEN, W.G., W.E.THIESSEN and P.R.RESMICK, *J. Amer. Chem. Soc.*, **84**, 2015 (1962).
601. DAVE, L.D., D.F.EVANS and G.WILKINSON, *J. Chem. Soc.*, 3684 (1959).
602. DAVID, H., G.MARTIN, G.MAVEL and G.STURTZ, *Bull. Soc. Chim. France*, 1616 (1962).
603. DAVID, H., G.MARTIN, G.MAVEL and G.STURTZ, *Bull. Soc. Chim. France*, 4616 (1962).
604. DAVIES, D.W., *Nature*, **190**, 1102 (1961).
605. DAVIES, D.W., *Mol. Phys.*, **6**, 489 (1963).
606. DAVIES, D.W., *Chem. Comm.*, 258 (1965).

607. Davis, B.R. and P.D.Woodgate, *J. Chem. Soc.*, 5943 (1965).
608. Davis, D.R. and J.D.Roberts, *J. Amer. Chem. Soc.*, **84**, 2252 (1962).
609. Davis, D.R., R.P.Lutz and J.D.Roberts, *J. Amer. Chem. Soc.*, **83**, 246 (1961).
610. Davis, J.B., L.M.Jackman, P.T.Siddons and B.C.L.Weedon, *Proc. Chem. Soc.*, 261 (1961).
611. Davis, J.C. and K.S.Pitzer, *J. Phys. Chem.*, **64**, 886 (1960).
612. Davis, J.C., K.S.Pitzer and C.N.R.Rao, *J. Phys. Chem.*, **64**, 1744 (1960).
613. Davis, J.C. and T. van Auken, *J. Amer. Chem. Soc.*, **87**, 3900 (1965).
614. Davison, A., J.A.McCleverty and G.Wilkinson, *J. Chem. Soc.*, 1133 (1963).
615. Dear, R.E.A. and F.L.M.Pattison, *J. Amer. Chem. Soc.*, **85**, 622 (1963).
616. Declerck, F., R.Degroote, J. De Lannoy, R.Nasielski-Hinkens and J.Nasielski, *Bull. Soc. Chim. Belg.*, **74**, 119 (1965).
617. Dehl, R., W.R.Vaughan and R.S.Berry, *J. Org. Chem.*, **24**, 1616 (1959).
618. Dehlsen, D.C. and A.V.Robertson, *Aust. J. Chem.*, **19**, 269 (1966).
619. de Jongh, H.A.P. and H.Wynberg, *Tetrahedron*, **21**, 515 (1965).
620. de Koch, W.T., P.R.Enslin, K.B.Norton, D.H.R.Barton, B.Sklarz and A.A. Bothner-By, *Tetrahedron Letters*, 309 (1962).
621. Delamu, J. and C.Barbier, *J. Chem. Phys.*, **41**, 1106 (1964).
622. Delamu, J. and J.Duplan, *Tetrahedron Letters*, 559 (1966).
623. Delpuech, J.J., *Bull. Soc. Chim. France*, 2697 (1964).
624. Demiel, A., *J. Org. Chem.*, **27**, 3500 (1962).
625. Dempsey, C.B., D.M.Donnelly and R.A.Laidlaw, *Chem. and Ind.*, 491 (1963).
626. Deno, N.C., *Chem. Eng. News*, **42**, 88 (1964).
627. Deno, N.C., D.B.Boyd, J.D.Hodge, C.U.Pittman and J.O.Turner, *J. Amer. Chem. Soc.*, **86**, 1745 (1964).
628. Deno, N.C., N.Friedman, J.D.Hodge and J.J.Houser, *J. Amer. Chem. Soc.*, **85**, 2995 (1963).
629. Deno, N.C. and J.J.Houser, *J. Amer. Chem. Soc.*, **86**, 1741 (1964).
630. Deno, N.C., C.U.Pittman and M.J.Wisotsky, *J. Amer. Chem. Soc.*, **86**, 4370 (1964).
631. Deno, N.C., H.G.Richey, Jr., N.Friedman, J.D.Hodge, J.J.Houser and C.U. Pittman Jr., *J. Amer. Chem. Soc.*, **85**, 2991 (1963).
632. Deno, N.C., H.G.Richey, J.D.Hodge and M.J.Wisotsky, *J. Amer. Chem. Soc.*, **84**, 1498 (1962).
633. Deno, N.C., H.G.Richey, J.S.Liu, J.D.Hodge, J.J.Houser and M.J.Wisotsky, *J. Amer. Chem. Soc.*, **84**, 2016 (1962).
634. Deno, N.C., H.G.Richey, J.S.Liu, D.N.Lincoln and J.O.Turner, *J. Amer. Chem. Soc.*, **87**, 4533 (1965).
635. de Selms, R.C. and C.M.Combs, *J. Org. Chem.*, **28**, 2206 (1963).
636. Dessy, R.E., T.J.Flautt, H.H.Jaffe and G.F.Reynolds, *J. Chem. Phys.*, **30**, 1422 (1959).
637. Dessy, R.E., F.Kaplan, G.R.Coe and R.M.Salinger, *J. Amer. Chem. Soc.*, **85**, 1191 (1963).
638. de Villepin, J., *J. Chim. Phys.*, **59**, 901 (1963).
639. Deutsch, I. and K.Deutsch, *Tetrahedron Letters*, 1849 (1966).
640. de Vries, G., *Rec. Trav. Chim.*, **84**, 1327 (1965).
641. Dewar, M.J.S. and R.C.Fahey, *J. Amer. Chem. Soc.*, **84**, 2012 (1962).
642. Dewar, M.J.S. and R.C.Fahey, *J. Amer. Chem. Soc.*, **85**, 2245 (1963).
643. Dewar, M.J.S. and R.C.Fahey, *J. Amer. Chem. Soc.*, **85**, 2248 (1963).
644. Dewar, M.J.S. and R.C.Fahey, *J. Amer. Chem. Soc.*, **85**, 2704 (1963).
645. Dewar, M.J.S., R.C.Fahey and P.J.Grisdale, *Tetrahedron Letters*, 343 (1963).
646. Dewar, M.J.S., G.J.Gleicher and B.P.Robinson, *J. Amer. Chem. Soc.*, **86**, 5698 (1964).
647. Dewhirst, K.C. and C.A.Reilly, *J. Org. Chem.*, **30**, 2870 (1965).
648. Dewhurst, B.B., J.S.E.Holker, A.Lablache-Combier, M.R.G.Leeming, J.Le-visalles and J.P.Pete, *Bull. Soc. Chim. France*, 3259 (1964).
649. DeWolf, M.Y. and J.D.Baldeschwieler, *J. Mol. Spec.*, **13**, 344 (1964).
650. Deyrup, J.A. and R.B.Greenwald, *J. Amer. Chem. Soc.*, **87**, 4138 (1965).

651. DICKINSON, W.C., *Phys. Rev.*, **81**, 717 (1951).
652. DIDRY, J.R. and J.GUY, *Compt. Rend.*, **253**, 422 (1961).
653. DIDRY, R.R., J.GUY and F.CABARET, *Compt. Rend.*, **257**, 1466 (1963).
654. DIEHL, P., *Helv. Chim. Acta*, **44**, 829 (1961).
655. DIEHL, P., *Helv. Chim. Acta*, **45**, 568 (1962).
656. DIEHL, P., *J. Chim. Phys.*, **61**, 199 (1964).
657. DIEHL, P., *Helv. Chem. Acta*, **47**, 1 (1964).
658. DIEHL, P., *Helv. Chim. Acta*, **48**, 567 (1965).
659. DIEHL, P., H.A.CHRIST and F.B.MALLORY, *Helv. Chim. Acta*, **45**, 504 (1962).
660. DIEHL, P. and R.FREEMAN, *Mol. Phys.*, **4**, 39 (1961).
661. DIEHL, P. and T.LEIPERT, *Helv. Chim. Acta*, **47**, 545 (1964).
662. DIEHL, P. and J.A.POPLE, *Mol. Phys.*, **3**, 557 (1960).
663. DIEHL, P. and G.SVEGLIADO, *Helv. Chim. Acta*, **46**, 461 (1963).
664. DIEKMANN, H., G.ENGLERT and K.WALLENFELS, *Tetrahedron*, **20**, 281 (1964).
665. DIEKMANN, J., *J. Org. Chem.*, **28**, 2880 (1963).
666. DIETRICH, M.W. and R.E.KELLER, *Anal. Chem.*, **36**, 258 (1964).
667. DIETRICH, M.W. and A.C.WAHL, *J. Chem. Phys.*, **38**, 1591 (1963).
668. DIMAIO, G., P.A.TARDELLA and C.IAVARONE, *Tetrahedron Letters*, 2825 (1966).
669. DONKT, E.V., R.H.MARTIN and F.GREERTS-EVRARD, *Tetrahedron*, **20**, 1495 (1964).
670. DISCHLER, B., *Z. Naturforsch.*, **19a**, 887 (1964).
671. DISCHLER, B., *Z. Naturforsch.*, **20a**, 888 (1965).
672. DISCHLER, B. and G.ENGLERT, *Z. Naturforsch.*, **16a**, 1180 (1961).
673. DISCHLER, B. and W.MAIER, *Z. Naturforsch.*, **16a**, 318 (1961).
674. DITTMER, D.C. and M.E.CHRISTY, *J. Org. Chem.*, **26**, 1324 (1961).
675. DIX, D.T., G.FRAENKEL, H.A.KARNES and M.S.NEWMAN, *Tetrahedron Letters*, 517 (1966).
676. DIXON, J.A., P.A.GWINNER and D.C.LINI, *J. Amer. Chem. Soc.*, **87**, 1379 (1965).
677. DJERASSI, C., C.D.ANTONACCIO, H.BUDZIKIEWICZ, J.M.WILSON and B.GILBERT, *Tetrahedron Letters*, 1001 (1962).
678. DJERASSI, C., B.F.BURROWS, C.G.OVERBERGER, T.TAKEKOSHI, C.D.GUTSCHE and C.T.CHANG, *J. Amer. Chem. Soc.*, **85**, 949 (1963).
679. DJERASSI, C., J.C.KNIGHT and H.BROCKMANN, *Ber.*, **97**, 3118 (1964).
680. DJERASSI, C., M.ISHIKAWA, H.BUDZIKIEWICZ, J.N.SHOOLERY and L.F.JOHNSON, *Tetrahedron Letters*, 383 (1961).
681. DODSON, R.M. and G.KLOSE, *Chem. and Ind.*, 450 (1963).
682. DOERING, W.E. and W.R.ROTH, *Angew. Chem.*, **75**, 27 (1963).
683. DOERING, W.E. and W.R.ROTH, *Tetrahedron*, **19**, 715 (1963).
684. DOERING, W.E., M.SAUNDERS, H.G.BOYTON, H.W.EARHART, E.F.WADLEY, W.R. EDWARDS and G.LABER, *Tetrahedron*, **4**, 178 (1958).
685. DOERING, W.E., M.R.WILLCOTT and M.JONES, *J. Amer. Chem. Soc.*, **84**, 1224 (1962).
686. DONNALLY, B. and T.M.SANDERS, *Rev. Sci. Instr.*, **31**, 977 (1960).
687. DORKO, E.A., *J. Amer. Chem. Soc.*, **87**, 5518 (1965).
688. DOSKOCILOVA, D. and B.SCHNEIDER, *Coll. Chech. Chem. Comm.*, **29**, 2290 (1964).
689. DOUGLAS, A.W. and J.H.GOLDSTEIN, *J. Mol. Spec.*, **16**, 1 (1965).
690. DOUGLASS, D.C. and A.FRATIELLO, *J. Chem. Phys.*, **39**, 3161 (1963).
691. DOWNS, A.V. and E.A.V.EBSWORTH, *J. Chem. Soc.*, 3516 (1960).
692. DRAGO, R.S. and D.BAFUS, *J. Phys. Chem.*, **65**, 1066 (1961).
693. DRAKE, J.E. and W.L.JOLLY, *J. Chem. Phys.*, **38**, 1033 (1963).
694. DREESKAMP, H., *Z. Phys. Chem.*, **38**, 121 (1963).
695. DREESKAMP, H. and E.SACKMANN, *Z. Phys. Chem.*, **34**, 261 (1962).
696. DREESKAMP, H. and E.SACKMANN, *Z. Phys. Chem.* **34**, 273 (1962).
697. DREESKAMP, H. and E.SACKMANN, *Z. Phys. Chem.*, **34**, 283 (1962).
698. DREESKAMP, H. and E.SACKMANN, *Ber. Bunsenges. Phys. Chem.*, **67**, 847 (1963).
699. DREESKAMP, H., E.SACKMANN and G.STEGMEIER, *Ber. Bunsenges.*, **67**, 860 (1963).
700. DUDEK, G.O., *Spectrochim. Acta*, **19**, 691 (1963).
701. DUDEK, G.O., *J. Amer. Chem. Soc.*, **85**, 694 (1963).
702. DUDEK, G.O., *J. Org. Chem.*, **30**, 548 (1965).

703. DUDEK, G.O. and E.P.DUDEK, *J. Amer. Chem. Soc.*, **86,** 4283 (1964).
704. DUDEK, G.O. and E.P.DUDEK, *Chem. Comm.*, 464 (1965).
705. DUDEK, G.O. and E.P.DUDEK, *J. Amer. Chem. Soc.*, **88,** 2407 (1966).
706. DUDEK, G.O. and R.H.HOLM, *J. Amer. Chem. Soc.*, **83,** 2099 (1961).
707. DUDEK, G.O. and R.H.HOLM, *J. Amer. Chem. Soc.*, **83,** 3914 (1961).
708. DURHAM, L.J., J.STUDEBAKER and M.J.PERKINS, *Chem. Comm.*, 456 (1965).
709. DUDEK, G.O. and G.P.VOLPP, *J. Amer. Chem. Soc.*, **85,** 2697 (1963).
710. DUDEK, G.O. and G.P.VOLPP, *J. Org. Chem.*, **30,** 50 (1965).
711. DUDEK, G.O. and F.H.WESTHEIMER, *J. Amer. Chem. Soc.*, **81,** 2641 (1959).
712. DÜRR, H., *Ber. Bunsenges.*, **69,** 641 (1965).
713. DÜRR, H., G.OURISON and B.WAEGELL, *Ber.*, **98,** 1858 (1965).
714. DVOLAITZKY, M. and A.S.DREIDING, *Helv. Chim. Acta*, **48,** 1988 (1965).
715. DYER, J., *Proc. Chem. Soc.*, 275 (1963).
716. DYER, J.R., *Applications of Absorption Spectroscopy of Organic Compounds,* Prentice-Hall, N.Y., 1965.
717. DYALL, L.K., *Aust. J. Chem.*, **17,** 419 (1964).
718. EADES, R.G., G.J.JENKS and A.BRADBURY, *J. Sci. Instr.*, **38,** 210 (1961).
719. EATON, P.E., *J. Amer. Chem. Soc.*, **84,** 2344 (1962).
720. EATON, D.R., A.D.JOSEY, R.E.BENSON, W.D.PHILLIPS and T.L.CAIRNS, *J. Amer. Chem. Soc.*, **84,** 4100 (1962).
721. EATON, D.R., A.D.JOSEY, W.D.PHILLIPS and R.E.BENSON, *Disc. Farad. Soc.*, **34,** 77 (1962).
722. EATON, D.R., A.D.JOSEY, W.D.PHILLIPS and R.E.BENSON, *J. Chem. Phys.*, **37,** 347 (1962).
723. EATON, D.R., A.D.JOSEY, W.D.PHILLIPS and R.E.BENSON, *Mol. Phys.*, **5,** 407 (1962).
724. EATON, D.R., A.D.JOSEY, W.D.PHILLIPS and R.E.BENSON, *J. Chem. Phys.*, **39,** 3513 (1963).
725. EATON, D.R., A.D.JOSEY and W.A.SHEPPARD, *J. Amer. Chem. Soc.*, **85,** 2689 (1963).
726. EATON, D.R., E.A.LALANCETTE, R.E.BENSON and W.D.PHILLIPS, *J. Amer. Chem. Soc.*, **84,** 3968 (1962).
727. EATON, D.R. and W.D.PHILLIPS, in *Advances in Magnetic Resonance* (ed. J.S. WAUGH), Academic Press, N.Y., 1965, p.103.
728. EATON, D.R., W.D.PHILLIPS, R.E.BENSON and A.D.JOSEY, *Proc. Xth Colloq.Spectroscopicum Internationale*, p.665. Spin Density Distribution in Conjugated Molecules by NMR contact shifts.
729. EATON, D.R., W.D.PHILLIPS and D.J.CALDWELL, *J. Amer. Chem. Soc.*, **85,** 397 (1963).
730. EATON, D.R. and W.A.SHEPPARD, *J. Amer. Chem. Soc.*, **85,** 1310 (1963).
731. EATON, P.E. and T.W.COLE, *J. Amer. Chem. Soc.*, **86,** 3157 (1964).
732. EBERSOLE, S., S.CASTELLANO and A.A.BOTHNER-BY, *J. Phys. Chem.*, **68,** 3430 (1964).
733. EBSWORTH, E.A.V. and S.G.FRANKISS, *J. Chem. Soc.*, 661 (1963).
734. EBSWORTH, E.A.V. and S.G.FRANKISS, *Trans. Faraday. Soc.*, **59,** 1518 (1963).
735. EBSWORTH, E.A.V. and S.G.FRANKISS, *J. Amer. Chem. Soc.*, **85,** 3516 (1963).
736. EBSWORTH, E.A.V. and J.J.TURNER, *J. Phys. Chem.*, **67,** 805 (1963).
737. EBERSON, L. and S.WINSTEIN, *J. Amer. Chem. Soc.*, **87,** 3506 (1965).
738. EDWARDS, O.E. and G.FENIAK, *Canad. J. Chem.*, **40,** 2416 (1962).
739. EDWARDS, O.E., G.FODOR and L.MARION, *Canad. J. Chem.*, **44,** 13 (1966).
740. EDWARDS, O.E. and M.LESAGE, *Canad. J. Chem.*, **41,** 1592 (1963).
741. ELGUERO, J., R.JACQUIER and G.TARRAGO, *J. Chim. Phys.*, **61,** 616 (1964).
742. ELGUERO, J., R.JACQUIER, G.TARRAGO and H.C.N.T.DUC, *Bull. Soc. Chim. France.* 293 (1966).
743. ELIEL, E.L., *Chem. and Ind.*, 568 (1959).
744. ELIEL, E.L., *Angew. Chem. Int. Ed.*, **4,** 761 (1965).
745. ELIEL, E.L., E.W.DELLA and T.H.WILLIAMS, *Tetrahedron Letters*, 831 (1963).
746. ELIEL, E.L. and M.H.GIANNI, *Tetrahedron Letters*, 97, (1962).
747. ELIEL, E.L., E.W.DELLA and T.H.WILLIAMS, *Tetrahedron Letters*, 831 (1963).
748. ELIEL, E.L., M.H.GIANNI, T.H.WILLIAMS and J.B.STOTHERS, *Tetrahedron Letters*, 741 (1962).

750. ELIEL, E.L., L.A.PILATO and V.C.BADDING, *J. Amer. Chem. Soc.*, **84**, 2377 (1962).
751. ELIEL, E.L. and B.P.THILL, *Chem. and Ind.*, 88 (1963).
752. ELLEMAN, D.D., L.C.BROWN and D.WILLIAMS, *J. Mol. Spec.*, **7**, 307 (1961).
753. ELLEMAN, D.D., L.C.BROWN and D.WILLIAMS, *J. Mol. Spec.*, **7**, 322 (1961).
754. ELLEMAN, D.D. and S.L.MANATT, *J. Mol. Spec.*, **9**, 477 (1962).
755. ELLEMAN, D.D. and S.L.MANATT, *J. Chem. Phys.*, **36**, 1945 (1962).
756. ELLEMAN. D.D. and S.L.MANATT, *J. Chem. Phys.*, **36**, 2346 (1962).
757. ELLEMAN, D.D., S.L.MANATT and C.D.PEARCE, *J. Chem. Phys.*, **42**, 650 (1965).
758. ELLIOTT, I.W. and H.C.DUNATHAN, *Tetrahedron*, **19**, 833 (1963).
759. ELS, H., G.ENGLERT, M.MULLER and A.FURST, *Helv.*, **48**, 989 (1965).
760. ELVIDGE, J.A., *Chem. Comm.*, 160 (1965).
761. ELVIDGE, J.A., *J. Chem. Soc.*, 474 (1959).
762. ELVIDGE, J.A. and R.G.FOSTER, *J. Chem. Soc.*, 590 (1963).
763. ELVIDGE, J.A. and R.G.FOSTER, *J. Chem. Soc.*, 981 (1964).
764. ELVIDGE, J.A. and L.M.JACKMAN, *Proc. Chem. Soc.*, 89 (1959).
765. ELVIDGE, J.A. and L.M.JACKMAN, *J. Chem. Soc.*, 859 (1961).
766. ELVIDGE, J.A., G.T.NEWBOLD, I.R.SENCIALL and T.G.SYMES, *J. Chem. Soc.*, 4157 (1964).
767. ELVIDGE, J.A. and P.D.RALPH, *J. Chem. Soc.*, (B), 243 (1966).
768. ELVIDGE, J.A. and P.D.RALPH, *J. Chem. Soc.*, (B), 249 (1966).
769. ELVIDGE, J.A. and P.D.RALPH, *J. Chem. Soc.*, (C) 387 (1966).
770. ELVIDGE, J.A. and R.STEVENS, *J. Chem. Soc.*, 2251 (1965).
771. EMERSON, G.F., J.E.MAHLER, R.KOCHHAR and R.PETTIT, *J. Org. Chem.*, **29**, 3620 (1964).
772. EMERSON, M.T., E.GRUNWALD and R.A.KROMHOUT, *J. Chem. Phys.*, **33**, 547 (1960).
773. EMERSON, M.T., E.GRUNWALD, M.L.KAPLAN and R.A.KROMHOUT, *J. Amer. Chem. Soc.*, **82**, 6307 (1960).
774. EMERSON, M.T., E.GRUNWALD, M.L.KAPLAN and R.A.KROMHOUT, *J. Amer. Chem. Soc.*, **82**, 6307 (1960).
775. ENGLERT, G., *Z. Naturforsch.*, **16b**, 413 (1961).
776. EPSTEIN, M. and S.A.BUCKLER, *J. Amer. Chem. Soc.*, **83**, 3279 (1961).
777. ERICKSON, L.E., *J. Amer. Chem. Soc.*, **87**, 1867 (1965).
778. ERNST, R.R. and H.PRIMAS, *Helv. Phys. Acta*, **36**, 583 (1963).
779. ERSKINE, R.L. and S.A.KNIGHT, *Chem. and Ind.*, 1160 (1960).
780. EVANEGA, G.R., W.BERGMANN and J.ENGLISH JR., *J. Org. Chem.*, **27**, 13 (1962).
781. EVANS, B.A. and R.E.RICHARDS, *J. Sci. Instr.*, **37**, 353 (1960).
782. EVANS, D.F., *J. Chem. Soc.*, 2003 (1959).
783. EVANS, D.F., *J. Chem. Soc.*, 877 (1960).
784. EVANS, D.F., *Mol. Phys.*, **5**, 183 (1962).
785. EVANS, D.F., *Mol. Phys.*, **6**, 179 (1963).
786. EVANS, D.F., S.L.MANATT and D.D.ELLEMAN, *J. Amer. Chem. Soc.*, **85**, 238 (1963).
787. EVANS, D.F., *J. Chem. Soc.*, 5575 (1963).
788. EVLETH, E.M., J.A.BERSON and S.L.MANATT, *Tetrahedron Letters*, 3087 (1964).
789. EWART, G., D.S.PAYNE, A.L.PORTE and A.P.LANE, *J. Chem. Soc.*, 3984 (1962).
790. FALES, H.M. and T.LUUKKAINEN, *Anal. Chem.*, **37**, 955 (1965).
791. FALES, H.M. and A.V.ROBERTSON, *Tetrahedron Letters*, 111 (1962).
792. FARGES, G. and A.S.DREIDING, *Helv. Chim. Acta*, **49**, 552 (1966).
793. FARNUM, D.G., *J. Amer. Chem. Soc.*, **86**, 934 (1964).
794. FARNUM, D.G., M.A.T.HEYBEY and B.WEBSTER, *J. Amer. Chem. Soc.*, **86**, 673 (1964).
795. FAWCETT, F.S., C.W.TULLOCK and D.D.COFFMAN, *J. Amer. Chem. Soc.*, **84**, 4275 (1962).
796. FAY, R.C. and T.S.PIPER, *J. Amer. Chem. Soc.*, **85**, 500 (1963).
797. FAY, R.C. and T.S.PIPER, *J. Amer. Chem. Soc.*, **84**, 2303 (1962).
798. FAY, C.K., S.STERNHELL and P.W.WESTERMAN, unpublished results.
799. FEENEY, J., A.LEDWITH and L.H.SUTCLIFFE, *J. Chem. Soc.*, 2021 (1962).
800. FEENEY, J. and L.H.SUTCLIFFE, *Trans. Faraday Soc.*, **56**, 1559 (1960).

801. FEENEY, J. and L.H.SUTCLIFFE, *J. Phys. Chem.*, **65**, 1894 (1961).
802. FEENEY, J. and L.H.SUTCLIFFE, *J. Chem. Soc.*, 1123 (1962).
803. FEENEY, J. and L.SUTCLIFFE, *Proc. Chem. Soc.*, 118 (1961).
804. FELTKAMP, H., N.C.FRANKLIN and Collaborators, *Ann.*, **683**, 55, 64, 75 (1965).
805. FELTKAMP, H. and N.C.FRANKLIN, *J. Amer. Chem. Soc.*, **87**, 1616 (1965).
806. FELTKAMP, H. and N.C.FRANKLIN, *Tetrahedron*, **21**, 1541 (1965).
807. FELTKAMP, H., N.L.FRANKLIN, M.HANACK and K.W.HEINZ, *Tetrahedron Letters*, 3535 (1964).
808. FENTON, D.E. and A.G.MASSEY, *J. Inorg. Nucl. Chem.*, **27**, 329 (1965).
809. FERGUSON, R.C., *J. Phys. Chem.*, **68**, 1594 (1964).
810. FERGUSON, R.C. and D.W.MARQUARDT, *J. Chem. Phys.*, **41**, 2087 (1964).
811. FERRIER, R.J. and N.R.WILLIAMS, *Chem. and Ind.*, 1696 (1964).
812. FERSTANDIG, L.L., *J. Amer. Chem. Soc.*, **84**, 3553 (1962).
813. FESSENDEN, R.W. and J.S.WAUGH, *J. Chem. Phys.*, **30**, 944 (1959).
814. FESSENDEN, R.W. and J.S.WAUGH, *J. Chem. Phys.*, **31**, 996 (1959).
815. FESSENDEN, R.W. and J.S.WAUGH, *J. Chem. Phys.*, **37**, 1466 (1962).
816. FETIZON, M., M.GOLFIER and P.LASZLO, *Bull. Soc. Chim. France*, 3205 (1965).
817. FIESER, L.F., T.GOTO and B.K.BHATTACHARYYA, *J. Amer. Chem. Soc.*, **82**, 1700 (1960).
818. FILLER, R. and S.M.NAGVI, *J. Org. Chem.*, **26**, 2571 (1961).
819. FINAR, I.L. and E.F.MOONEY, *Spectrochim. Acta*, **20**, 1269 (1964).
820. FINEGOLD, H., *Proc. Chem. Soc.*, 213 (1962).
821. FINEGOLD, H., *J. Chem. Phys.*, **41**, 1808 (1964).
822. FINEGOLD, H. and H.KWART, *J. Org. Chem.*, **27**, 2361 (1962).
823. FINGER, G.C., L.D.STARR, D.R.DICKERSON, H.S.GUTOWSKY and J.HAMER, *J. Org. Chem.*, **28**, 1666 (1963).
824. FINNEGAN, R.A. and R.S.McNEES, *J. Org. Chem.*, **29**, 3234 (1964).
825. FINNEGAN, W.G. and W.P.NORRIS, *J. Org. Chem.*, **28**, 1139 (1963).
826. FISHER, F. and D.E.APPLEQUIST, *J. Org. Chem.*, **30**, 2089 (1965).
827. FISCHER, A., R.M.GOLDING and W.C.TENANT, *J. Chem. Soc.*, 6032 (1965).
828. FITCH, S.J. and K.MOEDRITZER, *J. Amer. Chem. Soc.*, **84**, 1876 (1962).
829. FITZGERALD, J.S., *Aust. J. Chem.*, **18**, 589 (1965).
830. FIXMAN, M., *J. Chem. Phys.*, **35**, 679 (1961).
831. FLAUTT, T.J. and W.F.ERMAN, *J. Amer. Chem. Soc.*, **85**, 3212 (1963).
832. FLITCROFT, N. and H.D.KAESZ, *J. Amer. Chem. Soc.*, **85**, 1377 (1963).
833. FLYGARE, W.H., *J. Chem. Phys.*, **41**, 793 (1964).
834. FLYGARE, W.H., *J. Chem. Phys.*, **42**, 1563 (1965).
835. FLYNN, G.W., M.MATSUSHIMA, J.D.BALDESCHWIELER and N.C.CRAIG, *J. Chem. Phys.*, **38**, 2295 (1963).
836. FOOTE, C.S., *Tetrahedron Letters*, 579 (1963).
837. FORBES, W.F., *Canad. J. Chem.*, **40**, 1891 (1962).
838. FORMAN, S.E., A.J.DURBETAK, M.V.COHEN and R.A.OLOFSON, *J. Org. Chem.*, **30**, 169 (1965).
839. FORSEN, S., *Acta Chem. Scand.*, **13**, 1472 (1959).
840. FORSEN, S., *Acta Chem. Scand.*, **14**, 231 (1960).
841. FORSEN, S., *Arkiv Kemi*, **20**, 25 (1963).
842. FORSEN, S., *J. Phys. Chem.*, **67**, 1740 (1963).
843. FORSEN, S. and B.AKERMARK, *Acta Chem. Scand.*, **17**, 1712 (1963).
844. FORSEN, S. and B.AKERMARK, *Acta Chem. Scand.*, **17**, 1907 (1963).
845. FORSEN, S., T.ALM, B.GESTBLOM, S.RODMAR and R.A.HOFFMAN, *J. Mol. Spec.*, **17**, 13 (1965).
846. FORSEN, S., B.AKERMARK and T.ALM, *Acta Chem. Scand.*, **18**, 2313 (1964).
847. FORSEN, S., B.GESTBLOM, S.GRONOWITZ and R.A.HOFFMAN, *Acta. Chem. Scand.*, **18**, 313 (1964).
848. FORSEN, S., B.GESTBLOM, R.A.HOFFMAN and S.RODMAR, *Acta Chem. Scand.*, **19**, 503 (1965).
849. FORSEN, S. and R.A.HOFFMAN, *Acta. Chem. Scand.*, **17**, 1787 (1963).
850. FORSEN, S. and R.A.HOFFMAN, *J. Chem. Phys.*, **39**, 2892 (1963).

851. FORSEN, S. and R.A.HOFFMAN, *Acta Chem. Scand.*, **18**, 249 (1964).
852. FORSEN, S.R. and R.A.HOFFMAN, *J. Chem. Phys.*, **40**, 1189 (1964).
853. FORSEN, S., F.MERENYI and M.NILSSON, *Acta. Chem. Scand.*, **18**, 1208 (1964).
854. FORSEN, S. and M.NILSSON, *Acta Chem. Scand.*, **13**, 1383 (1959).
855. FORSEN, S. and M.NILSSON, *Acta Chem. Scand.*, **14**, 1333 (1960).
856. FORSEN, S., M.NILSSON, J.A.ELVIDGE, J.S.BURTON and R.STEVENS, *Acta Chem. Scand.*, **18**, 513 (1964).
857. FORSEN, S. and T.NORIN, *Tetrahedron Letters*, 2845 (1964).
858. FORSEN, S. and A.RUPPRECHT, *J. Chem. Phys.*, **33**, 1888 (1960).
859. FORT, A.W., *J. Amer. Chem. Soc.*, **84**, 4979 (1962).
860. FORT, R.C., G.W.H.CHEESEMAN and E.C.TAYLOR, *J. Org. Chem.*, **29**, 2440 (1964).
861. FORT, R.C. and P.R.SCHLEYER, *Chem. Revs.*, **64**, 277 (1964).
862. FORT, R.C. and P.R.SCHLEYER, *J. Org. Chem.*, **30**, 789 (1965).
863. FOSTER, A.B., A.H.HAINES, J.HOMER, J.LEHMANN and L.F.THOMAS, *J. Chem. Soc.*, 5005 (1961).
864. FOSTER, H., *Anal. Chem.*, **34**, 255R (1962).
865. FOSTER, H., *Anal. Chem.*, **36**, 266R (1964).
866. FOX, I.R., P.L.LEVINS and R.W.TAFT JR., *Tetrahedron Letters*, 249 (1961).
867. FRAENKEL, G., D.G.ADAMS and J.WILLIAMS, *Tetrahedron Letters*, 767 (1963).
868. FRANKEL, E.N., E.A.EMKEN, H.PETERS, V.L.DAVISON and R.O.BUTTERFIELD, *J. Org. Chem.*, **29**, 3292 (1964).
869. FRAENKEL, G., *J. Chem. Phys.*, **39**, 1614 (1963).
870. FRAENKEL, G. and W.BURLANT, *J. Chem. Phys.*, **42**, 3724 (1965).
871. FRAENKEL, G. and D.T.DIX, *J. Amer. Chem. Soc.*, **88**, 979 (1966).
872. FRAENKEL, G., R.E.CARTER, A.MCLACHLAN and J.H.RICHARDS, *J. Amer. Chem. Soc.*, **82**, 5846 (1960).
873. FRAENKEL, G., D.T.DIX and D.G.ADAMS, *Tetrahedron Letters*, 3155 (1964).
874. FRAENKEL, G., A.LOEWENSTEIN and S.MEIBOOM, *J. Phys. Chem.*, **65**, 700 (1961).
875. FRAENKEL, G., P.D.RALPH and J.P.KIM, *Canad. J. Chem.*, **43**, 674 (1965).
876. FRANCIS, S.A. and E.D.ARCHER, *Anal. Chem.*, **35**, 1363 (1963).
877. FRANCK-NEUMANN, M. and J.M.LEHN, *Mol. Phys.*, **7**, 197 (1963–64).
878. FRANCONI, C., *Z. Electrochem.*, **65**, 645 (1961).
879. FRANCONI, C. and G.FRAENKEL, *Rev. Sci. Instr.*, **31**, 657 (1960).
880. FRAENKEL, G., Y.ASSAHI, M.J.MITCHELL and M.P.CAVA, *Tetrahedron*, **20**, 1179 (1964).
881. FRANKISS, S.O., *J. Phys. Chem.*, **67**, 752 (1963).
882. FRANKLIN, N.C. and H.FELTKAMP, *Angew. Chem. Int. Ed.*, **4**, 774 (1965).
883. FRANZEN, W., *Rev. Sci. Instr.*, **33**, 933 (1962).
884. FRANZUS, B. and B.E.HUDSON JR., *J. Org. Chem.*, **28**, 2238 (1963).
885. FRASER, M., A.MALERA, B.B.MOLLOY and D.H.REID, *J. Chem. Soc.*, 3288 (1962).
886. FRASER, R.R., *Canad. J. Chem.*, **38**, 2226 (1960).
887. FRASER, R.R., *Canad. J. Chem.*, **38**, 505 (1961).
888. FRASER, R.R., *Canad. J. Chem.*, **40**, 78 (1962).
889. FRASER, R.R., *Canad. J. Chem.*, **40**, 1483 (1962).
890. FRASER, R.R., R.U.LEMIEUX and J.D.STEVENS, *J. Amer. Chem. Soc.*, **83**, 3901 (1961).
891. FRASER, R.R. and D.MCGREER, *Canad. J. Chem.*, **39**, 505 (1961).
892. FRASER, R.R. and S.O'FARELL, *Tetrahedron Letters*, 1143 (1962).
893. FRATIELLO, A. and J.P.LUONGO, *J. Amer. Chem. Soc.*, **83**, 3072 (1963).
894. FRANZ, J.E., C.OSUCH and M.W.DIETRICH, *J. Org. Chem.*, **29**, 2922 (1964).
895. FREEDMAN, H.H., A.E.YOUNG and V.R.SANDEL, *J. Amer. Chem. Soc.*, **86**, 4722 (1964).
896. FREEMAN, J.P., *J. Org. Chem.*, **27**, 2881 (1962).
897. FREEMAN, J.P., *J. Org. Chem.*, **28**, 2508 (1963).
898. FREEMAN, R., *Mol. Phys.*, **3**, 435 (1960).
899. FREEMAN, R., *Mol. Phys.*, **5**, 499 (1962).
900. FREEMAN, R., *Mol. Phys.*, **6**, 535 (1963).
901. FREEMAN, R., *J. Chem. Phys.*, **40**, 3571 (1964).

902. FREEMAN, R., *J. Chem. Phys.*, **43**, 3087 (1965).
903. FREEMAN, R. and W. A. ANDERSON, *J. Chem. Phys.*, **39**, 806 (1963).
903a. FREEMAN, R. and W. A. ANDERSON, *J. Chem. Phys.*, **37**, 2053 (1962).
904. FREEMAN, R. and W. A. ANDERSON, *J. Chem. Phys.*, **42**, 1199 (1965).
905. FREEMAN, R. and N. S. BHACCA, *J. Chem. Phys.*, 38, 1088 (1963).
906. FREEMAN, R., N. S. BHACCA and C. A. REILLY, *J. Chem. Phys.*, 38, 293 (1963).
907. FREEMAN, R., K. A. MCLAUCHLAN, J. I. MUSHER and K. G. R. PACHLER, *Mol. Phys.*, **5**, 321 (1962).
908. FREEMAN, R., K. A. MCLAUCHLAN, J. I. MUSHER and K. G. R. PACHLER, *Mol. Phys.*, **5**, 321 (1962).
909. FREEMAN, R. and K. PACHLER, *Mol. Phys.*, **5**, 85 (1962).
910. FREEMAN, R. and R. V. POUND, *Rev. Sci. Intr.*, **31**, 103 (1960).
911. FREEMAN, R. and D. H. WHIFFEN, *Mol. Phys.*, **4**, 321 (1961).
912. FREEMAN, R. and D. H. WHIFFEN, *Proc. Phys. Soc.*, **79**, 794 (1962).
913. FREI, K. and H. J. BERNSTEIN, *J. Chem. Phys.*, **37**, 1891 (1962).
914. FREI, K. and H. J. BERNSTEIN, *J. Chem. Phys.*, **38**, 1216 (1963).
915. FREYMANN, M. and R. FREYMANN, *Compt. Rend.*, **250**, 3638 (1960)
916. FREYMANN, R., M. DVOLAITZKY and J. JACQUES, *Compt. Rend.*, **253**, 1436 (1961).
917. FRIEBOLIN, H., S. KABUSS, W. MAIER and A. LÜTTRINGHAUS, *Tetrahedron Letters*, 683 (1962).
918. FRIEBOLIN, H. and W. MAIER, *Z. Naturforschung*, **16 A**, 640 (1961).
919. FRIEBOLIN, H., R. MECKE, S. KABUSS and A. LÜTTRINGHAUS, *Tetrahedron Letters*, 1929 (1964).
920. FRIEDEL, R. A. and H. L. RETCOFSKY, *J. Amer. Chem. Soc.*, **85**, 1300 (1963).
921. FRIEDRICH, H. J., *Angew. Chem.*, **76**, 496 (1964).
922. FRITZ, H. P. and H. KELLER, *Z. Naturforsch.*, **16b**, 231 (1961).
923. FRITZ, H. P. and C. G. KREITER, *Chem. Ber.*, **96**, 2008 (1963).
924. FRITZ, H. P. and C. G. KREITER, *Chem. Ber.*, **97**, 1398 (1964).
925. FRITZ, H. and W. LÖWE, *Angew. Chem. Int. Ed.*, **1**, 592 (1962).
926. FRITZ, H. P. and K. E. SCHWARZHANS, *Chem. Ber.*, **97**, 1390 (1964).
927. FUCHS, B. and R. G. HABER, *Tetrahedron Letters*, 1323 (1966).
928. FUJII, S. and S. SHIDA, *Bull. Chem. Soc. Japan*, **24**, 242 (1951).
929. FUJIWARA, S. and Y. ARATA, *Bull. Chem. Soc. Japan*, **36**, 578 (1963).
930. FUJIWARA, S. and H. SHIMIZU, *J. Chem. Phys.*, **32**, 1636 (1960).
931. FUJIWARA, S., H. SHIMIZU, Y. ARATA and S. AKAHORI, *Bull. Chem. Soc. Japan*, **33**, 428 (1960).
932. FUJIWARA, Y. and S. FUJIWARA, *Bull. Chem. Soc. Japan*, **36**, 574 (1963).
933. FUJIWARA, Y. and S. FUJIWARA, *Bull. Chem. Soc. Japan*, **37**, 1005 (1964).
934. FUKOTO, T. R., E. O. HORNIG, R. L. METCALF and M. Y. WINTON, *J. Org. Chem.*, **26**, 4620 (1961).
935. FURBERG, S. and B. PEDERSON, *Acta Chem. Scand.*, **17**, 1160 (1963).
936. GABBAI, A., A. MELERA, D. JANJICK and T. POSTERNAK, *Helv. Chim. Acta*, **49,** 168 (1966).
937. GAGNAIRE, D. and E. PAYO-SUBIZA, *Bull. Soc. Chim. France*, 2627 (1963).
938. GAGNAIRE, D. and E. PAYO-SUBIZA, *Bull. Soc. Chim. France*, 2633 (1963).
939. GAGNAIRE, D., E. PAYO-SUBIZA and A. ROUSSEAU, *J. Chim. Phys.*, **62**, 42 (1965).
940. GAGNAIRE, D., A. ROUSSEAU and P. SERVOZ-GAVIN, *J. Chim. Phys.*, **61**, 1207 (1964).
941. GAGNAIRE, D. and P. VOTTERO, *Bull. Soc. Chim. France*, 2779 (1963) and private communication.
942. GAINES, D. F. and R. SCHAEFFER, *J. Amer. Chem. Soc.*, **85**, 395 (1963).
943. GAINES, D. F. and R. SCHAEFFER, *J. Amer. Chem. Soc.*, **85**, 3592 (1963).
944. GALANTAY, E., A. SZABO and J. FRIED, *J. Org. Chem.*, **28**, 98 (1963).
945. GANS, R. and B. MROWKA, *Schriften Königsberg. Gelehrten Ges. Naturw.*, Ke, **12**, 1 (1935).
946. GAONI, Y., A. MELERA, F. SONDHEIMER and R. WOLOVSKY, *Proc. Chem. Soc.*, 397 (1964).
947. GARBISCH, E. W., *J. Org. Chem.*, **27**, 4243, 4249 (1962).
948. GARBISCH, E. W., *J. Amer. Chem. Soc.*, **85**, 927 (1963).

949. GARBISCH, E.W., *J. Amer. Chem. Soc.*, **85**, 1696 (1963).
950. GARBISCH, E.W., *J. Amer. Chem. Soc.*, **86**, 5561 (1964).
951. GARBISCH, E.W., *Chem. and Ind.*, 1715 (1964).
952. GARBISCH, E.W., *J. Amer. Chem. Soc.*, **86**, 1780 (1964).
953. GARBISCH, E.W., *J. Amer. Chem. Soc.*, **87**, 505 (1965).
954. GARBISCH, E.W., *J. Org. Chem.*, **30**, 2109 (1965).
955. GARBISCH, E.W. and D.B.PATTERSON, *J. Amer. Chem. Soc.*, **85**, 3228 (1963).
956. GARRATT, S., *J. Org. Chem.*, **28**, 1886 (1963).
957. GASSMAN, P.G. and D.C.HECKERT, *J. Org. Chem.*, **30**, 2859 (1965).
958. GATES, P.N. and E.F.MOONEY, *J. Chem. Soc.*, 4648 (1964).
959. GATLIN, L. and J.C.DAVIS, *J. Amer. Chem. Soc.*, **84**, 4464 (1962).
960. GAWER, A.H. and B.P.DAILEY, *J. Chem. Phys.*, **42**, 2658 (1965).
961. GAWRON, O., A.J.GLAID and T.P.FONDY, *J. Amer. Chem. Soc.*, **83**, 3634 (1961).
962. GEORGIAN, V., L.GEORGIAN, A.V.ROBERTSON and L.F.JOHNSON, *Tetrahedron*, **19**, 1219 (1963).
963. GERRARD, W., J.B.LEANE, E.F.MOONEY and R.G.REES, *Spectrochim. Acta*, **19**, 1964 (1963).
964. GESNER, B.D., *Tetrahedron Letters*, 3559 (1965).
965. GESTBLOM, B., *Acta Chem. Scand.*, **17**, 280 (1963).
966. GESTBLOM, B., S.GRONOWITZ, R.A.HOFFMAN and B.MATHIASSON, *Arkiv Kemi*, **23**, 517 (1965).
967. GESTBLOM, B., S.GRONOWITZ, R.A.HOFFMAN, B.MATHIASSON and S.RODMAR, *Arkiv Kemi*, **23**, 483 (1965).
968. GESTBLOM, B., S.GRONOWITZ, R.A.HOFFMAN, B.MATHIASSON and S.RODMAR, *Arkiv Kemi*, **23**, 501 (1965).
969. GESTBLOM, B., R.A.HOFFMAN and S.RODMAR, *Mol. Phys.*, **8**, 425 (1964).
970. GESTBLOM, B., R.A.HOFFMAN and S.RODMAN, *Acta Chem. Scand.*, **18**, 1222 (1964).
971. GESTBLOM, B. and B.MATHIASSON, *Acta Chem. Scand.*, **18**, 1905 (1964).
972. GESTBLOM, B. and S.RODMAR, *Acta Chem. Scand.*, **18**, 1767 (1964).
973. GHOSH, S.K. and S.K.SINHA, *J. Chem. Phys.*, **36**, 737 (1962).
974. GIANNI, M.H., E.L.STOGRYN and C.M.ORLANDO, *J. Phys. Chem.*, **67**, 1385 (1963).
975. GIBBONS, W.A. and H.FISCHER, *Tetrahedron Letters*, 43 (1964).
976. GIBBONS, W.A. and V.M.S.GIL, *Mol. Phys.*, **9**, 163 and 167 (1965).
977. GIESSNER-PETTRE, C., *Compt. Rend.*, **250**, 2547 (1960).
978. GIESSNER-PETTRE, C., *Compt. Rend.*, **252**, 3238 (1961).
979. GIL, V.M.S., *Mol. Phys.*, **9**, 97 (1965).
980. GIL, V.M.S. and W.A.GIBBONS, *Mol. Phys.*, **8**, 199 (1964).
981. GIL, V.M.S. and J.N.MURRELL, *Trans. Farad. Soc.*, **60**, 248 (1964).
982. GILBY, A.R. and D.F.WATERHOUSE, *Aust. J. Chem.*, **17**, 1311 (1964).
983. GILCHRIST, T., R.HODGES and A.L.PORTE, *J. Chem. Soc.*, 1780 (1962).
984. GILLESPIE, R.J. and T.BIRCHALL, *Canad. J. Chem.*, **41**, 148 (1963).
985. GILMAN, H. and W.H.ATWELL, *J. Amer. Chem. Soc.*, **87**, 2678 (1965).
986. GLASEL, J.A., L.M.JACKMAN and D.W.TURNER, *Proc. Chem. Soc.*, 426 (1961).
987. GLAZKOV, V.I., *Optics and Spectroscopy (U.S.S.R.)*, **9**, 217 (1960).
988. GLICK, R.E., *J. Phys. Chem.*, **65**, 1871 (1961).
989. GLICK, R.E. and A.A.BOTHNER-BY, *J. Chem. Phys.*, **25**, 362 (1956).
990. GLICK, R.E., D.F.KATES and S.J.EHRENSON, *J. Chem. Phys.*, **31**, 567 (1959).
991. GOETZ, H., F.NERDEL and K.REHSE, *Ann.*, **681**, 1 (1965).
992. GOFF, E. LE and R.B. LA COUNT, *J. Amer. Chem. Soc.*, **85**, 1354 (1963).
993. GOLDING, B.T. and R.W.RICKARDS, *Chem. and Ind.*, 1081 (1963).
994. GOLDSMITH, D.J. and J.A.HARTMAN, *J. Org. Chem.*, **29**, 3520 (1964).
995. GOLDSTEIN, J.H., G.S.REDDY and L.MANDELL, *J. Amer. Chem. Soc.*, **83**, 1300 (1961).
996. GOLDSTEIN, J.H., L.MANDELL and E.B.WHIPPLE, *J. Chem. Phys.*, **30**, 1109 (1959).
997. GOLDSTEIN, J.H. and G.S.REDDY, *J. Amer. Chem. Soc.*, **83**, 5020 (1961).
998. GOLDSTEIN, J.H. and G.S.REDDY, *J. Chem. Phys.*, **36**, 2644 (1962).
999. GOMPPER, R. and E.KUTTER, *Chem. Ber.*, **98**, 2825 (1965).

1000. GOOD, C.D. and D.M.RITTER, *J. Amer. Chem. Soc.*, **84**, 1162 (1962).
1001. GOODWIN, S., J.N.SHOOLERY and L.F.JOHNSON, *Proc. Chem. Soc.*, 306 (1958).
1002. GOODWIN, S., J.N.SHOOLERY and L.F.JOHNSON, *J. Amer. Chem. Soc.*, **81**, 3065 (1959).
1003. GORMAN, W.G., R.K.KULLNIG and F.C.NACHOD, *Appl. Spec.*, **17**, 77 (1963).
1004. GORODETSKY, M. and Y.MAZUR, *J. Amer. Chem. Soc.*, **86**, 5218 (1964).
1005. GORDON, S.L. and J.D.BALDESCHWIELER, *J. Chem. Phys.*, **41**, 571 (1965).
1006. GORDON, S. and B.P.DAILEY, *J. Chem. Phys.*, **34**, 1084 (1961).
1007. GORE, E.S., D.J.BLEARS and S.S.DANYLUK, *Canad. J. Chem.*, **43**, 2135 (1965).
1008. GRAHAM, D.M. and C.E.HOLLOWAY, *Canad. J. Chem.*, **41**, 2114 (1963).
1009. GRAHAM, J.D. and M.T.ROGERS, *J. Amer. Chem. Soc.*, **84**, 2249 (1962).
1010. GRANACHER, I., *Helv. Phys. Acta*, **34**, 272 (1961).
1011. GRANACHER, I., H.SUCR, A.ZENHAUSERN and H.ZOLLINGER, *Helv. Chim. Acta*, **44**, 313 (1961).
1012. GRANGER, P., *J. Chim. Phys.*, **62**, 594 (1965).
1013. GRANT, D.M., *Ann. Rev. Phys. Chem.*, **15**, 489 (1964).
1014. GRANT, D.M. and H.S.GUTOWSKY, *J. Chem. Phys.*, **34**, 699 (1961).
1015. GRANT, D.M., R.C.HIRST and H.S.GUTOWSKY, *J. Chem. Phys.*, **38**, 470 (1963).
1016. GRANT, D.M. and W.M.LITCHMAN, *J. Amer. Chem. Soc.*, **87**, 3994 (1965).
1017. GRASHEY, R., R.HUISGEN, K.K.SUN and R.M.MORIARTY, *J. Org. Chem.*, **30**, 74 (1965).
1018. GRAY, P. and L.REEVES, *J. Chem. Phys.*, **32**, 1878 (1960).
1019. GREEN, M.C.H., J.A.McCLEVERTY and L.PRATT and G.WILKINSON, *J. Chem. Soc.*, 4854 (1961).
1020. GRIESBAUM, K., W.NAEGELE and G.G.WANLESS, *J. Amer. Chem. Soc.*, **87**, 3151 (1965).
1021. GRIESBAUM, K., A.A.OSWALD and B.E.HUDSON JR., *J. Amer. Chem. Soc.*, **85**, 1969 (1963).
1022. GRIESBAUM, K., A.A.OSWALD and B.E.HUDSON, *J. Amer. Chem. Soc.*, **85**, 1969 (1963).
1023. GRIFFIN, C.E., *Tetrahedron*, **20**, 2399 (1964).
1024. GRIFFIN, G.W. and L.I.PETERSON, *J. Amer. Chem. Soc.*, **85**, 2268 (1963).
1025. GRIFFITH, D.L. and J.D.ROBERTS, *J. Amer. Chem. Soc.*, **87**, 4089 (1965).
1026. GRIMLEY, T.B., *Mol. Phys.*, **6**, 329 (1963).
1027. GRIMME, W., H.HOFFMANN and E.VOGEL, *Angew. Chem. Int. Ed.*, **4**, 354 (1965).
1028. GRIOT, R.G. and A.J.FREY, *Tetrahedron*, **19**, 1661 (1963).
1029. GRISWOLD, A.A. and P.S.STARCHER, *J. Org. Chem.*, **30**, 1687 (1965).
1030. GRONOWITZ, S. and B.GESTBLOM, *Arkiv Kemi*, **18**, 513 (1961).
1031. GRONOWITZ, S., *Arkiv Kemi*, **13**, 295 (1959).
1032. GRONOWITZ, S., B.GESTBLOM and R.A.HOFFMAN, *Acta Chem. Scand.*, **15**, 1201 (1961).
1033. GRONOWITZ, S., B.GESTBLOM and B.MATHIASSON, *Arkiv Kemi*, **20**, 407 (1963).
1034. GRONOWITZ, S. and R.A.HOFFMAN, *Acta Chem. Scand.*, **13**, 1687 (1959).
1035. GRONOWITZ, S. and R.A.HOFFMAN, *Arkiv Kemi*, **15**, 499 (1960).
1036. GRONOWITZ, S. and R.A. HOFFMAN, *Arkiv Kemi*, **16**, 459 (1961).
1037. GRONOWITZ, S. and R.A.HOFFMAN, *Arkiv Kemi*, **16**, 539 (1961).
1038. GRONOWITZ, S., A.B.HORNFELDT, B.GESTBLOM and R.A.HOFFMAN, *Arkiv Kemi*, **18**, 133 (1961).
1039. GRONOWITZ, S., A.B.HORNFELDT, B.GESTBLOM and R.A.HOFFMAN, *J. Org. Chem.*, **26**, 2615 (1961).
1040. GRONOWITZ, S., P.MOSES and A.B.HORNFELDT, *Arkiv Kemi*, **17**, 237 (1961).
1041. GRONOWITZ, S., B.NORRMAN, B.GESTBLOM, B.MATHIASSON and R.A.HOFFMAN, *Arkiv Kemi*, **22**, 65 (1964).
1042. GRONOWITZ, S., U.RUDEN and B.GESTBLOM, *Arkiv Kemi*, **20**, 297 (1963).
1043. GRONOWITZ, S., G.SÖRLIN, B.GESTBLOM and R.A.HOFFMAN, *Arkiv Kemi*, **19**, 483 (1963).
1044. GROVING, N. and A.HOLM, *Acta Chem. Scand.*, **19**, 443 (1955).

1045. GRUEN, L. C. and P. T. McTIGUE, *J. Chem. Soc.*, 5217, 5224 (1963).
1046. GRUNWALD, E., *J. Phys. Chem.*, **67**, 2208 (1963).
1047. GRUNWALD, E., *J. Phys. Chem.*, **67**, 2211 (1963).
1048. GRUNWALD, E. and C. F. JUMPER, *J. Amer. Chem. Soc.*, **85**, 2051 (1963).
1049. GRUNWALD, E., C. F. JUMPER and S. MEIBOOM, *J. Amer. Chem. Soc.*, **84**, 4664 (1962).
1050. GRUNWALD, E., C. F. JUMPER and S. MEIBOOM, *J. Amer. Chem. Soc.*, **84**, 4666 (1962).
1051. GRUNWALD, E., C. F. JUMPER and S. MEIBOOM, *J. Amer. Chem. Soc.*, **85**, 522 (1963).
1052. GRUNWALD, E., G. J. KARABATSOS, R. A. KROMHOUT and E. L. PURLEE, *J. Chem. Phys.*, **33**, 556 (1960).
1053. GRUNWALD, E., A. LOWENSTEIN and S. MEIBOOM, *J. Chem. Phys.*, **27**, 630 (1957).
1054. GRUNWALD, E. and S. MEIBOOM, *J. Amer. Chem. Soc.*, **85**, 2047 (1963).
1055. GRUNWALD, E. and E. PRICE, *J. Amer. Chem. Soc.*, **87**, 3139 (1965).
1056. GRUTZNER, J. B., L. M. JACKMAN and J. M. LAWLOR, Unpublished results.
1057. GÜNTHER, H., *Angew. Chem. Int. Ed.*, **4**, 702 (1965).
1058. GÜNTHER, H. and H. H. HINRICHS, *Tetrahedron Letters*, 787 (1966).
1059. GUTSCHE, C. D. and T. D. SMITH, *J. Amer. Chem. Soc.*, **82**, 4067 (1960).
1060. GUSTAFSON, D. H. and W. F. ERMAN, *J. Org. Chem.*, **30**, 1665 (1965).
1061. GUTOWSKY, H. S., *Techniques of Organic Chemistry*, Weissberger, Vol. I, Part 4, p. 2663 (Interscience 1960).
1062. GUTOWSKY, H. S., *J. Chem. Phys.*, **37**, 2196 (1962).
1063. GUTOWSKY, H. S., *Pure Appl. Chem.*, **7**, 93 (1963).
1064. GUTOWSKY, H. S., G. G. BELFORD and P. E. McMAHON, *J. Chem. Phys.*, **36**, 3353 (1962).
1065. GUTOWSKY, H. S. and C. H. HOLM, *J. Chem. Phys.*, **25**, 1228 (1956).
1066. GUTOWSKY, H. S. and J. JONAS, *Inorg. Chem.*, **4**, 430 (1965).
1067. GUTOWSKY, H. S. and C. JUAN, *J. Chem. Phys.*, **37**, 120 (1962).
1068. GUTOWSKY, H. S. and C. S. JUAN, *J. Amer. Chem. Soc.*, **84**, 308 (1962).
1069. GUTOWSKY, H. S., M. KARPLUS and D. M. GRANT, *J. Chem. Phys.*, **31**, 1278 (1959).
1070. GUTOWSKY, H. S., D. W. McCALL and C. P. SLICHTER, *Phys. Rev.*, **84**, 589 (1951).
1071. GUTOWSKY, H. S., D. W. McCALL and C. P. SLICHTER, *J. Chem. Phys.*, **21**, 279 (1953).
1072. GUTOWSKY, H. S. and V. D. MOCHEL, *J. Chem. Phys.*, **39**, 1195 (1963.
1073. GUTOWSKY, H. S., V. D. MOCHEL and B. G. SOMMERS, *J. Chem. Phys.*, **36**, 1153 (1962).
1074. GUY, J. and J. TILLIEU, *J. Chem. Phys.*, **24**, 1117 (1956).
1075. GWYNN, D. E., G. M. WHITESIDES and J. D. ROBERTS, *J. Amer. Chem. Soc.*, **87**, 2862 (1965).
1076. HAAKE, P. and W. B. MILLER, *J. Amer. Chem. Soc.*, **85**, 4044 (1963).
1077. HABRAKEN, C. L. and J. A. MOORE, *J. Org. Chem.*, **30**, 1892 (1965).
1078. HAHN, E. L. and D. E. MAXWELL, *Phys. Rev.*, **84**, 1246 (1951).
1079. HAHN, E. L. and D. E. MAXWELL, *Phys. Rev.*, **88**, 1070 (1952).
1080. HAIGH, C. W., M. H. PALMER and B. SEMPLE, *J. Chem. Soc.*, 6004 (1965).
1081. HALL, G. G. and A. HARDISSON, *Proc. Roy. Soc.* (London, A **268**, 328 (1962).
1082. HALL, G. G., A. HARDISSON and L. M. JACKMAN, *Disc. Farad. Soc.*, **34**, 15 (1962).
1083. HALL, G. G., A. HARDISSON and L. M. JACKMAN, *Tetrahedron*, **19**, Suppl. 2, 101 (1963).
1084. HALL, L. D., *Chem. and Ind.*, 950 (1963).
1085. HALL, L. D., *Tetrahedron Letters*, 1457 (1964).
1085a. HALL, L. D., *J. Org. Chem.*, **29**, 297 (1964).
1086. HALL, L. D. and L. HOUGH, *Proc. Chem. Soc.*, 382 (1962).
1087. HALL, L. D. and L. F. JOHNSON, *Tetrahedron*, **20**, 883 (1964).
1088. HALL, L. D. and J. F. MANVILLE, *Chem. and Ind.*, 991 (1965).
1089. HAMEKA, H. F., *Mol. Phys.*, **2**, 64 (1959).
1090. HAMEKA, H. F., *J. Chem. Phys.*, **37**, 3008 (1962).
1091. HAMEKA, H. F., *Rev. Mod. Phys.*, **34**, 87 (1962).
1092. HATTON, H. V. and R. E. RICHARDS, *Mol. Phys.*, **3**, 253 (1960); **5**, 139 (1962).
1093. HAMLOW, H. P., S. OKUDA and N. NAKAGAWA, *Tetrahedron Letters*, 2553 (1964).
1094. HAMMAKER, R. M., *J. Chem. Phys.*, **43**, 1843 (1965).

1095. HAMMAKER, R.M., *Canad. J. Chem.*, **43**, 2916 (1965).
1096. HAMMOND, G.S. and R.C.NEUMAN, *J. Phys. Chem.*, **67**, 1655 (1963).
1097. HAMMOND, G.S., N.J.TURRO and A.A.FISCHER, *J. Amer. Chem. Soc.*, **83**, 4674 (1961).
1098. HAMMOND, P.R., *J. Chem. Soc.*, 1365 (1962).
1099. HAMMOND, P.R., *J. Chem. Soc.*, 1370 (1962).
1100. HAMMOND, P.R., *J. Chem. Soc.*, 2565 (1963).
1101. HAMPEL, B. and J.M.KRÄMER, *Chem. Ber.*, **98**, 3255 (1965).
1102. HAMPEL, B. and J.M.KRÄMER, *Tetrahedron*, **22**, 1601 (1966).
1103. HAND, E.S. and T.COHEN, *J. Amer. Chem. Soc.*, **87**, 133 (1965).
1104. HAND, E.S. and R.M.HOROWITZ, *J. Org. Chem.*, **29**, 3088 (1964).
1105. HANNA, M.W. and A.L.ASHBAUGH, *J. Phys. Chem.*, **68**, 811 (1964).
1106. HANNA, M.W. and J.K.HARRINGTON, *J. Phys. Chem.*, **67**, 940 (1963).
1107. HANSON, J.R., *J. Chem. Soc.*, 5036 (1965).
1108. HARDMAN, G.E.G., *Phillips Technical Review*, **24**, No.6, p.206 (1963).
1109. HARRIS, R.K., *J. Phys. Chem.*, **66**, 768 (1962).
1110. HARRIS, R.K., *J. Mol. Spec.*, **10**, 309 (1963).
1111. HARRIS, R.K., *Canad. J. Chem.*, **42**, 2275 (1964).
1112. HARRIS, R.K., *J. Mol. Spec.*, **15**, 100 (1965).
1113. HARRIS, R.K., A.R.KATRITZKY, S.OKSNE, A.S.BAILEY and W.G.PATERSON, *J. Chem. Soc.*, 197 (1963).
1114. HARRIS, R.K. and N.SHEPPARD, *Trans. Farad. Soc.*, **59**, 606 (1963).
1115. HART, H. and F.FREEMAN, *Chem. and Ind.*, 332 (1963).
1116. HART, H. and F.FREEMAN, *J. Org. Chem.*, **28**, 1220 (1963).
1117. HASEK, R.H., P.G.GOTT, R.H.MEEN and J.C.MARTIN, *J. Org. Chem.*, **28**, 2496 (1963).
1118. HASEK, R.H., R.H.MEEN and J.C.MARTIN, *J. Org. Chem.*, **30**, 1495 (1965).
1119. HASHIMOTO, M. and Y.TSUDA, *International Symposium on Nuclear Magnetic Resonance*, Tokyo, 1965.
1120. HASSNER, A.F. and M.J.HADDADIN, *J. Org. Chem.*, **28**, 224 (1963).
1121. HASSNER, A. and C.HEATHCOCK, *J. Org. Chem.*, **29**, 1350 (1964).
1122. HASSNER, A. and M.S.MICHELSON, *J. Org. Chem.*, **27**, 298 (1962).
1123. HASSNER, A. and M.S.MICHELSON, *J. Org. Chem.*, **27**, 3974 (1962).
1124. HATTON, J.V., *J. Chem. Phys.*, **40**, 933 (1964).
1125. HATTON, J.V. and R.E.RICHARDS, *Mol. Phys.*, **3**, 253 (1960).
1126. HATTON, J.V. and R.E.RICHARDS, *Trans. Farad. Soc.*, **56**, 315 (1960).
1127. HATTON, J.V. and R.E.RICHARDS, *Mol. Phys.*, **5**, 139 (1962).
1128. HATTON, J.V. and R.E.RICHARDS, *Mol. Phys.*, **5**, 153 (1962).
1129. HATTON, J.V. and W.G.SCHNEIDER, *Canad. J. Chem.*, **40**, 1285 (1962).
1130. HATTON, J.V., W.G.SCHNEIDER and W.SIEBRAND, *J. Chem. Phys.*, **39**, 1330 (1963).
1131. HAUGWITZ, R.D., P.W.JEFFS and E.WENKERT, *J. Chem. Soc.*, 2001 (1965).
1132. HAUTH, H., D.STAUFFACHER, P.NIKLAUS and A.MELERA, *Helv. Chim. Acta*, **48**, 1087 (1965).
1133. HAWDON, A.R. and I.J.LAWRENSON, *Chem. and Ind.*, 1690 (1963).
1134. HAWTHORNE, J.O., E.L.MIHELIC, M.S.MORGAN and M.H.WILT, *J. Org. Chem.*, **28**, 2831 (1963).
1135. HAY, R.W. and P.P.WILLIAMS, *J. Chem. Soc.*, 2270 (1964).
1136. HAYASHI, T., M.IGARASHI, S.HAYASHI and H.MIDORIKAWA, *Bull. Chem. Soc., Japan*, **38**, 2063 (1965).
1137. HAYES, S., *Bull. Soc. Chim. France*, 2715 (1964).
1138. HAYNES, L.J. and K.L.STUART, *J. Chem. Soc.*, 1789 (1963).
1139. HAYTER, R.G., *J. Amer. Chem. Soc.*, **85**, 3120 (1963).
1140. HEAP, N. and G.H.WHITHAM, *J. Chem. Soc.*, (B)164 (1966).
1141. HEATHCOCK, C., *Canad. J. Chem.*, **40**, 1865 (1962).
1142. HEACOCK, R.A., O.HUTZINGER, B.D.SCOTT, J.W.DALY and B.WITKOP, *J. Amer. Chem. Soc.*, **85**, 1825 (1963).

1143. HECHT, A.M., *J. Phys.* (Paris), **26,** 167A (1965).
1144. HECHT, H.G., *Theoretica Chim. Acta,* **3,** 202 (1965).
1145. HEEL, H. and W.ZEIL, *Z. Electrochem.,* **64,** 962 (1960).
1146. HEFFERNAN, M.L. and A.A.HUMFFRAY, *Mol. Phys.,* **7,** 527 (1963–4).
1147. HELLYER, R.O., *Aust. J. Chem.,* **17,** 1418 (1964).
1148. HELLYER, R.O., I.R.C.BICK, R.G.NICHOLLS and H.ROTTENDORF, *Aust. J. Chem.,* **16,** 703 (1963).
1149. HENDERSON, R., R.McCRINDLE, K.H.OVERTON and A.MELERA, *Tetrahedron Letters,* 3969 (1964).
1150. HENDRICKSON, J.B., *J. Amer. Chem. Soc.,* **83,** 2018 (1961).
1151. HENDRICKSON, J.B., M.L.MADDOX, J.J.SIMS and H.D.KAESZ, *Tetrahedron,* **20,** 449 (1964).
1152. HENRICK, C.A. and P.R.JEFFERIES, *Chem. and Ind.,* 1801 (1963).
1153. HENRICK, C.A. and P.R.JEFFERIES, *Aust. J. Chem.,* **18,** 2005 (1965).
1154. HERBISON-EVANS, D. and R.E.RICHARDS, *Trans. Farad. Soc.,* **58,** 845 (1962).
1155. HERBISON-EVANS, D. and R.E.RICHARDS, *Mol. Phys.,* **8,** 19 (1964).
1156. HERSHENSON, H.M., *NMR and ESR spectra Index for 1958–1963,* Academic Press, New York 1965.
1157. HERZ, W. and L.A.GLICK, *J. Org. Chem.,* **28,** 2970 (1963).
1158. HERZ, W., H.WATANABE, M.MIYAZAKI and Y.KISHIDA, *J. Amer. Chem. Soc.,* **84,** 2601 (1962).
1159. HESSE, M., W. VON PHILIPSBORN, D.SCHUMANN, G.SPITELLER, M.SPITELLER-FRIEDMANN, W.I.TAYLOR, H.SCHMID and P.KARRER, *Helv. Chim. Acta,* **47,** 878 (1964).
1160. HEWGILL, F.R., B.R.KENNEDY and D.KILPIN, *J. Chem. Soc.,* 2904 (1965).
1161. HEWGILL, F.R. and B.S.MIDDLETON, *J. Chem. Soc.,* 2914 (1965).
1162. HIGHAM, P. and R.E.RICHARDS, *Proc. Chem. Soc.,* 128 (1959).
1163. HIGHET, R.J. and T.J.BATTERHAM, *J. Org. Chem.,* **29,** 475 (1964).
1164. HIGHET, R.J. and P.F.HIGHET, *J. Org. Chem.,* **30,** 902 (1965).
1165. HILL, R.K. and T.H.CHAN, *Tetrahedron,* **21,** 2015 (1965).
1166. HINE, J., R.P.BAYER and G.G.HAMMER, *J. Amer. Chem. Soc.,* **84,** 1751 (1962).
1167. HINE, J. and J.G.HOUSTON, *J. Org. Chem.,* **30,** 1328 (1965).
1168. HINE, J., J.G.HOUSTON and J.H.JENSEN, *J. Org. Chem.,* 1184 (1965).
1169. HIMMAN, R.L. and S.THEODOROPULOS, *J. Org. Chem.,* **28,** 3052 (1963).
1170. HINMAN, R.L. and E.B.WHIPPLE, *J. Amer. Chem. Soc.,* **84,** 2534 (1962).
1171. HIROIKE, E., *J. Phys. Soc., Japan,* **15,** 270 (1960).
1172. HIRAGA, K., *Chem. Pharm. Bull.,* **13,** 1300 (1965).
1173. HIRST, R.C. and D.M.GRANT, *J. Amer. Chem. Soc.,* **84,** 2009 (1962).
1174. HIRST, R.C. and D.M.GRANT, *J. Chem. Phys.,* **40,** 1909 (1964).
1175. HJEDS, H., K.P.HANSEN and B.JERSLEV, *Acta Chem. Scand.,* **19,** 2166 (1965).
1176. HOARAU, J., *Ann. Chim. (France),* **13,** 544 (1956).
1177. HOARAU, J., N.LUMBROSO and A.PACAULT, *Compt. Rend.,* **242,** 1702 (1956).
1178. HOBGOOD, R.T. and J.H.GOLDSTEIN, *J. Mol. Spec.,* **12,** 76 (1964).
1179. HOBGOOD, R.T., R.E.MAYO and J.H.GOLDSTEIN, *J. Chem. Phys.,* **39,** 2501 (1963).
1180. HOBGOOD, R.T., G.S.REDDY and J.H.GOLDSTEIN, *J. Phys. Chem.,* **67,** 110 (1963).
1181. HOFMANN, A., W. VON PHILIPSBORN and C.H.EUGSTER, *Helv. Chim. Acta,* **48,** 1322 (1965).
1182. HOFFMANN, E.G., *Ber. Bunsenges. Phys. Chem.,* **64,** 144 (1960).
1183. HOFFMANN, E.G., *Trans. Farad. Soc.,* **58,** 642 (1962).
1184. HOFFMANN, E.G., *Bull. Soc. Chim. France,* 1467 (1963).
1185. HOFFMAN, R.A., *Mol. Phys.,* **1,** 326 (1958).
1186. HOFFMAN, R.A., *J. Chem. Phys.,* **33,** 1256 (1960).
1187. HOFFMAN, R.A., *Arkiv Kemi,* **17,** 1 (1961).
1188. HOFFMAN, R.A., B.GESTBLOM and S.FORSEN, *J. Mol. Spec.,* **13,** 221 (1964).
1189. HOFFMAN, R.A., B.GESTBLOM, S.GRONOWITZ and S.FORSEN, *J. Mol. Spec.,* **11,** 454 (1963).

1190. HOFFMAN, R.A., B.GESTBLOM and S.FORSEN, *J. Chem. Phys.*, **39**, 486 (1963).
1191. HOFFMAN, R.A., B.GESTBLOM and S.FORSEN, *J. Chem. Phys.*, **40**, 3734 (1964).
1192. HOFFMAN, R.A. and S.GRONOWITZ, *Acta Chem. Scand.*, **13**, 1477 (1959).
1193. HOFFMAN, R.A. and S.GRONOWITZ, *Arkiv Kemi*, **15**, 45 (1960).
1194. HOFFMAN, R.A. and S.GRONOWITZ, *Arkiv Kemi*, **16**, 471 (1961).
1195. HOFFMAN, R.A. and S.GRONOWITZ, *Arkiv Kemi*, **16**, 501 (1961).
1196. HOFFMAN, R.A. and S.GRONOWITZ, *Arkiv Kemi*, **16**, 515 (1961).
1197. HOFFMAN, R.A. and S.GRONOWITZ, *Arkiv Kemi*, **16**, 563 (1961).
1198. HOFFMAN, R.A. and S.GRONOWITZ, *J. Amer. Chem. Soc.*, **83**, 3910 (1961)
1199. HOFFMAN, R.W. and H.HÄUSER, *Tetrahedron Letters*, 197 (1964).
1200. HOFMAN, W., L.STEFANIAK, T.URBANSKI and M.WITANOWSKI, *J. Amer. Chem. Soc.*, **86**, 554 (1964).
1201. HOGEVEEN, H., G.MACCAGNANI and F.TADDEI, *Rec. Trav. Chim.*, **83**, 937 (1964).
1202. HOLMES, E.T. and H.R.SNYDER, *J. Org. Chem.*, **29**, 2155 (1964) and following papers.
1203. HOLMES, J.R. and H.D.KAESZ, *J. Amer. Chem. Soc.*, **83**, 3903 (1961).
1204. HOLMES, J.R. and D.KIVELSON, *J. Amer. Chem. Soc.*, **83**, 2959 (1961).
1205. HOLMES, E.T. and H.R.SNYDER, *J. Org. Chem.*, **29**, 2725 (1964).
1206. HOLMES, R.R. and J.A.FORSTNER, *J. Amer. Chem. Soc.*, **82**, 5509 (1960).
1207. HOOGZAND, C. and W.HÜBEL, *Tetrahedron Letters*, 637 (1961).
1208. HOOPER, D.L. and R.KAISER, *Canad. J. Chem.*, **43**, 2363 (1965).
1209. HOOVER, F.W. and D.D.COFFMAN, *J. Org. Chem.*, **29**, 3567 (1964).
1210. HOOVER, F.W. and H.S.ROTHROCK, *J. Org. Chem.*, **28**, 2082 (1963).
1211. HÖRNFELDT, A.B., *Arkiv Kemi*, **22**, 211 (1964).
1212. HÖRNFELDT, A.B. and S.GRONOWITZ, *Acta Chem. Scand.*, **16**, 789 (1962).
1213. HÖRNFELDT, A.B. and S.GRONOWITZ, *Arkiv Kemi*, **21**, 239 (1965).
1214. HOUSE, H.O., R.G.CARLSON, H.MÜLLER, A.W.NOLTES and C.D.SLATER, *J. Amer. Chem. Soc.*, **84**, 2614 (1962).
1215. HOUSE, H.O. and W.F.GILMORE, *J. Amer. Chem. Soc.*, **83**, 3972 (1961).
1216. HOUSE, H.O. and H.BABAD, *J. Org. Chem.*, **28**, 90 (1963).
1217. HOUSE, H.O. and W.F.BERKOWITZ, *J. Org. Chem.*, **28**, 307 (1963).
1218. HOUSE, H.O. and V.KRAMAR, *J. Org. Chem.*, **28**, 3362 (1963).
1219. HOUSE, H.O., R.W.MAGIN and H.W.THOMPSON, *J. Org. Chem.*, **85**, 2403 (1963).
1220. HOUSE, H.O. and G.H.RASMUSSON, *J. Org. Chem.*, **28**, 27 (1963).
1221. HOUSE, H.O. and H.W.THOMPSON, *J. Org. Chem.*, **28**, 164 (1963).
1222. HOUSE, H.O., D.D.TRAFICANTE and R.A.EVANS, *J. Org. Chem.*, **28**, 348 (1963).
1223. HOUSE, H.O., P.P.WICKHAM and H.C.MULLER, *J. Amer. Chem. Soc.*, **84**, 3139 (1962).
1224. HOWARD, B.B., B.LINDER and M.T.EMERSON, *J. Chem. Phys.*, **36**, 485 (1962).
1225. HRUSKA, F., E.BOCK and T.SCHAEFER, *Canad. J. Chem.*, **41**, 3034 (1963).
1226. HRUSKA, F., H.M.HUTTON and T.SCHAEFER, *Canad. J. Chem.*, **43**, 1942 (1965).
1227. HRUSKA, F., H.M.HUTTON and T.SCHAEFER, *Canad. J. Chem.*, **43**, 2392 (1965).
1228. HRUSKA, F., G.KOTOWYCZ and T.SCHAEFER, *Canad. J. Chem.*, **43**, 2827 (1965).
1229. HRUSKA, F., T.SCHAEFER and C.A.REILLY, *Canad. J. Chem.*, **42**, 697 (1964).
1230. HUISGEN, R., L.FEILER and G.BINSCH, *Angew. Chem.*, **76**, 892 (1964).
1231. HUISGEN, R. and G.SZEIMIES, *Chem. Ber.*, **98**, 1153 (1965).
1232. HUITRIC, A.C. and J.B.CARR, *J. Org. Chem.*, **26**, 2648 (1961).
1233. HUITRIC, A.C., J.B.CARR, W.F.TRAGER and B.J.NIST, *Tetrahedron*, **19**, 2145 (1963).
1234. HUITRIC, A.C., W.S.STAVROPOULOS and B.J.NIST, *J. Org. Chem.*, **28**, 1539 (1963).
1235. HUITRIC, A.C. and W.F.TRAGER, *J. Org. Chem.*, **27**, 1926 (1962).
1236. HUNECK, S. and J.M.LEHN, *Bull. Soc. Chim. France*, 1702 (1963).
1237. HÜTTEL, R., H.CHRIST and K.HERZOG, *Chem. Ber.*, **97**, 2710 (1964).
1238. HUTTON, H.M., W.F.REYNOLDS and T.SCHAEFER, *Canad. J. Chem.*, **40**, 1758(1962).
1239. HUTTON, H.M. and T.SCHAEFER, *Canad. J. Chem.*, **41**, 684 (1963).
1240. HUTTON, H.M. and T.SCHAEFER, *Canad. J. Chem.*, **41**, 1623 (1963).
1241. HUTTON, H.M. and T.SCHAEFER, *Canad. J. Chem.*, **41**, 1625 (1963).

1242. HUTTON, H.M. and T.SCHAEFER, *Canad. J. Chem.*, **41**, 2429 (1963).
1243. HUTTON, H.M. and T.SCHAEFER, *Canad. J. Chem.*, **41**, 2774 (1963).
1244. HUTTON, H.M. and T.SCHAEFER, *J. Phys. Chem.*, **68**, 1602 (1964).
1245. HUTTON, H.M. and T.SCHAEFER, *Canad. J. Chem.*, **43**, 3116 (1965).
1246. HYNE, J.B. and R.WILLIS, *J. Amer. Chem. Soc.*, **85**, 3650 (1963).
1247. INAMOTO, N., S.MASUDA, Y.NAGAI and O.SIMAMURA, *J. Chem. Soc.*, 1433 (1963).
1248. ITO, K., *J. Amer. Chem. Soc.*, **80**, 3502 (1958).
1249. ITO, S. and I.MIURA, *Bull. Chem. Soc. Japan*, **38**, 2197 (1965).
1250. ITOH, T., K.OHNO and H.YOSHIZUMI, *J. Phys. Soc. Japan*, **10**, 103 (1955).
1251. ITOH, J. and S.SATO, *J. Phys. Soc. Japan*, **14**, 851 (1959).
1252. JACKMAN, L.M., Unpublished results.
1252a. JACKMAN, L.M., *Fortschr. Chem. Organ. Natur*, **23**, 315 (1965).
1253. JACKMAN, L.M., *J. Chem. Soc.*, 4585 (1961).
1254. JACKMAN, L.M. and J.W.LOWN, *J. Chem. Soc.*, 3776 (1962).
1255. JACKMAN, L.M., Q.N.PORTER and G.R.UNDERWOOD, *Aust. J. Chem.*, **18**, 1221 (1965).
1256. JACKMAN, L.M., Q.N.PORTER and G.R.UNDERWOOD, *International Symposium on Nuclear Magnetic Resonance*, Tokyo, 1965.
1257. JACKMAN, L.M., L.RADOM and S.STERNHELL, Unpublished data.
1258. JACKMAN, L.M., F.SONDHEIMER, Y.AMIEL, D.A.BEN EFRIAM, Y.GAONI, R.WO-LOVSKY and A.A.BOTHNER-BY, *J. Amer. Chem. Soc.*, **84**, 4307 (1962).
1259. JACKMAN, L.M. and R.H.WILEY, *J. Chem. Soc.*, 2881 (1960).
1260. JACKMAN, L.M. and R.H.WILEY, *J. Chem. Soc.*, 2886 (1960).
1261. JACOBS, H.A.M., M.H.BERG, L.BRANDSMA and J.F.ARENS, *Rec. Trav. Chim.*, **84**, 1113 (1965).
1262. JACOBSEN, H.J. and S.LAWESSON, *Tetrahedron*, **21**, 3331 (1966).
1263. JACQUESY, J.C., R.JACQUESY, J.LEVISALLES, J.P.PETE and H.RUDLER, *Bull. Soc. Chim. France*, 2224 (1964).
1264. JAQUESY, J.C. and J.LEVISALLES, *Bull. Soc. Chim. France*, 1866 (1962).
1265. JANOT, M.M., P.LONGEVIALLE and R.GOUTAREL, *Bull. Soc. Chim. France*, 2158 (1964).
1266. JARDETZKY, C.D., *J. Amer. Chem. Soc.*, **83**, 2919 (1961).
1267. JARDETZKY, C.D. and O.JARDETZKY, *J. Amer. Chem. Soc.*, **82**, 222 (1960).
1268. JARDETZKY, O., *J. Amer. Chem. Soc.*, **85**, 1823 (1963).
1269. JARDETZKY, O., P.PAPPAS and N.G.WADE, *J. Amer. Chem. Soc.*, **85**, 1657 (1963).
1270. JARDETZKY, O., N.G.WADE and J.J.FISCHER, *Nature*, **197**, 183 (1963).
1271. JARDINE, R.V. and R.K.BROWN, *Canad. J. Chem.*, **41**, 2067 (1963).
1272. JEFFERIES, P.R., R.S.ROSICH and D.E.WHITE, *Tetrahedron Letters*, 1853 (1963).
1273. JEFFORD, C.W., J.GUNSHER and K.RAMEY, *J. Amer. Chem. Soc.*, **87**, 4384 (1965).
1274. JEFFORD, C.W., S.MAHAJAN, J.WASLYN and B.WAEGELL, *J. Amer. Chem. Soc.*, **87**, 2183 (1965).
1275. JEFFORD, C.W., B.WAEGELL and K.RAMEY, *J. Amer. Chem. Soc.*, **87**, 2191 (1965).
1276. JENKINS, J.M. and B.L.SHAW, *Proc. Chem. Soc.*, 279 (1963).
1277. JENNY, E.F. and J.DRUEY, *J. Amer. Chem. Soc.*, **82**, 3111 (1960).
1278. JENSEN, F.R., D.S.NOYCE, C.H.SEDERHOLM and A.J.BERLIN, *J. Amer. Chem. Soc.*, **82**, 1256 (1960).
1279. JENSEN, F.R., D.S.NOYCE, C.H.SEDERHOLM and A.J.BERLIN, *J. Amer. Chem. Soc.*, **84**, 386 (1962).
1280. JENSEN, F.R. and L.A.SMITH, *J. Amer. Chem. Soc.*, **86**, 956 (1964).
1281. JHA, S., *Proc. Indian Acad. Sci.*, **54A**, 13 (1961).
1282. HOCHSTRASSER, G., G.BENE and R.EXTERMANN, *Compt. Rend.*, **248**, 218 (1959).
1283. JOHNS, S.R. and J.A.LAMBERTON, *Chem. Comm.*, 458 (1965).
1284. JOHNSON, C.E. and F.A.BOVEY, *J. Chem. Phys.*, **29**, 1012 (1958).
1285. JOHNSON, C.S., *J. Chem. Phys.*, **41**, 3277 (1964).
1286. JOHNSON, C.S., in *Advances in Magnetic Resonance*, Vol. I (Ed. J.S.WAUGH) Academic Press, N.Y., 1965, p.33.
1287. JOHNSON, C.S., M.A.WEINER, J.S.WAUGH and D.SEYFERTH, *J. Amer. Chem. Soc.*, **83**, 1306 (1961).

1288. JOHNSON, F. and J.P.HEESCHEN, *J. Org. Chem.*, **29**, 3252 (1964).

1289. JOHNSON, F., W.D.GUROWITZ and N.A.STARKOVSKY, *Tetrahedron Letters*, 1173 (1962).

1290. JOHNSON, F., N.A.STARKOVSKY and W.D.GUROWITZ, *J. Amer. Chem. Soc.*, **87**, 3492 (1965).

1291. JOHNSON, F.P., A.MELERA and S.STERNHELL, *Aust. J. Chem.*, **19**, 1523 (1966).

1292. JOHNSON, L.F. and N.S.BHACCA, *J. Amer. Chem. Soc.*, **85**, 3700 (1963).

1293. JOHNSON, L.F., A.V.ROBERTSON, W.R.J.SIMPSON and B.WITKOP, *Aust. J. Chem.*, **19**, 115 (1966).

1294. JOHNSON, L.F. and J.N.SHOOLERY, *Anal. Chem.*, **34**, 1136 (1962).

1295. JOHNSTON, T.P., C.R.STRINGFELLOW and A.GALLAGHER, *J. Org. Chem.*, **27**, 4068 (1962).

1296. JONAS, J., A.ALLERHAND and H.S.GUTOWSKY, *J. Chem. Phys.*, **42**, 3396 (1965).

1297. JONAS, J., W.DERBYSHIRE and H.S.GUTOWSKY, *J. Phys. Chem.*, **69**, 1 (1965).

1298. JONAS, J. and H.S.GUTOWSKY, *J. Chem. Phys.*, **42**, 140 (1965).

1299. JONATHAN, N., S.GORDON and B.P.DAILEY, *J. Chem. Phys.*, **36**, 2443 (1962).

1300. JONES, E.R.H. and D.A.WILSON, *J. Chem. Soc.*, 2933 (1965).

1301. JONES, R., A.J.RYAN, S.STERNHELL and S.E.WRIGHT, *Tetrahedron*, **19**, 1497 (1963).

1302. JONES, R.A.Y. and A.R.KATRITZKY, *Angew. Chem. Int. Edit.*, **1**, 32 (1962).

1303. JONES, R.A.Y., A.R.KATRITZKY and J.M.LAGOWSKI, *Chem. and Ind.*, 870 (1960).

1304. JONES, R.A.Y., A.R.KATRITZKY, J.N.MURRELL and N.SHEPPARD, *J. Chem. Soc.*, 2576 (1962).

1305. JONES, R.G., R.C.HIRST and H.J.BERNSTEIN, *Canad. J. Chem.*, **43**, 683 (1965).

1306. JOOP, N. and H.ZIMMERMANN, *Ber. Bunsenges. Phys. Chem.*, **66**, 440 (1962).

1307. JOOP, N. and H.ZIMMERMANN, *Ber. Bunsenges. Phys. Chem.*, **66**, 541 (1963).

1308. JOSHIMURA, J. and F.W.LICHTENTHALER, *Angew. Chem.*, **77**, 740 (1965).

1309. JOUVE, M.P., *Compt. Rend.*, **256**, 1987 (1963).

1310. JOUVE, M.P., Private communication 1966.

1311. JOUVE, M.P. and M.P.SIMONNIN, *Compt. Rend.*, **257**, 121 (1963).

1312. JUAN, C. and H.S.GUTOWSKY, *J. Chem. Phys.*, **37**, 2198 (1962).

1313. JUNG, A. and M.BRINI, *Bull. Soc. Chim. France*, 693 (1964).

1314. JUNG, A and M.BRINI, *Bull. Soc. Chim. France*, 587 (1965).

1315. JUNG, D. and A.A.BOTHNER-BY, *J. Amer. Chem. Soc.*, **86**, 4025 (1964).

1316. JUNGNICKEL, J.L. and J.W.FORBES, *Anal. Chem.*, **35**, 938 (1963).

1317. JUNGNICKEL, J.L. and C.A.REILLY, *J. Mol. Spec.*, **16**, 135 (1965).

1318. JUTZ, C. and H.AMSCHLER, *Chem. Ber.*, **96**, 2100 (1963).

1319. KABUS, S.S., H.FRIEBOLIN and H.SCHMID, *Tetrahedron Letters*, 469 (1965).

1320. KAGAN, H.B., J.J.BASSELIER and J.L.LUCHE, *Tetrahedron Letters*, 941 (1964).

1321. KAISER, R., *Rev. Sci. Instr.*, **31**, 963 (1960).

1322. KAISER, R., *Rev. Sci. Instr.*, **31**, 963 (1960).

1323. KAISER, R., *J. Chem. Phys.*, **39**, 2435 (1963).

1324. KAISER, R. and D.L.HOPPER, *Mol. Phys.*, **8**, 403 (1964).

1325. KAMPMEIER, J.A. and G.CHEN, *J. Amer. Chem. Soc.*, **87**, 2608 (1965).

1326. KAMEI, H., *Bull. Chem. Soc. Japan*, **38**, 1212 (1965).

1327. KAN, R.O., *J. Amer. Chem. Soc.*, **86**, 5180 (1964).

1328. KAPADI, A.H. and SUKH DEV, *Tetrahedron Letters*, 1171 (1964).

1329. KAPLAN, F. and G.K.MELLOY, *Tetrahedron Letters*, 2427 (1964).

1330. KAPLAN, F. and J.D.ROBERTS, *J. Amer. Chem. Soc.*, **83**, 4666 (1961).

1331. KAPLAN, F., G.SINGH and H.ZIMMER, *J. Phys. Chem.*, **67**, 2509 (1963).

1332. KARABATSOS, G.J., *J. Amer. Chem. Soc.*, **83**, 1230 (1961).

1333. KARABATSOS, G.J. and J.D.GRAHAM, *J. Amer. Chem. Soc.*, **82**, 5250 (1960).

1334. KARABATSOS, G.J., J.D.GRAHAM and F.M.VANE, *J. Amer. Chem. Soc.*, **83**, 2778 (1961); *J. Phys. Chem.*, **65**, 1657 (1961).

1335. KARABATSOS, G.J., J.D.GRAHAM and F.M.VANE, *J. Amer. Chem. Soc.*, **84**, 37 (1962).

1336. KARABATSOS, G.J., J.D.GRAHAM and F.M.VANE, *J. Amer. Chem. Soc.*, **84**, 753 (1962).

1337. KARABATSOS, G.J. and H.HSI, *J. Amer. Chem. Soc.*, **87**, 2864 (1965).

1338. KARABATSOS, G.J. and C.E.ORZECH, *J. Amer. Chem. Soc.*, **86**, 3574 (1964).

1339. KARABATSOS, G.J. and C.E.ORZECH, *J. Amer. Chem. Soc.*, **87**, 560 (1965).

1340. KARABATSOS, G.J. and R.A.TALLER, *J. Amer. Chem. Soc.*, **85**, 3624 (1963).

1341. KARABATSOS, G.J., R.A.TALLER and F.M.VANE, *J. Amer. Chem. Soc.*, **85**, 2326 (1963).

1342. KARABATSOS, G.J., R.A.TALLER and F.M.VANE, *J. Amer. Chem. Soc.*, **85**, 2327 (1963).

1343. KARABATSOS, G.J. and R.A.TALLER, *J. Amer. Chem. Soc.*, **86**, 4373 (1964).

1344. KARABATSOS, G.J., R.A.TALLER and F.M.VANE, *Tetrahedron Letters*, 1081 (1964).

1345. KARABATSOS, G.J. and F.M.VANE, *J. Amer. Chem. Soc.*, **85**, 3886 (1963).

1346. KARABATSOS, G.J., F.M.VANE and S.MEYERSON, *J. Amer. Chem. Soc.*, **85**, 733 (1963).

1347. KARABATSOS, G.J., F.M.VANE, R.A.TALLER and N.HSI, *J. Amer. Chem. Soc.*, **86**, 3351 (1964).

1348. KARPLUS, M., *J. Chem. Phys.*, **30**, 11 (1959).

1349. KARPLUS, M., *J. Chem. Phys.*, **33**, 316 (1960).

1350. KARPLUS, M., *J. Chem. Phys.*, **33**, 941 (1960).

1351. KARPLUS, M., *J. Chem. Phys.*, **33**, 1842 (1960).

1352. KARPLUS, M., *J. Amer. Chem. Soc.*, **82**, 4431 (1960).

1353. KARPLUS, M., *J. Amer. Chem. Soc.*, **84**, 2458 (1962) and references therein.

1354. KARPLUS, M., *J. Amer. Chem. Soc.*, **85**, 2870 (1963).

1355. KARPLUS, M. and H.J.KOLKER, *J. Chem. Phys.*, **38**, 1263 (1963).

1356. KARPLUS, M. and J.A.POPLE, *J. Chem. Phys.*, **38**, 2803 (1963).

1357. KASIWAGI, H., N.NAKAWAGA and J.NIWA, *Bull. Soc. Chem. Japan*, **36**, 410 (1963).

1358. KASIWAGI, H. and J.NIWA, *Bull. Chem. Soc. Japan*, **36**, 405 (1963).

1359. KATAYAMA, M., S.FUJIWARA, H.SUZUKI, Y.NAGAI and O.SIMANURA, *J. Mol. Spec.*, **5**, 85 (1960).

1360. KATAYMAM, M., S.FUJIWARA, H.SUZUKI, Y.NAGAI and O.SIMANURA, *J. Mol. Spec.*, **5**, 85 (1960).

1361. KATRITZKY, A.R. and R.A.Y.JONES, *Proc. Chem. Soc.*, 313 (1960).

1362. KATRITZKY, A.R. and R.A.Y.JONES, *Chem. and Ind.*, 722 (1961).

1363. KATRITZKY, A.R. and F.W.MAINE, *Tetrahedron*, **20**, 299 (1964).

1364. KATRITZKY, A.R., S.ØKSNE and A.J.BOULTON, *Tetrahedron*, **18**, 777 (1962).

1365. KATRITZKY, A.R., S.ØKSNE and A.J.BOULTON, *Tetrahedron*, **18**, 779 (1962).

1366. KATRITZKY, A.R. and R.E.REAVILL, *J. Chem. Soc.*, 753 (1963).

1367. KATRITZKY, A.R. and R.E.REAVILL, *Rec. Trav. Chim.*, **83**, 1230 (1964).

1368. KATRITZKY, A.R. and R.E.REAVILL, *J. Chem. Soc.*, 3825 (1965).

1369. KATRITZKY, A.R., R.E.REAVILL and F.J.SWINBOURNE, *J. Chem. Soc.*, (B)351 (1966).

1370. KATRITZKY, A.R. and F.J.SWINBOURNE, *J. Chem. Soc.*, 6707 (1965).

1371. KATRITZKY, A.R., B.TERNAI and G.J.T.TIDDY, *Tetrahedron Letters*, 1713 (1966).

1372. KATRITZKY, A.R. and B.WALLIS, *Chem. and Ind.*, 2025 (1964).

1373. KATRITZKY, A.R. and A.J.WARING, *Chem. and Ind.*, 695 (1962).

1374. KATRITZKY, A.R. and A.J.WARING, *J. Chem. Soc.*, 3046 (1963).

1375. KATRITZKY, A.R. and A.J.WARING, *J. Chem. Soc.*, 1523 (1964).

1376. KATZ, T.J., *J. Amer. Chem. Soc.*, **82**, 3785 (1960).

1377. KATZ, T.J. and P.J.GARRATT, *J. Amer. Chem. Soc.*, **85**, 2852 (1963).

1378. KATZ, T.J. and P.J.GARRATT, *J. Amer. Chem. Soc.*, **86**, 5194 (1964).

1379. KATZ, T.J. and E.H.GOLD, *J. Amer. Chem. Soc.*, **86**, 1600 (1964).

1380. KATZ, T.J., M.YOSHIDA and L.C.SIEW, *J. Amer. Chem. Soc.*, **87**, 4516 (1965).

1381. KATZ, T.J., M.ROSENBERGER and R.K.O'HARA, *J. Amer. Chem. Soc.*, **86**, 249 (1964).

1382. KATZ, J.J., M.R.THOMAS and H.H.STRAIN, *J. Amer. Chem. Soc.*, **84**, 3587 (1962).

1383. KAWAZOE, Y., Y.SATO, M.NATSUME, H.HASEGAWA, T.OKAMOTO and K.TSUDA, *Chem. Pharm. Bull.* (Tokyo), **10**, 338 (1962).

1384. KAWASAKI, Y., K.KAWAKAMI and T.TANAKA, *Bull. Chem. Soc. Japan*, **38**, 1102 (1965).

1385. KAWASAKI, Y., R.UEEDA and T.TANAKA, *International Symposium on Nuclear Magnetic Resonance, Tokyo*, Sept. 1965.
1386. KAZAN, J. and F.D.GREENE, *J. Org. Chem.*, **28**, 2965 (1963).
1387. ITO, K., T.ISOBE and K.SONE, *J. Chem. Phys.*, **31**, 861 (1959).
1388. KIEFER, E.F. and W.L.WATERS, *J. Amer. Chem. Soc.*, **87**, 4402 (1965).
1389. KELLER, C.E. and R.PETTIT, *J. Amer. Chem. Soc.*, **88**, 604 (1966).
1390. KENDALL, C.E. and Z.G.HAJOS, *J. Amer. Chem. Soc.*, **82**, 3219 (1960).
1391. KENDE, A.S. and P.T.IZZO, *J. Amer. Chem. Soc.*, **86**, 3587 (1964).
1392. KERN, C.W. and W.N.LIPSCOMB, *Phys. Rev. Letters*, **7**, 19 (1961).
1393. KERN, C.W. and W.N.LIPSCOMB, *J. Chem. Phys.*, **37**, 260 (1962).
1394. KEUNING, R., *Rev. Sci. Instr.*, **31**, 839 (1960).
1395. KEVILL, D.N. and N.H.CROMWELL, *J. Phys. Chem.*, **67**, 2876 (1963).
1396. KICE, J.L. and T.S.CANTRELL, *J. Amer. Chem. Soc.*, **85**, 2298 (1963).
1397. KICE, J.L. and E.H.MORKVED, *J. Amer. Chem. Soc.*, **85**, 3472 (1963).
1398. KIMURA, M., S.MATSUOKA, S.HATTORI and K.SENDA, *J. Phys. Soc. Japan*, **14**, 684 (1959).
1399. KING, R.B., *J. Amer. Chem. Soc.*, **85**, 1922 (1963).
1400. KING, R.W., Private communication (1965).
1401. KING, T.J. and J.P.YARDLAY, *J. Chem. Soc.*, 4308 (1961).
1402. KINGSBURY, C.A. and W.B.THORNTON, *J. Org. Chem.*, **31**, 1000 (1966).
1403. KITAHARA, Y., M.C.CASERIO, F.SCARDIGLIA and J.D.ROBERTS, *J. Amer. Chem. Soc.*, **82**, 3106 (1960).
1404. KITAHARA, Y., I.MURATA, M.FUNAMIZU and T.ASANO, *Bull. Chem. Soc. Japan*, **37**, 1399 (1964).
1405. KITAHARA, Y., I.MURATA, K.SHIRAHATA, S.KITAGIRI and M.AZUMI, *Bull. Chem. Soc. Japan*, **38**, 780 (1965).
1406. KITAHARA, Y. and A.YOSHIKOSHI, *Bull. Soc. Chem. Japan*, **37**, 890 (1964).
1407. KLEIN, M.P. and G.W.BARTON, *Rev. Sci. Instr.*, **34**, 754 (1963).
1408. KLINCK, R.E. and J.B.STOTHERS, *Canad. J. Chem.*, **40**, 1071 (1962).
1409. KLINCK, R.E. and J.B.STOTHERS, *Canad. J. Chem.*, **40**, 2329 (1962).
1410. KLINCK, R.E. and J.B.STOTHERS, *Canad. J. Chem.*, **44**, 45 (1966).
1411. KLOSE, G., *Ann. Physik.*, **9**, 262 (1962).
1412. KLOSE, G., *Ann. Physik.*, **10**, 391 (1963).
1413. KLOSE, G., *Mol. Phys.*, **6**, 585 (1963).
1414. KLUGER, R., *J. Org. Chem.*, **29**, 2045 (1964).
1415. KNOX, G.R., P.L.PAUSON and G.V.D.TIERS, *Chem. and Ind.*, 1046 (1950).
1416. KNOX, C.H., E.V.VELARDE, S.M.BERGER and D.H.CUADNIELLO, *Chem. and Ind.*, 860 (1962).
1417. KNOX, L.H., E.VELARDE, S.BERGER, D.CUADRIELLO and A.D.CROSS, *Tetrahedron Letters*, 1249 (1962).
1418. KNOX, L.H., E.VELARDE, S.BERGER, D.CUADRIELLO and A.D.CROSS, *J. Org. Chem.*, **29**, 2187 (1964).
1419. KNOX, L.H., E.VELARDE, S.BERGER, D.CUADRIELLO, P.W.LANDIS and A.D.CROSS, *J. Amer. Chem. Soc.*, **85**, 1851 (1963).
1420. KNOX, L.H., E.VELARDE and A.D.CROSS, *J. Amer. Chem. Soc.*, **85**, 2533 (1963).
1421. KOCH, W. and H.ZOLLINGER, *Helv. Chim. Acta*, **48**, 1791 (1965).
1422. KOIDE, S. and E.DUVAL, *J. Chem. Phys.*, **41**, 315 (1964).
1423. KOKKO, J.P. and J.H.GOLDSTEIN, *Spectrochim. Acta*, **19**, 1119 (1963).
1424. KOKKO, J.P., L.MANDELL and J.H.GOLDSTEIN, *J. Amer. Chem. Soc.*, **84**, 1042 (1962).
1425. KOLKER, H.J. and M.KARPLUS, *J. Chem. Phys.*, **41**, 1259 (1964).
1426. KOTOWYCZ, G., T.SCHAEFER and E.BOCK, *Canad. J. Chem.*, **42**, 2541 (1964).
1427. KONDO,M., *Bull. Soc. Chem. Japan*, **38**, 1271 (1965).
1428. KONDO, Y., T.TAKEMOTO and K.YASUDA, *Chem. Pharm. Bull.*, **12**, 976 (1964).
1429. KORNBLUM, N., R.A.BROWN, *J. Amer. Chem. Soc.*, **85**, 1359 (1963).
1430. KORNBLUM, N. and R.A.BROWN, *J. Amer. Chem. Soc.*, **86**, 2681 (1964).
1431. KORSCH, B.H. and N.V.RIGGS, *Tetrahedron Letters*, 523 (1964).

1432. KORVER, P.K., P.J. VAN DER HAAK, N.STEINBERG and T.J. DE BOER, *Rec. Trav. Chim.*, **84**, 129 (1965).
1433. KOSOWER, E.M. and T.J.SORENSEN, *J. Org. Chem.*, **28**, 687 (1963).
1434. KOSOWER, E.M. and T.J.SORENSEN, *J. Org. Chem.*, **28**, 687 (1963).
1435. KOSSANYI, J., *Bull. Soc. Chim. France*, 704 (1965).
1436. KOSTER, D.F. and A.DANTI, *J. Phys. Chem.*, **69**, 486 (1965).
1437. KOTOWYCZ, G., T.SCHAEFER and E.BOCK, *Canad. J. Chem.*, **42**, 2541 (1964).
1438. KOWALEWSKI, V.J. and D.G.KOWALEWSKI, *J. Chem. Phys.*, **32**, 1272 (1960).
1439. KOWALEWSKI, V.J. and D.G.KOWALEWSKI, *J. Chem. Phys.*, **33**, 1793 (1960).
1440. KOWALEWSKI, D.G. and V.J.KOWALEWSKI, *J. Phys. Radium*, **22**, 129 (1961).
1441. KOWALEWSKI, D.G. and V.J.KOWALEWSKI, *Arkiv Kemi*, **16**, 373 (1961).
1442. KOWALEWSKI, V.J. and D.G.KOWALEWSKI, *J. Chem. Phys.*, **36**, 266 (1962).
1443. KOWALEWSKI, D.G. and V.J.KOWALEWSKI, *J. Chem. Phys.*, **37**, 1009 (1962).
1444. KOWALEWSKI, V.J. and D.G.KOWALEWSKI, *J. Chem. Phys.*, **37**, 2603 (1962).
1445. KOWALEWSKI, D.G. and V.J.KOWALEWSKI, *Mol. Phys.*, **8**, 93 (1964).
1446. KOWALEWSKI, D.G. and V.J.KOWALEWSKI, *Mol. Phys.*, **9**, 319 (1965).
1447. KRAKOWER, E. and L.W.REEVES, *Spectrochim. Acta*, **20**, 71 (1964).
1448. KRAKOWER, E. and L.W.REEVES, *Spectrochim. Acta*, **20**, 71 (1964).
1449. KRAMER, D.N., R.M.GAMSON and F.M.MILLER, *J. Org. Chem.*, **27**, 1848 (1962).
1450. KRAMER, F.A. and R.WEST, *J. Phys. Chem.*, **69**, 673 (1965).
1451. KRAMER, H.E.A. and R.GOMPPER, *Tetrahedron Letters*, 969 (1963).
1452. KRAMER, H.E.A. and R.GROMPPER, *Z. Phys. Chemie*, **43**, 292 (1964).
1453. KRAUCH, C.H., S.FARID and G.O.SCHENK, *Chem. Ber.*, **98**, 3102 (1965).
1454. KRAUS, W., *Chem. Ber.*, **97**, 2719 (1964).
1455. KREEVOY, M.M., H.B.CHARMAN and D.R.VINARD, *J. Amer. Chem. Soc.*, **83**, 1978 (1961).
1456. KREILICK, R.W. and S.I.WEISSMAN, *J. Amer. Chem. Soc.*, **84**, 306 (1962).
1457. KRESGE, A.J., *J. Chem. Phys.*, **39**, 1360 (1963).
1458. KRESGE, A.J., G.W.BARRY, K.R.CHARLES and Y.CHIANG, *J. Amer. Chem. Soc.*, **84**, 4343 (1962).
1459. KRESPAN, C.G., *J. Org. Chem.*, **27**, 3588 (1962).
1460. KROMHOUT, R.A., *Bull. Amer. Phys. Soc.*, **4**, 392 (1959).
1461. KROPP, P.J., *J. Amer. Chem. Soc.*, **86**, 4053 (1964).
1462. KREUGER, G.L., F.KAPLAN, M.ORCHIN and W.H.FAUL, *Tetrahedron Letters*, 3979 (1965).
1463. KUHN, S.J. and J.S.MCINTYRE, *Canad. J. Chem.*, **44**, 105 (1966).
1464. KUIVILA, H.G., W.RAHMAN and R.M.FISH, *J. Amer. Chem. Soc.*, **87**, 2853 (1965).
1465. KUIVILA, H.G. and C.R.WARNER, *J. Org. Chem.*, **29**, 2845 (1964).
1466. KULA, R.J., D.T.SAWYER, S.I.CHAN and C.M.FINLEY, *J. Amer. Chem. Soc.*, **85**, 2930 (1963).
1467. KULLNIG, R.K. and F.C.NACHOD, *J. Phys. Chem.*, **67**, 1361 (1863).
1468. KUMLER, N.D., E.KUN and J.N.SHOOLERY, *J. Org. Chem.*, **27**, 1165 (1962).
1469. KUN, K.A. and H.G.CASSIDY, *J. Org. Chem.*, **26**, 3223 (1961).
1470. KUNTZ, JR., I.D., P.R.SCHLEYER and A.ALLERHAND, *J. Chem. Phys.*, **35**, 1533 (1961).
1471. KUNTZLER, J.E., *Endeavour*, **23**, 114 (1964).
1472. KUPCHAN, S.M., W.S.JOHNSON and S.RAJAGOPLAN, *Tetrahedron*, **7**, 47 (1959).
1473. KURATH, P., W.COLE, J.TADANIER, M.FREIFELDER and E.V.SCHUBER, *J. Org. Chem.*, **28**, 2189 (1963).
1474. KURIYAMA, K., E.KONDO and K.TORI, *Tetrahedron Letters*, 1485 (1963).
1475. KURONO, M., Y.MAKI, K.NAKANISHI, M.OHASHI, K.UEDA, S.UYEO, M.C.WOODS and Y.YAMAMOTO, *Tetrahedron Letters*, 1917 (1965).
1476. KURTZ, A.N., W.E.BILLUPS, R.B.GREENLEE, H.F.HAMIL and W.T.PACE, *J.Org. Chem.*, **30**, 3141 (1965).
1477. KURYLA, W.C. and D.G.LEIS, *J. Org. Chem.*, **29**, 2773 (1964).
1478. KUTNEY, J.P., *Steroids*, **2**, 225 (1963).
1479. LABLACHE-COMBIER, A., J.LEVISALLES, J.P.PETE and H.RUDLER, *Bull. Soc. Chim. France*, 1689 (1963).

1480. LACHER, J.R., J.W.POLLOCK and J.D.PARK, *J. Chem. Phys.*, **20**, 1047 (1952).
1481. LA LANCETTE, E.A. and R.E.BENSON, *J. Amer. Chem. Soc.*, **85**, 2853 (1963).
1482. LA LANCETTE, E.A. and R.E.BENSON, *J. Amer. Chem. Soc.*, **87**, 1941 (1965).
1483. LALONDE, R.T., S.EMMI and R.R.FRASER, *J. Amer. Chem. Soc.*, **86**, 5548 (1964).
1484. LAMATY, G., C.TAPIERO and R.WYLDE, *Bull. Soc. Chim. France*, 3085 (1965).
1485. LAMBERT, J.B., *Tetrahedron Letters*, 1901 (1963).
1486. LAMBERT, J.B., L.J.DURHAM, P.LEPOUTERE and J.D.ROBERTS, *J. Amer. Chem. Soc.*, **87**, 3896 (1965).
1487. LAMBERT, J.B. and R.G.KESKE, *J. Amer. Chem. Soc.*, **88**, 620 (1966).
1488. LAMBERT, J.B., W.L.OLIVER and J.D.ROBERTS, *J. Amer. Chem. Soc.*, **87**, 5085 (1965).
1489. LAMBERT, J.B. and J.D.ROBERTS, *J. Amer. Chem. Soc.*, **85**, 3710 (1963).
1490. LAMBERT, J.B. and J.D.ROBERTS, *J. Amer. Chem. Soc.*, **87**, 3884 (1965).
1491. LAMBERT, J.B. and J.D.ROBERTS, *J. Amer. Chem. Soc.*, **87**, 3891 (1965).
1492. LAMBERT, J.B. and J.D.ROBERTS, *J. Amer. Chem. Soc.*, **87**, 4087 (1965).
1493. LANGENBUCHER, F., R.MECKE and E.D.SCHMID, *Ann.*, **669**, 11 (1963).
1494. LANGENBUCHER, F., E.D.SCHMID and R.MECKE, *J. Chem. Phys.*, **39**, 1901 (1963).
1495. LANSBURY, P.T. and J.F.BIERON, *J. Org. Chem.*, **28**, 3564 (1963).
1496. LANSBURY, P.T., V.A.PATTISON, W.A.CLEMENT and J.DSIDLER, *J. Amer. Chem. Soc.*, **86**, 2247 (1964).
1497. LAPLANCHE, L.A. and M.T.ROGERS, *J. Amer. Chem. Soc.*, **85**, 3728 (1963).
1498. LAPLANCHE, L.A. and M.T.ROGERS, *J. Amer. Chem. Soc.*, **86**, 337 (1964).
1499. LARSEN, D.W. and A.L.ALLRED, *J. Amer. Chem. Soc.*, **87**, 1216 and 1219 (1965).
1500. LASZLO, P., *Thesis, Université de Paris* (1965).
1501. LASZLO, P., *Sciences*, 58 (1963).
1502. LASZLO, P., *Bull. Soc. Chim. France*, 85 (1964).
1503. LASZLO, P., *Bull. Soc. Chim. France*, 2658 (1964).
1504. LASZLO, P., *Bull. Soc. Chim. France*, 1131 (1966).
1505. LASZLO, P. and H.J.T.BOS, *Tetrahedron Letters*, 1325 (1965).
1506. LASZLO, P. and J.I.MUSHER, *Bull. Soc. Chim. France*, 2558 (1964).
1507. LASZLO, P. and P. VON R.SCHLEYER, *J. Amer. Chem. Soc.*, **85**, 2017 (1963).
1508. LASZLO, P. and P.R.SCHLEYER, *J. Amer. Chem. Soc.*, **85**, 2709 (1963).
1509. LASZLO, P. and P.R.SCHLEYER, *J. Amer. Chem. Soc.*, **86**, 1171 (1964).
1510. LASZLO, P. and P.R.SCHLEYER, *Bull. Soc. Chim. France*, 87 (1964).
1511. LASZLO, P. and D.H.WILLIAMS, *J. Amer. Chem. Soc.*, **88**, 2799 (1966).
1511a. LAUTERBUR, P.C., *Phys. Rev. Letters*, **1**, 343 (1958).
1512. LAUTERBUR, P.C., *Tetrahedron Letters*, 274 (1961).
1513. LAUTERBUR, P.C., *J. Amer. Chem. Soc.*, **83**, 1846 (1961).
1514. LAUTERBUR, P.C., *J. Amer. Chem. Soc.*, **84**, 1192 (1962).
1515. LAUTERBUR, P.C. and R.J.KURLAND, *J. Amer. Chem. Soc.*, **84**, 3405 (1962).
1516. LAVIE, D., B.S.BENJAMINOV and Y.SHVO, *Tetrahedron*, **20**, 2585 (1964).
1517. LAVIE, D., Y.SHVO and E.GLOTTER, *Tetrahedron*, **19**, 2255 (1963).
1518. LAWRENSON, I.J., *J. Chem. Soc.*, 1117 (1965).
1519. LAWSON, K.D., W.S.BREY and L.B.KIER, *J. Amer. Chem. Soc.*, **86**, 463 (1964).
1520. LEANE, J.B. and R.E.RICHARDS, *Trans. Farad. Soc.*, **55**, 518 (1959).
1521. LEANE, J.B. and R.E.RICHARDS, *Trans. Farad. Soc.*, **55**, 707 (1959).
1522. LEDAAL, T., *Tetrahedron Letters*, 1653 (1966).
1523. LEE, J. and K.G.ORRELL, *J. Chem. Soc.*, 582 (1965).
1524. LEE, J. and L.H.SUTCLIFFE, *Trans. Farad. Soc.*, **55** 880 (1959).
1525. LEE, W.W., A.BENTTEZ, C.D.ANDERSON, L.GOODMAN and B.R.BAKER, *J. Amer. Chem. Soc.*, **83**, 1906 (1961).
1526. LEFEVRE, C.G. and R.J.W.LEFEVRE, *Rev. Pure Appl. Chem.*, **5**, 261 (1955).
1527. LEFEVRE, R.J.W., P.H.WILLIAMS and J.M.ECKERT, *Aust. J. Chem.*, **18**, 1133 (1965).
1528. LEGOFF, E. and R.B.LACOUNT, *J. Amer. Chem. Soc.*, **85**, 1354 (1963).
1529. LEHN, J.M., *Bull. Soc. Chim. France*, 1832 (1962).
1530. LEHN, J.M. and M.FRANCK-NEUMANN, *J. Chem. Phys.*, **43**, 1421 (1965).
1531. LEHN, J.M., J.LEVISALLES and G.OURISSON, *Bull. Soc. Chim. France*, 1096 (1963).
1532. LEHN, J.M. and G.OURISSON, *Bull. Soc. Chim. France*, 1137 (1962).

1533. LEHN, J.M. and G.OURISSON, *Bull. Soc. Chim. France*, 1113 (1963).
1534. LEHN, J.M. and J.J.RIEHL, *Mol. Phys.*, **8**, 33 (1964).
1535. LEHN, J.M. and J.J.RIEHL, *J. Chim. Phys.*, **62**, 573 (1965).
1536. LEHN, J.M. and A.VYSTRCIL, *Tetrahedron*, **19**, 1733 (1963).
1537. LEMAL, D.M., F.MENGER and G.W.CLARK, *J. Amer. Chem. Soc.*, **85**, 2529 (1963).
1538. LEMIEUX, R.U. and J.HOWARD, *Canad. J. Chem.*, **41**, 393 (1963).
1539. LEMIEUX, R.U., R.K. KULLNIG, H.J.BERNSTEIN and W.G.SCHNEIDER, *J. Amer. Chem.Soc.*, **80**, 6098 (1958).
1540. LEMIEUX, R.U., D.R.LINEBACK, M.L.WOLFROM, F.B.MOODY, E.G.WALLACE and F.KOMITSKY, *J. Org. Chem.*, **30**, 1092 (1965).
1541. LEMIEUX, R.U. and J.W.LOWN, *Canad. J. Chem.*, **41**, 889 (1963).
1542. LEMIEUX, R.U. and J.W.LOWN, *Tetrahedron Letters*, 1229 (1963).
1543. LEMIEUX, R.U. and J.W.LOWN, *Canad. J. Chem.*, **42**, 893 (1964).
1544. LEMIEUX, R.U., J.S.MARTIN and J.HAYAMI, *International Symposium on Nuclear Magnetic Resonance*, Tokyo, 1965.
1545. LEMIEUX, R.U. and A.R.MORGAN, *J. Amer. Chem. Soc.*, **85**, 1889 (1963).
1546. LEMIEUX, R.U. and R.NAGARAJAN, *Canad. J. Chem.*, **42**, 1270 (1964).
1547. LEMIEUX, R.U. and J.D.STEVENS, *Canad. J. Chem.*, **43**, 2059 (1965).
1548. LEMIEUX, T.U. and J.D.STEVENS, *Canad. J. Chem.*, **44**, 249 (1966).
1549. LEMIEUX, R.U., J.D.STEVENS and R.R.FRASER, *Canad. J. Chem.*, **40**, 1955 (1962).
1550. LENZ, R.W. and J.P.HEESCHEN, *J. Polymer. Sci.*, **51**, 247 (1961).
1551. LEONARD, N.J. and K.JANN, *J. Amer. Chem. Soc.*, **84**, 4806 (1963).
1552. LEONARD, N.J. and R.A.LAURSEN, *J. Amer. Chem. Soc.*, **85**, 2026 (1963).
1553. LEONARD, N.J. and J.V.PAUKSTELIS, *J. Org. Chem.*, **28**, 3021 (1963).
1554. LEPAGE, M., R.MUMMA and A.A.BENSON, *J. Amer. Chem. Soc.*, **82**, 3713 (1960).
1555. LEPLEY, A.R. and R.H.BECKER, *Tetrahedron*, **21**, 2365 (1965).
1556. LEVENBERG, M.I. and J.H.RICHARDS, *J. Amer. Chem. Soc.*, **86**, 2634 (1964).
1557. LEVINE, S.G., N.H.EUDY and E.C.FARTHING, *Tetrahedron Letters*, 1517 (1963).
1558. LEUSINK, A.J., J.W.MARSMAN and H.A.BUDDING, *Rec. Trav. Chim.*, **84**, 689 (1965).
1559. LEWIN, A.H. and S.WINSTEIN, *J. Amer. Chem. Soc.*, **84**, 2484 (1962).
1560. LEWIN, A.H., *J. Amer. Chem. Soc.*, **86**, 2303 (1964).
1561. LEWIN, A.H., J.LIPOWITZ and T.COHEN, *Tetrahedron Letters*, 1241 (1965).
1562. LEWIS, W.C. and B.E.NORCROSS, *J. Org. Chem.*, **30**, 2866 (1965).
1563. LICHTENTHALER, F.W., *Tetrahedron Letters*, 775 (1963).
1564. LICHTENTHALER, F.W., *Chem. Ber.*, **96**, 845 (1963).
1565. LICHTENTHALER, F.W., *Chem. Ber.*, **96**, 2047 (1963).
1566. LICHTENTHALER, F.W. and H.O.L.FISCHER, *J. Amer. Chem. Soc.*, **83**, 2005 (1961).
1567. LIN, W.C., *J. Chinese Chem. Soc.*, **11**, 36 (1964).
1568. LINN, W.J., O.W.WEBSTER and R.E.BENSON, *J. Amer. Chem. Soc.*, **85**, 2032 (1963).
1569. LINDSAY, W.S., P.STOKES, G.HUMBER and V.BOCKELHEIDE, *J. Amer. Chem. Soc.*, **83**, 943 (1961).
1570. LIPPERT, E., *Z. Electrochem.*, **67**, 267 (1963).
1571. LIPPERT, E. and H.PRIGGE, *Ber. Bunsenges. Physik Chem.*, **67**, 415 (1963).
1572. LIPPERT, VON E., D.SAMUEL and E.FISCHER, *Ber. Bunsenges.*, **69**, 155 (1965).
1573. LOAN, L.D., R.W.MURRAY and P.R.STORY, *J. Amer. Chem. Soc.*, **87**, 737 (1965).
1574. LOEWENSTEIN, A., *J. Phys. Chem.*, **67**, 1728 (1963).
1575. LOEWENSTEIN, A. and T.M.CONNOR, *Ber. Bunsenges.*, **67**, 280 (1963).
1576. LOEWENSTEIN, A., J.F.NEUMER and J.D.ROBERTS, *J.Amer.Chem.Soc.*, **82**, 3599(1960).
1577. LOGAN, T.J. and T.J.FLAUTT, *J. Amer. Chem. Soc.*, **82**, 3446 (1960).
1578. LOHR, L.L. and W.N.LIPSCOMB, *Inorg. Chem.*, **3**, 22 (1964).
1579. LOMBARDI, E. and P.B.SOGO, *J. Chem. Phys.*, **32**, 635 (1960).
1580. LONDON, F., *J. Chem. Phys.*, **5**, 837 (1937).
1581. LONGUET-HIGGINS, H.C. and L.SALEM, *Proc. Roy. Soc.* (London), A**257**, 445(1960).
1582. LONGONE, D.T. and C.L.WARREN, *J. Amer. Chem. Soc.*, **84**, 1507 (1962).
1583. LOWE, I.J. and D.E.BARNAAL, *Rev. Sci. Instr.*, **34**, 143 (1963).
1584. LOWRY, J.B. and N.V.RIGGS, *Tetrahedron Letters*, 2911 (1964) and references therein.

418   *References*

1585. LUMBROSO, N., T.K.WU and B.P.DAILY, *J. Phys. Chem.*, **67**, 2469 (1963).
1586. LUSTIG, E., *J. Phys. Chem.*, **65**, 491 1961).
1587. LUSTIG, E., *J. Chem. Phys.*, **37**, 2725 (1962).
1588. LUSTIG, E. and W.B.MONITZ, *Anal. Chem.*, **38**, 331 R (1966).
1589. LUSTIG, E. and R.M.MORIARTY, *J. Amer. Chem. Soc.*, **87**, 3252 (1965).
1590. LUTTRINGHAUS, A., E.FUTTERER and H.PRINZBACH, *Tetrahedron Letters*, 1209 (1963).
1591. LUTTRINGHAUS, A., S.KABUSS, W.MAIER and H.FRIEBOLIN, *Z. Naturforsch.*, **16b**, 761 (1961).
1592. LUZ, Z., D.GILL and S.MEIBOOM, *J. Chem. Phys.*, **30**, 1540 (1959).
1593. LUZ, Z. and S.MEIBOOM, *J. Chem. Phys.*, **39**, 366 (1963).
1594. LUZ, Z. and B.SILVER, *J. Amer. Chem. Soc.*, **84**, 1095 (1962).
1595. LYLE, R.E., *Chem. Eng. News*, 72 (1966).
1596. LYNCH, B.M., *Canad. J. Chem.*, **41**, 2380 (1963).
1597. LYNDEN-BELL, R.M., *Trans. Farad. Soc.*, **57**, 888 (1961).
1598. LYNDEN-BELL, R.M., *Mol. Phys.*, **6**, 537 (1963).
1599. LYNDEN-BELL, R.M., *Mol. Phys.*, **6**, 601 (1963).
1600. LYNDEN-BELL, R.M., *Proc. Roy. Soc.*, A **286**, 337 (1965).
1601. LYNDEN BELL, R.M. and N.SHEPPARD, *Proc. Roy. Soc.*, **269**, 385 (1962).
1602. MA, J.C.N. and E.W.WARNHOFF, *Canad. J. Chem.*, **43**, 1849 (1965).
1603. McADAMS, D.R., *J. Chem. Phys.*, **36**, 1948 (1962).
1604. McALPINE, J.B., Private communication 1966.
1605. McBEE, E.T., W.L.DILLING and H.P.BRAENDLIN, *J. Org. Chem.*, **27**, 2704 (1962).
1606. McBRIDE, D.W., E.DUDEK and F.G.A.STONE, *J. Chem. Soc.*, 1752 (1964).
1607. McCANN, A.P., F.SMITH, J.A.S.SMITH and J.D.THWAITES, *J. Sci. Inst.*, **39**, 349 (1962).
1608. McCAPRA, F., A.I.SCOTT, P.DELMOTTE, J.DELMOTTE-PLAQUEE and N.S.BHACCA, *Tetrahedron Letters*, 869 (1964).
1609. McCASLAND, G.E., S.FURUTA, L.F.JOHNSON and J.N.SHOOLERY, *J. Amer. Chem. Soc.*, **83**, 2335 (1961).
1610. McCASLAND, G.G., S.FURUTA, L.F.JOHNSON and J.N.SHOOLERY, *J. Org. Chem.*, **28**, 894 (1963).
1611. McCLELLAN, W.R., *J. Amer. Chem. Soc.*, **83**, 1598 (1961).
1612. McCLELLEN, W.M., H.H.HOEHN, H.N.CRIPPS, E.L.MUETTERTIES and B.W.HOWK, *J. Amer. Chem. Soc.*, **83**, 1601 (1961).
1613. McCONNELL, H.M., *J. Chem. Phys.*, **24**, 460 (1956).
1614. McCONNELL, H.M., *J. Mol. Spec.*, **1**, 11 (1957); *J. Chem. Phys.*, **30**, 26 (1959).
1615. McCONNELL, H.M., *J. Chem. Phys.*, **27**, 226 (1957).
1616. McCONNELL, H.M., *J. Chem. Phys.*, **28**, 430 (1958).
1617. McCONNELL, H.M., A.D.McLEAN and C.A.REILLY, *J. Chem. Phys.*, **23**, 1152 (1955).
1618. McCOY, C.R. and A.L.ALLRED, *J. Amer. Chem. Soc.*, **84**, 912 (1962).
1619. McCOY, C.R. and A.L.ALLRED, *J. Inorg. Nucl. Chem.*, **25**, 1219 (1963).
1620. MACDONALD, C.G., J.S.SHANNON and S.STERNHELL, *Aust. J. Chem.*, **17**, 38 (1964) and references therein.
1621. MACDONALD, C.G., J.S.SHANNON and S.STERNHELL, *Aust. J. Chem.*, **19**, 1527 (1966).
1622. McFARLANE, W. and S.O.GRIM, *J. Organometal. Chem.*, **5**, 147 (1966).
1623. McGARVEY, B.R. and G.SLOMP JR., *J. Chem. Phys.*, **30**, 1586 (1959).
1624. McGRADY, M.M. and R.S.TOBIAS, *Inorg. Chem.*, **3**, 1157 (1964).
1625. McGREER, D.E. and M.M.MOCEK, *J. Chem. Ed.*, **40**, 358 (1963).
1626. MACHLEIDT, H. and R.WESSENDORF, *Ann.*, **674**, 1 (1964).
1627. MACIEL, G.E. and R.V.JAMES, *J. Amer. Chem. Soc.*, **86**, 3893 (1964).
1628. MACKOR, E.L. and C.MACLEAN, *Pure Appl. Chem.*, **8**, 393 (1964).
1628a. McLACHLAN, A.D., *J. Chem. Phys.*, **32**, 1263 (1960).
1629. McLACHLAN, A.D. and M.R.BAKER, *Mol. Phys.*, **5**, 255 (1961).
1630. McLAUCHLAN, K.A. and T.SCHAEFER, *Canad. J. Chem.*, **44**, 321 (1966).

1631. McLAUCHLAN, K.A. and D.H.WHIFFEN, *Proc. Chem. Soc.*, 144 (1962).
1632. McLAUGHLIN, D.E., M.TAMRES and S.SEARLES, *J. Amer. Chem. Soc.*, **82,** 5621 (1960).
1633. MacLEAN, C. and E.L.MACKOR, *Mol. Phys.*, **3,** 223 (1960).
1634. MacLEAN, C. and E.L.MACKOR, *Mol. Phys.*, **4,** 241 (1961).
1635. MacLEAN, C. and L.E.MACKOR, *J. Chem. Phys.*, **34,** 2207 (1961).
1636. MacLEAN, C. and E.L.MACKOR, *Disc. Farad. Soc.*, **34,** 165 (1962).
1637. MacLEAN, C., J.H. VAN DER WAALS and E.L.MACKOR, *Mol. Phys.*, **1,** 247 (1958).
1638. McLEAN, S., *Canad. J. Chem.*, **38,** 2278 (1960).
1639. McLEAN, S. and P.HAYNES, *Canad. J. Chem.*, **41,** 1231 (1963).
1640. McLEAN, S. and P.HAYNES, *Canad. J. Chem.*, **41,** 1233 (1963).
1641. McMAHON, P.E., *J. Mol. Spec.*, **16,** 221 (1965).
1642. McMAHON, P.E. and W.C.TINCHER, *J. Mol. Spec.*, **15,** 180 (1965).
1643. MacNICOL, D.D., R.WALLACE and J.C.D.BRAND, *Trans. Farad. Soc.*, **61,** 1 (1965).
1644. McWEENY, R., *Proc. Phys. Soc.*, **65A,** 839 (1952).
1645. McWEENY, R., *Mol. Phys.*, **1,** 311 (1958).
1646. MADDOX, I.J. and R.McWEENY, *J. Chem. Phys.*, **36,** 2353 (1962).
1647. MAERCKER, A. and J.D.ROBERTS, *J. Amer. Chem. Soc.*, **88,** 1742 (1966).
1648. MAGNUSSON, G., *Acta Chem. Scand.*, **17,** 273 (1963).
1649. MAHER, J.P. and D.F.EVANS, *Proc. Chem. Soc.*, 176 (1963).
1650. MAHER, J.P. and D.F.EVANS, *J. Chem. Soc.*, 5534 (1963).
1651. MAHER, J.P. and D.F.EVANS, *J. Chem. Soc.*, 673 (1965).
1652. MAHLER, J.E., D.H.GIBSON and R.PETTIT, *J. Amer. Chem. Soc.*, **85,** 3959 (1963).
1653. MAHLER, J.E., D.A.K.JONES and R.PETTIT, *J. Amer. Chem. Soc.*, **86,** 3589 (1964).
1654. MAIER, G., *Angew. Chem. Int. Ed.*, **2,** 621 (1963).
1655. MAIER, G., *Chem. Ber.*, **98,** 2438 (1965).
1656. MAKI, A.H. and R.J.VOLPICELLI, *Rev. Sci. Instr.*, **36,** 325 (1965).
1657. MAKISUMI, Y., H.WATANABE and K.TORI, *Chem. Pharm. Bull.*, **12,** 204 (1964).
1658. MALINOWSKI, E.R., *J. Amer. Chem. Soc.*, **83,** 4479 (1961).
1659. MALINOWSKI, E.R., M.S.MANHAS, G.H.MÜLLER and A.K.BOSE, *Tetrahedron Letters*, 1161 (1963).
1660. MALINOWSKI, E.R., L.Z.POLLARA and J.P.LARMANN, *J. Amer. Chem. Soc.*, **84,** 2649 (1962).
1661. MALINOWSKI, E.R. and T.VLADIMIROFF, *J. Amer. Chem. Soc.*, **86,** 3575 (1964).
1662. MOLLOY, B.B., D.H.REID and S.McKENZIE, *J. Chem. Soc.*, 4368 (1965).
1663. MANN, F.G., B.P.TONG and V.P.WYSTRACH, *J. Chem. Soc.*, 1155 (1963).
1664. MANATT, S.L. and D.D.ELLEMAN, *J. Amer. Chem. Soc.*, **84,** 1305 (1962).
1665. MANATT, S.L. and D.D.ELLEMAN, *J. Amer. Chem. Soc.*, **84,** 1579 (1962).
1666. MANATT, S.L., D.D.ELLEMAN and S.J.BROIS, *J. Amer. Chem. Soc.*, **87,** 2220 (1965).
1667. MANATT, S.L., G.L.JUVINALL and D.D.ELLEMAN, *J. Amer. Chem. Soc.*, **85,** 2664 (1963).
1668. MANATT, S.L., M.VOGEL, D.KNUTSON and J.D.ROBERTS, *J. Amer. Chem. Soc.*, **86,** 2645 (1964).
1669. MANNSCHRECK, A., W.SEITZ and H.A.STAAB, *Ber. Bunsenges.*, **67,** 470 (1963).
1670. MANNSCHRECK, A., H.A.STAAB and D.WURMB-GERLICH, *Tetrahedron Letters*, 2003 (1963).
1671. MARCUS, E., J.K.CHAN and C.B.STROW, *J. Org. Chem.*, **31,** 1369 (1966).
1672. MARCUS, S.H. and S.I.MILLER, *J. Phys. Chem.*, **68,** 331 (1964).
1673. MARGRAFT, J.H., T.BACHMANN and D.P.HOLLIS, *J. Org. Chem.*, **30,** 3472 (1965).
1674. MARKGRAF, J.H., B.A.HESS, C.W.NICHOLS and R.W.KING, *J. Org. Chem.*, **29,** 1499 (1964).
1675. MARKGRAF, J.H. and R.W.KING, *J. Org. Chem.*, **29,** 3094 (1964).
1676. MARK, V., *J. Amer. Chem. Soc.*, **85,** 1884 (1963).
1677. MARKHAM, K.R. and I.D.RAE, *Aust. J. Chem.*, **18,** 1497 (1965).
1678. MARKI, F., A.V.ROBERTSON and B.WITKOP, *J. Amer. Chem. Soc.*, **83,** 3341 (1961).
1679. MARSHALL, T.W. and J.A.POPLE, *Mol. Phys.*, **1,** 199 (1958).
1680. MARSHALL, T.W. and J.A.POPLE, *Mol. Phys.*, **3,** 339 (1960).

1681. MARTIN, G., B. CASTRO and M. MARTIN, *Compt. Rend.*, **261**, 395 (1965).
1682. MARTIN, G. J. and M. L. MARTIN, *J. Chim. Phys.*, **61**, 1222 (1964).
1683. MARTIN, G. and G. MAVEL, *Compt. Rend.*, **253**, 2523 (1961).
1684. MARTIN, J. S. and B. P. DAILEY, *J. Chem. Phys.*, **37**, 2594 (1962).
1685. MARTIN, J. S. and B. P. DAILEY, *J. Chem. Phys.*, **39**, 1722 (1963).
1686. MARTIN, M., *J. Chim. Phys.*, **59**, 736 (1962).
1687. MARTIN, M. L., C. ANDRIEU and G. J. MARTIN, *Tetrahedron Letters*, 921 (1966).
1688. MARTIN, M. L., R. MANTIONE and G. J. MARTIN, *Tetrahedron Letters*, 3185 (1965).
1689. MARTIN, M. and G. MARTIN, *Compt. Rend.*, **249**, 884 (1959).
1690. MARTIN, M. and G. MARTIN, *Compt. Rend.*, **256**, 403 (1963).
1691. MARTIN, M. L., G. J. MARTIN and P. CAUBERE, *Bull. Soc. Chim. France*, 3066 (1964).
1692. MARTIN, M. and M. QUILBEUF, *Compt. Rend.*, 4151 (1961).
1693. MARTIN, M. M. and G. J. GLEICHER, *J. Amer. Chem. Soc.*, **86**, 238 (1964).
1694. MARTIN, R. H., *Tetrahedron*, **20**, 897 (1964).
1695. MARTIN, R. H., N. DEFAY and F. GREETS-EVRARD, *Tetrahedron*, **20**, 1091 (1964).
1696. MARTIN, R. H., N. DEFAY and F. GREETS-EVRARD, *Tetrahedron*, **20**, 1505 (1964).
1697. MARTIN, R. H., N. DEFAY and F. GREETS-EVRARD, *Tetrahedron*, **21**, 2421 (1965).
1698. MARTIN, R. H., N. DEFAY and F. GREETS-EVRARD, *Tetrahedron*, **21**, 2435 (1965).
1699. MARTIN, R. H., N. DEFAY, F. GREETS-EVRARD and S. DELAVARENNE, *Bull. Soc. Chim. Belgium*, **73**, 189 (1964).
1699a. MARTIN, R. H., N. DEFAY, F. GREETS-EVRARD and S. DELAVARENNE, *Tetrahedron*, **20**, 1073 (1964).
1700. MARTIN, R. H., N. DEFAY, F. GREETS-EVRARD and H. FIGEYS, *Bull. Soc. Chim. Belgium*, **73**, 199 (1964).
1700a. MARTIN, R. H., N. DEFAY, F. GREETS-EVRARD, P. H. GIVEN, J. R. JONES and R. W. WEDEL, *Tetrahedron*, **21**, 1833 (1965).
1701. MARTIN, R. H., R. FLAMMANG and M. ARBAOUI, *Bull. Soc. Chim. Belgium*, **74**, 418 (1965).
1702. MARTIN, R. H., J. P. VAN TRAPPEN, N. DEFAY and J. F. W. McOMIE, *Tetrahedron*, **20**, 2373 (1964).
1703. MARTIN-SMITH, M., S. T. REID and S. STERNHELL, *Tetrahedron Letters*, 2393 (1965) and unpublished results.
1704. MARVELL, E. N. and S. PROVANT, *J. Org. Chem.*, **29**, 3084 (1964).
1705. MASAMUNE, S., *Tetrahedron Letters*, 945 (1965).
1706. MASAMUNE, S. and N. T. CASTELLUCCI, *J. Amer. Chem. Soc.*, **84**, 2452 (1962).
1707. MASSEY, A. G., E. W. RANDALL and D. SHAW, *Chem. and Ind.*, 1244 (1963).
1708. MASSEY, A. G., E. W. RANDALL and D. SHAW, *Chem. and Ind.*, 1246 (1963).
1709. MASSICOT, J. and J. P. MARTHE, *Bull. Soc. Chim. France*, 1962 (1962).
1710. MASSICOT, J., J. P. MARTHE and S. HEITZ, *Bull. Soc. Chim. France*, 2712 (1963).
1711. MATEOS, J. L. and D. J. CRAM, *J. Amer. Chem. Soc.*, **81**, 2756 (1959).
1712. MATHIAS, A., *Tetrahedron*, **21**, 1073 (1965).
1713. MATHIS, C. T. and J. H. GOLDSTEIN, *J. Phys. Chem.*, **68**, 571 (1964).
1714. MATHIS, C. T. and J. H. GOLDSTEIN, *Spectrochim. Acta*, **20**, 871 (1964).
1715. MATHUR, R., E. D. BECKER, R. B. BRADLEY and N. C. LI, *J. Phys. Chem.*, **67**, 2190 (1963).
1716. MATSUURA, S. and T. GOTO, *Tetrahedron Letters*, 1499 (1963).
1717. MATSUURA, S. and T. GOTO, *J. Chem. Soc.*, 1773 (1963).
1718. MATSUURA, S. and T. GOTO, *J. Chem. Soc.*, 623 (1965).
1719. MATWIYOFF, N. A. and R. S. DRAGO, *J. Chem. Phys.*, **38**, 2583 (1963).
1720. MAUGER, A. B., F. IRREVERRE and B. WITKOP, *J. Amer. Chem. Soc.*, **88**, 2019 (1966).
1721. MAVEL, G., *Compt. Rend.*, **248**, 3699 (1959).
1722. MAVEL, G., *Compt. Rend.*, **250**, 1477 (1960).
1723. MAVEL, G., *J. Chim. Phys.*, **59**, 683 (1962).
1724. MAVEL, G., *J. Chim. Phys.*, **61**, 182 (1964).
1725. MAVEL, G. and G. MARTIN, *Compt. Rend.*, **254**, 260 (1962).
1726. MAVEL, G. and G. MARTIN, *Compt. Rend.*, **257**, 1703 (1963).
1727. MAVEL, G. and M. MARTIN, *J. Chim. Phys.*, **57**, 445 (1960).

1728. MAYO, D.W., P.J.SAPIENZA, R.R.LORD and W.D.PHILIPS, *J. Org. Chem.*, **29**, 2682 (1964).
1729. MAYO, R.E. and J.H.GOLDSTEIN, *J. Mol. Spec.*, **14**, 173 (1964).
1730. MAYO, R.E. and J.H.GOLDSTEIN, *Rev. Sci. Instr.*, **35**, 1231 (1964).
1731. MAYO, R.E. and J.H.GOLDTSEIN, *Mol. Phys.*, **10**, 301 (1966).
1732. MAZUR, R.H., *J. Org. Chem.*, **28**, 248 (1963).
1733. MAZUR, R.H., W.N.WHITE, D.A.SEMENOW, C.C.LEE, M.S.SILVER and J.D.RO-BERTS, *J. Amer. Chem. Soc.*, **81**, 4390 (1959).
1734. MEIBOOM, S., *Z. Electrochem.*, **64**, 60 (1960).
1735. MEIBOOM, S., *J. Chem. Phys.*, **34**, 375 (1961).
1736. MEIER, W., D.MEUCHE and E.HEILBRONNER, *Helv. Chim. Acta*, **46**, 1929 (1963).
1737. MEINWALD, J. and Y.C.MEINWALD, *J. Amer. Chem. Soc.*, **85**, 2514 (1963).
1738. MEINWALD, J., Y.C.MEINWALD and T.N.BAKER, *J. Amer. Chem. Soc.*, **86**, 4074 (1964).
1739. MEMORY, J.D., *Bull. Amer. Phys. Soc.*, **8**, 532 (1963).
1740. MEMORY, J.D., *J. Chem. Phys.*, **38**, 1341 (1963).
1741. MEMORY, J.D. and T.B.COBB, *J. Chem. Phys.*, **38**, 1453 (1963).
1742. MEMORY, J.D. and T.B.COBB, *J. Chem. Phys.*, **39**, 2316 (1963).
1743. MEMORY, J.D. and T.B.COBB, *J. Chem. Phys.*, **39**, 2386 (1963).
1744. MEUCHE, D. and E.HEILBRONNER, *Helv. Chim. Acta*, **45**, 1965 (1962).
1745. MEUCHE, D., B.B.MALLOY, D.H.REID and E.HEILBRONNER, *Helv. Chim. Acta*, **46**, 2483 (1963).
1746. MEUCHE, D., M.NEUENSCHWANDER, H.SCHALTEGGER and H.U.SCHLUNEGGER, *Helv. Chim. Acta*, **47**, 1211 (1964).
1747. MEYER, W.L., D.L.DAVIS, L.FOSTER, A.S.LEVINSON, V.L.SAWIN, D.C.SHEW and R.F.WEDDLETON, *J. Amer. Chem. Soc.*, **87**, 1573 (1965).
1748. MEYER, W.L., H.R.MAHLER and R.H.BAKER JR., *Biochem. Biophys. Acta*, **64**, 353 (1962).
1749. MEYER, W.L. and A.S.LEVINSON, *J. Org. Chem.*, **28**, 2859 (1963).
1750. MEYER, W.L. and R.B.MEYER, *J. Amer. Chem. Soc.*, **85**, 2170 (1963).
1751. MEYERS, O.E. and J.C.SHEPPARD, *J. Amer. Chem. Soc.*, **83**, 4739 (1961).
1752. MEZZADROLI, G., *Bollettino Scientifico della Facolta di Chimica Industriale Bologna*, **21**, 217 (1963).
1753. MICHEL, G.W. and H.R.SNYDER, *J. Org. Chem.*, **27**, 2034 (1962).
1754. MIDDAUGH, R.L. and R.S.DRAGO, *J. Amer. Chem. Soc.*, **85**, 2575 (1963).
1755. MIDDLETON, W.S. and W.H.SHARKEY, *J. Amer. Chem. Soc.*, **81**, 803 (1959).
1756. MILES, H.T., *J. Amer. Chem. Soc.*, **85**, 1007 (1963).
1757. MILES, M.L., T.M.HARRIS and C.R.HAUSER, *J. Amer. Chem. Soc.*, **85**, 3884 (1963).
1758. MILLER, J.D. and R.H.PRINCE, *J. Chem. Soc.*, 3185 (1965).
1759. MILLER, J.M. and M.ONYSZCHUK, *Canad. J. Chem.*, **42**, 1518 (1964).
1760. MILLER, R.G. and M.STILES, *J. Amer. Chem. Soc.*, **85**, 1798 (1963).
1761. MILNE, J.W., J.S.SHANNON and S.STERNHELL, *Aust. J. Chem.*, **18**, 139 (1965).
1762. MIRRINGTON, R.N., E.RITCHIE, C.W.SHOPPEE, W.C.TAYLOR and S.STERNHELL, *Tetrahedron Letters*, 365 (1964).
1763. MISLOW, K., *Tetrahedron Letters*, 1415 (1964).
1764. MISLOW, K., M.A.W.GLASS, H.B.HOPPS, E.SIMON and G.H.WAHL, *J. Amer. Chem. Soc.*, **86**, 1710 (1964).
1765. MISLOW, K., R.GRAEVE, A.J.GORDON and G.H.WAHL, *J. Amer. Chem. Soc.*, **85**, 1199 (1963).
1766. MOFFETT, R.B., *J. Org. Chem.*, **28**, 2885 (1963).
1767. MOHACSI, E., *J. Chem. Ed.*, **41**, 38 (1964).
1768. MOHACSI, E., *The Analyst*, **91**, 57 (1966).
1769. MOLE, T., *Aust. J. Chem.*, **16**, 801 (1963).
1770. MOLE, T. and J.R.SURTEES, *Aust. J. Chem.*, **17**, 310 (1963).
1771. MONIZ, W.M. and J.A.DIXON, *J. Amer. Chem. Soc.*, **83**, 1671 (1961).
1772. MONIZ, W.B., C.F.PORANSKI and T.N.HALL, *J. Amer. Chem. Soc.*, **88**, 190 (1966).
1773. MOODIE, R.B., T.M.CONNOR and R.STEWART, *Canad. J. Chem.*, **37**, 1402 (1959).

1774. MOODIE, R.B., T.M.CONNOR and R.STEWART, *Canad. J. Chem.*, **38**, 626 (1960).
1775. MOORE, J.A. and C.L.HABRAKEN, *J. Amer. Chem. Soc.*, **86**, 1456 (1964).
1776. MOORE, J.A., F.S.MARASCIA, R.W.MEDEIROS and E.WYSS, *J. Amer. Chem. Soc.*, **84**, 3022 (1962).
1777. MOORE, D.W., *J. Chem. Phys.*, **34**, 1470 (1961).
1778. MOORE, D.W., H.B.JONASSEN, T.B.JOYNER and A.J.BERTRAND, *Chem. and Ind.*, 1304 (1960).
1779. MOORE, D.W. and A.G.WHITTAKER, *J. Amer. Chem. Soc.*, **82**, 5007 (1960).
1780. MOORE, H.W. and H.R.SNYDER, *J. Org. Chem.*, **28**, 297 (1963).
1781. MOORE, H.W. and H.R.SNYDER, *J. Org. Chem.*, **28**, 535 (1963).
1782. MOORE, W.R., W.R.MOSER and J.E.LAPRADE, *J. Org. Chem.*, **28**, 2200 (1963).
1783. MORGAN, L.O., J.MURPHY and P.F.COX, *J. Amer. Chem. Soc.*, **81**, 5043 (1959).
1784. MORI, N., S.OMURA, O.YAMAMATO, T.SUZUKI and Y.TSUZUKI, *Bull. Chem. Soc. Japan*, **36**, 1047 (1963).
1785. MORIARTY, R.M., *J. Org. Chem.*, **28**, 1296 (1963).
1786. MORIARTY, R.M., *J. Org. Chem.*, **29**, 2748 (1964).
1787. MORIARTY, R.M., *Tetrahedron Letters*, 509 (1964).
1788. MORIARTY, R.M., *J. Org. Chem.*, **30**, 600 (1965).
1789. MORIARTY, R.M. and J.M.KLIEGMAN, *Tetrahedron Letters*, 891 (1966).
1790. MORIARTY, R.M. and T.D.D'SILVA, *J. Org. Chem.*, **28**, 2445 (1963).
1791. MORISAWA, Y., *Chem. Pharm. Bull.*, **12**, 1060 (1964).
1792. MORISAWA, Y., *Chem. Pharm. Bull.*, **12**, 1066 (1964).
1793. MORITZ, A.G. and N.SHEPPARD, *Mol. Phys.*, **5**, 361 (1962).
1794. MORROW, D.F., M.E.BROKKE and G.W.MOERSCH, *Chem. and Ind.*, 1655 (1962).
1795. MORTIMER, F.S., *J. Mol. Spec.*, **3**, 335 (1959).
1796. MORTIMER, F.S., *J. Mol. Spec.*, **5**, 199 (1960).
1797. MOSS, R.A., *Tetrahedron Letters*, **711** (1966).
1798. MOTCHANE, J.L., *Compt. Rend.*, **254**, 1614 (1962).
1799. MOUSSEBOIS, C. and J.F.M.OTH, *Helv. Chim. Acta*, **47**, 942 (1964).
1800. MOYNEHAN, T.M., K.SCHOFIELD, R.A.Y.JONES and A.R.KATRITZKY, *Proc. Chem. Soc.*, 218 (1961).
1801. MOYNEHAN, T.M., K.SCHOFIELD, R.A.Y.JONES and A.R.KATRITZKY, *J. Chem. Soc.*, 2673 (1962).
1802. MULLER, J.C., *Bull. Soc. Chim. France*, 1815 (1964).
1803. MULLER, J.C., *Bull. Soc. Chim. France*, 2027 (1964).
1804. MULLER, N., *J. Chem. Phys.*, **36**, 359 (1962).
1805. MULLER, N., *J. Chem. Phys.*, **37**, 2729 (1962).
1806. MULLER, N. and D.T.CARR, *J. Phys. Chem.*, **67**, 112 (1963).
1807. MULLER, N. and D.E.PRITCHARD, *J. Chem. Phys.*, **31**, 768 (1959).
1808. MULLER, N. and D.E.PRITCHARD, *J. Chem. Phys.*, **31**, 1471 (1959).
1809. MULLER, N. and D.E.PRITCHARD, *J. Amer. Chem. Soc.*, **82**, 248 (1960).
1810. MULLER, N. and P.I.ROSE, *J. Amer. Chem. Soc.*, **84**, 3973 (1962).
1811. MULLER, N. and P.J.SCHULTZ, *J. Phys. Chem.*, **68**, 2026 (1964).
1812. MULLER, N. and W.C.TOSCH, *J. Chem. Phys.*, **37**, 1167 (1962).
1813. MURAYAMA, K. and K.NUKADA, *Bull. Chem. Soc. Japan*, **36**, 1233 (1963).
1814. MURDOCH, H.D., *J. Organometal. Chem.*, **4**, 119 (1965).
1815. MURRELL, J.N. and V.M.S.GIL, *Trans. Farad. Soc.*, **61**, 402 (1965).
1816. MURRELL, J.N., V.M.S.GIL and F.B.VAN DUIJNEVELDT, *Rec. Trav. Chim.*, **84**, 1399 (1965).
1817. MURTHY, S.V., *J. Sci. Ind. Res (India)*, **21B**, 333 (1962).
1818. MUSHER, J.I., *J. Chem. Phys.*, **34**, 594 (1961).
1819. MUSHER, J.I., *J. Chem. Phys.*, **35**, 1159 (1961).
1820. MUSHER, J.I., *J. Chem. Phys.*, **36**, 1086 (1962).
1821. MUSHER, J.I., *J. Chem. Phys.*, **37**, 34 (1962).
1822. MUSHER, J.I., *J. Chem. Phys.*, **37**, 192 (1962) and earlier papers.
1823. MUSHER, J.I., *J. Chem. Phys.*, **40**, 983 (1964).
1824. MUSHER, J.I., *J. Chem. Phys.*, **43**, 4081 (1965).

1825. MUSHER, J.I., *J. Amer. Chem. Soc.*, **83**, 1146 (1961).
1826. MUSHER, J.I., *Bull. Amer. Phys. Soc.*, **8**, 35 (1963).
1827. MUSHER, J.I., *Mol. Phys.*, **6**, 93 (1963).
1828. MUSHER, J.I., *Mol. Phys.*, **6**, 93 (1963).
1829. MUSHER, J.I., *Spectrochim. Acta*, **16**, 835 (1960).
1830. MUSHER, J.I. and E.J. COREY, *Tetrahedron*, **18**, 791 (1962).
1831. MUSHER, J.I. and R.G. GORDON, *J. Chem. Phys.*, **36**, 3097 (1962).
1832. MUETTERTIES, E.L. and W.D. PHILLIPS, *J. Amer. Chem. Soc.*, **81**, 1084 (1959).
1833. NAGAI, Y., J. HOOZ and R.A. BENKESER, *Bull. Chem. Soc. Japan*, **37**, 53 (1964).
1834. NAGARJAN, V., *J. Sci. Ind. Res. (India)*, **18B**, 84 (1959).
1835. NAGASAMPAGI, B.A., R.C. PANDEY, V.S. PANSARE, J.R. PRAHLAD and SUKH DEV, *Tetrahedron Letters*, 411 (1964).
1836. NAGATA, W., T. TERASAWA and K. TORI, *J. Amer. Chem. Soc.*, **86**, 3746 (1964).
1837. NAGESWARA RAO, B.D. and J.D. BALDESCHWIELER, *J. Chem. Phys.*, **37**, 2473 (1962).
1838. NAGESWARA RAO, B.D., J.D. BALDESCHWIELER and J.I. MUSHER, *J. Chem. Phys.*, **37**, 2480 (1962).
1839. NAGESWARA RAO, B.D. and P. VENKATESWARLU, *Indian Academy Sci. Proc.*, **54A**, 1 (1961).
1840. NAIR, M.D. and R. ADAMS, *J. Amer. Chem. Soc.*, **82**, 3786 (1960).
1841. NAIR, M.D. and R. ADAMS, *J. Amer. Chem. Soc.*, **83**, 922 (1961).
1842. NAIR, P.M., G. GOPAKUMAR and T. FAIRWELL, *International Symposium on Nuclear Magnetic Resonance*, Tokyo, (1965).
1843. NAIR, P.M. and G. GOPAKUMAR, *Tetrahedron Letters*, 709 (1964).
1844. NAKAGAWA, N., *International Symposium on Nuclear Magnetic Resonance*, Tokyo, 1965.
1845. NAKAGAWA, N. and S. FUJIWARA, *Bull. Chem. Soc. Japan*, **34**, 143 (1961).
1846. NAKANISHI, K., Y. TAKAHASHI and M. BUDZIKIEWICZ, *J. Org. Chem.*, **30**, 1729 (1965).
1847. NARASIMHAN, P.T. and M.T. ROGERS, *J. Chem. Phys.*, **31**, 1302 (1959).
1848. NARASIMHAN, P.T. and M.T. ROGERS, *J. Chem. Phys.*, **31**, 1428 (1959).
1849. NARASIMHAN, P.T. and M.T. ROGERS, *J. Chem. Phys.*, **33**, 727 (1960).
1850. NARASIMHAN, P.T. and M.T. ROGERS, *J. Chem. Phys.*, **34**, 1049 (1961).
1851. NARASIMHAN, P.T. and M.T. ROGERS, *J. Phys. Chem.*, **63**, 1388 (1959).
1852. NARASIMHAN, P.T. and M.T. ROGERS, *J. Amer. Chem. Soc.*, **82**, 34 (1960).
1853. NARASIMHAN, P.T. and M.T. ROGERS, *J. Amer. Chem. Soc.*, **82**, 5983 (1960).
1854. NARAYANAN, C.R. and K.N. IYER, *Tetrahedron Letters*, 3741 (1965).
1855. NARAYANAN, C.R. and R.V. PACHAPURKAR, *Chem. and Ind.*, 322 (1964).
1856. NARAYANAN, C.R. and N.K. VENKATASUBRAMANIAN, *Tetrahedron Letters*, 3639 (1965).
1857. NEALE, R.S. and E.B. WHIPPLE, *J. Amer. Chem. Soc.*, **86**, 3130 (1964).
1858. NIEDRICH, R.A., D.M. GRANT and M. BARFIELD, *J. Chem. Phys.*, **42**, 3733 (1965).
1859. NEIKAM, W.C. and B.P. DAILEY, *J. Chem. Phys.*, **38**, 445 (1963).
1860. NELSON, F.A. and H.E. WEAVER, *Science*, **146**, 223 (1964).
1861. NEUENSCHWANDER, M., D. MEUCHE and H. SCHALTEGGER, *Helv. Chim. Acta*, **46**, 1760 (1963).
1862. NEUENSCHWANDER, M., D. MEUCHE and H. SCHALTEGGER, *Helv. Chim. Acta*, **47**, 1022 (1964).
1863. NEUMAN, R.C. and G.S. HAMMOND, *J. Phys. Chem.*, **67**, 1659 (1963).
1864. NEUMAN, R.C., G.S. HAMMOND and T.J. DOUGHERTY, *J. Amer. Chem. Soc.*, **84**, 1506 (1962).
1865. NEUMAN, R.C. and L.B. YOUNG, *J. Phys. Chem.*, **69**, 1777 (1965).
1866. NEWMAN, M.S., *J. Org. Chem.*, **27**, 323 (1962).
1867. NEWMAN, M.S., G. FRAENKEL and W.N. KIRN, *J. Org. Chem.*, **28**, 1851 (1963).
1868. NEWMARK, R.A. and C.H. SEDERHOLM, *J. Chem. Phys.*, **39**, 3131 (1963).
1869. NEWMARK, R.A. and C.H. SEDERHOLM, *J. Chem. Phys.*, **43**, 602 (1965).
1870. NEWSOROFF, G.P. and S. STERNHELL, *Tetrahedron Letters*, 3499 (1964) and unpublished data.
1871. NG, S. and C.H. SEDERHOLM, *J. Chem. Phys.*, **40**, 2090 (1964).

1872. NG, S., J.TANG and C.H.SEDERHOLM, *J. Chem. Phys.*, **42**, 79 (1965).
1873. NICKON, A., M.A.CASTLE, R.HARADA, C.E.BERKOFF and R.O.WILLIAMS, *J.Amer. Chem. Soc.*, **85**, 2185 (1963).
1874. NILSSON, K. and S.STERNHELL, *Acta Chem. Scand.*, **19**, 2441 (1965).
1875. NIXON, J.F., *Chem. and Ind.*, 1555 (1963).
1876. NOGGLE, J.H., *Rev. Sci. Instr.*, **35**, 1166 (1964).
1877. NORDLANDER, J.E. and J.D.ROBERTS, *J. Amer. Chem. Soc.*, **81**, 1769 (1959).
1878. NORDLANDER, J.E., W.G.YOUNG and J.D.ROBERTS, *J. Amer. Chem. Soc.*, **83**, 494 (1961).
1879. NORIN, T., *Acta Chem. Scand.*, **17**, 738 (1963).
1880. NORRIS, R.K. and S.STERNHELL, *Aust. J. Chem.*, **19**, 617 (1966).
1881. NORRIS, R.K. and S.STERNHELL, *Aust. J. Chem.*, **19**, 841 (1966).
1882. NOZOE, T. and K.MATSUI, *Bull. Chem. Soc. Japan*, **34**, 616 (1961).
1883. NUKADA, K., O. YAMANOTO, T.SUZUKI, M.TAKEUCHI and M.OHNISHI, *Anal. Chem.*, **35**, 1892 (1963).
1884. O'CONNOR, R., *J. Org. Chem.*, **26**, 4375 (1961).
1885. ODAJIMA, A., J.A.SAUER and A.E.WOODWARD, *J. Phys. Chem.*, **66**, 718 (1962).
1886. OGATA, M., H.KANO and K.TORI, *Chem. Pharm. Bull.*, **11**, 1527 (1963).
1887. OGATA, M., H.WATANABE, K.TORI and H.KANO, *Tetrahedron Letters*, 19 (1964).
1888. OGLIARUSO, M., R.RIEKE and S.WINSTEIN, *J. Amer. Chem. Soc.*, **88**, 4731 (1966).
1889. OHLINE, R.W., A.L.ALLRED and F.G.BORDWELL, *J. Amer. Chem. Soc.*, **86**, 4641 (1964).
1890. OHASHI, O., M.MASHIMA and M.KUBO, *Canad. J. Chem.*, **42**, 970 (1964).
1891. OKI, M. and H.IWAMURA, *Bull. Chem. Soc. Japan*, **36**, 1 (1963).
1892. OKI, M. and H.IWAMURA, *International Symposium on Nuclear Magnetic Resonance*, Tokyo 1965.
1893. OKI, M., H.IWAMURA and N.HAYAKAMA, *Bull. Chem. Soc. Japan*, **37**, 1865 (1964).
1894. OKOMOTO, T. and Y.KAWAZOE, *Chem. Pharm. Bull. (Tokyo)*, **11**, 643 (1963).
1895. OLAH, G.A., *J. Amer. Chem. Soc.*, **86**, 932 (1964).
1896. OLAH, G.A., *J. Amer. Chem. Soc.*, **87**, 1103 (1965).
1897. OLAH, G.A., E.B.BAKER and M.B.COMISAROW, *J. Amer. Chem. Soc.*, **86**, 1265 (1964).
1898. OLAH, G.A., E.B.BAKER, J.C.EVANS, W.S.TOLGYESI, J.S.McINTYRE and I.J. BASTIEN, *J. Amer. Chem. Soc.*, **86**, 1360 (1964).
1899. OLAH, G.A. and M.B.COMISAROW, *J. Amer. Chem. Soc.*, **86**, 5682 (1964).
1900. OLAH, G.A., C.A.CUPAS and M.B.COMISAROW, *J. Amer. Chem. Soc.*, **88**, 362 (1966).
1901. OLAH, G.A., S.J.KUHN, W.S.TOLGYESI and E.B.BAKER, *J. Amer. Chem. Soc.*, **84**, 2733 (1962).
1902. OLAH, G.A. and C.U.PITTMAN, *J. Amer. Chem. Soc.*, **87**, 3507 (1965).
1903. OLAH, G.A. and C.U.PITTMAN, *J. Amer. Chem. Soc.*, **87**, 3509 (1965).
1904. OLAH, G.A., C.U.PITTMAN and T.S.SORENSEN, *J. Amer. Chem. Soc.*, **88**, 2331 (1966).
1905. OLAH, G.A. and W.S.TOLGYESI, *J. Amer. Chem. Soc.*, **83**, 5031 (1961).
1906. OLAH, G.A., W.S.TOLGYESI, S.J.KUHN, M.E.MOFFAT, I.J.BASTIEN and E.B.BAKER, *J. Amer. Chem. Soc.*, **85**, 1328 (1963).
1907. OLOFSON, R.A. and J.S.MICHELMAN, *J. Org. Chem.*, **30**, 1854 (1965) and private communication.
1908. OLOMUCKI, M., G.DESVAGES, NGUYEN-VAN THOAI and J.ROCHE, *Compt. Rend.*, **260**, 4519 (1965).
1909. O'LONAE, J.K., C.M.COMBS and R.L.GRIFFITH, *J. Org. Chem.*, **29**, 1730 (1964).
1910. OLSEN, R.K. and H.R.SNYDER, *J. Org. Chem.*, **28**, 3050 (1963).
1911. OLSEN, R.K. and H.R.SNYDER, *J. Org. Chem.*, **30**, 184 (1965).
1912. ONSAGER, L., *J. Amer. Chem. Soc.*, **58**, 1486 (1936).
1913. O'REILLY, D.E., *J. Chem. Phys.*, **36**, 855 (1962).
1914. OSAWA, Y. and M.NEEMAN, *J. Amer. Chem. Soc.*, **85**, 2856 (1963).
1915. OSTERCAMP, D.L., *J. Org. Chem.*, **30**, 1169 (1965).
1916. OSWALD, A.A., K.GRIESBAUM and B.E.HUDSON JR., *J. Org. Chem.*, **28**, 1262 (1963).

1917. OSWALD, A.A., K.GRIESBAUM and W.NAEGELE, *J. Amer. Chem. Soc.*, **86**, 3791 (1964).
1918. OSWALD, A.A., K.GRIESBAUM, W.A.THALER and B.E.HUDSON, JR., *J. Amer. Chem. Soc.*, **84**, 3897 (1962).
1919. OTTAVI, M.H. and M.L.DE BROGLIE, *Compt. Rend.*, **253**, 430 (1961).
1920. OUELLETTE, R.J., *J. Amer. Chem. Soc.*, **86**, 4378 (1964).
1921. OUELLETTE, R.J., *Canad. J. Chem.*, **43**, 707 (1965).
1922. OUELLETTE, R.J., G.E.BOOTH and K.LIPTAK, *J. Amer. Chem. Soc.*, **87** 3436 (1965).
1923. OUELLETTE, R.J., K.LIPTAK and G.E.BOOTH, *J. Org. Chem.*, **31**, 546 (1966).
1924. OVERBERGER, C.G. and J.P.ANSELME, *Chem. and Ind.*, 280 (1964).
1925. OVERBERGER, C.G., J.P.ANSELME and J.R.HALL, *J. Amer. Chem. Soc.*, **85**, 2752 (1963).
1926. OVERBERGER, C.G. and A.DRUCKER, *J. Org. Chem.*, **29**, 360 (1964).
1927. PACHLER, K.G.R., *Spectrochim. Acta*, **19**, 2085 (1963).
1928. PADWA, A., *Tetrahedron Letters*, 813 (1964).
1929. PAKE, G.E., *Amer. J. Phys.*, **18**, 438, 473 (1950).
1930. PALA, G., *Nature*, **204**, 1190 (1964).
1931. PAQUETTE, L.A., *J. Amer. Chem. Soc.*, **86**, 4096 (1964).
1932. PAQUETTE, L.A., J.H.BARRETT, R.P.SPITZ and R.PITCHER, *J. Amer. Chem. Soc.*, **87**, 3417 (1965).
1933. PARELLO, J., *Bull. Soc. Chim. France*, 2033 (1964).
1934. PARELLO, J., P.LONGEVIALLE, W.VETTER and J.A.MCCLOSKEY, *Bull. Soc. Chim. France*, 2787 (1963).
1935. PARHAM, W.E. and L.D.HUESTIS, *J. Amer. Chem. Soc.*, **84**, 813 (1962).
1936. PARHAM, W.E. and R.KONCOS, *J. Amer. Chem. Soc.*, **83**, 4034 (1961).
1937. PARKER, G.W. and J.D.MEMORY, *J. Chem. Phys.*, **43**, 1388 (1965).
1938. PARKHURST, R.H.J.O.RODIN and R.M.SILVERSTEIN, *J. Org. Chem.*, **28**, 120 (1963).
1939. PARMIGIANI, A., A.PEROTTI and V.RIGANTI, *Gazz. Chim. Italiana*, **91**, 1148 (1961).
1940. PASCAL, Y., J.P.MARIZUR and J.WIEMANN, *Bull. Soc. Chim. France*, 2211 (1965).
1941. PASCUAL, C., J.MEIER and W.SIMON, *Helv. Chim. Acta*, **49**, 164 (1966).
1942. PASCUAL, C., D.WEGMANN, U.GRAF, R.SCHEFFOLD, P.F.SOMMER and W.SIMON, *Helv. Chim. Acta*, **47**, 213 (1964).
1943. PAASIVIRTA, J., *Suomen. Kem.*, **36B**, 76 (1963).
1944. PASTO, D.J. and J.L.MIESEL, *J. Amer. Chem. Soc.*, **85**, 2118 (1963).
1945. PATCHETT, A.A., F.HOFFMAN, F.F.GIARRUSSO, H.SCHWAM and G.E.ARTH, *J. Org. Chem.*, **27**, 3822 (1962).
1946. PATEL, D.J., M.E.H.HOWDEN and J.D.ROBERTS, *J. Amer. Chem. Soc.*, **85**, 3218 (1963).
1947. PATERSON, W.G. and G.BIGAM, *Canad. J. Chem.*, **41**, 1841 (1963).
1948. PATERSON, W.G. and H.SPEDDING. *Canad. J. Chem.*, **41**, 2706 (1963).
1949. PATERSON, W.G. and N.R.TIPMAN, *Canad. J. Chem.*, **40**, 2122 (1962).
1950. PATERSON, W.G. and E.J.WELLS, *J. Mol. Spec.*, **14**, 101 (1964).
1951. PAUDLER, W.W. and H.L.BLEWITT, *Tetrahedron*, **21**, 353 (1965).
1952. PAUDLER, W.W. and H.L.BLEWITT, *J. Org. Chem.*, **31**, 1295 (1966).
1953. PAUDLER, W.W. and J.E.KUDER, *J. Org. Chem.*, **31**, 809 (1966).
1954. PAUDLER, W.W. and S.WAGNER, *Chem. and Ind.*, 1693 (1963).
1955. PAUL, E.G. and D.M.GRANT, *J. Amer. Chem. Soc.*, **85**, 1701 (1963).
1956. PAULING, L., *J. Chem. Phys.*, **4**, 673 (1936).
1957. PAULSEN, P.L. and W.D.COOKE, *Anal. Chem.*, **35**, 1560 (1963).
1958. PAVIA, A., J.WYLDE, R.WYLDE and E.ARNAL, *Bull. Soc. Chim. France*, 2709, 2718 (1965).
1959. PEARSON, R.G., J.PALMER, M.M.ANDERSON and A.L.ALLRED, *Z. Electrochem.*, **64**, 110 (1960).
1960. PEARSON, R.G. and R.D.LANIER, *J. Amer. Chem. Soc.*, **86**, 765 (1964).
1961. PECHER, J., R.H.MARTIN, N.DEFAY, M.KAISIN, J.PEETERS, G.VAN BINST, H.VERZELA and F.ALDERWEIRELDT, *Tetrahedron Letters*, 270 (1961).
1962. PEDERSEN, B.F. and B.PEDERSEN, *Tetrahedron Letters*, 2995 (1965).

1963. PEJKOVIC-TADIC, I., M.HRANISAVLJECIC-JAKOVLJEVIC, S.NESIC, C.PASCUAL and W.SIMON, *Helv. Chim. Acta*, **48**, 1157 (1965).
1964. PERKAMPUS, H.H. and E.BAUMGARTEN, *Angew. Chem.*, **76**, 965 (1964).
1965. PERLIN, A.S., *Canad. J. Chem.*, **44**, 539 (1966).
1966. PETRAKIS, L. and H.J.BERNSTEIN, *J. Chem. Phys.*, **37**, 2731 (1962).
1967. PETRAKIS, L. and C.H.SEDERHOLM, *J. Chem. Phys.*, **34**, 1174 (1961).
1968. PETRAKIS, L. and C.H. SEDERHOLM, *J. Chem. Phys.*, **35**, 1243 (1961).
1969. PETRAKIS, L. and C.H.SEDERHOLM, *J. Chem. Phys.*, **36**, 1087 (1962).
1970. PETROV, A.A. and V.B.LEBEDEV, *J. Gen. Chem. USSR*, **32**, 653 (1962).
1971. PETTIT, G.R., I.B.DOUGLASS and R.A.HILL, *Canad. J. Chem.*, **42**, 2357 (1964).
1972. PHILLIPS, W.D., H.C.MILLER and E.L.MUETTERTIES, *J. Amer. Chem. Soc.*, 4496 (1959).
1973. PIERRE, J.L., P.CHAUTEMPS and P.ARNAUD, *Compt. Rend.*, **261**, 4025 (1965).
1974. PIERRE, J.L. and P.ARNAUD, *Bull. Soc. Chim. France*, 1040 (1966).
1975. PIETTE, L.H. and W.A.ANDERSON, *J. Chem. Phys.*, **30**, 899 (1959).
1976. PIKE, J.E., M.A.REBENSTORF, G.SLOMP and F.A.MACKELLER, *J. Org. Chem.*, **28**, 2499 (1963).
1977. PIKE, J.E., G.SLOMP and F.A.MACKELLER, *J. Org. Chem.*, 2502 (1963).
1978. PINHEY, J.T. personal communication.
1979. PINHEY, J.T. and S.STERNHELL, *Tetrahedron Letters*, 275 (1963).
1980. PINHEY, J.T. and S.STERNHELL, *Aust. J. Chem.*, **18**, 543 (1965).
1981. PIOZZI, F., A.QUILICO, T.AJELLO, V.SPRIO and A.MELERA, *Tetrahedron Letters*, 1829 (1965).
1982. PITTMAN, C.U. and G.A.OLAH, *J. Amer. Chem. Soc.*, **87**, 5123 (1965).
1983. PIZEY, J.S. and W.E.TRUCE, *J. Chem. Soc.*, 865 (1964).
1984. PLIENINGER, H., H.BAUR, A.R.KATRITZKY and U.LERCH, *Ann.*, **654**, 165 (1962).
1985. POHLAND, A.E., R.C.BADGEE and N.H.CROMWELL, *Tetrahedron Letters*, 4369 (1965).
1986. POMMIER, J., *Compt. Rend.*, **257**, 1699 (1963).
1987. POPLE, J.A., *Proc. Roy. Soc. (London)*, A**239**, 550 (1957).
1988. POPLE, J.A., *Mol. Phys.*, **1**, 1 (1958).
1989. POPLE, J.A., *Mol. Phys.*, **1**, 168 (1958).
1990. POPLE, J.A., *Mol. Phys.*, **1**, 170 (1958).
1991. POPLE, J.A., *Mol. Phys.*, **1**, 175 (1958).
1992. POPLE, J.A., *Mol. Phys.*, **1**, 216 (1958).
1993. POPLE, J.A., *Mol. Phys.*, **7**, 301 (1963–4).
1994. POPLE, J.A. and D.P.SANTRY, *Mol. Phys.*, **8**, 1 (1964).
1995. POPLE, J.A., *J. Chem. Phys.*, **24**, 1111 (1956).
1996. POPLE, J.A., *J. Chem. Phys.*, **37**, 53 (1962).
1997. POPLE, J.A., *J. Chem. Phys.*, **37**, 60 (1962).
1998. POPLE, J.A., *J. Chem. Phys.*, **37**, 3009 (1962).
1999. POPLE, J.A., *J. Chem. Phys.*, **38**, 1276 (1963).
2000. POPLE, J.A., *J. Chem. Phys.*, **41**, 2559 (1964).
2001. POPLE, J.A. and A.A.BOTHNER-BY, *J. Chem. Phys.*, **42**, 1339 (1965).
2002. POPLE, J.A. and D.P.SANTRY, *Mol. Phys.*, **9**, 311 (1965).
2003. POPLE, J.A. and T.SCHAEFER, *Mol. Phys.*, **3**, 547 (1960).
2004. POPLE, J.A., W.G.SCHNEIDER and H.J.BERNSTEIN, *High Resolution NMR*, McGraw-Hill (1959).
2005. PORTE, A.L. and H.S.GUTOWSKY, *J. Org. Chem.*, **28**, 3216 (1963).
2006. PORTE, A.L., H.S.GUTOWSKY and I.M.HUNSBERGER, *J. Amer. Chem. Soc.*, **82**, 5057 (1960).
2007. POTTS, K.T., *J. Org. Chem.*, **28**, 543 (1963).
2008. POUND, R.V. and R.FREEMAN, *Rev. Sci. Instr.*, **31**, 96 (1960).
2009. POUND, R.V. and R.FREEMAN, *Rev. Sci. Instr.*, **31**, 103 (1960).
2010. POUZARD, G., L.PUJOL, J.ROGGERO and E.VINCENT, *J. Chim. Phys.*, **61**, 612 (1964).
2011. POWLES, J.G. and J.H.STRANGE, *Mol. Phys.*, **5**, 329 (1962).
2012. POZIOMEK, E.J., D.N.KRAMER, W.A.MOSHER and H.O.MICHEL, *J. Amer. Chem. Soc.*, **83**, 3916 (1961).

2013. PRELOG, V., *Pure Appl. Chem.*, **7**, 551 (1963).

2014. PREMUZIC, E. and L.W.REEVES, *Canad. J. Chem.*, **40**, 1870 (1962).

2015. PREMUZIC, E. and L.W.REEVES, *Canad. J. Chem.*, **42**, 1498 (1964).

2016. PREMUZIC, E. and L.W.REEVES, *J. Chem. Soc.*, 4817 (1964).

2017. PRIMAS, H., *Spectrochim. Acta*, **14**, 17 (1959).

2018. PRIMAS, H., R.ARNDT and R.ERNST, *Z. Instrum. Kde.*, **68**, 55 (1960).

2019. PRIMAS, H., R.ARNDT and R.ERNST, *Advances in Molecular Spectroscopy*, Ed. by A.MANGINI (Pergamon Press, 1962). Vol.III. p.1246.

2020. PRINZBACH, H. and E.DRUCKREY, *Tetrahedron Letters*, 2959 (1965).

2021. PRINZBACH, H., H.HAGEMANN, J.H.HARTENSTEIN and R.KITZING, *Chem. Ber.*, **98**, 2201 (1965).

2022. PRINZBACH, H. and J.H.HARTENSTEIN, *Angew. Chem.*, **75**, 639 (1963).

2023. PRITCHARD, J.G. and P.C.LAUTERBUR, *J. Amer. Chem. Soc.*, **83**, 2105 (1961).

2024. PRITCHARD, J.G. and R.L.VOLLMER, *J. Org. Chem.*, **28**, 1545 (1963).

2025. PROSSER, F. and L.GOODMAN, *J. Chem. Phys.*, **38**, 374 (1963).

2026. PUMPHREY, N.W. and M.J.T.ROBINSON, *Tetrahedron Letters*, 741 (1963).

2027. PURCELL, J.M. and J.A.CONNELY, *Anal. Chem.*, **37**, 1181 (1965).

2028. PURCELL, J.M. and H.SUSI, *Appl. Spec.*, **19**, 105 (1965).

2029. QUINN, H.W., J.S.MCINTYRE and D.J.PETERSON, *Canad. J. Chem.*, 2896 (1965).

2030. RADLICK, P. and W.ROSEN, *J. Amer. Chem. Soc.*, **88**, 3461 (1966).

2031. RADLICK, P. and S.WINSTEIN, *J. Amer. Chem. Soc.*, **85**, 344 (1963).

2032. RAE, I.D., *Canad. J. Chem.*, **44**, 1334 (1966).

2033. RAE, I.D., *Aust. J. Chem.*, **18**, 1807 (1965).

2034. RAE, I.D., *Aust. J. Chem.*, **19**, 409 (1966).

2035. RAE, I.D. and L.K.DYALL, *Aust. J. Chem.*, **19**, 835 (1966).

2036. RANEY, K.C. and J.MESSICK, *Tetrahedron Letters*, 4423 (1965).

2037. RAMIREZ, F. and N.B.DESAI, *J. Amer. Chem. Soc.*, **85**, 3252 (1963).

2038. RAMIREZ, F., N.B.DESAI and N.RAMANATHAN, *J. Amer. Chem. Soc.*, **85**, 1874 (1963).

2039. RAMIREZ, F., N.B.DESAI and N.RAMANATHAN, *Tetrahedron Letters*, 323 (1963).

2040. RAMIREZ, F., O.P.MADAN, N.B.DESAI, S.MEYERSON and E.M.BANAS, *J. Amer. Chem. Soc.*, **85**, 2681 (1963).

2041. RAUSCH, M.D. and V.MARK, *J. Org. Chem.*, **28**, 3225 (1963).

2042. RAMIREZ, F. and A.V.PATWARDHAN, *J. Org. Chem.*, **30**, 2575 (1965).

2043. RAMIREZ, F., A.V.PATWARDHAN, N.B.DESAI, N.RAMANATHAN and C.V.GRECO, *J. Amer. Chem. Soc.*, **85**, 3056 (1963).

2044. RAMIREZ, F., N.RAMANATHAN and N.B.DESAI, *J. Amer. Chem. Soc.*, **85**, 3465 (1963).

2045. RAMSEY, N.F., *Phys. Rev.*, **91**, 303 (1953).
    RAMSEY, N.F. and E.M.PURCELL, *ibid*, **85**, 143 (1952).

2046. RAMSEY, N.F., *Phys. Rev.*, **78**, 699 (1950).

2047. RANDALL, E.W. and J.D.BALDESCHWIELER, *J. Mol. Spec.*, **8**, 365 (1962).

2048. RANDALL, J.C., J.J.MCLESKEY, P.SMITH and M.E.HOBBS, *J. Amer. Chem. Soc.*, **86**, 3229 (1964).

2049. RANDALL, J.C., R.L.VAULX, M.E.HOBBS and C.R.HAUSER, *J. Org. Chem.*, **30**, 2035 (1965).

2050. RANFT, J., *Ann. Phys.*, **8**, 322 (1961).

2051. RANFT, J., *Ann. Phys.*, **9**, 124 (1962).

2052. RANFT, J., *Ann. Phys.*, **9**, 279 (1962).

2053. RANFT, J., *Ann. Phys.*, **10**, 1 (1963).

2054. RANFT, J., *Ann. Phys.*, **10**, 399 (1963).

2055. RANFT, J. and S.DÄHNE, *Helv. Chim. Acta*, **47**, 1160 (1964).

2056. RAO, B.D.N., *Mol. Phys.*, **7**, 307 (1963–4).

2057. RAO, B.D.N. and J.D.BALDESCHWIELER, *J. Mol. Spec.*, **11**, 440 (1964).

2058. RAO, B.D.N. and P.VENKATESWARLU, *Proc. Indian Acad. Sci.*, **52A**, 109 (1960).

2059. RAO, B.D.N. and P.VENKATESWARLU, *Proc. Indian Acad. Sci.*, **54A**, 305 (1961).

2060. RAPER, A.H. and E.ROTHSTEIN, *J. Chem. Soc.*, 1027 (1963).

2061. RAPPE, C., *Acta Chem. Scand.*, **19**, 31 (1965).

2062. RASSAT, A., C.W.JEFFORD, J.M.LEHN and B.WAEGELL, *Tetrahedron Letters*, 233 (1964).
2063. FRATIELLO, A., *J. Chem. Phys.*, **41**, 2204 (1964).
2064. RAUK, A., E.BUNCEL, R.Y.MOIR and S.WOLFE, *J. Amer. Chem. Soc.*, **87**, 5498 (1965).
2065. RAUTENSTRAUCH, V. and F.WINGLER, *Tetrahedron Letters*, 4703 (1965).
2066. RAYNES, W.T., A.D.BUCKINGHAM and H.J.BERNSTEIN, *J. Chem. Phys.*, **36**, 3481 (1962).
2067. REDDY, G.S., C.E.BOOZER, J.H.GOLDSTEIN, *J. Chem. Phys.*, **34**, 700 (1961).
2068. REDDY, G.S. and J.H.GOLDSTEIN, *J. Amer. Chem. Soc.*, **83**, 2045 (1961).
2069. REDDY, G.S. and J.H.GOLDSTEIN, *J. Amer. Chem. Soc.*, **84**, 583 (1962).
2070. REDDY, G.S. and J.H.GOLDSTEIN, *J. Mol. Spec.*, **8**, 475 (1962).
2071. REDDY, G.S. and J.H.GOLDSTEIN, *J. Chem. Phys.*, **38**, 2736 (1963).
2072. REDDY, G.S. and J.H.GOLDSTEIN, *J. Chem. Phys.*, **38**, 2736 (1963).
2073. REDDY, G.S. and J.H.GOLDSTEIN, *J. Chem. Phys.*, **39**, 3509 (1963).
2074. REDDY, G.S., R.T.HOBGOOD JR., and J.H.GOLDSTEIN, *J. Amer. Chem. Soc.*, **84**, 336 (1962).
2075. REDDY, G.S., L.MANDELL and J.H.GOLDSTEIN, *J. Chem. Soc.*, 1414 (1963).
2076. REDDY, G.S., L.MANDELL and J.H.GOLDSTEIN, *J. Amer. Chem. Soc.*, **83**, 4729 (1961).
2077. REEVES, L.W., *Canad. J. Chem.*, **38**, 748 (1960).
2078. REEVES, L.W., *Canad. J. Chem.*, **39**, 1711 (1961).
2079. REEVES, L.W. in *Advances in Physical Organic Chemistry*, Vol.3 (Ed. by V.GOLD) Academic Press, N.Y. 1965, p.187.
2080. REEVES, L.W., E.A.ALLEN and K.D.STROMME, *Canad. J. Chem.*, **38**, 1249 (1960).
2081. REEVES, L.W. and W.G.SCHNEIDER, *Canad. J. Chem.*, **35**, 251 (1957).
2082. REEVES, L.W. and K.O.STROMME, *Canad. J. Chem.*, **38**, 1241 (1960).
2083. REEVES, L.W. and K.O.STROMME, *Canad. J. Chem.*, **39**, 2318 (1961).
2084. REEVES, L.W. and K.O.STROMME, *Trans. Farad. Soc.*, **57**, 390 (1961).
2085. REEVES, L.W. and E.J.WELLS, *Canad. J. Chem.*, **41**, 2698 (1963).
2086. REILLY, C.A., *Anal. Chem.*, **32**, 221R (1960).
2087. REILLY, C.A., *J. Chem. Phys.*, **37**, 456 (1962).
2088. REILLY, C.A. and J.D.SWALEN, *J. Chem. Phys.*, **32**, 1378 (1960).
2089. REILLY, C.A. and J.D.SWALEN, *J. Chem. Phys.*, **33**, 617 (1960).
2090. REILLY, C.A. and J.D.SWALEN, *J. Chem. Phys.*, **33**, 1257 (1960).
2091. REILLY, C.A. and J.D.SWALEN, *J. Chem. Phys.*, **34**, 980 (1961).
2092. REILLY, C.A. and J.D.SWALEN, *J. Chem. Phys.*, **35**, 1522 (1961).
2093. REINECKE, M.G., *J. Org. Chem.*, **28**, 3574 (1963).
2094. REINECKE, M.G., J.W.JOHNSON and J.F.SEBASTIAN, *Tetrahedron Letters*, 1183 (1963).
2095. REINECKE, M.G., J.W.JOHNSON and J.F.SEBASTIAN, *J. Amer. Chem. Soc.*, **85**, 2859 (1963).
2096. REINECKE, M.G., J.W.JOHNSON and J.F.SEBASTIAN, *Chem. and Ind.*, 151 (1964).
2097. REINHARD, R.R., *Rev. Sci. Instr.*, **36**, 549 (1965).
2098. RENK, E., P.R.SCHAFER, W.H.GRAHAM, R.H.MAZUR and J.D.ROBERTS, *J. Amer. Chem. Soc.*, **83**, 1987 (1961).
2099. REUBEN, J., D.SAMUEL and B.L.SILVER, *J. Amer. Chem. Soc.*, **85**, 3093 (1963).
2100. REUBEN, J., A.TZALMONA and D.SAMUEL, *Proc. Chem. Soc.*, 353 (1962).
2101. REYNOLDS, W.F. and T.SCHAEFER, *Canad. J. Chem.*, **41**, 540 (1963).
2102. REYNOLDS, W.F. and T.SCHAEFER, *Canad. J. Chem.*, **41**, 2339 (1963).
2103. REYNOLDS, W.F. and T.SCHAEFER, *Canad. J. Chem.*, **42**, 2119 (1964).
2104. REYNOLDS, W.F. and T.SCHAEFER, *Canad. J. Chem.*, **42**, 2641 (1964).
2105. RICHARDS, J.H. and W.F.BEACH, *J. Org. Chem.*, **26**, 623 (1961).
2106. RICHARDS, J.H., G.FRAENKEL, R.E.CARTER and A.MCLACHLAN, *J. Amer. Chem. Soc.*, **82**, 5846 (1960).
2107. RICHARDS, R.E., *Proc. Roy. Soc.*, **255A**, 72 (1960).
2108. RICHARDS, R.E. and T.SCHAEFER, *Mol. Phys.*, **1**, 331 (1958).
2109. RICHARDS, R.E. and T.SCHAEFER, *Proc. Roy. Soc.*, **A246**, 429 (1958).
2110. RICHARDS, R.E. and J.W.WHITE, *Proc. Roy. Soc.*, **269**, 287, 301 (1962).

2111. RICHEY, H. G. and R. K. LUSTGARTEN, *J. Amer. Chem. Soc.*, **88,** 3136 (1966).
2112. RICHEY, H. G., J. C. PHILIPS and L. E. RENNICK, *J. Amer. Chem. Soc.*, **87,** 1381 (1965).
2113. RICHEY, H. G., L. E. RENNICK, A. S. KUSHNER, J. M. RICHEY and J. C. PHILIPS, *J. Amer. Chem. Soc.*, **87,** 4017 (1965).
2114. RICKBORN, B., D. A. MAY and A. A. THELEN, *J. Org. Chem.*, **29,** 91 (1964).
2115. RIEHL, J. J., J. M. LEHN and F. HEMMERT, *Bill. Soc. Chim. France*, 224 (1963).
2116. RIGGS, N. V., *Aust. J. Chem.*, **16,** 521 (1963).
2117. RINEHART, K. L. JR., D. E. BUBLITZ and D. H. GUSTAFSON, *J. Amer. Chem. Soc.*, **85,** 970 (1963).
2118. RINEHART, K. L., A. K. FRERICHS, P. A. KITTLE, L. F. WESTMAN, D. H. GUSTAFSON, R. L. PRUETT and J. E. MCMAHON, *J. Amer. Chem. Soc.*, **82,** 4111 (1960).
2119. RITCHIE, E., W. C. TAYLOR and S. T. K. VAUTIN, *Aust. J. Chem.*, **18,** 2021 (1965).
2120. ROBERTS, B. W., J. B. LAMBERT and J. D. ROBERTS, *J. Amer. Chem. Soc.*, **87,** 5439 (1965).
2121. ROBERTS, J. D., *An Introduction to Spin–Spin Splitting in High Resolution Nuclear Magnetic Resonance Spectra*, W. A. Benjamin, N.Y., 1961.
2122. ROBERTSON, A. V., *Aust. J. Chem.*, **16,** 451 (1963).
2123. ROBERTSON, A. V., J. E. FRANCIS and B. WITKOP, *J. Amer. Chem. Soc.*, **84,** 1709 (1962).
2124. ROBINSON, C. H., N. F. BRUCE and E. P. OLIVETO, *J. Org. Chem.*, **28,** 975 (1963).
2125. ROBINSON, M. J. T., *Tetrahedron Letters*, 1685 (1965).
2126. ROBINSON, S. D. and B. L. SHAW, *Z. Naturforsch.* **18b,** 507 (1963).
2127. RODMAR, S., S. FORSEN, B. GESTBLOM, S. GRONOWITZ and R. A. HOFFMAN, *Acta Chem. Scand.*, **19,** 485 (1965).
2128. ROGERS, M. T., *J. Chem. Phys.*, **31,** 1430 (1959).
2129. ROGERS, M. T. and J. L. BURDETT, *Canad. J. Chem.*, **43,** 1516 (1965).
2130. ROGERS, M. T. and J. D. GRAHAM, *J. Amer. Chem. Soc.*, **84,** 3666 (1962).
2131. ROGERS, M. T. and J. C. WOODBREY, *J. Phys. Chem.*, **66,** 540 (1962).
2132. ROGERS, M. T. and J. C. WOODBREY, *J. Phys. Chem.*, **66,** 542 (1962).
2133. ROLL, D. B., B. J. NIST and A. C. HUITRIC, *Tetrahedron*, **20,** 2851 (1964).
2134. ROSENBAUM, J. and M. C. R. SYMONS, *Mol. Phys.*, **3,** 205 (1960).
2135. ROSEN, W. E., L. DORFMAN and M. P. LINFIELD, *J. Org. Chem.*, **29,** 1723 (1964).
2136. ROSENBERG, J. L. JR., J. E. MAHLER and R. PETTIT, *J. Amer. Chem. Soc.*, **84,** 2842 (1962).
2137. ROSENBERG, J. L. JR., J. E. MAHLER and R. PETTIT, *J. Amer. Chem. Soc.*, **84,** 2842 (1962).
2138. ROSENBLUM, M., V. NAYAK, S. K. DAS GUPTA and A. LONGROY, *J. Amer. Chem. Soc.*, **85,** 3874 (1963).
2139. ROSENKRANZ, R. E., K. ALLNER, R. GOOD, W. VON PHILIPSBORN and C. H. EUGSTER, *Helv. Chim. Acta.* **46,** 1259 (1963).
2140. ROTH, W. R., *Ann.*, **671,** 25 (1964).
2141. ROTTENDORF, H. and S. STERNHELL, *Aust. J. Chem.*, **16,** 647 (1963).
2142. ROTTENDORF, H. and S. STERNHELL, *Tetrahedron Letters*, 1289 (1963).
2143. ROTTENDORF, H. and S. STERNHELL, *Aust. J. Chem.*, **17,** 1315 (1964).
2144. ROTTENDORF, H., S. STERNHELL and J. R. WILMSHURST, *Aust. J. Chem.*, **18,** 1759 (1965).
2145. ROUSSELOT, M. M., *Compt. Rend.*, **262C,** 26 (1966).
2146. ROWLAND, R. L. and D. L. ROBERTS, *J. Org. Chem.*, **28,** 1165 (1963).
2147. RUDRUM, M. and D. F. SHAW, *J. Chem. Soc.*, 52, (1965).
2148. RUIDISCH, I. and M. SCHMIDT, *Chem. Ber.*, **96,** 1424 (1963).
2149. TUITE, R. J., H. R. SNYDER, A. L. PORTE and H. S. GUTOWSKY, *J. Phys. Chem.*, **65,** 187 (1961).
2150. RUSSELL, G. A. and A. ITO, *J. Amer. Chem. Soc.*, **85,** 2983 (1963).
2151. RYSCHEWITSCH, G. E., W. S. BREY and A. SAJI, *J. Amer. Chem. Soc.*, **83,** 1010 (1961).
2152. RYAN, M. T. and W. L. LEHN, *J. Organometal. Chem.*, **4,** 455 (1965).
2153. SACKMANN, E., *Z. Physik Chem.*, **34,** 283 (1962).
2154. SAIKA, A., *J. Amer. Chem. Soc.*, **82,** 3540 (1960).

2155. SAIKA, A. and C.P. SLICHTER, *J. Chem. Phys.*, **22**, 26 (1954).

2156. SAITO, H. and K. NUKADA, *J. Mol. Spec.*, **18**, 1 (1965).

2157. SAITO, H. and K. NUKADA, *J. Mol. Spec.*, **18**, 355 (1965).

2158. SAITO, H. and K. NUKADA, *Tetrahedron Letters*, 2117 (1965).

2159. SALINGER, R.M. and R.E. DESSY, *Tetrahedron Letters*, 729 (1963).

2160. SANDEL, V.R. and H.H. FREEDMAN, *J. Amer. Chem. Soc.*, **85**, 2328 (1963).

2161. SANDOVAL, A., F. WALLS, J.N. SHOOLERY, J.M. WILSON, H. BUDZIKIEWICZ and C. DJERASSI, *Tetrahedron Letters*, 409 (1962).

2162. SASSON, M., A. TZALMONA and A. LOEWENSTEIN, *J. Sci. Instr.*, **40**, 133 (1963).

2163. SATO, T. and Y. MIKAMI, *J. Chem. Soc. Japan.*, Ind. Chem. Section, **68**, 1401 (1965).

2164. SAUER, J., A. MIELERT, D. LANG and D. PETER, *Chem. Ber.*, **98**, 1435 (1965).

2165. SAUERS, R.R. and P.E. SONNET, *Chem. and Ind.*, 786 (1963).

2166. SAUNDERS, M., Private communications 1963.

2167. SAUNDERS, M., *Tetrahedron Letters*, 1699 (1963).

2168. SAUNDERS, M., J. PLOSTNIEKS, P.S. WHARTON and H.H. WASSERMAN, *J. Chem. Phys.*, **32**, 317 (1960).

2169. SAUNDERS, M., P. R. SCHLEYER and G.A. OLAH, *J. Amer. Chem. Soc.*, **86**, 5680 (1964).

2170. SAUNDERS, M. and F. YAMADA, *J. Amer. Chem. Soc.*, **85**, 1882 (1963).

2171. SAUPE, A., *Z. Naturforsch.*, **19a**, 161 (1964).

2172. SAUPE, A., *Z. Naturforsch.*, **20a**, 572 (1965).

2173. SAUPE, A. and G. ENGLERT, *Phys. Rev. Letters*, **11**, 462 (1963).

2174. SAUPE, A. and G. ENGLERT, *Z. Naturforsch.*, **19a**, 172 (1964).

2175. SAVERS, R.R. and P.E. SONNET, *Chem. and Ind.*, 786 (1963).

2176. SAVITSKY, G.B., *J. Phys. Chem.*, **67**, 2723 (1963).

2177. SAVITSKY, G.B. and K. NAMIKAWA, *J. Phys. Chem.*, **67**, 2430 (1963).

2178. SAVITSKY, G.B., R.M. PEARSON and K. NANIKAWA, *J. Phys. Chem.*, **69**, 1425 (1965).

2179. SCALA, A.A. and E.I. BECKER, *J. Org. Chem.*, **30**, 3491 (1965).

2180. SCHAEFER, T., *Canad. J. Chem.*, **37**, 882 (1959).

2181. SCHAEFER, T., *Canad. J. Chem.*, **37**, 884 (1959).

2182. SCHAEFER, T., *Canad. J. Chem.*, **40**, 1 (1962).

2183. SCHAEFER, T., *Canad. J. Chem.*, **40**, 431 (1962).

2184. SCHAEFER, T., *Canad. J. Chem.*, **40**, 1678 (1962).

2185. SCHAEFER, T., *J. Chem. Phys.*, **36**, 2235 (1962).

2186. SCHAEFER, T., W.F. REYNOLDS and Y. YONEMOTO, *Canad. J. Chem.*, **41**, 2969 (1963).

2187. SCHAEFER, T. and W.G. SCHNEIDER, *J. Chem. Phys.*, **32**, 1218 (1960).

2188. SCHAEFER, T. and W.G. SCHNEIDER, *J. Chem. Phys.*, **32**, 1224 (1960).

2189. SCHAEFER, T. and W.G. SCHNEIDER, *Canad. J. Chem.*, **41**, 966 (1963).

2190. SCHAEFER, T. and W.G. SCHNEIDER, *Canad. J. Chem.*, **37**, 2078 (1959).

2191. SCHAEFER, T. and W.G. SCHNEIDER, *Canad. J. Chem.*, **38**, 2066 (1960).

2192. SCHAEFER, T. and W.G. SCHNEIDER, *Canad. J. Chem.*, **41**, 966 (1963).

2193. SCHAEFER, T. and T. YONEMOTO, *Canad. J. Chem.*, **42**, 2318 (1964).

2194. SCHAFER, P.R., D.R. DAVIS, M. VOGEL, K. NAGARAJAN and J.D. ROBERTS, *Proc. Natl. Acad. Sci. USA*, **47**, 49 (1961).

2195. SCHARF, H.D. and F. KORTE, *Chem. Ber.*, **98**, 3672 (1965).

2196. SCHEIDEGGER, U., K. SCHAFFNER and O. JEGER, *Helv. Chim. Acta*, **45**, 400 (1962).

2197. SHEINBLATT, M., *J. Chem. Phys.*, **36**, 3103 (1962).

2198. SHEINBLATT, M., *J. Chem. Phys.*, **39**, 2005 (1963).

2199. SHEINBLATT, M. and Z. LUZ, *J. Phys. Chem.*, **66**, 1535 (1962).

2200. SHECHTER, H., W.J. LINK and G.V.D. TIERS, *J. Amer. Chem. Soc.*, **85**, 1601 (1963).

2201. SCHLEYER, P.R., R.C. FORT, W.E. WATTS, M.B. COMISAROW and G.A. OLAH, *J. Amer. Chem. Soc.*, **86**, 4195 (1964).

2202. SCHLEYER, P.R., D.C. KLEINFELTER and H.G. RICHEY, *J. Amer. Chem. Soc.*, **85** 479 (1963).

2203. SCHLEYER, P.R., W.E. WATTS, R.C. FORT, M.B. COMISAROW and G.A. OLAH, *J. Amer. Chem. Soc.*, **86**, 5679 (1964).

2204. SCHMIDBAUR, H., *J. Amer. Chem. Soc.*, **85**, 2336 (1963).

2205. SCHMIDBAUR, H. and F.SCHINDLER, *J. Organometal. Chem.*, **2**, 466 (1965).
2206. SCHMIDBAUR, H. and W.SIEBERT, *Z. Naturforsch.*, **20b**, 596 (1965).
2207. SCHMIDPETER, A., *Tetrahedron Letters*, 1421 (1963).
2208. SCHMUTZLER, R., *Angew. Chem. Int. Ed.* **4**, 496 (1965).
2209. SCHNEIDER, W.G., *J. Phys. Chem.*, **66**, 2653 (1962).
2210. SCHNEIDER, W.G., H.J.BERNSTEIN and J.A.POPLE, *Ann. N.Y. Acad. Sci.*, **70**, 806 (1958); *Canad. J. Chem.*, **35**, 1487 (1957).
2211. SCHNEIDER, W.G., H.J.BERNSTEIN and J.A.POPLE, *J. Amer. Chem. Soc.*, **80**, 3497 (1958).
2212. SCHNEIDER, W.G., H.J.BERNSTEIN and J.A.POPLE, *J. Chem. Phys.*, **28**, 601 (1958).
2213. SCHOLLKOPF, U. and collaborators, *Chem. Ber.*, **97**, 636 (1964); *ibid.*, **96**, 2266 (1963); *Tetrahedron Letters*, 105 (1963).
2214. SCHÖLLKOPF, U. and J.PAUST, *Chem. Ber.*, **98**, 2221 (1965).
2215. SCHREIER, E., *Helv. Chim. Acta*, **47**, 1529 (1964).
2216. SCHRÖDER, G., *Angew. Chem.*, **75**, 722 (1963).
2217. SCHRÖDER, G., *Chem. Ber.*, **97**, 3140, 3150 (1964).
2218. SCHRÖDER, G., R.MERENYI and J.F.H.OTH, *Tetrahedron Letters*, 773 (1964).
2219. SCHRÖDER, G., J.F.M.OTH and R.MERENYI, *Angew. Chem. Int. Ed.*, **4**, 752 (1965).
2220. SCHROETER, S.H. and E.L.ELIEL, *J. Org. Chem.*, **30**, 1 (1965).
2221. SCHUBERT, W.M. and R.H.QUACCHIA, *J. Amer. Chem. Soc.*, **85**, 1278 (1963).
2222. SCHUG, J.C., *J. Chem. Phys.*, **39**, 2798 (1963).
2223. SCHUG, J.C. and J.C.DECK, *J. Chem. Phys.*, **37**, 2618 (1962).
2224. SCHUG, J.C., P.E.MCMAHON and H.S.GUTOWSKY, *J. Chem. Phys.*, **33**, 843 (1960).
2225. SCHULZE, J. and F.A.LONG, *J. Amer. Chem. Soc.*, **86**, 322 (1964).
2226. SCHUT, R.N., W.G.STRYCKER and T.M.H.LIU, *J. Org. Chem.*, **28**, 3046 (1963).
2227. SCHWEIZER, E.E. and R.SCHEPERS, *Tetrahedron Letters*, 979 (1963).
2228. SCHWEIZER, M.P., S.I.CHAN, G.K.HELMKAMP and P.O.P.Ts'O, *J. Amer. Chem. Soc.*, **86**, 696 (1964).
2229. SCRIBE, P., *Compt. Rend.*, **261**, 160 (1965).
2230. SEEBACH, D., *Chem. Ber.*, **96**, 2723 (1963).
2231. SEEBACH, D., *Chem. Ber.*, **97**, 2953 (1964).
2232. SEEBACH, D., *Angew. Chem. Int. Ed.*, **4**, 121 (1965).
2233. SEIFFERT, W., *Angew. Chem.*, **74**, 250 (1962).
2234. SEIFFERT, W., H.ZIMMERMANN and G.SCHEIBE, *Angew. Chem.*, **74**, 249 (1962).
2235. SEITZ, L.M. and T.L.BROWN, *J. Amer. Chem. Soc.*, **88**, 2174 (1966).
2236. SELTZER, S., *J. Amer. Chem. Soc.*, **87**, 1534 (1965).
2237. SELWOOD, P.W., in *Physical Methods of Organic Chemistry* (Ed. by A.WEISS-BERGER), Vol.1, Interscience, N.Y., 1960, p.2873.
2238. SERIVS, K.L., *J. Amer. Chem. Soc.*, **87**, 5495 (1965).
2239. SERIVS, K.L. and J.D.ROBERTS, *J. Phys. Chem.*, **67**, 2885 (1963).
2240. SERVIS, K.L. and J.D.ROBERTS, *J. Amer. Chem. Soc.*, **87**, 1339 (1965).
2241. SEYDEN-PENNE, J. and T.STRZALKO, *Compt. Rend.*, **260**, 5059 (1965).
2242. SEYDEN-PENNE, J., T.STRZALKO and M.PLAT, *Tetrahedron Letters*, 4597 (1965).
2243. SEYFERTH, D., H.YAMAZAKI and D.L.ALLESTON, *J. Org. Chem.*, **28**, 703 (1963).
2244. SEYFERTH, D., *Rec. Chem. Prog.*, **26**, 87 (1965).
2245. SEYFERTH, D. and L.G.VAUGHAN, *J. Organometal. Chem.*, **1**, 201 (1963).
2246. SHABTAI, J. and E.GIL-AV, *J. Org. Chem.*, **28**, 2893 (1963).
2247. SHAMMA, M. and J.B.MOSS, *J. Amer. Chem. Soc.*, **84**, 1739 (1962).
2248. SHAPIRO, B.L., S.J.EBERSOLE, G.J.KARABATSOS, F.M.VANE and S.L.MANNATT, *J. Amer. Chem. Soc.*, **85**, 4041 (1963).
2249. SHAPIRO, B.L., S.J.EBERSOLE and R.M.KOPCHIK, *J. Mol. Spec.*, **11**, 326 (1963).
2250. SHAPIRO, B.L., R.M KOPCHIK and S.J.EBERSOLE, *J. Chem. Phys.*, **39**, 3154 (1963).
2251. SHAPIRO, B. L., G. J. KARABATSOS, F.M. VANE, J.S.FLEMING and S.S.RATKA, *J. Amer. Chem. Soc.*, **85**, 2784 (1963).
2252. SHARTS, C.M. and J.D.ROBERTS, *J. Amer. Chem. Soc.*, **83**, 871 (1961).
2253. SHAW, B.L., *Chem. and Ind.*, 1190 (1962).
2254. SHEFFER, H.E. and J.A.MOORE, *J. Org. Chem.*, **28**, 129 (1963).

2255. SHEPPARD, N. and R.K.HARRIS, *Proc. Chem. Soc.*, 418 (1961).

2256. SHEPPARD, N. and J.J.TURNER, *Proc. Roy. Soc.*, **252 A**, 506 (1959).

2257. SHEPPARD, N. and J.J.TURNER, *Mol. Phys.*, **3**, 168 (1960).

2258. SHER, A. and R.E.NORBERG, *Rev. Sci. Instr.*, **31**, 508 (1960).

2259. SHERMAN, C., *Rev. Sci. Instr.*, **30**, 568 (1959).

2260. SHIMIZU, H., *J. Chem. Phys.*, **40**, 3357 (1964).

2261. SHIMIZU, H. and S.FUJIWARA, *Bull. Chem. Soc. Japan*, **33**, 920 (1960).

2262. SHIMIZU, H., M.KATAYAMA, S.FUJIWARA, *Bull. Chem. Soc. Japan*, **32**, 419 (1959).

2263. SHIMIZU, H. and Y.HAMRA, *Bull. Chem. Soc. Japan*, **37**, 763 (1964).

2264. SHIMIZU, M. and H.SHIMIZU, *J. Chem. Phys.*, **41**, 2329 (1964).

2265. SHINE, H.J., L.T.FANG, H.E.MALLORY, N.C.CHAMBERLAIN and F.STEHLING, *J. Org. Chem.*, **28**, 2326 (1963).

2266. SHINER, V.J. and B.MARTIN, *J. Amer. Chem. Soc.*, **84**, 4824 (1962).

2267. SHONO, T., T.MORIKAWA, A.OKU and R.ODA, *Tetrahedron Letters*, 791 (1964).

2268. SHOOLERY, J.N., No 2. *Technical Information Bulletin*, Varian Associates, Palo Alto, USA (1959).

2269. SHOOLERY, J.N., *J. Chem. Phys.*, **31**, 1427 (1959).

2270. SHOLERY, J.N., *Disc. Farad. Soc.*, **34**, 104 (1962).

2271. SHOOLERY, J.N., *Disc. Farad. Soc.*, **34**, 104 (1962).

2272. SHOOLERY, J.N., *J. Chem. Phys.* **31**, 1427 (1959).

2273. SHOOLERY, J.N., L.F.JOHNSON and W.A.ANDERSON, *J. Mol. Spec.*, **5**, 110 (1960).

2274. SHOOLERY, J.N., L.F.JOHNSON, S.FURUTA and G.E.MCCASLAND, *J. Amer. Chem. Soc.*, **83**, 4243 (1961).

2275. SHOOLERY, J.N. and M.T.ROGERS, *J. Amer. Chem. Soc.*, **80**, 5121 (1958).

2276. SHOPPEE, C.W., T.E.BELLAS, R.E.LACK and S.STERNHELL, *J. Chem. Soc.*, 2483 (1965).

2277. SHOPPEE, C.W., F.P.JOHNSON, R.E.LACK and S.STERNHELL, *Tetrahedron Letters*, 2319 (1964).

2278. SHOPPEE, C.W., F.P.JOHNSON, R.E.LACK and S.STERNHELL, *Chem. Comm.*, 347 (1965) and unpublished work.

2279. SHOPPEE, C.W., F.P.JOHNSON, R.E.LACK, R.L.RAWSON and S.STERNHELL, *J. Chem. Soc.*, 2476 (1965).

2280. SHOPPEE, C.W., F.P.JOHNSON, R.E.LACK and S.STERNHELL, *J. Chem. Soc.*, 2489 (1965).

2281. SHULGIN, A.T. and A.W.BAKER, *J. Org. Chem.*, **28**, 2468 (1963).

2282. SIDDALL, T.H. and C.A.PROHASKA, *J. Amer. Chem. Soc.*, **84**, 2502 (1962).

2283. SIDDALL, T.H. and C.A.PROHASKA, *J. Amer. Chem. Soc.*, **84**, 3467 (1962).

2284. SIDDALL, T.H., C.A.PROHASKA and W.E.SHULER, *Nature*, **190**, 903 (1961).

2285. SILVER, B. and Z.LUZ, *J. Amer. Chem. Soc.*, **83**, 786 (1961).

2286. SILVERSTEIN, R.M. and G.C.BASSLER, *Spectrometric Identification of Organic Compounds*, Wiley, 1963.

2287. SIMONNIN, M.P., *Compt. Rend.*, **257**, 1075 (1963).

2288. SIMONNIN, M.P., *J. Organometal. Chem.*, **5**, 155 (1966).

2289. SINGER, L.A. and D.J.CRAM, *J. Amer. Chem. Soc.*, **85**, 1080 (1963).

2290. SINGH, G. and H.ZIMMER, *J. Org. Chem.*, **30**, 417 (1965).

2291. SINHA, S.K. and A.MAKHERJI, *J. Chem. Phys.*, **32**, 1652 (1960).

2292. SKELL, P.S. and G.P.BEAN, *J. Amer. Chem. Soc.*, **84**, 4655 (1962).

2293. SKULSKI, L., *Bull. Acad. Sci. Polon.*, *Ser. Sci. Chem.*, **12**, 299 (1964).

2294. SKULSKI, L., G.C.PALMER and M.CALVIN, *Tetrahedron Letters*, 1773 (1963).

2295. SLOMP, G., *J. Amer. Chem. Soc.*, **84**, 673 (1962).

2296. SLOMP, G., *J. Amer. Chem. Soc.*, **84**, 673 (1962).

2297. SLOMP, G. and B.R.MCGARVEY, *J. Amer. Chem. Soc.*, **81**, 2200 (1959).

2298. SLOMP, G. and F.A.MACKELLAR, *J. Amer. Chem. Soc.*, **82**, 999 (1960).

2299. SLOMP, G. and F.A.MACKELLAR, *J. Amer. Chem. Soc.*, **84**, 204 (1962).

2300. SLOMP, G., F.A.MACKELLAR and L.A.PAQUETTE, *J. Amer. Chem. Soc.*, **83**, 4472 (1961).

2301. SLOMP, G. and W.J.WECHTER, *Chem. and Ind.*, 41 (1962).

2302. SMITH, G.V. and H.KRILOFF, *J. Amer. Chem. Soc.*, **85**, 2016 (1963).
2303. SMITH, G.V. and P.J.TROTTER, *J. Org. Chem.*, **28**, 2450 (1963).
2304. SMITH, G.W., *J. Chem. Phys.*, **39**, 2031 (1963).
2305. SMITH, G.W., *J. Chem. Phys.*, **42**, 435 (1965).
2306. SMITH, G.W., *J. Mol. Spec.*, **12**, 146 (1964).
2307. SMITH, I.C. and W.G.SCHNEIDER, *Canad. J. Chem.*, **39**, 1158 (1961).
2308. SMITH, J.A.S., *J. Chem. Soc.*, 4736 (1962).
2309. SMITH, L.L., *Steroids*, **4**, 395 (1964).
2310. SMITH, L.L., M.MARX, H.MENDELSOHN, T.FOELL and J.J.GOODMAN, *J. Amer. Chem.. Soc.*, **84**, 1265 (1962).
2311. SMITH, P. and J.J.MCLESKEY, *Canad. J. Chem.*, **43**, 2418 (1965).
2312. SMITH, R.C., *Rev. Sci. Instr.*, **34**, 296 (1963).
2313. SMITH, S.L. and R.H.COX, *J. Mol. Spec.*, **16**, 216 (1965).
2314. SMITH, W.B., *J. Chem. Ed.*, **41**, 97 (1964).
2315. SMITH, W.B., *J. Phys. Chem.*, **67**, 2841 (1963).
2316. SMITH, W.B., *J. Phys. Chem.*, **67**, 2843 (1963).
2317. SMITH, W.B. and B.A.SHOULDERS, *J. Amer. Chem. Soc.*, **86**, 3118 (1964).
2318. SMITH, W.B. and B.A.SHOULDERS, *J. Phys. Chem.*, **69**, 579 (1965).
2319. SMITH, W.B. and B.A.SHOULDERS, *J. Phys. Chem.*, **69**, 2022 (1965).
2320. SMOLINSKY, G., *J. Amer. Chem. Soc.*, **83**, 4483 (1961).
2321. SNYDER, E.I., *J. Amer. Chem. Soc.*, **85**, 2624 (1963).
2322. SNYDER, E.I., L.J.ALTMAN and J.D.ROBERTS, *J. Amer. Chem. Soc.*, **84**, 2004 (1962).
2323. SNYDER, E.I. and B.FRANZUS, *J. Amer. Chem. Soc.*, **86**, 1166 (1964).
2324. SNYDER, E.I. and J.D.ROBERTS, *J. Amer. Chem. Soc.*, **84**, 1582 (1962).
2325. SNYDER, L. and R.G.PARR, *J. Chem. Phys.*, **34**, 837 (1961).
2326. SNYDER, L.C., *Bull. Am. Phys. Soc.*, Ser., II, **10**, 358 (1965).
2327. SNYDER, L.C., *J. Chem. Phys.*, **43**, 4041 (1965).
2328. SNYDER, L.C. and E.W.ANDERSON, *J. Amer. Chem. Soc.*, **86**, 5023 (1964).
2329. SNYDER, L.C. and E.W.ANDERSON, *J. Chem. Phys.*, **42**, 3336 (1965).
2330. SOMERS, B.G. and H.S.GUTOWSKY, *J. Amer. Chem. Soc.*, **85**, 3065 (1963).
2331. SOWDEN, J.C., C.H.BOWERS, L.HOUGH and S.H.SHUTE, *Chem. and Ind.*, 1827 (1962).
2232. SONDHEIMER, F., *Pure Appl. Chem.*, **7**, 363 (1963).
2333. SONDHEIMER, F., Y.GAONI, L.M.JACKMAN, N.A.BAILEY and R.MASON, *J. Amer. Chem. Soc.*, **84**, 4595 (1962).
2334. SONE, T. and Y.MATSUKI, *Bull. Chem. Soc. Japan*, **37**, 1235 (1964).
2335. SORENSEN, T.S., *Canad. J. Chem.*, **42**, 2768 (1964).
2336. SORENSEN, T.S., *J. Amer. Chem. Soc.*, **87**, 5075 (1965).
2337. SOUTHWICK, P.L., J.A.FITZGERALD and G.E.MILLIMAN, *Tetrahedron Letters*, 1247 (1965).
2338. SPECKAMP, W.N., U.K.PANDIT and H.O.HUISMAN, *Tetrahedron Letters* 3279 (1964).
2339. SPEZIALE, A.J. and C.C.TUNG, *J. Org. Chem.*, **28**, 1353 (1963).
2340. SPIESECKE, H. and W.G.SCHNEIDER, *J. Chem. Phys.*, **35**, 722 (1961).
2341. SPIESECKE, H. and W.G.SCHNEIDER, *J. Chem. Phys.*, **35**, 731 (1961).
2342. SPIESECKE, H. and W.G.SCHNEIDER, *Tetrahedron Letters*, 468 (1961).
2343. SPINNER, E., *J. Phys. Chem.*, **64**, 275 (1960).
2344. SRINIVASAN, R., *J. Amer. Chem. Soc.*, **83**, 4923 (1961).
2345. STAAB, H.A. and F.BINNIG, *Tetrahedron Letters*, 319 (1964).
2346. STAAB, H.A. and H.BRÄUNLING, *Tetrahedron Letters*, 45 (1965).
2347. STAAB, H.A. and A.MANNSCHRECK, *Angew. Chemie Int. Ed.*, **2**, 216 (1963).
2347a. STAAB, H.A. and A.MANNSCHRECK, *Chem. Ber.*, **98**, 1111 (1965).
2348. STAAB, H.A., F.VOGTLE and A.MANNSCHRECK, *Tetrahedron Letters*, 697 (1965).
2349. STAIFF, D.C. and A.C.HUITRIC, *J. Org. Chem.*, **28**, 3531 (1963).
2350. STAIFF, D.C. and A.C.HUITRIC, *J. Org. Chem.*, **29**, 3106 (1964).
2351. STACEY, F.W. and J.F.HARRIS, *J. Amer. Chem. Soc.*, **85**, 963 (1963).
2352. STACY, G.W., A.J.PAPA, F.W.VILLAESCUSSA and S.C.RAY, *J. Org. Chem.*, **29**, 607 (1964).

434 *References*

2353. STAFFORD, S.L. and J.D.BALDESCHWIELER, *J. Amer. Chem. Soc.*, **83**, 4473 (1961).
2354. STEELE, J.A., L.A.COHEN and E.MOSETTIG, *J. Amer. Chem. Soc.*, **85**, 1134 (1963).
2355. STEHLING, F.C., *Anal. Chem.*, **35**, 773 (1963).
2356. STEPHEN, M.J., *Mol. Phys.*, **1**, 223 (1958).
2357. STEPHEN, M.J., *J. Chem. Phys.*, **34**, 484 (1964).
2358. STERNHELL, S., *Revs. Pure Appl. Sci.*, **14**, 15 (1964).
2359. STEVENS, R.M., R.M.PITZER and W.N.LIPSCOMP, *J. Chem. Phys.*, **38**, 550 (1963).
2360. STEVENS, R.M., R.M.PITZER and W.N.LIPSCOMB, *J. Chem. Phys.*, **38**, 550 (1963).
2361. STEWART, F.H.C. and N.DANIELI, *Chem. and Ind.*, 1926 (1963).
2362. STORY, P.R. and M.SAUNDERS, *J. Amer. Chem. Soc.*, **84**, 4876 (1962).
2363. STORY, R.R., L.C.SNYDER, D.C.DOUGLASS, E.W.ANDERSON and R.L.KORNEGAY, *J. Amer. Chem. Soc.*, **85**, 3630 (1963).
2364. STOTHERS, J.B. *Quart. Revs.*, **19**, 144 (1965).
2365. STOTHERS, J.B., J.D.TALMAN and R.R.FRASER, *Canad. J. Chem.*, **42**, 1530 (1964).
2366. SUBRAMANIAN, P.M., M.T.EMERSON and N.A.LEBEL, *J. Org. Chem.*, **30**, 2642 (1965).
2367. SUCHR, H., *Ber. Bunsen Ges.*, **68**, 169 (1964).
2368. SUGIYAMA, H., S.ITO and T.NOZOE, *Tetrahedron Letters*, 179 (1965).
2369. SUHR, H., *Chem. Ber.*, **96**, 1720 (1963).
2370. SUHR, H., *Z. Elektrochem.*, **66**, 466 (1962).
2371. SUNNERS, B., L.H.PIETTE and W.G.SCHNEIDER, *Canad. J. Chem.*, **38**, 681 (1960).
2372. SWALEN, J.D. and C.A.REILLY, *J. Chem. Phys.*, **37**, 21 (1962).
2373. SZAREK, W.A., S.WOLFE and J.K.N.JONES, *Tetrahedron Letters*, 2743 (1964).
2374. TADANIER, J., *J. Org. Chem.*, **28**, 1744 (1963).
2375. TADANIER, J. and W.COLE, *J. Org. Chem.*, **27**, 4610 (1962).
2376. TAFT, R.W.JR., *J. Phys. Chem.*, **64**, 1805 (1960).
2377. TAFT, R.W., S.EHRENSON, I.C.LEWIS and R.E.GLICK, *J. Amer. Chem. Soc.*, **81**, 5352 (1959).
2378. TAFT, R.W., E.PRICE, I.R.FOX, I.C.LEWIS, K.K.ANDERSON and G.T.DAVIS, *J. Amer. Chem. Soc.*, **85**, 3146 (1963).
2379. TAFT, R.W., F.PROSSER, L.GOODMAN and G.T.DAVIS, *J. Chem. Phys.*, **38**, 380 (1963).
2380. TAKAHASHI, K., *Bull. Chem. Soc. Japan*, **37**, 291 (1964).
2381. TAKAHASHI, K., *Bull. Chem. Soc. Japan*, **37**, 963 (1964).
2382. TAKAHASHI, K. and G.HAZATO, *Bull. Chem. Soc. Japan*, **38**, 1807 (1965).
2383. TAKAHASHI, K., T.KANDA and Y.MATSUKI, *Bull. Chem. Soc. Japan*, **37**, 768 (1964).
2384. TAKAHASHI, K., T.KANDA and Y.MATSUKI, *Bull. Chem. Soc. Japan*, **38**, 1799 (1965).
2385. TAKAHASHI, K., T.KANDA, F.SHOJI and Y.MATSUKI, *Bull. Chem. Soc. Japan*, **38**, 508 (1965).
2386. TAKAHASHI, K., T.SONE, Y.MATSUKI and G.HAZATO, *Bull. Chem. Soc. Japan*, **36**, 108 (1963).
2387. TAKAHASHI, K., T.SONE, Y.MATSUKI and G.HAZATO, *Bull. Chem. Soc. Japan*, **38**, 1041 (1965).
2388. TAKAHASHI, M., D.R.DAVIS and J.D.ROBERTS, *J.Amer. Chem. Soc.*, **84**, 2935 (1962).
2389. TAKAHASHI, T., *Tetrahedron Letters*, 565 (1964).
2390. TAKAMIZAWA, A. and Y.HAMASHIMA, *Chem. and Pharm. Bull.*, **13**, 142 (1965).
2391. TAKAMIZAWA, A. and K.HIRAI, *J. Org. Chem.*, **30**, 2290 (1965).
2392. TAKEDA, K. and M.IKUTA, *Tetrahedron Letters*, 277 (1964).
2393. TAKEDA, K., M.IKUTA and M.MIYAWAKI, *Tetrahedron*, **20**, 2991 (1964).
2394. TANIGUCHI, H., I.M.MATHAI and S.J.MILLER, *Tetrahedron*, **22**, 867 (1966).
2395. TAURINS, A. and W.G.SCHNEIDER, *Canad. J. Chem.*, **38**, 1237 (1960).
2396. TAYLOR, T.G., *Chem. and Ind.*, 649 (1963).
2397. TEMPLE, C. and J.A.MONTGOMERY, *J. Amer. Chem. Soc.*, **86**, 2946 (1964).
2398. TEMPLE, C. and J.A.MONTGOMERY, *J. Org. Chem.*, **30**, 826 (1965).
2399. TENSMEYER, L.G. and C.AINSWORTH, *J. Org. Chem.*, **31**, 1878 (1966).
2400. THEOBALD, J.G. and J.UBERSFELD, *Compt. Rend.*, 255 (1962).
2401. THOMPSON, D.S., R.A.NEWMARK and C.H.SEDERHOLM, *J. Chem. Phys.*, **37**, 411 (1962).

2402. THOMPSON, J.E., *J. Org. Chem.*, **30**, 4276 (1965).
2403. TIERS, G.V.D., *J. Phys. Chem.*, **62**, 1151 (1958) and *Characteristic NMR Shielding Values for Hydrogen in Organic Structures*, (Minnesota Mining and Manufacturing Company, 1958).
2404. TIERS, G.V.D., *J. Phys. Chem.*, **66**, 1192 (1962).
2405. TIERS, G.V.D., *J. Phys. Chem.*, **67**, 1373 (1963).
2406. TIERS, G.V.D., *Proc. Chem. Soc.*, 389 (1960).
2407. TIERS, G.V.D., *J. Amer. Chem. Soc.*, **84**, 3972 (1962).
2408. TIERS, G.V.D. and F.A.BOVEY, *J. Phys. Chem.*, **63**, 302 (1959).
2409. TIERS, G.V.D., C.A.BROWN, R.A.JACKSON and T.N.LAHR, *J. Amer. Chem. Soc.*, **86**, 2526 (1964).
2410. TIERS, G.V.D. and R.I.COON, *J. Org. Chem.*, **26**, 2097 (1961).
2411. TIERS, G.V.D. and D.R.HOTCHKISS, *J. Phys. Chem.*, **66**, 560 (1962).
2412. TIERS, G.V.D. and P.C.LAUTERBUR, *J. Chem. Phys.*, **36**, 1110 (1962).
2413. TIERS, G.V.D., S.PLOVAN and S.SEARLES, *J. Org. Chem.*, **25**, 285 (1960).
2414. TILLIEU, J., *Ann. Phys.*, **2**, 471, 631 (1957).
2415. TIMMERMAN, H., R.F.REKKER and W.T.NAUTA, *Rec. Trav. Chim.*, **84**, 1348 (1965).
2416. TIMMONS, C.J., *Chem. Comm.*, 576 (1965).
2417. TOBLER, E. and D.J.FOSTER, *J. Org. Chem.*, **29**, 2839 (1964).
2418. TOMOEDA, M., M.INUZUKA, T.FURUTA and K.TAKAHASHI, *Tetrahedron Letters*, 1233 (1964).
2419. TOKI, S., K.SHIMA and H.SAKURAI, *Bull. Chem. Soc. Japan*, **38**, 760 (1965).
2420. TORI, K., *Chem. Pharm. Bull.*, **12**, 1439 (1964).
2421. TORI, K. and K.AONO, *Ann. Repts. Shionogi Res. Labs.*, No.14, p.136 (1964).
2422. TORI, K., K.AONO, Y.HATA, R.MUNEYUKI, T.TSUJI and H.TANIDA, *Tetrahedron Letters*, 9 (1966).
2423. TORI, K., Y.HAMASHIMA and A.TAKAMIZAWA, *Chem. Pharm. Bull.*, **12**, 924 (1964).
2424. TORI, K., Y.HATA, R.MUNEYUKI, Y.TAKANO, T.TSUJI and H.TANIDA, *Canad. J. Chem.*, **42**, 926 (1964).
2425. TORI, K. and K.KITAHONOKI, *J. Amer. Chem. Soc.*, **87**, 386 (1965).
2426. TORI, K., K.KITAHONOKI, Y.TAKANO, H.TANIDA and T.TSUJI, *Tetrahedron Letters*, 559 (1964).
2427. TORI, K., K.KITAHONOKI, Y.TAKANO, H.TANIDA and T.TSUJI, *Tetrahedron Letters*, 869 (1965).
2428. TORI, K., T.KOMENO and T.NAKAGAWA, *J. Org. Chem.*, **29**, 1136 (1964).
2429. TORI, K. and E.KONDO, *Tetrahedron Letters*, 645 (1963).
2430. TORI, K. and E.KONDO, *Steroids*, **4**, 713 (1964).
2431. TORI, K. and E.KONDO, *Steroids*, **4**, 715 (1964).
2432. TORI, K. and K.KURIYAMA, *Chem. and Ind.*, 1525 (1963).
2433. TORI, K. and K.KURIYAMA, *Tetrahedron Letters*, 3939 (1964).
2434. TORI, K., R.MUNEYUKI and H.TANIDA, *Canad. J. Chem.*, **41**, 3142 (1963).
2435. TORI, K. and T.NAKAGAWA, *J. Phys. Chem.*, **68**, 3163 (1964).
2436. TORI, K., M.OGATA and H.KANO, *Chem. Pharm. Bull.*, **11**, 235 (1963).
2437. TORI, K., M.OGATA and H.KANO, *Chem. Pharm. Bull.*, **11**, 681 (1963).
2438. TORI, K., Y.TAKANO and K.KITAHONOKI, *Chem. Ber.*, **97**, 2798 (1964).
2439. TRAGER, W.F. and A.C.HUITRIC, *J. Org. Chem.*, **30**, 3257 (1965).
2440. TRAGER, W.F. and A.C.HUITRIC, *Tetrahedron Letters*, 825 (1966).
2441. TRAGER, W.F., B.J.NIST and A.C.HUITRIC, *Tetrahedron Letters*, 267 (1965).
2442. TRAYNHAM, J.G. and M.T.YANG, *Tetrahedron Letters*, 575 (1965).
2443. TROFIMENKO, S., *J. Org. Chem.*, **28**, 2755 (1963).
2444. TROFIMENKO, S., *J. Org. Chem.*, **28**, 3242 (1963).
2445. TROST, B.M., *J. Amer. Chem. Soc.*, **88**, 853 (1966).
2446. TRUCE, W.E. and B.GROTEN, *J. Org. Chem.*, **27**, 128 (1962).
2447. TRUCE, W.E., H.G.KLEIN and R.B.KRUSE, *J. Amer. Chem. Soc.*, **83**, 4636 (1961).
2448. TUITE, R.J., H.R.SNYDER, A.L.PORTE and H.S.GUTOWSKY, *J. Phys. Chem.*, **65**, 187 (1961).
2449. TUNG, C.C. and A.J.SPEZIALE, *J. Org. Chem.*, **28**, 1521 (1963).

436 *References*

2450. Tung, C.C., A.J.Speciale and H.W.Frazier, *J. Org. Chem.*, **28**, 1514 (1963).
2451. Turner, D.W., *J. Chem. Soc.*, 847 (1962).
2452. Turner, J.J., *Mol. Phys.*, **3**, 417 (1960).
2452a. Turro, N.J. and W.B.Hammond, *J. Amer. Chem. Soc.*, **88**, 3672 (1966).
2453. Uebel, J.J. and J.C.Martin, *J. Amer. Chem. Soc.*, **86**, 4618 (1964).
2454. Ulbricht, T.L.V. *Tetrahedron Letters*, 1027 (1963).
2455. Ullman, E.F., *J. Amer. Chem. Soc.*, **81**, 5386 (1959).
2456. Ungnade, H.E., L.W.Kissinger, A.Narath and D.C.Barham, *J. Org. Chem.*, **28**, 134 (1963).
2457. Untch, K.G. and R.J.Kurland, *J. Amer. Chem. Soc.*, **85**, 346 (1963).
2458. Untch, K.G. and R.J.Kurland, *J. Mol. Spec.*, **14**, 156 (1964).
2459. Van Auken, T.V. and K.L.Rinehart, *J. Amer. Chem. Soc.*, **84**, 3736 (1962).
2460. Van Binst, G., J.C.Nouls and R.H.Martin, *Bull. Soc. Chim. Belgium*, **73**, 226 (1964).
2461. Van der Haak, P.J. and T.J.de Boer, *Rec. Trav. Chim.*, **83**, 186 (1964).
2462. Van der Kelen, G.P., *Bull. Soc. Chim. Belgium*, **72**, 644 (1963).
2463. Van der Kelen, G.P. and Z.Eeckhaut, *J. Mol. Spec.*, **10**, 141 (1963).
2464. Van der Vlies, C., *Rev. Trav. Chim.*, **84**, 1289 (1965).
2465. Van Dort, H.M. and T.J.Sekuur, *Tetrahedron Letters*, 1303 (1963).
2466. Van Dyke, C.H. and A.G.MacDiarmid, *Inorg. Chem.*, **3**, 1071 (1964).
2467. Van Geet, A.L. and D.N.Hume, *Anal. Chem.*, **37**, 978, 983 (1965).
2468. Van Meurs, N., *Z. Anal. Chem.*, **205**, 194 (1964).
2469. Van Vleck, J.H. *The Theory of Electric and Magnetic Susceptibilities*, Oxford University Press, 1932, p.22.
2470. Varian Associates, *NMR amd EPR Spectroscopy*, Pergamon Press, 1960.
2471. Varian Associates, *NMR at Work*. No.23.
2472. Varian Associates, *NMR at Work*. No.87.
2473. Varian Associates, *NMR at Work*. No.92.
2474. Vasil'ev, A.M., *Soviet Phys.*, JETP, **16**, 822 (1963).
2475 Veber, D.F. and W.Lwowski, *J. Amer. Chem. Soc.*, **85**, 646 (1963).
2476. Veigele, W.J. and A.W.Bevan, *Rev. Sci. Instr.*, **34**, 1158 (1963); *ibid*, **34**, 21 (1963).
2477. Veillard, A., *J. Chim. Phys.*, **59**, 1056 (1962).
2478. Villotti, R., A.Cervantes and A.D.Cross, *J. Chem. Soc.*, 3621 (1964).
2479. Vogel, E., W.A.Böll and H.Günther, *Tetrahedron Letters*, 609 (1965).
2480. Vogel, E., R.Erb, G.Lenz and A.A.Bothner-By, *Ann.*, **682**, 1 (1965).
2481. Vogel, E., W.Grimme and S.Kate, *Tetrahedron Letters*, 3625 (1965).
2482. Vogel, E., H.Kiefer and W.R.Roth, *Angew. Chem.*, **76**, 432 (1964).
2483. Vogel, E., W.Pretzer and W.A.Böll, *Tetrahedron Letters*, 3613 (1965).
2484. Vogel, E. and H.D.Roth, *Angew. Chem.*, **76**, 145 (1964).
2485. Vogel, E., D.Wendisch and W.R.Roth, *Angew. Chem.*, **76**, 432 (1964).
2486. Vogel, E., W.Wiedemann, H.Kiefer and W.F.Harrison, *Tetrahedron Letters*, 673 (1963).
2487. Vogel, E., W.Wiedemann, H.Kiefer and W.F.Harrison, *Tetrahedron Letters*, 673 (1963).
2488. Vo-Quang, L. and M.P.Simonnin, *Bull. Soc. Chim. France*, 1534 (1965).
2489. Waack, R. and M.A.Doran, *J. Amer. Chem. Soc.*, **85**, 4042 (1963).
2490. Waack, R., M.A.Doran, E.B Baker and G.A.Olah, *J. Amer. Chem. Soc.*, **88**, 1272 (1966).
2491. Waegel, B., *Bull. Soc. Chim. France*, 855 (1964).
2492. Waegel, B. and W.Jefford, *Bull. Soc. Chim. France*, 844 (1964).
2493. VanWazer, J.R. and D.Grant, *J. Amer. Chem. Soc.*, **86**, 1450 (1964).
2494. Wagniere, G. and M.Gouterman, *Mol. Phys.*, **5**, 621 (1962).
2495. Wamser, C.A. and B.B.Stewart, *Rev. Sci. Instr.*, **36**, 397 (1965).
2496. Warner, H.R. and W.E.M.Lands, *J. Amer. Chem. Soc.*, **85**, 1359 (1963).
2497. Warrener, R.M., Private communication, 1965.
2498. Wasserman, H.H. and E.V.Dehmlow, *J. Amer. Chem. Soc.*, **84**, 3786 (1962).

2499. WATANABE, H., T.TOTANI, K.TORI and T.NAKAGAWA, *Proc. XIIIth Colloque Ampere*, Sept. 1964, p.374.
2500. WATSON, R.F. and J.F.EASTHAM, *J. Amer. Chem. Soc.*, **87**, 664 (1965).
2501. WATTS, V.S. and J.H.GOLDSTEIN, *J. Chem. Phys.*, **42**, 228 (1965).
2502. WATTS, V.S., J.LOEMKER and J.H.GOLDSTEIN, *J. Mol. Spec.*, **17**, 348 (1965).
2503. WATTS, V.S., G.S.REDDY and J.H.GOLDSTEIN, *J. Mol. Spec.*, **11**, 325 (1963).
2504. WAUGH, J.S. and S.CASTELLANO, *J. Chem. Phys.*, **35**, 1900 (1961).
2505. WAUGH, J.S. and F.A.COTTON, *J. Phys. Chem.*, **65**, 562 (1961).
2506. WAUGH, J.S. and F.W.DOBBS, *J. Chem. Phys.*, **31**, 1235 (1959).
2507. WAUGH, J.S. and R.W.FESSENDEN, *J. Amer. Chem. Soc.*, **79**, 846 (1957).
2508. WAUGH, J.S. and E.L.WEI, *J. Chem. Phys.*, **43**, 2308 (1965).
2509. WEBB, D.L. and H.H.JAFFE, *J. Amer. Chem. Soc.*, **86**, 2419 (1964).
2510. WECHTER, W.J., G.SLOMP, F.A.MACKELLAR, R.WIECHERT and U.KERB, *Tetrahedron*, **21**, 1625 (1965).
2511. WEIDLER, A.M., *Acta Chem. Scand.*, **17**, 2742 (1963).
2512. WEINBERGER, M.A. and R.GREENHALGH, *Canad. J. Chem.*, **41**, 1038 (1963).
2513. WEINBERGER, M.A., R.M.HEGGIE and H.L.HOLMES, *Canad. J. Chem.*, **43**, 2585 (1965).
2514. WEINMAYR, V., *J. Org. Chem.*, **28**, 492 (1963).
2515. WEINSTEIN, B. and A.H.FENSELAU, *J. Org. Chem.*, **27**, 4094 (1962).
2516. WEISS, E., K.STARK, J.E.LANCASTER and H.D.MURDOCH, *Helv. Chim. Acta*, **46**, 288 (1963).
2517. WEITKAMP, H., U.HASSERODT and F.KORTE, *Chem. Ber.*, **95**, 2280 (1962).
2518. WEITKAMP, H. and F.KORTE, *Chem. Ber.*, **95**, 2896 (1962).
2519. WEITKAMP, H. and F.KORTE, *Tetrahedron*, **20**, 2125 (1964) and references therein.
2520. WELLINGTON, C.A. and W.D.WALTERS, *J. Amer. Chem. Soc.*, **83**, 4888 (1961).
2521. WELLMAN, K.M. and F.G.BORDWELL, *Tetrahedron Letters*, 1703 (1963).
2522. WELLS, P.R., *J. Chem. Soc.*, 1967 (1963).
2523. WELLS, P.R., *Aust. J. Chem.*, **16**, 165 (1963).
2524. WELLS, P.R., *Aust. J. Chem.*, **17**, 967 (1964).
2525. WELLS, P.R. and P.G.E.ALCORN, *Aust. J. Chem.*, **16**, 1108 (1963).
2526. WELLS, P.R. and W.KITCHING, *Tetrahedron Letters*, 1531 (1963).
2527. WELLS, P.R. and W.KITCHING, *Aust. J. Chem.*, **17**, 1204 (1964).
2528. WELLS, P.R., W.KITCHING and R.F.HENZELL, *Tetrahedron Letters*, 1029 (1964).
2529. WENKERT, E., A.AFONSO, P.BEAK, R.W.J.CARNEY, P.W.JEFFS and J.D.McCHESNEY, *J. Org. Chem.*, **30**, 713 (1965).
2530. WERNER, R.P.M. and S.A.MANASTYRSKY, *J. Amer. Chem. Soc.*, **83**, 2023 (1961).
2531. WESSELY, F. and S.STERNHELL, unpublished data.
2532. WEST, R., H.Y.NIU and M.ITO, *J. Amer. Chem. Soc.*, **85**, 2584 (1963).
2533. WESTERHOF, P., J.HARTOG and J.S.JALKES, *Rec. Trav. Chim.*, **84**, 863 (1965).
2534. WEYGAND, F., W.STEGLICH, D.MAYER and W.VON PHILIPSBORN, *Chem. Ber.* **97**, 2023 (1964).
2535. WHARTON, P.S. and T.I.BAIR, *J. Org. Chem.*, **30**, 1681 (1965).
2536. WHIPPLE, E.B. and Y.CHIANG, *J. Chem. Phys.*, **40**, 713 (1964).
2537. WHIPPLE, E.B., Y.CHIANG and R.L.HINMAN, *J. Amer. Chem. Soc.*, **85**, 26 (1963).
2538. WHIPPLE, E.B., J.H.GOLDSTEIN and L.MANDELL, *J. Amer. Chem. Soc.*, **82**, 3010 (1960).
2539. WHIPPLE, E.B., J.H.GOLDSTEIN, L.MANDELL, G.S.REDDY and G.R.McCLURE, *J. Amer. Chem. Soc.*, **81**, 1321 (1959).
2540. WHIPPLE, E.B., J.H.GOLDSTEIN and W.E.STEWART, *J. Amer. Chem. Soc.*, **81**, 4761 (1959).
2541. WHIPPLE, E.B., J.H.GOLDSTEIN and G.R.McCLURE, *J. Amer. Chem. Soc.*, **82**, 3811 (1960).
2542. WHIPPLE, E.B., J.H.GOLDSTEIN and L.MANDELL, *J. Amer. Chem. Soc.*, **82**, 3010 (1960).
2543. WHIPPLE, E.B. and M.RUTA, *J. Amer. Chem. Soc.*, **87**, 3060 (1965).
2544. WHITE, E.H. and H.C.DUNATHAN, *J. Amer. Chem. Soc.*, **86**, 453 (1964).

438 *References*

2545. WHITE, R.F.M., in *Physical Methods in Heterocyclic Chemistry*, Ed. by A.R. KAT-RITZKY, Academic Press, New York, 1963, Vol.II, Chapter 9.
2546. WHITESIDES, C.M., BEAUCHAMP and J.D.ROBERTS, *J. Amer. Chem. Soc.*, **85**, 2665 (1963).
2547. WHITESIDES, G.M., J.J.GROCKI, D.HOLTZ, H.STEINBERG and J.D.ROBERTS, *J. Amer. Chem. Soc.*, **87**, 1058 (1965).
2548. WHITESIDES, G.M., D.HOLTZ and J.D.ROBERTS, *J. Amer. Chem. Soc.*, **86**, 2628 (1964).
2549. WHITESIDES, G.M., F.KAPLAN, K.NAGARAJAN and J.D.ROBERTS, *Proc. Natl. Acad. Sci. USA*, **48**, 1112 (1962).
2550. WHITESIDES, G.M., F.KAPLAN and J.D.ROBERTS, *J. Amer. Chem. Soc.*, **85**, 2167 (1963).
2551. WHITESIDES, G.M., J.E.NORDLANDER and J.D.ROBERTS, *J. Amer. Chem. Soc.*, **84**, 2010 (1962).
2552. WHITESIDES, G.M. and J.D.ROBERTS, *J. Phys. Chem.*, **68**, 1583 (1964).
2553. WHITESIDES, G.M. and J.D.ROBERTS, *J. Amer. Chem. Soc.*, **87**, 4878 (1965).
2554. WHITESIDES, G.M., M.WITANOWSKI and J.D.ROBERTS, *J. Amer. Chem. Soc.*, **87**, 2854 (1965).
2555. WHITMAN, D.R., *J. Chem. Phys.*, **36**, 2085 (1962).
2556. WHITMAN, D.R., *J. Mol. Spec.*, **10**, 250 (1963).
2557. WHITTAKER, A.G., D.W.MOORE and S.SIEGEL, *J. Phys. Chem.*, **68**, 3431 (1964).
2558. WHITTAKER, A.G. and S.SIEGEL, *J. Chem. Phys.*, **42**, 3320 (1965).
2559. WHITTAKER, A.G. and S.SIEGEL, *J. Chem. Phys.*, **43**, 1575 (1965).
2560. WIBERG, K.B. and H.W.HOLMQUIST, *J. Org. Chem.*, **24**, 578 (1959).
2561. WIBERG, K.B., G.M.LAMPMAN, R.P.CIULA, D.S.CONNOR, P.SCHERTLER and J.LAVANISH, *Tetrahedron*, **21**, 2749 (1965).
2562. WIBERG, K.B., B.R.LOWRY and T.H.COLBY, *J. Amer. Chem. Soc.*, **83**, 3998 (1961).
2563. WIBERG, K.B., B.R.LOWRY and B.J.NIST, *J. Amer. Chem. Soc.*, **84**, 1594 (1962).
2564. WIBERG, K.B. and B.J.NIST, *Interpretation of NMR Spectra*, W.A.Benjamin, New York (1962).
2565. WIBERG, K.B. and B.J.NIST, *J. Amer. Chem. Soc.*, **83**, 1226 (1961).
2566. WIBERG, K.B. and B.J.NIST, *J. Amer. Chem. Soc.*, **85**, 2788 (1963).
2567. WIBERG, K.B. and B.J.NIST, *J. Amer. Chem. Soc.*, **85**, 2790 (1963).
2568. WIEMANN, J., N.THOAI and F.WEISBUCH, *Bull. Soc. Chim. France*, 2187 (1964).
2569. WILCOX, C.F. and D.L.NEALY, *J. Org. Chem.*, **28**, 3446 (1963).
2570. WILEY, D.W. and H.E.SIMMONS, *J. Org. Chem.*, **29**, 1876 (1964).
2571. WILEY, R.H. and T.H.CRAWFORD, *J. Polymer Sci.*, **3**, 829 (1965).
2572. WILEY, R.H., T.H.CRAWFORD and N.F.BRAY, *J. Polymer. Sci.*, **3**, 99 (1965).
2573. WILEY, R.H., C.E.STAPLES and T.H.CRAWFORD, *J. Org. Chem.*, **29**, 2986 (1964).
2574. WILKING, S., *Z. Physics*, **157**, 401 (1959).
2575. WILLIAMS, D.H., *Tetrahedron Letters*, 2305 (1965).
2576. WILLIAMS, D.H. and N.S.BHACCA, *Chem. and Ind.*, 506 (1965).
2577. WILLIAMS, D.H. and N.S.BHACCA, *Tetrahedron*, **21**, 1641 (1965).
2578. WILLIAMS, D.H. and N.S.BHACCA, *Tetrahedron*, **21**, 2021 (1965).
2579. WILLIAMS, D.H. and N.S.BHACCA, *J. Amer. Chem. Soc.*, **85**, 2861 (1963).
2580. WILLIAMS, D.H. and N.S.BHACCA, *J. Amer. Chem. Soc.*, **86**, 2742 (1964).
2581. WILLIAMS, D.H., N.S.BHACCA and C.DJERASSI, *J. Amer. Chem. Soc.*, **85**, 2810 (1963).
2582. WILLIAMS, D.H. and D.A.WILSON, *J. Chem. Soc. (B)*, 144, (1966).
2583. WILLIAMS, G.A. and H.S.GUTOWSKY, *J. Chem. Phys.*, **30**, 717 (1959).
2584. WILLIAMS, J.K., *J. Org. Chem.*, **29**, 1377 (1964).
2585. WILLIAMS, J.K., D.W.WILEY and B.C.McKUSICK, *J. Amer. Chem. Soc.*, **84**, 2210 (1962).
2586. WILLIAMS, J.K. and W.H.SHARKEY, *J. Amer. Chem. Soc.*, **81**, 4269 (1959).
2587. WILLIAMS, R.E., S.G.GIBBINS and I.SHAPIRO, *J. Amer. Chem. Soc.*, **81**, 6164 (1959).
2588. WILLIAMSON, K.L., *J. Amer. Chem. Soc.*, **85**, 516 (1963).
2589. WILLIAMSON, K.L. and J. C. FENSTERMAKER, *International Symposium on Nuclear Magnetic Resonance, Tokyo, 1965*.

2590. WILLIAMSON, K.L., N.C.JACOBUS and K.T.SOUCY, *J. Amer. Chem. Soc.*, **86**, 4021 (1964).
2591. WILLIAMSON, K.L. and W.S.JOHNSON, *J. Amer. Chem. Soc.*, **83**, 4623 (1961).
2592. WILLIAMSON, K.L., C.A.LANDFORD and C.R.NICHOLSON, *J. Amer. Chem. Soc.*, **86**, 762 (1964).
2593. WILLIAMSON, K.L., L.R.SLOAN, T.HOWELL and T.A.SPENCER, *J. Org. Chem.*, **31**, 436 (1966).
2594. WILLIAMSON, K.L. and T.A.SPENCER, *Tetrahedron Letters*, 3267 (1965).
2595. WILZBACH, K.E. and L.KAPLAN, *J. Amer. Chem. Soc.*, **87**, 4004 (1965).
2596. WILSON, D.J., V.BOEKELHEIDE and R.W.GRIFFITH JR., *J.Amer.Chem. Soc.*, **82**, 6302 (1960).
2597. WILSON, E.B., *J. Chem. Phys.*, **27**, 60 (1957).
2598. WINKHAUS, G., L.PRATT and G.WILKINSON, *J. Chem. Soc.*, 3807 (1961).
2599. WINSTEIN, S., *J. Amer. Chem. Soc.*, **87**, 381 (1965).
2600. WINSTEIN, S., P.CARTER, F.A.L.ANET and A.J.R.BOURN, *J. Amer. Chem. Soc.*, **87**, 5247 (1965).
2601. WINTERFELDT, E. and H.PREUSS, *Chem. Ber.*, **99**, 450 (1966).
2602. WINTERSTEINER, O., M.MOORE and A.I.COHEN, *J. Org. Chem.*, **29**, 1325 (1964).
2603. WILT, J.W. and W.J.WAGNER, *Chem. and Ind.*, 1389 (1964).
2604. WIERZCHOWSKI, K.L., D.SHUGAR and A.R.KATRITZKY, *J. Amer. Chem. Soc.*, **85**, 827 (1963).
2605. WISHNIA, A. and M.SAUNDERS, *J. Amer. Chem. Soc.*, **84**, 4235 (1962).
2606. WITANOWSKI, M. and J.D.ROBERTS, *J. Amer. Chem. Soc.*, **88**, 737 (1966).
2607. WITIAK, D.T. and B.B.CHAUDHARI, *J. Org. Chem.*, **30**, 1467 (1965).
2608. WITTSTRUCK, T.A., S.K.MALHOTRA and H.J.RINGOLD, *J. Amer. Chem. Soc.*, **85**, 1699 (1963).
2609. WITTSTRUCK, T.A., S.K.MALHOTRA, H.J.RINGOLD and A.D.CROSS, *J. Amer. Chem. Soc.*, **85**, 3038 (1963).
2610. WOHL, R.A., *Chimia*, **18**, 219 (1964).
2611. WOLFF, M.E., J.F.KERWIN, F.F.OWINGS, B.B.LEWIS and B.BLANK, *J. Org. Chem.*, **28**, 2729 (1963).
2612. WOLFROM, M.L., F.KOMITSKY, G.FRAENKEL, J.H.LOOKER, E.E.DICKEY, P.MCWAIN A.THOMPSON, P.M.MUNDELL and O.M.WINDRATH, *Tetrahedron Letters*, 749 (1963).
2613. WOLOVSKY, R., *J. Amer. Chem. Soc.*, **87**, 3638 (1965).
2614. WONG, W.C. and C.C.LEE, *Canad. J. Chem.*, **42**, 1245 (1964).
2615. WOO, P.W.K., H.W.DION and L.F.JOHNSON, *J. Amer. Chem. Soc.*, **84**, 1066 (1962).
2616. WU, T.K. and B.P.DAILEY, *J. Chem. Phys.*, **41**, 2796 (1964).
2617. WU, T.K. and B.P.DAILEY, *J. Chem. Phys.*, **41**, 3307 (1964).
2618. WULFMAN, D.S., L.DURHAM and C.E.WULFMAN, *Chem. and Ind.*, 859 (1962).
2619. WYNBERG, H., A.DE GROOT and D.W.DAVIES, *Tetrahedron Letters*, 1083 (1963).
2620. YAMUGUCHI, I., *Bull. Chem. Soc. Japan*, **34**, 451 (1961).
2621. YAMAGUCHI, I., *Bull. Chem. Soc. Japan*, **34**, 744 (1961).
2622. YAMAGUCHI, I., *Mol. Physics*, **6**, 105 (1963).
2623. YAMAGUCHI, I., and S.BROWNSTEIN, *J. Phys. Chem.*, **68**, 1572 (1964).
2624. YAMAGUCHI, I. and N.HAYAKAWA, *Bull. Chem. Soc. Japan*, **33**, 1128 (1960).
2625. YAMAGUCHI, S., S.OKUDA and N.NAKAGAWA, *Chem. Pharm. Bull.*, **11**, 1465 (1963).
2626. YATES, P. and R.S.DEWEY, *Tetrahedron Letters*, 847 (1962).
2627. YATES, P. and E.S.HAND, *Tetrahedron Letters*, 669 (1961).
2628. YATSIV, S., *Phys. Rev.*, **113**, 1522 (1959).
2629. YEW, F.F., R.J.KURLAND and B.J.MAIR, *Anal. Chem.*, **36**, 843 (1964).
2630. INOUE, Y., N.FURUTACHI and K.NAKANISHI, *J. Org. Chem.*, **31**, 175 (1966).
2631. YONEMOTO, T., W.F.REYNOLDS, H.M.HUTTON and T.SCHAEFER, *Canad. J. Chem.*, **43**, 2668 (1965).
2632. YONEZAWA, T., K.FUKUI, H.KATO, H.KITANO, S.HATTORI and S.MATSUOKA, *Bull. Chem. Soc. Japan*, **34**, 707 (1961).
2633. YONEZAWA, T., I.MORISHIMA, M.FUJII and K.FUKUI, *Bull. Chem. Soc. Japan*, **38**, 1224 (1965).

2634. YONEZAWA, T., H. SAITO, S. MATSUOKA and K. FUKUI, *Bull. Chem. Soc. Japan*, **38,** 1431 (1965).
2635. ZALKOW, L.H. and N.N. GIROTRA, *J. Org. Chem.*, **28,** 2033 and 2037 (1963).
2636. ZAUGG, H.E. and R.W. DE NET, *J. Amer. Chem. Soc.*, **84,** 4574 (1962).
2637. ZEIL, W. and H. BUCHERT, *Z. Phys. Chem.*, **38,** 47 (1963).
2638. ZIMMERMANN, H.E. and A. ZWEIG, *J. Amer. Chem. Soc.*, **83,** 1196 (1961).
2639. ZUPANCIC, I., *J. Sci. Inst.*, **39,** 621 (1962).
2640. ZÜRCHER, R.F., *Helv. Chim. Acta*, **44,** 1380 (1961).
2641. ZÜRCHER, R.F., *Helv. Chim. Acta*, **44,** 1755 (1961).
2642. ZÜRCHER, R.F., *Helv. Chim. Acta*, **46,** 2054 (1963).
2643. ZÜRCHER, R.F., *J. Chem. Phys.*, **37,** 2421 (1962).
2644. ZÜRCHER, R.F., in *Nuclear Magnetic Resonance in Chemistry*, Ed. by BIAGIO PESCE. Academic Press, N.Y., 1965, p.45.
2645. ZWEIG, A., J.E. LEHNSEN, J.E. LANCASTER and M.T. NEGLIA, *J. Amer. Chem. Soc.*, **86,** 3940 (1963).
2646. ZWEIG, A., J.E. LANCASTER, M.T. NEGLIA and W.H. JURA, *J. Amer. Chem. Soc.*, **86,** 4130 (1964).

INDEX

Absorption intensity *see* Intensity
Absorption signals 27
Acenaphthenes 197, 233
Acenaphthylene 188
Acetaldehyde 164
  diethylacetal 166
Acetamide 164
  axial and equatorial derivatives 240
  derivatives 176
  N,N-dimethyl 178
  N-ethyl 164
  N-methyl 164
  N-*iso*-propyl 164
  N-*n*-propyl 164
Acetates 176
  axial and equatorial 240
Acetic acid
  chemical shift 164
  methyl ester 164
Acetic anhydride 177
Acetone
  chemical shift 164
  cyanohydrin 166
  dimethylketal 166
  solvent effects of 106
Acetonitrile
  benzene solvent effect for 111
  chemical shift 164
  solvent effects for 106
Acetophenone 164
Acetylacetone, tautomerism 177
Acetyl bromide 177
Acetyl chloride 177
Acetylene
  analysis of ^{13}C-satellite spectrum 140
  $J_{^{13}C,H}$ 141
  long-range shielding in 73
Acetylenes
  chemical shifts 164, 193
  conjugated 187
  spin–spin coupling 328
Acid anhydrides, stereochemistry from solvent shifts 248
Acidinium ions 262
Acridines, chemical shifts 212
Acrylic acid, β, β-dimethyl derivatives 171
Acrylonitrile, analysis of spectrum of 143
Acylation shift 176

Acyl cations 262
Adamantane, 1-substituted, correlation of σ * with chemical shifts 66
Admantanes 196, 232
Adamantyl carbonium ion 251
Alcohols
  chemical shifts 164, 176, 179
  chemical shifts of OH 215
  spin–spin coupling of OH groups 298
Aldehydes
  aromatic, restricted rotation 362
  benzene solvent effects with 111
  chemical shifts 164, 191
  conformation from spin–spin coupling 284
  long-range spin–spin coupling 333
  α,β-unsaturated, spin–spin coupling 284, 324, 333
  vicinal spin–spin coupling 284
Aliphatic hydrocarbons, chemical shifts 168
Alkanes, chemical shifts 168
Alkenes
  chemical shifts 164, 171, 187
  conformation 364
  cyclic, chemical shifts 188
Alkenyl carbonium ions 252
Alkylamines, N-methyl 178
Alkyl carbonium ions 250
Alkyl derivatives, chemical shifts 164
Alkyl halides, chemical shifts 164
Alkyl nitrites 227
Alkyl zinc derivatives 265
Alkynyl carbonium ions 252
Allenes
  chemical shifts 190
  geminal spin–spin coupling 278
  long-range spin–spin coupling 328
  non-equivalence 378
  vicinal spin–spin coupling 284
Allyl carbonium ions 252
Allylic anions 263
Allylic groups, cyclic, chemical shifts 188
Allylic methylene and methine groups, chemical shifts 172
Allylic protons, chemical shifts 170

441

Amides
  chemical shifts 164, 215
  exchange with D_2O 360
  partial double bond character 361
  protonation 384
  spin–spin coupling 328, 360
  stereochemistry from solvent shifts 248
Amines
  acylation shift 180
  average spectra 259
  chemical shifts 164, 178, 180
  chemical shifts of NH 215, 359
  exchange of −NH 181
  nitrogen inversion 366
  protonation 180
  spin–spin coupling of NH groups 298
Analysis of spectra 122
  first order 124
  use of ^{13}C satellites 140
  use of contact shifts 144
  use of deuterium substitution 142
  use of field variation 139
  use of solvent effects 144
  use of spin decoupling 145
Androstanes, chemical shifts of angular
    methyl groups 242
Anhydrides, cyclic, $J_{13_{C,H}}$ 141
Aniline
  derivatives 203, 359
  N,N-dimethyl 178
  N,N-dimethyl-p-nitro 178
  N,N-dimethyl-p-nitroso 227
  N-methyl 178
  substituted, correlation of chemical
    shift with σ^- 66
Anisole
  chemical shifts 164
  derivatives 180, 203
[18]Annulene 97
Annulenes, spin–spin coupling 309
Anthracene, 9,10-dihydro 197
Anthracenes
  chemical shifts 205
  spin–spin coupling 306
Anthrone 197
  9,9-dimethyl 100
  9-spirocyclopropane, chemical shifts 100
Arenonium ions 254
Aromatic anions 265, 266
Aromatic cations 261
Aromatic compounds
  benzene solvent effects 111
  chemical shifts 201
  nonbenzoid, chemical shifts 205
  para-disubstituted 221
  solvent effects 204
  spin–spin coupling 305, 330, 338

Aromaticity and the chemical shift 98
Aryl carbonium ions 251
Asymmetry and spin–spin coupling 314
Axial and equatorial groups, chemical shifts
    240
Axial and equatorial protons
  chemical shifts 199, 200, 238
  spin–spin coupling 288, 338
Azaindolizine, chemical shifts 210
Azapolynuclear derivatives, chemical shifts
    214
Azetidine 199
  2-phenyl, spin–spin coupling 272
Aziridines
  chemical shifts 100
  inversion 366
  steroidal, chemical shifts 101
Azobenzenes 224
Azoxy compounds 178
Azulenes, chemical shifts 205
Azulenonium ions 254

Benzaldehyde, substituted, correlation of
    Hammett σ-constants with chemical
    shifts in 66
Benzazines, chemical shifts 212
Benzene, solvent effect of 111
Benzene derivatives
  chemical shifts 201
  correlation of Hammett σ-constants with
    chemical shifts in 66
  influence of substituents on chemical
    shifts 202
  $J_{13_{C,H}}$ 141
  polysubstituted 170
  solvent effects 106, 108
  spin–spin coupling 306
Benzenonium ions 254
Benzimidazole, N-acetyl 177
  spin–spin coupling 308
1 *H*-Benz[e]indene, 2,3-dihydro-1-oxo, che-
    mical shifts 91
Benzoates 164, 179
Benzocyclobutenes 224
  chemical shifts 235
Benzofuran, 2,2-dimethyldihydro 168
Benzofurans
  chemical shifts 209
  spin-spin coupling 308
Benzoic acid, methyl ester 164
Benzopyrazoles
  chemical shifts 210
  spin–spin coupling 308
p-Benzoquinone derivatives 189
Benzothiophens
  chemical shifts 210
  spin–spin coupling 308

Bentylacetone, spectrum of 136
Benzyl alcohol, spectra of 131
Benzyl chloride derivatives 175
Benzylic methyl groups, chemical shifts 173
Benzylic protons, chemical shifts 170, 173
Benzyllithium 265
Beyerol, 15β,16β-oxido-, chemical shifts 99
Biacetyl 177
Bicyclic compounds, chemical shifts 196
Bicyclo[2.2.1]heptanes
    chemical shifts 100, 229
    vicinal spin–spin coupling 289, 335
Bicyclo[2.2.1]heptanols, chemical shifts 79
Bicyclo[2.2.1]heptene
    effect of substituents on chemical shifts 64
    vicinal spin–spin coupling 289
Bicyclo[2.1.1]hexanes, spin–spin coupling 276, 288, 335
Bicyclo[2.2.2]octanes
    chemical shifts 229, 232
    spin–spin coupling 288, 335
Bicyclo[2.1.3]octanes, spin–spin coupling 335
Bicyclo[2.2.2]octenes, chemical shifts 232
Biphenylenes
    chemical shifts 206
    spin–spin coupling 309
Bisdehydro[14]annulene, chemical shifts 97
Bond order
    and allylic coupling 322
    and homoallylic coupling 325
    and side-chain coupling 331
    and vicinal spin–spin coupling 303
Bromine, solvent effects for 106
Bromochloromethane, solvent effects for 106
Bromocyclohexane, axial and equatorial protons 239
Bromoform, solvent effects for 106
N-Bromosuccinimide 200
Bromotrichloromethane, solvent effects for 106
Buckingham's theory of solvent effects 107
Bulk susceptibility, correction for 104
Bullvalene 362
Butadiene derivatives
    chemical shifts 187, 225
    spin–spin coupling 285, 341
Butanal 164

Butane
    2,3-diacetoxy 163
    2,3-dibromo 163
    2,3-dichloro 163
    2,3-diphenyl 163
n-Butanol
    3-methoxy 166
    spectrum of 143
tert-Butanol 164
But-1-ene 164
    3,3-dimethyl 164
    3-methyl 164
But-2-ene
    2-methyl 171
    solvent effects for *cis* and *trans* 106
Butenes 171
    spin–spin coupling 323
Butenolides 188
    spin–spin coupling 317
tert-Butyl acetate 164
tert-Butylamine 164
tert-Butylbenzene 164, 168
tert-Butyl benzoate 164
tert-Butyl bromide 164
tert-Butyl chloride 164
tert-Butyl cyanide 164
tert-Butyl derivatives, chemical shifts 164, 168
tert-Butyl iodide 164
tert-Butyl isocyanide 164
tert-Butyl mercaptan 164
tert-Butyl methyl ketone 164
But-1-yne 164
    3,3-dimethyl 164
    3-methyl 164
But-2-yne, solvent effects for 106
iso-Butyramide 164
n-Butyramide 164
iso-Butyric acid 164
n-Butyric acid 164
    2-bromo 166
    methyl ester 164
Butyrolactam 198
Butyrolactone 198

13C
    chemical shifts 64, 77
    coupling constants 77, 345
    satellites 140
Calibration of spectra 43
    side band technique 43
    wiggle beat technique 43
Camphor derivatives 233
Carbanions 262
Carbonates, cyclic, chemical shifts 235
Carbon disulphide, solvent effects for 106

Carbonium ions 249
  non-classical 256
Carbon tetrachloride, solvent effects for 106
Carbonyl compounds
  conformation 364
  cyclic 192
  derivatives, chemical shifts 226
  hydrates 384
  α,β-unsaturated, chemical shifts 170, 171, 187, 190, 192, 223
  α,β-unsaturated, coupling constants 285, 323
Carbonyl group, influence on chemical shifts 88, 207
Carboxylic acids, chemical shifts 164, 216, 359
CAT (computer of averaged transients) 41
Chemical exchange
  effect on line shapes 56
  measurement of rate by double resonance 157
Chemical shift 11
  additivity rules 160
  applications 159
  average over a number of conformations 60
  bar-graphs 161
  classification systems 162
  compilations 159
  highly accurate determination by spin decoupling 155
  limitations to correlations 159
Chlorobenzene, solvent effects for 106
Chlorocyclohexane, axial and equatorial protons 239
Chlorodibromomethane, solvent effects for 106
Chloroform, solvent effects for 106
Chromene 189
  2,2-dimethyl and derivatives 169
Chromones and analogs, chemical shifts 213
Chrysenes, chemical shifts 205
Cinnolines
  chemical shifts 212
  spin–spin coupling 309
Clerodin, spectrum of 156
Collision complex 109, 246
Complex formation, and vicinal spin–spin coupling 304
Computer of averaged transients 41
Configuration, *threo-erythro* from spin–spin coupling 291
Conformation
  ethane derivatives 368

from chemical shifts 245
from vicinal spin–spin coupling 289
Conformational inversion 60, 364
  and symmetry 365
Contact shifts 102
Cotton–Mouton effect 75
Coumarin 188
Coupling *see* Spin–spin coupling
Coupling path 269
Cresols 175
Cubane, chemical shift 71
Cumulenes, spin–spin coupling 328
Curvature 31
Cycl[3,2,2]azine
  chemical shifts 211
  spin–spin coupling 309
Cycling 32
Cycloalkanes 196
Cyclobutane, chemical shift 196
Cyclobutanes
  chemical shifts 235
  1,2-dione-3,4-dimethyl 172
  geminal spin–spin coupling 276
  long-range spin–spin coupling 336
  stereochemistry from chemical shifts 234
  vicinal spin–spin coupling 287
Cyclobutanone 198
  shielding by the carbonyl group in 91
Cyclobutene derivatives 188
  1-methyl 172
  spin–spin coupling in derivatives 287, 303, 320
Cyclodecene, vicinal spin–spin coupling 303
Cycloheptane, chemical shift 196
Cycloheptanone 198
  derivatives 169
Cycloheptatrienes 172, 188
Cycloheptane, vicinal spin–spin coupling 303
Cycloheptenone 192
Cyclohexa-1,3-diene, vicinal spin–spin coupling 285
Cyclohexadienes 188
Cyclohexadienones, vicinal spin–spin coupling 285
Cyclohexane
  axial and equatorial protons 239
  chemical shift 196
  conformation from chemical shifts 239
  4-*tert*-butyl derivatives 239
  1,2-diphenyl 224
  solvent effects for 106
Cyclohexane derivatives
  axial and equatorial protons from spin–spin coupling 288

Cyclohexane derivatives *(cont.)*
  conformational inversion 364
  geminal spin–spin coupling 275
  influence of substituents on chemical
    shifts 237
  long-range spin–spin coupling 337
  stereochemistry from chemical shifts
    234
  vicinal spin–spin coupling 290
Cyclohexane thiol axial and equatorial
    protons 239
Cyclohexanol
  axial and equatorial protons 239
  4-*t*-butyl, spin–spin coupling 290
  cis-4-*t*-butyl, chemical shifts 79
  trans-4-*t*-butyl, chemical shifts 79
  conformation from spin–spin coupling
    290
  cis-2-*cis*-6-dimethyl, chemical shifts 79
  trans-2-*trans*-6-dimethyl, chemical shifts
    79
  2,2-dimethyl-*cis*-4-*t*-butyl, chemical shifts
    79
  2,2-dimethyl-*trans*-4-*t*-butyl, chemical
    shifts 79
  3,3-dimethyl, chemical shifts 79
  trans-3-*trans*-5-dimethyl, chemical shifts
    79
  cis,cis-3,5-dimethyl, chemical shifts 79
  trans-2-methyl, chemical shifts 79
  cis-2-methyl-*cis*-4-*t*-butyl, chemical shifts
    79
  cis-2-methyl-*trans*-4-*t*-butyl, chemical
    shifts 79
  trans-2-methyl-*cis*-4-*t*-butyl, chemical
    shifts 79
  trans-2-methyl-*trans*-4-*t*-butyl, chemical
    shifts 79
  cis-3-methyl, chemical shifts 79
  3,3,5,5-tetramethyl, chemical shifts 79
  cis-3,3,5-trimethyl, chemical shifts 79
Cyclohexanone 198
  2-bromo, conformation from vicinal
    coupling 291
  derivatives 169, 241
Cyclohexene
  chemical shifts 188, 238
  2,4-dimethyl 172
  vicinal spin–spin coupling 303
Cyclohexenone 188, 192
  derivatives 168, 172
Cyclohexenyl carbonium ion 252
Cyclohexyl acetate, axial and equatorial
    protons 239
Cyclohexylamine
  axial and equatorial protons 239
  N,N-dimethyl 178

N-methyl, axial and equatorial protons
    239
Cyclohexyl derivatives 238
Cyclononene, vicinal spin–spin coupling
    303
Cyclooctane 196
Cyclooctanone 198
Cyclooctatetraene, spin–spin coupling 285
Cyclooctatetraenyl dianion 266
Cyclooctene 188
  vicinal spin–spin coupling 303
Cyclopentadiene
  chemical shifts 188
  vicinal spin–spin coupling 285
Cyclopentadienyl anion 266
Cyclopentane
  chemical shift 85, 196
  1,2-diphenyl 224
  solvent effects for 106
Cyclopentane derivatives
  conformation from spin–spin coupling
    287
  geminal spin–spin coupling 275
  stereochemistry from chemical shifts
    234
  vicinal spin–spin coupling 287
Cyclopentanone 198
  derivatives 168
Cyclopentene
  chemical shifts 83
  derivatives 188
  vicinal spin–spin coupling 303
Cyclopentenones, chemical shifts 168,
    172, 188, 192, 235
Cyclopentenyl carbonium ion 252
Cyclopropane
  chemical shift 196
  ring current in 98
Cyclopropane derivatives
  chemical shifts 197, 228
  geminal spin–spin coupling 276
  $J_{13_C,H}$ 141
  stereochemistry from chemical shifts 227
  vicinal spin–spin coupling 286
Cyclopropanone 198
Cyclopropene
  chemical shifts 101
  derivatives 188
  1-methyl 172
  vicinal spin–spin coupling 303
Cyclopropenium cations 262
Cyclopropyl carbonium ions 253

Dailey–Shoolery relation 65
Decalones, axial and equatorial protons
    241

1,4-Decamethylenebenzene, chemical shift 97
Deceptively simple spectra 136, 147, 150
Delta scale 48
Deuterium exchange 53, 215, 299, 359
Deuterochloroform, purification and fast exchange 299
Diacetylene, solvent effects for 106
Diamagnetic anisotropy 72
  and solvent effects 105
  determination 74
  determination from chemical shift 77
  of acetylenes 93
  of aromatic systems 94
  of benzene 96
  of carbon–carbon double bonds 83
  of carbon–carbon single bonds 78
  of carbon–halogen bonds 80
  of carbon–hydrogen bonds 78
  of the carbonyl group 88
  of cyclopropane 98
  of formaldehyde 88
  of nitrile groups 93
  of nitro groups 94
  of nitrogen (sp^2) atoms 82
  of nitrogen (sp^3) atoms 81
  of ether oxygen atoms 81
  of iodine 80
Diamagnetic shielding 61
Diaziridines, long-range shielding by 101
Dibenzo-[cf][1,2]-diazepin-11-one, spectrum of 24
Dibenzyl 182
1,1-Dibromo-2,2-dicyanoethane 58
1,2-Dibromoethylene-*cis* and *trans*, solvent effects on chemical shifts 109
Dibromemethane, chemical shift 160
Di-*tert*-butyl disulphide 164
Di-*tert*-butyl ether 164
β-Dicarbonyl derivatives, tautomerism 382
Dichloroacetaldehyde, double resonance spectra of 152
1,2-Dichloroethylene, *cis* and *trans*, solvent effects on chemical shifts 109
Dichloromethane, chemical shift 160
Dicyanoacetylene, solvent effects for 106
Dicyanomethane 182
Dielectric constant and polar solvent shifts 107
Diethyl disulphide 164
Diethyl ether 164
  solvent effects for 106
Difluoromethane, chemical shift 160
Dihedral angle, definition 280
1,4-Dihydrobenzenes, spin–spin coupling 321
Dihydrodeoxycodeine-D 82

2,5-Dihydrofuran
  chemical shifts 188
  spin–spin coupling 328
2,3-Dihydrofurans
  chemical shifts 188, 235
  spin–spin coupling 304, 317
Dihydroindoles, chemical shifts 236
Dihydroisoindoles, chemical shifts 236
2,3-Dihydropyran, vicinal spin–spin coupling 304
Dihydropyrans 188
Dihydropyridines
  spin–spin coupling 339
  vicinal spin–spin coupling 304
Dihydropyrone 169
2,5-Dihydropyrrole, spin–spin coupling 328
Diphydrothiophens 188
2,5-Dihydroxypropiophenone 103
  hydrogen bonding in 103
Diiodomethane, chemical shift 160
β-Diketones
  chemical shifts 177
  tautomerism 382
Dimethoxymethane 182
Dimethyl disulphide 164
Dimethyl ether 164
Dimethyl sulphate 177
Dimethyl sulphide 164
Dimethyl sulphite 177
Dimethyl sulphone 177
Dimethyl sulphoxide
  chemical shift 177
  solvent effects 111
  solvent effects for alcohols 299
Dinitromethane 182
2,4-Dinitrophenylhydrazones, chemical shifts 217
para-Dioxan 200
  solvent effects 10
meta-Dioxans
  chemical shifts 237, 240
  spin–spin coupling 273
1,2-Diphenylpropionic acid 143
Diphenyls, non-equivalence 378
Dipolar broadening 37
  in solids 8
Dipole moment, influence on chemical shift 68
Di-*iso*-propyl disulphide 164
Di-*n*-propyl disulphide 164
Di-*iso*-propyl ether 164
Di-*n*-propyl ether 164
Di-*iso*-propyl sulphide 164
Di-*n*-propyl sulphide 164
Dispersion signals 27
Disulphides, dialkyl, chemical shifts 164

Diterpenes 244
DOG technique 42
Double irradiation *see* Spin decoupling

Effective shielding coefficients 181
Electronegativity
  and spin–spin coupling to ^{13}C 347
  correlations of chemical shifts with 64, 66
  influence on geminal coupling constants 276, 278
  influence on proton chemical shifts 63
  influence on vicinal spin–spin coupling 283, 302
Electrostatic effects, influence on proton shielding 67, 79
Enamines, chemical shifts 191
Enol acetates, chemical shifts 187
Enols 216
Episulphides, steroidal, chemical shifts 101
Epoxides *see* Oxirane
Equivalence
  accidental, definition 119
  and internal rotation 368
  and symmetry 368 ff.
  chemical, definition 119
  magnetic, definition 119
Esters
  chemical shifts 164, 176, 179
  inorganic, chemical shifts 176
Ethane
  1-amino-1-phenyl 166
  chemical shift 164
  Cotton–Mouton effect in 80
  coupling constants 280
  diamagnetic anisotropy of 76
  1,1-dibromo 166
  1,1-diphenyl 182
  $J_{13C,H}$ 141
  monosubstituted, chemical shifts 64
Ethanol
  chemical shifts 164
  low resolution spectrum of 13
  proton exchange in 17
  spectra of mixtures with water 18, 19
  spectrum of 14
  spin–spin coupling 15, 299
Etherification shift 179
Ethers 179
  chemical shifts 164, 179
  cyclic, chemical shifts 189
Ethyl acetate 164
  solvent effects of 106
Ethyl acetoacetate, tautomerism 177
Ethylamine, chemical shifts 164

Ethylbenzene 164
  4-substituted 175
  substituted, correlation of chemical shifts with Hammett σ-constants 66
Ethyl benzoate 164
Ethyl bromide 164
Ethyl chloride 164
Ethyl cyanate 178
Ethyl derivatives, chemical shifts 164
Ethylene
  analysis of the spectrum of partially deuterated ethylene 143
  $J_{13C,H}$ 141
Ethylene carbonate 200
Ethyleneimine 199
Ethyleneimines, chemical shifts 229
Ethylene oxide 199
  chemical shift 100
Ethylene sulphide 199
  chemical shift 100
Ethylene sulphides, chemical shifts 229
Ethyl fluoride 164
Ethyl iodide 164
Ethyl isocyanate 178
Ethyl isocyanide 164
Ethyl isothiocyanate 178
Ethyl lithium 265
Ethyl mercaptan 164
Ethyl methyl ketone 164
Ethyl methyl sulphone 177
Ethyl nitrate, solvent effects for 106
Ethyl nitrite 177
Ethyl orthoformate 182
Ethyl phenyl ketone 164
Ethyl thiocyanate 177
Ethyl p-toluenesulphonate 164
Exchange processes 55, 59, 358
  of labile protons 215
External lock spectrometer 29
External references for spectra 44

Fast exchange in carbonium ions 253, 256
Ferrocenes
  chemical shifts 206
  spin–spin coupling 309
Ferromagnetic impurities, removal of 36
Field-frequency lock 29
Field gradients 31
Field sweep 26
First order analysis 125
First order spectra, definition 124
Fluorene 197
Fluorenyl anion 266
Fluorine derivatives, spin–spin coupling in 348

Formaldehyde, diamagnetic anisotropy of 88

Formamide, N-alkyl, benzene solvent effects 111

Formamides, chemical shifts 191

Formates 191
  chemical shifts 191
  long-range spin–spin coupling in 343

Formyl protons, chemical shifts 191

Frequency response 38

Frequency sweep 26

Fulvenes, chemical shifts 206

Functional groups, influence on α-protons 176

Furanoquinolines, chemical shifts 213

Furans
  chemical shifts 207, 209, 214
  $J_{13_{C,H}}$ 141
  methyl derivatives 173
  spin–spin coupling 306

Furazan, chemical shifts 209

Furenidones, spin–spin coupling 342

Furoxans, isomerization 364

Gans–Mrowka relation 76

Gases, chemical shifts for 105

Gauche and *trans* interactions 291

Geminal spin–spin coupling *see* Spin–spin coupling, geminal

β-*D*-Glucopyranosides, aryl tetra-O-acetyl 113

Grignard reagents
  chemical shifts 263
  ligand exchange 360

Gyromagnetic ratio 3

H—C—X, chemical shifts 174

H—C—C—X, chemical shifts 167

H—C—CO—X, chemical shifts 176

H—C—O—X, chemical shifts 176

Half-height width and vicinal coupling 288

Hamiltonian, spin 122

Hammett σ-constants, correlation with chemical shifts 66

Helmholtz coils 26

Heterocyclic compounds
  chemical shifts 207
  protonation 384
  saturated, chemical shifts 199, 240
  solvent effects 208
  spin–spin coupling 305, 331
  substituent effects on chemical shifts 208
  tautomerism 383

Heterocyclopropanes
  chemical shifts 197, 224
  geminal spin–spin coupling 276
  stereochemistry from chemical shifts 227
  vicinal spin–spin coupling 287

n-Hexane, solvent effects for 106

Hex-2-en-4-onal, spectrum of 148

Hindered rotation, effect on n.m.r. line shapes 58

Homoaromatic anions 266

Homogeneity 30
  automatic control of 30

Homotropylium cations 262

Hückel's rule 98

Hybridization
  from $J_{13_{C,H}}$ 347
  influence on chemical shifts 70

Hydrazine derivatives, chemical shifts 217

Hydrazones
  chemical shifts 226
  spin–spin coupling 344

Hydrido compounds, spin–spin coupling 344

Hydroaromatic compounds 197, 207

Hydrogen bonding 359
  effect on chemical shifts 103, 216

Hydrogen exchange 359

Hydroxyl groups
  averaged spectra 359
  axial and equatorial 240
  chemical shifts 215
  long-range spin–spin coupling 341
  spin–spin coupling 298

Hydroxymethylene compounds, tautomerism 383

Imidazole
  N-acetyl 177
  derivatives, chemical shifts 209
  derivatives, spin–spin coupling 307
  methyl derivatives 173

Indane 197

Indanones, chemical shifts 197, 223

Indazoles
  chemical shifts 210
  spin–spin coupling 308

Indene
  chemical shifts 188
  spin–spin coupling 317

Indenyl anion 266

Indole
  and derivatives, chemical shifts 209
  and derivatives, spin–spin coupling 308
  N-methyl 178
  methyl derivatives 174

Indolizines, spin–spin coupling 309
Inductive effects, influence on chemical shift 69
Integration of spectra 49
Intensity 13
  factors influencing the measurements of 52
  measurement of 49
Interatomic diamagnetic shielding 63
Internal lock spectrometer 29
Internal references for spectra 49
  for acidic systems 249
Intrinsic non-equivalence 373
Iodocyclohexane, axial and equatorial 239
Isobutane 164, 168
  chemical shifts 71
Isobutyric acid methyl ester 164
Isochronous nuclei 120
Isoeugenol, spectra of *cis* and *trans* 135
Isomerism
  cis–trans, from chemical shifts 224
  cis–trans, from coupling constants 301
  rapid, *cis–trans* 362
  syn–anti, from chemical shifts 226
Isonitriles
  chemical shifts 164
  spin–spin coupling to ^{14}N 354
Isopropyl chloride, spectrum of 128
Isoquinolines
  chemical shifts 212
  spin–spin coupling 308
Isothiazoles, chemical shifts 213
Isothiocyanates 178
Isoxazoles, chemical shifts 213

Karplus equation 281
  examples 294 ff.
  limitations 292
Ketals
  cyclic, chemical shifts 190, 236
  formation equilibria 384
Ketene dimer, spin–spin coupling 317
Ketimines 226
Ketones
  benzene and toluene solvent effects 112
  chemical shifts 164, 180, 195
  cyclic, chemical shifts 188, 190, 198, 207
  α-halogeno, conformation from spin–spin coupling 291
  stereochemistry from solvent shifts 248
  α,β-unsaturated, chemical shifts 170, 187, 190, 192, 223
  α,β-unsaturated, benzene solvent effects 111

Lactams 198
  N-acetyl 198
  N-methyl 198
  partial double bond character 362
β-Lactams
  geminal spin–spin coupling 276
  vicinal spin–spin coupling 287
Lactones 188, 198
  spin–spin coupling 317
  stereochemistry from solvent shifts 248
  Larmor precession 61
Lewis acids, complex formation with 181
Line shapes, theory of the effect of chemical exchange on 57
Liquid crystals for the study of diamagnetic anisotropy 75
Lithium alkyls 264
Local diamagnetic shielding 62
Local paramagnetic shielding 62
Longitudinal relaxation time, T_1 *see* Spin–spin lattice relaxation time
Long-range shielding 63
  by acetylene triple bonds 92
  by aromatic rings 94
  by carbon–carbon double bonds 83
  by carbon–carbon single bonds 78
  by carbon–hydrogen bonds 78
  by carbonyl groups 88
  by diazirine rings 101
  by ether oxygen atoms 80
  by furan nuclei 97
  by halogens 80
  by hydroxyl groups 80
  by nitrile groups 93
  by nitro groups 94
  by sp^2-nitrogen atoms 82
  by sp^3-nitrogen atoms 81
  by thiophene nuclei 97
  by three-membered rings 98
  in oximes 82
  in Schiff's bases 82
  in α,β-unsaturated acids and esters 91
  in α,β-unsaturated acid chlorides 91
  in α,β-unsaturated amides 91
  in α,β-unsaturated ketones 91
  point dipole approximation for 74
α-Lumicolchicine, chemical shifts 92

M-rule 334
Magnetic field
  homogeneity of 22
  stabilization of 22
Magnetic shielding 61
Magnets 22
Maleic anhydride 188
Meisenheimer complexes 266

Mesomeric effect, influence on chemical shifts 70
Methacrylic acid and derivatives 171
Methane
  chemical shift 164
  monosubstituted, chemical shifts 64
Methanethiol 164
Methanol, chemical shift 164
Methine protons, chemical shifts 165
Methoxybenzenes, correlation of Hammett σ-constants with chemical shifts 66
Methoxyl groups 180
  axial and equatorial 240
  long-range coupling to 243
Methylacetylene, $J_{13_{C,H}}$ 141
Methylacroleins, $J_{13_{C,H}}$ 141
Methylamine 164
Methyl bromide
  chemical shift 160, 164
  solvent effects for 106
Methyl chloride, chemical shift 160, 164
Methylcyclohexane 169
Methylene bromide, solvent effects for 106
Methylene chloride, solvent effects for 106
Methylenedioxy group
  chemical shifts 200
  spin–spin coupling 273, 277
Methylene group
  chemical shifts 165
  exocyclic, chemical shifts 190
  non-equivalence 372
Methylene iodide, solvent effects for 106
Methyl esters 180
Methyl ethers 180
Methyl fluoride, chemical shift 160, 164
Methyl groups
  angular, chemical shifts 244
  attached to nitrogen, chemical shifts 178
  attached to sulphur, chemical shifts 177
  axial and equatorial 240
  β in enones 223
  chemical shifts 164, 168
  chemical shifts of angular 244
  configuration from spin–spin coupling 298
  effect on chemical shifts in aromatic systems 208
  long-range coupling to 337
  virtual coupling 298
Methyl isocyanide 164
Methyl iodide
  chemical shift 160, 164
  solvent effects for 106
Methylisothiocyanate 178

Methyl *iso*-propyl ketone 164
Methyl *n*-propyl ketone 164
Methyl thioacetate 177
Methyl thiocyanate 177
Methylthiophenes, correlation of Hammett σ-constants with chemical shifts 66
Methyl vinyl sulphide, spin–spin coupling 342
Microcells 40
Molecular Zeeman effect and diamagnetic anisotropy 74
Monascin 153
Monosaccharides 241
Morpholine 200
Multiple irradiation *see* Spin decoupling

^{15}N Labels 383
NH groups
  averaged spectra 359
  chemical shifts 217
  spin–spin coupling 298, 355
Naphthalenes
  chemical shifts 205 ff.
  methyl derivatives, chemical shifts 173, 174
  α- and β-nitro, chemical shifts 94
  spin–spin coupling 306
Neighbouring paramagnetic shielding 63
Neohexylorganometallic compounds 263
Neopentane, solvent effects for 106
Neutron, magnetic properties of 1
Nickel(II) N,N′-di(6-quinolyl)-aminotroponeiminate, spectrum of 102
Nitriles, chemical shifts 164, 225
Nitroalkenes 187
Nitrobenzene, solvent effects for 106
Nitro compounds
  aromatic 203, 204
  chemical shifts 164
Nitrocyclohexane, axial and equatorial protons 239
Nitroethane 164
  solvent effects for 106
Nitrogen, spin–spin coupling to protons 354
Nitromethane 164
  solvent effects for 106
cis- and *trans*-1-Nitropropene 94
N-Nitrosoamines 226
Non-benzenoid aromatic compounds, chemical shifts 205
Non-equivalence 368 ff.
  solvent effects 221
Norbornadiene 188
Norbornadienyl cation 261

Norbornane, derivatives, chemical shifts 100, 230

2-Norbornanone 100

Norbornene, derivatives, chemical shifts 230

2-Norbornene, 7-methyl, *syn* and *anti*, chemical shifts 84

5-Norbornene-2-carbonitrile, *exo* and *endo*, chemical shifts 84

5-Norbornene-2-carboxylic acid
 exo and *endo*, chemical shifts 84
 2-methyl, *exo* and *endo*, chemical shifts 84

2-Norbornen-7-ol, *syn* and *anti*, chemical shifts 84

5-Norbornen-2-ol, *endo* and *exo*, chemical shifts 84

Norbonyl cation 260

Nuclear induction 27

Nuclear magnetic moment 1

Nuclear Overhauser effect 157

Nuclear quadrupole moment 1

^{17}O labels 384

Olefins
 chemical shifts 184
 cyclic, chemical shifts 188
 cyclic, spin–spin coupling 285, 303, 339
 geminal spin–spin coupling 277
 $J_{13C,H}$ 141
 long-range coupling 323
 vicinal spin–spin coupling 301

Onsager model, use for estimating polar solvent effects 107

Organometallic compounds, ligand exchange 360

Overhauser effect *see* Nuclear Overhauser effect

Oxazoles, chemical shifts 213

Oxetane 199

N-Oxides, heterocyclic, chemical shifts 213

Oximes, chemical shifts 191, 216, 226

Oxirane, chemical shift 100

Oxiranes
 chemical shifts 224, 228
 solvent effects of chloroform on chemical shifts of 101
 stereochemistry from chemical shifts 227
 steroidal, chemical shifts 101
 vicinal spin–spin coupling 287

4-Oxo-1,2,3,4-tetrahydrophenanthrene, chemical shifts 91

Oxygen, paramagnetic broadening by 31

Ozonides 200

Palmarin 157

Paramagnetic broadening 9
 by oxygen 31

Paramagnetic impurities, removal of 37

Paramagnetic substances 367

Partial double bond character 361

Pentalenyl dianion
 chemical shifts 266
 spin–spin coupling 309

neo-Pentane 168

Pent-3-ene-1-yne-4-methyl 171

Pent-1-yne 164

Peri effects 204, 208, 214, 225, 244

Phase sensitive detection 29

Phenanthrene
 derivatives, chemical shifts 204, 205
 derivatives, spin–spin coupling 306, 342
 9,10-dihydro 197
 4-oxo-1,2,3,4-tetrahydrophenanthrene, chemical shifts 91

Phenanthroline, spin–spin coupling 309

Phenetole 164

Phenol
 intermolecular hydrogen bonding in 103
 substituted, correlation of chemical shifts with σ^- 66

Phenolates 204

Phenols 203, 204, 216

Phenylacetylene, correlation of chemical shift with Taft's substituent parameter 67

Phenylcyclohexane, axial and equatorial protons 239

5-Phenyl-1,2,3,4,7,7-hexachlorobicyclo-[2.2.1]hept-2-ene, correlation of chemical shifts of 5- and 6-protons with Hammett σ-values 66, 69

Phenylhydrazine, N-methyl 178

Phenylmethyl carbanions 265

Phenyl *iso*-propyl ether 164

Phenyl *n*-propyl ether 164

Phenyl *iso*-propyl ketone 164

Phenyl *n*-propyl ketone 164

Phosphate esters, chemical shifts 179

Phosphine derivatives
 chemical shifts 179
 spin–spin coupling 300

Phosphorus, spin–spin coupling to protons 353

Phosphorus derivatives 181
 chemical shifts 179

Phthalazine, spin–spin coupling 309

Phthalimide, N-methyl 179

Picolines 174

Piperidine 199

Piperidine *(cont.)*
  N-acyl 199
  axial and equatorial protons 240
  chemical shifts 100
  N-methyl 178
Pivalic acid 164
  methyl ester 164
Pivalic aldehyde 164
Point dipole approximation for long-range
    shielding 74
Polyacetylenes, spin–spin coupling 330
Polyenes
  chemical shifts 170, 187, 225
  cyclic, chemical shifts 206
Polynuclear hydrocarbons, chemical shifts
    174, 204
Porphin, chemical shifts 97
Porphyrins, chemical shifts 97
Precessional motion of nucleus 3
Propanal, 2-methyl 164
Propane
  1-bromo-3-phenyl 166
  chemical shifts 164
  1-chloro-2,2-difluoro 166
  1-chloro-1-nitro 166
  coupling constants 280
  1,1-diphenyl 168
  1,3-diphenyl 166
  1-nitro 164
  2-nitro 164
iso-Propanol 164
n-Propanol 164
Propene
  1-acetoxy-2-methyl 171
  2-acetoxy 171
  2-bromo 171
  1-bromo-2-methyl 171
  chemical shifts 164
  2-chloro 171
  2-cyano 171
  derivatives 192
  derivatives, chemical shifts 170
  derivatives, spin–spin coupling 316,
    323
  2-methyl 171
  cis and *trans*-1-nitro 94
Propiolactam 198
Propiolactone 198
Propionaldehyde 164
Propionamide 164
Propionic acid 164
  2-chloro 166
  methyl ester 164
Propionitrile 164
iso-Propyl acetate 164
n-Propyl acetate 164
iso-Propylamine 164

n-Propylamine 164
iso-Propyl benzene 164
n-Propyl benzene 164
iso-Propyl benzoate 164
n-Propyl benzoate 164
iso-Propyl bromide 164
n-Propyl bromide 164
iso-Propyl chloride 164
n-Propyl chloride 164
iso-Propyl cyanide 164
n-Propyl cyanide 164
iso-Propyl derivatives, chemical shifts 164
n-Propyl derivatives, chemical shifts 164
iso-Propyl fluoride 164
n-Propyl fluoride 164
iso-Propyl groups, non-equivalence 375
iso-Propyl iodide 164
n-Propyl iodide 164
iso-Propyl isocyanide 164
n-Propyl isocyanide 164
iso-Propylisothiocyanate 178
iso-Propyl mercaptan 164
n-Propyl mercaptan 164
iso-Propyl thiocyanate 177
iso-Propyl *p*-toluenesulphonate 164
n-Propyl *p*-toluenesulphonate 164
Propyne
  chemical shifts 164
  solvent effects for 106
Proton exchange, effect on spectrum 17
Pteridines
  chemical shifts 213
  spin–spin coupling 309
Purines 208
  chemical shifts 210
  spin–spin coupling 309
Pyrane, tetrahydro-, chemical shifts 100
Pyrazines
  chemical shifts 211
  spin–spin coupling 307
Pyrazole
  derivatives 208
  derivatives, chemical shifts 209
  derivatives, spin–spin coupling 307
  N-methyl 179
  methyl derivatives 173
Pyrene
  chemical shifts 205
  2,7-diacetoxy-*trans*-15,16-dimethyl-15,
    16-dihydro, chemical shifts 97
  spin–spin coupling 306
Pyridazines 207
  chemical shifts 211
  spin–spin coupling 307
Pyridines
  chemical shifts 207, 211, 213
  long-range shielding by the lone pair 82

Pyridines *(cont.)*
spin–spin coupling   307
α-Peridones
chemical shifts   211
N-methyl   178
4-methyl, protonation   299
Pyrimidines   207
chemical shifts   211
methyl derivatives   174
spin–spin coupling   307
Pyrimidones, chemical shifts   213
γ-Pyrone, vicinal spin–spin coupling   304
Pyrone derivatives   172, 189
spin–spin coupling   340
Pyrrole
and derivatives   208
derivatives, chemical shifts   209
derivatives, spin–spin coupling   306
N-methyl   179
methyl derivatives   173
Pyrrolidine   199
N-acetyl   199
N-acetyl, derivatives   236
N-methyl   178
Pyrrolidone, N-methyl   178
Pyrrolines
N-acetyl   236
methyl derivatives   169
Pyrrolones, tautomerism   383
Pyrylium salts   309

Quadrupole broadening   10, 58, 142, 360
Quaternary ammonium salts   354
Quinazoline   207, 213
derivatives   207, 213
spin–spin coupling   309
Quinolines
chemical shifts   207, 211
spin–spin coupling   308
Quinolizidine, methyl derivatives, chemical shifts   82
Quinolizidone, chemical shifts   92
Quinones, stereochemistry from solvent shifts   248
Quinoxalines, spin–spin coupling   309

Radiciol   157
Radiofrequency oscillator   25
Radiofrequency phase   33
Reaction field, contribution to solvent effect   107
References
delta (δ) values of   49
internal references for intensity measurements   51
Resolution   313
criteria of   31

Restricted rotation   361
Rifomycin B, chemical shifts   97
Ring current
in aromatic systems   95
in cyclopropane   98
Ring size, from vicinal spin–spin coupling   303
Ringing patterns   35
Rotation, internal and equivalence   368
Rotenolone A and B, methyl ethers, chemical shifts in   91
Rotenone, 12-desoxy-6′,7′-dihydro, chemical shifts   92

Sample preparation   36
Santonins, stereochemistry from vicinal coupling   295
Satellites   345
Saturation   10
Saturation factor   10, 314
Schiff's bases   172, 226
Selenophene
chemical shifts   209
spin–spin coupling   307
Semicarbazones   336
Sensitivity, enhancement of   41
Shielding constants   182
Shielding effects, additivity of   181, 185
Shikimic acid, stereochemistry from vicinal coupling   296
Shims, current   23
Shoolery's rules   181
Side band modulation   39
Side-chain spin–spin coupling   331
Signal-to-noise ratio   38
Silane derivatives, spin–spin coupling   300
Silicon tetrachloride, solvent effects for   106
Simon's rules   184
Six-membered rings, configuration from spin–spin coupling   288
S-methyl groups   180
Solvent effects   53, 104, 181, 246
and stereochemistry   246
as an aid to analysis of spectra   144
correlation with heats of vaporization   105
Solvents, purification and fast exchange   299
Solvent shifts   246
Spin decoupling (multiple irradiation)   151
analysis of spectra by   145
determination of highly accurate chemical shifts by   155

Spin decoupling*(cont.)*
  determination of relative signs of coupling constant by   154
  field sweep method   152
  frequency sweep method   152
  heteronuclear   153
  homonuclear   153
  interrelation of two or more coupled groups of nuclei   157
  location of "hidden" absorption by   155
  measurement of rate of chemical exchange by   157
  simplification of complex spin multiplets by   153
  theory of   151
Spin Hamiltonian   122
Spin–lattice relaxation   6
Spin–lattice relaxation time, T_1 7
Spinning of sample and homogeneity   23
Spinning side bands   33
Spin numbers   1, 355
Spin–spin coupling   13
  applications   269
  across 4 single bonds   334
  across 5 bonds   333, 342
  across 6 bonds   342
  allylic   316
  and bond-order in aromatic systems   310
  and ring size in aromatic systems   310
  average over a number of conformations   61
  benzylic   330
  between ^{13}C and ^1H 345
  between ^{19}F and ^1H 348
  between nitrogen and protons   354, 383
  between ^{31}P and ^1H 353
  between protons and miscellaneous elements   354
  constant, J, definition of   17, 116
  effect of substituents in aromatic compounds   310
  geminal   270
    across a heteroatom   279
    and configuration   275
    in exocyclic $=CH_2$ groups   279
    in formaldehyde derivatives   277
    in unsaturated systems   277
    in vinyl derivatives   278
    influence of π-bonds   273, 278
    influence of H—C—H angle   273
    influence of ring size   275
    influence of substituents   276
    measurement by partial deuteration   143
    relative signs   270
    small values of   277
  solvent effects   277, 279
    in X = CH_2 groups   277
  homoallylic   316, 325
  inter-ring in aromatic compounds   333
  long-range   312
  mechanism of   115
  of OH, NH and SH groups   298
  relative sign of   117, 154
  signs in aromatic systems   305
  theoretical calculation of   118
  vicinal   280
    and bond order   303
    and conformation   289
    and fast exchange   299
    axial–equatorial   288
    axial–axial   288
    equatorial–equatorial   288
    gauche and *trans*   291
    in five-membered rings   287
    in flexible systems   289
    in H—C—X—H   298
    in H—C—C—H and H—C—C—H   284
    in unsaturated systems   301
    in vinyl derivatives   278
    influence of dihedral angle   281
    influence of ring size in saturated systems   286
    influence of substituents in saturated systems   283
    of methyl groups   298
    relative signs   280, 285
    solvent effects   304
Spin–spin multiplet   115
  analysis of   122
  exact analysis of   123
  recognition of   115
Spin–spin relaxation time, T_2 8
Spin systems
  AA'BB'   134
  AA'XX'   134
  AB   129
  AB$_2$ 130
  A$_2$B$_2$ 134
  ABC   133
  ABX   133
  ABX$_n$ 134
  AMX   132
  AX$_2$ 130
  A$_2$X$_2$ 134
  classification of   121
  miscellaneous   139
Stannic chloride, solvent effects for   106
Stereochemistry
  cis–trans and vicinal coupling   286, 301
  cis–trans from chemical shifts   222

Stereochemistry *(cont.)*
   from allylic coupling   322
   from chemical shifts   219
   from solvent shifts   246
Stereoisomers, chemical shifts in *meso* and
   dl forms   163
Steroids
   chemical shifts   92
   chemical shifts of angular methyl groups
      242
   cyano substituted, chemical shifts   94
   3α,5α-cyclo, chemical shifts   99
   episulphides, chemical shifts   99
   epoxides, chemical shifts   99
   11-keto-5α-, chemical shifts   91
   6α- and 6β-nitro, chemical shifts   94
   Δ⁴-3-oxo, chemical shifts   245
Stilbenes   224
   oxides   224
   sulphides   224
Strained systems, spin–spin copuling   335
Structure determination, and vicinal coupling   300
Styrene
   derivatives, chemical shifts   170, 187, 224
   α-methyl   171
   oxide, spectrum of   126
   oxide, spin–spin coupling   272
   sulphide, spin–spin coupling   272
Subspectra   123
α-substituents, influence on chemical shifts
   165
β-substituents, influence on chemical shifts
   165
Succinic anhydrides   200
   chemical shifts   235
Succinimide   200
Sulphides   180
Sulphinamides, non-equivalence   377
Sulphinates   180
   non-equivalence   377
Sulphites, non-equivalence   377
Sulphonates   180
Sulphones   177, 180
Sulphoxides, non-equivalence   377
Sulphur derivatives   180
Sulphydryl groups
   averaged spectra   359
   spin–spin coupling   298
Symmetry
   and equivalence   368 ff.
   and fast exchange   221
   and rapid inversion   365
Syn–anti isomerism   226

Taft's σ*-constants, correlation with chemical shifts   66, 70

Tau scale   48
Tautomerism   380
   and spin–spin coupling   299
Temperature dependence of spectra   368
Tetrahydrofuran   199
Tetrahydropyran   199
1,2,5,6-Tetrahydropyridine, spectrum of
   146
Tetrahydrothiophene   199
   dioxide   199
Tetrahydrothiopyran   199
Tetralin   197
α-Tetralone   197
   chemical shifts   91
Tetramethylsilane (TMS) as an internal
   reference   46
Tetrazoles, chemical shifts   213
Thiapyran, tetrahydro-, chemical shifts
   100
Thiapyrones, vicinal spin–spin coupling
   304
Thiazoles
   chemical shifts   209
   methyl derivatives   173
   spin–spin coupling   307
Thienopyrroles
   chemical shifts   213
   spin–spin coupling   309
Thiethane   199
Thiirane, chemical shift   100
Thioacetic acid   177
Thiocyanates   177
Thioethers, chemical shifts   164
Thiols
   chemical shifts   164, 215
   spin–spin coupling of SH groups   298
Thiophens   207, 214
   chemical shifts   209
   $J_{13_{C,H}}$ 141
   methyl derivatives   173, 175
   spin–spin coupling   307
Thiopyrones, vicinal spin–spin coupling
   304
Thiosulphites, cyclic, chemical shifts   236
Thiosulphonates   180, 220
Threo and *erythro* forms
   from chemical shifts   163
   from spin–spin coupling   291
β-Thujone, chemical shifts   92
Time-dependent phenomena   17
   applications   358
Toluene
   chemical shifts   164
   derivatives, chemical shifts   170
   nitro derivatives   175
   solvent effects of   112
p-Toluenesulphonic acid, methyl ester   164

Transverse relaxation time, T_2 *see* Spin–spin relaxation time
Tribromomethane, chemical shift 160
Triazines, chemical shifts 211
Triazoles, chemical shifts 213
1,1,1-Trichloroethane 177
1,1,2-Trichloroethane 166
Trichloromethane, chemical shift 160
Tricyclo[2.2.1.0^{2,6}]heptane, chemical shifts 100
Tricyclo[3.1.1.0^{6,7}]heptane 99
Tricyclo[2.2.1.0^{2,6}]heptan-3-one
   chemical shifts 100
   1,7,7-trimethyl 100
Tricyclo[2.1.1.0^{5,6}]hexane, 1-methyl, chemical shifts 99
Tricyclo[1.1.1.0^{4,5}]pentane,1,3-dimethyl, chemical shifts 99
Triethylamine, solvent effects for 106
Trifluoroacetates 179
Trifluoroacetic acid as solvent for amines 359
Trifluoromethane, chemical shift 160
Triiodomethane, chemical shift 160
β-Triketones, tautomerism 382
Trimethylene oxide 199
2,4,4-Trimethyl-2-pentene 171
Trimethylsilyl ethers 240
Trinitromethane 182
Trioxans 200, 237
Triphenylene, chemical shifts 205
Triphenylmethane 182
Triphenylmethyl anion 263
Triphenylmethyl cation 263
Triptycene 197
Triterpenes 244
   chemical shifts 92
Trithians 200

Tropolones 204
   spin–spin coupling 309
Tropylium cation 262

UCP non-equivalence, definition 374
Urea derivatives 178
Urethanes 178

Valance-bond tautomerism 362
Valerolactam 198
Valerolactone 198
van der Waal interactions
   and chemical shifts 105
   and solvent effects 71
van der Waals' shifts 204
Variable temperature studies and time-dependent phenomena 358
Vicinal spin–spin coupling *see* Spin–spin coupling, vicinal
Vinyl derivatives
   chemical shifts 187
   $J_{13C,H}$ 141
   spin–spin coupling 278
Vinyl ethers, chemical shifts 187
Vinylic protons 184
Virtual coupling 133, 147
Viscosity, effect on spin–lattice relaxation 8
Volume susceptibility, correction for 4

Width at half-height (W_H) 288, 314
Wiggle-beats 35, 313
W-rule 334

Xylenes 170, 175, 203

OTHER TITLES IN THE SERIES IN
ORGANIC CHEMISTRY

Vol. 1. WATERS—*Vistas in Free-Radical Chemistry*

Vol. 2. TOPCHIEV et al.—*Boron Fluoride and its Compounds as Catalysts in Organic Chemistry*

Vol. 3. JANSSEN—*Synthetic Analgesics—Part I: Diphenylpropylamines*

Vol. 4. WILLIAMS—*Homolytic Aromatic Substitution*

Vol. 5. JACKMAN—*Applications of Nuclear Magnetic Resonance Spectroscopy in Organic Chemistry*

Vol. 6. GEFTER—*Organophosphorus Monomers and Polymers*

Vol. 7. SCOTT—*Interpretation of the Ultra-violet Spectra of Natural Products*

Vol. 8. HELLERBACH et al.—*Synthetic Analgesics—Part IIA: Morphinans. Part IIB: 6,7-Benzomorphans*

Vol. 9. HANSON—*The Tetracyclic Diterpenes*